T5-AFE-758

Industrial Hazard
and Safety Handbook

Industrial Hazard and Safety Handbook

(Revised impression)

RALPH W KING, BSc, CEng, FIChemE
JOHN MAGID, BSc, MInstInfSci

Butterworth Scientific
London – Boston – Sydney – Wellington – Durban – Toronto

First published 1979
  Reprinted, 1980
  Third (Revised) impression, 1982

© Butterworth & Co (Publishers) Ltd, 1979

British Library Cataloguing in Publication Data

King, Ralph W
  Industrial hazard and safety handbook.
  1. Industrial safety
  I. Title II. Magid, John
  614.8'52   T55   78-40474

  ISBN 0-408-00304-9

Photoset by Amos Typesetters, Hockley, Essex.
Printed in England by J. W. Arrowsmith Ltd., Bristol

Foreword

by Sir Bernard Braine, D.L., M.P.,

This is a most useful and timely book. It is useful because it demonstrates with great clarity how knowledge, understanding and training can change attitudes and prevent practices which lead to industrial ill-health and accident. It is timely because, with the new regulations about safety at work which came into force in October, 1978, a large army of safety representatives and joint management/trade union safety committees are coming into existence in order to comply with the provisions of the Health and Safety at Work Act 1974, and management can expect many more questions to be asked about health and safety in the work place than have been asked up till now.

By and large Great Britain compares favourably with other countries with regard to the prevention of industrial ill-health and accidents, but the toll of avoidable death and injury is still far too high for complacency and in some occupations it remains stubbornly and disgracefully high. What is more, the relentless advance of technology, the introduction of new processes and the use of new materials often bring new hazards to health and safety that have not been foreseen or are only dimly perceived.

The importance of the new approach is that it is designed to persuade managers and employees to work sensibly together rather than to bludgeon everyone into improvement. In any event, it would be impossible and undesirable in our complex society to enforce safety regulations solely by a new kind of police force. There are, after all, only 900 factory inspectors in the country and even if there were more of them they cannot be everywhere all the time.

What is needed is a self-regulatory system which encourages people to be vigilant, to think constructively about health and safety, and to act voluntarily in their own joint interest. There is room here for a new kind of partnership. As Wordsworth said of an earlier generation:

'The discipline of slavery is unknown among us;
Hence the more do we require the discipline of virtue.'

All this, however, requires knowledge and understanding of a high order. It requires a determination on the part of management to observe an approved code of practice. It requires a new breed of alert, practical and persuasive safety officers who possess a recognised diploma in safety management. It requires all safety representatives from the shop floor to have had some training so that they know what they are talking about and what to look out for. It also requires monitoring by the Health and Safety

Commission, continuous research into new, potential hazards and the dissemination of information and guidance in a form that will be readily understood by all concerned.

This will certainly be expensive, but in the end it will save money. For lives that are lost or blighted by someone's carelessness or ignorance, and the consequent cost of hospitalisation and working days lost as a result of industrial accident or ill-health add up to a massive waste of human resources. Add to this the cost of mechanical failure and damage to productive equipment and buildings from fire and explosion caused by human error and one can readily see that investment in better safety procedures is money well spent.

Improved health and safety at work makes sense all round. This book tells what must be done to achieve it.

Preface

In this revised impression, only two years after the first printing, statistics and references to standards in the field of safety have been up-dated and the authors have taken the opportunity to introduce a few new ideas. The understanding of hazards often needs deep technical appreciation, without which even the most safety conscious firm may have its disasters. This book is an attempt to identify and warn of the main hazards found in industry and to provide appropriate references for further study. Here within a single volume, it is possible to find quickly the information needed on any hazard without having to thumb through numerous publications. However, as no single volume can provide all the information needed at all times, we have given a comprehensive list of references at the end of each section.

One omission, which came to light in time for this second impression, was polyurethane foam. Although this is not a serious industrial hazard, nevertheless, because of its widespread use, it was thought that some mention should be made. Accordingly, we have added a short appendix on polyurethane foams, which should provide a general background to the problems associated with these materials.

The book has been written for safety specialists, representatives and students, for managers and engineers responsible for design, production, inspection and maintenance in industry, as well as insurers and lawyers whose work is concerned with industrial accidents and their consequences.

The emphasis is on hazards to health and safety, how they arise, how they often pass unnoticed, and how to spot, eliminate or control them. The book contains six main chapters, each comprising a number of sections and sub-sections, and three appendices. The titles of the chapters and main sub-sections should provide a clear understanding of the form and objectives of each part of the book and render further explanation unnecessary. Although the book is intended mainly for reference, it is hoped that the reader will be tempted to browse until eventually he is familiar with most of the contents.

We believe that the book is unusual in attempting to deal with a number of common hazards which are rarely discussed in this context. These include human stresses brought on by various conditions, such as insomnia, as well as unsatisfactory working environments. The hazards caused or hidden by secrecy about industrial processes are also discussed frankly. A special section (5.2) on that common complaint of our era 'The Slipped Disc' has been written in consultation with the leading pioneer in this field,

Dr James Cyriax of St. Thomas' Hospital, London, to whom we owe a special debt of gratitude.

Discussions of hazards, particularly after a serious accident, are often bedevilled by considerations of guilt and liability. Thus expensive exercises launched to find the causes of serious accidents and point out lessons for the future have a habit of confusing the vital issues. Before progress can be made, it is sometimes necessary to take a cool, hard look both at our own attitudes and the attitudes of those most involved or liable to be held responsible. We are all human and too often it is a case of 'do as I say, not as I do.'

This was forcibly brought home to one of the authors when, with both hands swathed in plaster and limping from an injured knee, he visited a company who saw fit to employ him as an adviser on certain hazards. Their sympathy quickly turned to laughter when he confessed that the injuries were caused in a moment of aberration through taking excessive liberties with his wife's bicycle. This should be a salutary reminder not to be too harsh or critical of others mistakes.

In this book we have had to refer on many occasions to the person whose special responsibility it is within an organisation to be concerned with problems of safety. Sometimes it is the safety manager, or the safety officer or engineer, the safety superintendent, or sometimes the training and personnel manager. In seeking an acceptable and non-discriminating common denominator for this person we have finally settled on the term 'safety specialist.' To possible critics we would add that both 'safety' and 'specialist' are somewhat relative terms, so that the expression allows a certain and necessary degree of elasticity. Throughout this book we have used the masculine forms of the personal pronouns to indicate both male and female; we trust this will not cause offence.

Many firms, organisations and individuals have helped us greatly with illustrations, information and friendly criticism. We are painfully aware that we have at times been so overwhelmed that even the civilised courtesy of a 'Thank you' was forgotten. To these we offer our apologies and a belated thanks. We also express our thanks to all those closely concerned with this book, Sue Williams, who typed the manuscript so well, Harry Ibbetson, formerly administrative director of the International Wool Secretariat who read the manuscript and checked the proofs, and the staff of the Publishers. Not least we should thank our wives and families, who learnt what it is to have an author in the front sitting room, and who helped, encouraged and even disciplined us when our footsteps began to wander.

We should particularly like to acknowledge the help we have received from two organisations, on whom we have drawn extensively for illustrations and information. These are the Health and Safety Executive who through Her Majesty's Stationery Office, London, allowed us to reproduce several illustrations, and the National Safety Council in Chicago.

We will gratefully appreciate any comments from readers which we hope one day will assist us when preparing a second edition.

RALPH KING
JOHN MAGID

Contents

5 Common Industrial Hazards 477

6 Special Industrial Hazards 605

Appendices 755

Index 803

1 PEOPLE, PROFITS AND SAFETY

1

People, profits and safety

1.1 INTRODUCTION

Despite the growing interest in industrial health and safety, progress in reducing the toll of death, injury, industrial disease and property loss is slow. Before getting immersed in detail, it is worth trying to examine our own attitudes to safety, and to its opposite — danger. Many writers on safety assume that it is universally desired, and by implication that danger is something everyone wants to avoid. Yet this assumption is at best only superficially true and often demonstrably false.

Many popular sports such as motor racing, boxing and steeplechasing are very hazardous. This is nothing new. The appeal of hazard, danger and sudden death is instinctive and part of man's historical heritage. Accidents and disasters hold an immense fascination for most people which a good safety record sadly lacks. To quote Nietzsche, 'A heart full of cheerfulness and courage needs a little danger from time to time or the world gets unbearable',[1] or again Prince Charles, 'There is enormous satisfaction in achieving something which is potentially hazardous and which requires concentration and self discipline'.[2]

There is thus an ambivalence in man's quest for safety. While man with one hand consciously endeavours to remove hazards from the things he knows, with the other he creates, perhaps unconsciously, newer and even nastier hazards than the ones he is removing. The attitude of most people towards questions of safety reveal split[3] personalities—a Dr Jekyll and Mr Hyde situation. At heart, most of us realise that even if we totally banished industrial accidents and injuries, there would be other dangers. To quote Jung:

'In view of the rapidly increasing avalanche of world population, man has already begun to seek ways and means of keeping the rising flood at bay. But nature may anticipate all our attempts by turning man against his own creative mind. The H-bomb for instance would put an effective stop to over population. In spite of our proud domination of nature, we are still her victims for we have not yet learned to control our own nature. Slowly but it appears inevitably we are courting disaster'.[3]

While the immediate causes of most industrial accidents are external, the solutions must come from within our own minds. Even dreams can help us, as Jung pointed out. Perhaps the main reason for this is that we tend to forget hazards which cause accidents — particularly if we suspect we are in some way connected with them. As Nietzsche said, 'Where pride is insistent enough, memory prefers to give way'.[4] The knowledge however remains in the subconscious mind, which may possibly even be able to make a few useful deductions without its owner being consciously aware of it. Thus a warning can be conveyed in a dream when our conscious defences are down.

There are many stories of people being warned of dangers to themselves (such as the Titanic disaster) in dreams which they thereby avoided, or of harm befalling a friend (which generally happened).

Dreams also can to a skilled analyst reveal suicidal tendencies which if allowed full rein could lead to disaster. This is discussed further in section 1.7 'Human factors'. Lest the reader thinks this is an exaggeration, consider first that until recently, when our gas was made by coal and coke gasification and contained much carbon monoxide, there were about 1000 suicides a year in Britain from gas poisoning.[5] The poisonous town's gas clearly fulfilled a 'demand'. We now have non-toxic natural gas in the UK, but the 'demand' is still there. It would be naive to imagine that this fortuitously vanished when the supply ceased. Has the 'demand' sought alternatives?

Speaking in a parliamentary debate in 1970 on the safety at work bill, Leo Abse, M.P. for Pontypool, who had been acting professionally for many years in the sphere of industrial accidents said,

'It is jejune and naive to imagine that accidents occur only because of mechanical hazards. There is an unconscious motivation, and I have dealt with too much clinical material during the past twenty years not to have become keenly conscious of the fact'.[6]

Before leaving this theme, it is as well to recognise that many apparent accidents, especially those involving fire and explosion are caused deliberately. Motives may vary considerably and include financial gain (often the hope of collecting insurance on a loss making business), politically motivated sabotage — or sometimes the settling of a personal grudge.

Risks and life are inseparable. In order to survive, man has acquired through the ages instinctive habits and reflexes to protect him from the more common hazards. A man picking up a hot stone instictively drops it before serious injury to his hand can occur.

Modern life and industry have created many fresh hazards against which man's primitive instincts offer no protection. His 'natural' behaviour may even increase his injuries. A man grasping a bare electrical cable carrying 400 V is an example. The current passing through his hand paralyses the nerves causing the muscles to contract and he grips the cable tighter until heart failure occurs.

To guard against industrial hazards requires knowledge, understanding and training. Mere theoretical knowledge is not enough. New habits must be formed to suit the situation, and these must be so deeply ingrained that they are not forgotten in times of tension. Nobody can learn to drive a car safely merely by reading books, without practice under the supervision of a competent instructor. The training of men for battle or any hazardous occupation is similar.

Our attitudes towards safety and profit are changing slowly but perceptibly. Most of us would agree that our activities as a whole should show a profit. The digging of holes merely to fill them in again with nothing to show for the exercise is taken as an example of a pointless and unprofitable operation. The conflicts that arise over profits are more concerned with how they should be shared rather than whether they are necessary. Unprofitable or uncompetitive activities thus tend to phase themselves out of existence when society recognises that there is no good reason for subsidising them.

Many successes in improving safety have occurred when it was believed there was a clear economic incentive to do so. The distinction between personal safety and the safety of material goods, ships, buildings, etc is only one that has appeared fairly recently in human history. The first attempts at organised safety probably came about through elementary forms of insurance — particularly of ships and their cargos at sea through the spreading of individual risks among a group of merchants.

Loans on security of ships carrying cargoes are described by Demosthenes, the Athenian orator in the 4th century B.C.[7] The cargoes included slaves who were often as valuable as the other goods carried. Both goods and slaves would be covered by 'respondentia bonds' while the vessels were covered by 'bottomry bonds'. Although the main object of such mutual insurance was the reduction of risk, the experience and capital so gained resulted slowly in improved standards of ship construction and the development of astronomy and navigational aids.

Improvements in safety are often the sequel to some traumatic experience. One such was the Great Fire of London in 1666 which drew attention to the high fire risks inherent in the construction and spacing of buildings of that time, the absence of any co-ordinated method of fighting fires, and the needs for some form of building insurance.

Following the Great Fire, the first fire insurance office the Phoenix was founded in 1680 and the first body of fire fighters composed of Thames watermen was organised at about the same time (*Figure 1.1.1*). But for a long time each insurance company had its own separate 'fire brigade'. These finally merged in 1833 as the London Fire Engine Establishment. In 1865 it became the Metropolitan Fire Brigade.

In the UK in the nineteenth century, improvements in industrial safety were slow and generally the result of legislation, much of which lacked teeth. The early Factory Acts (dating from 1833) were concerned with the textile trades, but gradually their provisions were extended to other industries until by 1878 a consolidated act was passed and the duties of certifying surgeons were defined. The first Medical Inspector of Factories was appointed in 1898 and the Factories and Workshops Consolidation Act 1901 produced a comprehensive code for factory health and safety enforced by a centralised inspectorate. Prosecutions however were generally few, penalties low, the inspectorate understaffed and overworked, and companies defence in the Courts far better financed than the prosecution.

Figure 1.1.1.   Early fire brigade (GLC, London Fire Brigade)

Losses resulting from property damage may well have been the main spur to improvements in factory safety in the first half of the nineteenth century. Public opinion was particularly aroused by the frequent boiler explosions which occurred at the time — mostly the result of inadequate design, construction and working conditions. In 1854 the Manchester Steam Users' Association was formed to provide periodical inspection. This was followed by the Steam Boiler Assurance Company in 1858 which provided insurance based on satisfactory inspection, and the Boiler Explosions Act 1882 which provided for investigation of all boiler explosions and severe penalties where negligence was proved. Those following the history of industrial safety legislation in Britain must be impressed by the prominent part played by Trade Unions in enforcing minimum standards on the laggards. Enforcement is, however, not the only approach.

The Robens Committee was appointed in 1970 to review the legal and other aspects of industrial safety in a comprehensive way. Reporting in 1972,[8] it found that the responsibility for industrial safety was far too fragmented. A number of authorities with ill-defined areas of responsibility under a number of different Ministries or Departments had the burden of enforcing a tangled web of legislation. This was couched in language so incomprehensible to the ordinary man that a lawyer was generally needed to interpret it. Much of it was obsolete and referred to processes and factory conditions of fifty years or more ago. Over-reliance on legal regulations paradoxically encouraged apathy.

Even with up-to-date legislation and a unified agency to enforce it, it was considered that such a legalistic approach could only lead to limited improvements. The primary responsibility for improving the current health and safety record lay with those who created the risks and those who worked with them.

The outcome of the Robens report was the Health and Safety at Work etc Act 1974.[9] This created the Health and Safety Commission to take over the responsibilities for health and safety from the fragmented official authorities which previously existed. The Health and Safety Commission consists of a Chairman and six to nine members, appointed by the Secretary of State for Employment after consulting all interested parties. The Commission reports to the Secretary of State for Employment on most matters, and to the Secretary of State for the Environment on the control of emissions and to the Secretary of State for Energy on licensing nuclear installations (see *Figure 1.1.2*).

Under the Act the Health and Safety Commission were directed to set up a three man Executive to act as its operational arm. The Executive controls a unified inspectorate whose staff includes the former inspectors of factories, mines and quarries, nuclear installations, alkali and clean air, and explosives. It also controls the Employment Medical Advisory Service which had recently been set up. The Executive has a major responsibility for research, information, education and advice in health and safety at work. Both the Commission and Executive have greater power and autonomy than the authorities they replaced and can bring civil proceedings in Courts.

The Act gives wide powers to the Employment Secretary (acting on behalf of the Commission) to make new regulations which may repeal or modify existing provisions thereby improving standards where necessary and otherwise maintaining previous standards. Eventually most of the then existing safety legislation is expected to disappear, to be replaced by these new regulations. Legal aspects of Health and Safety are dealt with in more detail in section 1.11.

Alongside the official inspectorates, various voluntary organisations have grown up with the object of improving safety on a broad front by education, training and campaigning on specific issues. The most broadly based of these, and probably the best known, are RoSPA, the Royal Society for the Prevention of Accidents, and BSC, the British Safety Council.

RoSPA[10] began in 1916 as the London 'Safety First' Council under the wartime instigation of Mr. H. E. Blain (later Sir Herbert Blain) as an attempt to reduce traffic accidents in London streets. In 1924, the National 'Safety First' Association was formed. This incorporated the British Industrial 'Safety First' Association and various local organisations. In 1941 the name of the parent organisation was changed to RoSPA.

RoSPA has four functional divisions, Road Safety, Home Safety, Industrial Safety and Agricultural Safety. From its early years the organisation has attracted support from industrial leaders and leading public figures and as its name implies it enjoys royal patronage. RoSPA undertakes research, publication of accident statistics, education and training of safety officers and the public, and plans campaigns on safety aspects requiring special attention. It has long been a pioneer in the formulation and publication of safety codes, many of which have since become accepted nationally.

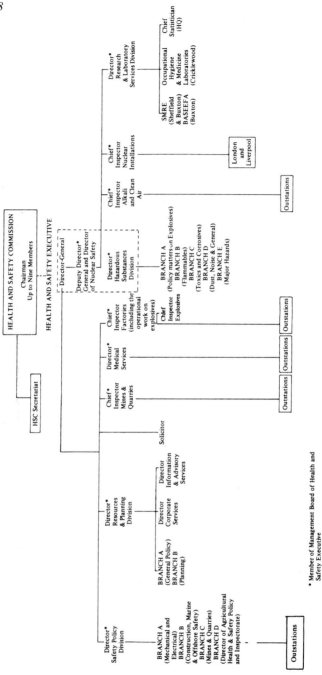

Figure 1.1.2   Health and Safety Commission. Outline organisation Chart (HMSO)

In 1925 the Society produced its first road safety code which was the forerunner of the Highway Code. In 1966, RoSPA took over the running of the Industrial Safety training Centre at Birmingham. Specialised safety engineering codes written in conjunction with leading British firms such as ICI Ltd. have been published by RoSPA for many years. These tend to be in advance of the minimum requirements set out in official codes, and to give more detailed and explicit technical guidance.

The British Safety Council[11] was formed in 1957 to conduct propaganda and education among factory employers and workers on lines similar to those of the National Safety Council in the USA. Although its work runs parallel to that of RoSPA in many ways, it seems to employ a 'hard selling' tactic and has acted as a pressure group in a number of campaigns. It claims responsibility for initiating legislation and technical improvements in the design and manufacture of domestic paraffin heaters, articulated vehicles, children's nightwear and ships' life jackets, the introduction of the kiss of life method of artificial respiration and car seat belts.[11]

While differing in style, both RoSPA and the BSC work in the same fields, in competition with each other. The main criticism that has been levelled at both organisations in the past is that their activities have not drawn in workers and trade unions into direct involvement.[11]

Both these organisations however, have undoubtedly been responsible for initiating improvements and for training, particularly of industrial safety engineers. (BSC prefers the term 'safety engineer' to 'safety officer', due to 'snobbish' implications which the latter have in their eyes.[11] In the absence of any generally agreed title we prefer here to use the expression 'safety specialist'.)

Other voluntary organisations (some of which are no longer in existence) concerned with particular aspects of industrial health and safety include:

The Association of Certifying Factory Surgeons (1887).
The British Institute of Management (1947).
The British Medical Association Occupational Health Committee (1948).
The Industrial Welfare Society (1918).
The National Institute of Industrial Psychology (1921).
The Royal Society for the Promotion of Health (1876).
The Association of Industrial Medical Officers (1955).
The Institution of Industrial Safety Officers (1952).
The Institute of Occupational Health.
The Institution of Public Health Engineers.
The Council for Science and Society (especially of high technology).
The British Society for Social Responsibility in Science (concerned mainly with worker's education).

Since the emphasis of these organisations is mainly on human life, health and safety, we should mention other voluntary organisations which are primarily concerned with property losses from industrial accidents, particularly fires. These include:

The Fire Brigades Union.
The Fire Offices Committee.
The Fire Protection Association.
The Fire Research Organisation.
The Insurance Technical Bureau.
The National Association of Fire Officers.
The Society of Fire Protection Engineers.
The Society for the Protection of Life from Fire.

There is scarcely any industrial professional or trade union association which does not have some special interest in industrial health and safety. This interest is usually catered for by a standing committee or by a special department.

Many of these associations which have overlapping interests have, in the apparent interests of their members, produced codes of practice relating for instance to the design, construction and operation of particular types of installation. The result is, in many cases, a proliferation of codes covering the same subject, all differing in one or more details from the others, which are revised and brought up to date at infrequent intervals. Such fragmentation is similar to that found by the Robens Committee in safety legislation upon which it commented so pointedly.

A similar fragmented situation existed in the USA prior to 1970. Up to that year, legislation and enforcement in matters of industrial safety had been largely left to the States, and standards varied greatly. This situation was changed dramatically with the passage of the very comprehensive Occupational Safety and Health Act of 1970. This covers approximately 55 million employees and gives powers to a number of federal enforcement agencies to develop federal codes in the fields with which they are specially concerned. Many of these codes, e.g. in matters of industrial hygiene, are now followed in the UK and other countries, and are referred to in this book. A good account of the American Act of 1970 is given in reference 12.

One obvious and sometimes forgotten point is that industrial safety is usually a complex technical subject and depends to a large extent on the technical knowledge and competence of designers, constructors and operators. A plant designed by a technically competent and safety conscious team of engineers throughout is usually a safe plant, but one flaw in the team's knowledge and expertise can result in hidden dangers which the safety specialist is frequently incapable of recognising.

Industrial hazards may be caused by human error or by physical or mechanical malfunctions, but it is very often possible to eliminate the worst consequences of human error by engineering modification.

To give an example: late at night a man runs to jump on a moving bus. He misses his foothold, falls in the road and is run over and killed by a following lorry. The direct cause of the accident was the unsafe act of the man. The underlying cause of his recklessness was his anxiety to catch the last bus home. If he missed it he knew he would have to walk five miles and explain his late arrival to a sceptical wife.

Future accidents of this kind could be prevented by discipline and persuasion. A regulation could be issued by the bus company instructing their conductors to eject any passenger attempting to board the bus while in motion, and another to their drivers instructing them to halt after such an event and not to proceed until the man had been ejected. This would be accompanied by warning notices on all buses forbidding passengers to board a bus whilst in motion and explaining the consequences.

Another and more effective remedy would be for the bus owner to have doors fitted to the bus which are remotely controlled by the driver and interlocked with the brake so that the bus could not be started until the door was closed. This would make it obvious to all would-be passengers that the bus could not be boarded whilst in motion.

Thus we see that an injury caused by human error in a particular circumstance could have been prevented by an engineering modification which would have made it far less likely for the particular accident to occur. A great deal of safety engineering is concerned with cases of this kind and requires judgement and experience to determine whether the expense of the modification is justified. The modification also needs checking very thoroughly to ensure that *it* has not introduced some new and unsuspected hazard.

REFERENCES

1. Nietzsche, F., quoted by Flesh, R., *A Dictionary of unusual quotations,* 3rd ed., Cassell, London, 4 (1968).
2. H.R.H. Prince Charles quoted by Davis, W., 'High life' *British Airways,* January (1977).
3. Jung, C. G., *Man and his symbols,* 1st ed., Aldus Books Ltd., London, 111 (1964).
4. Nietzsche, F., quoted by Jung, C. G. in Reference 3.
5. Morton, F., *Report of the inquiry into the safety of natural gas as a fuel,* prepared for the Ministry of Technology, HMSO (1970).
6. Leo Abse quoted by Tye, J., *Safety uncensored,* 1st ed., Corgi Books, London, 56 (1971).
7. Chartered Insurance Institute, *Elements of insurance,* No. 51, Wilmer Bros, Birkenhead, 9 (1974).
8. Lord Robens Committee, *Safety and health at work,* HMSO, London, 7 (1972).
9. Fife, I., and Machin, E. A., *Redgrave's Health and Safety in Factories,* 1st ed., Butterworth, London, 530 et seq (1976).
10. RoSPA, *50 RoSPA,* Birmingham (1966).
11. Tye, J., and Ullyett, K., *Safety uncensored,* Corgi, London (1971).
12. National Safety Council, 'The Occupational Safety & Health Act of 1970', Chapter 2 in *Accident Prevention Manual for Industrial Operations,* 7th Ed., N.S.C. Chicago. 20-47 (1977).

1.2 DEFINITIONS

Contents

'When I use a word', Humpty Dumpty said in a rather scornful tone, 'it means just what I choose it to mean, – neither more nor less'.[1]

Many common words such as accident, damage, hazard, injury and risk are used repeatedly in this book. So far as is possible they are used consistently and unambiguously, but to avoid misunderstandings their definitions are discussed below.

1.2.1 Accidents

Norman Crane wrote an article[2] on various meanings of the word 'accident' which illustrates the confusion that often arises from its different uses. Many people associate accidents exclusively with injury or damage and this seems to be a fairly recent tendency. It is probably a spin-off from section 80, part V of the Factories Act 1961[3] which begins:

Notification of Accidents. (1) Where an accident (a) in a factory (b)
(a) causes loss of life to a person employed in the factory, or
(b) disables any such person for more than three days from earning full wages at the work at which he was employed;
written notice of the accident in the prescribed form (c) . . . shall forthwith be sent to the inspector (d) . . .

The Act states that the word 'accident' must be given its 'ordinary meaning' and refers to Fenton v. Thorley (1903) A.C. 443 for guidance. But the Act's own meaning is no ordinary definition but a qualified one, i.e. 'a notifiable accident'. The Act does not say that a non-notifiable accident is not an accident!

Most dictionaries define an accident as an 'unintentional event'. A few of us harbour stranger notions, such as Kafka's[4] that 'Accidents exist only in our heads'.

One difficulty with equating an accident with injury or damage is that it leaves one to find another word for other unintentional events. The definition preferred here is that adopted by H.W. Heinrich[5] and later Frank Bird[6] in their classical studies on 'accident prevention'. 'An accident is an unplanned event which has a probability of causing personal injury or property damage'. This excludes many 'unintentional events' under the wider definition, yet includes a far wider range than the narrow class of notifiable accidents. When referring to an accident, we should therefore qualify it according to its type or class. Let us take some simple examples. They are all caused by the same hazard, a banana skin on a pavement. Let us try to decide which of the following are accidents and typify them.

Pedestrian (1) seeing the banana skin steps over it carefully while maintaining his pace, so as not to collide with other walkers.

Pedestrian (2) steps on the banana skin, slips, but recovers her balance and proceeds without collision, injury or damage.

Pedestrian (3), who was carrying a bottle of wine, steps on the banana skin, slips, drops the bottle which breaks, but recovers her balance and walks on uninjured.

Pedestrian (4) steps on the banana skin, slips and falls, cutting his hand slightly.

Pedestrian (5) steps on the banana skin, slips and falls, breaking a wrist and tearing his trousers.

According to our adopted definition, pedestrians (2), (3), (4) and (5) were all involved in an accident — *slipping on a banana skin*.

Pedestrian (1) was not involved in an accident but had a 'near miss'.

Pedestrian (2) had a 'non-injury, no damage accident' which we could call a minor mishap.

Pedestrian (3) had a 'non-injury accident with material damage'.

Pedestrian (4) had a 'minor injury accident with no material damage'.

Pedestrian (5) had a 'major (disabling) injury accident with material damage'.

Many people would say that pedestrian (2) had a 'near miss' or a 'near miss accident'. This is strictly incorrect as she did not miss but trod on the banana skin and slipped. Her 'near miss' referred to injury or damage, but it is preferable to relate 'near misses' only to accidents or the hazards that cause them.

1.2.1.1 Symbolic classification of accidents

If for the moment we ignore the distinction between minor and major injuries, we see that accidents can be split into four types as shown in Table 1.2.1.

Table 1.2.1 TYPES OF ACCIDENT

	A *Injury*	$\bar{A}$ *Not injury*
B Property damage	AB Injury and property damage	B$\bar{A}$ Property damage and not injury
$\bar{B}$ not property damage	A$\bar{B}$ Injury and not property damage	$\bar{A}\bar{B}$ Not injury and not property damage

This table serves to introduce a further concept, that of logical symbolism which is used later in hazard analysis. The symbols A, $\bar{A}$ and B, $\bar{B}$ are borrowed from Boolean Algebra[7] which is basic to symbolic logic. Each symbol can have one of two values only, 0 or 1. What is not A is $\bar{A}$ and what is not B is $\bar{B}$. The theme is developed further in section 2.6.

Personal injuries can also be subdivided into minor injuries, usually those having no permanent effects and leading to less than three days off work, and major injuries, leading to three or more days off work. 'Major injuries' and 'minor injuries' are 'sub-sets' of injuries, just as injuries and property damage are sub-sets of 'losses'. There are various shades of major injury including fatal injuries, although economically a fatal injury usually represents a less serious loss than one causing a serious and permanent disability, hence the truism 'it is cheaper to kill than to maim'.

Insurers often try to calculate the 'maximum probable loss' in any insurance situation while safety engineers study the 'maximum credible accident'. They are nearly, but not quite, the same thing.

1.2.1.2 Emotional recognition of accidents

Before leaving the subject of accidents, it is worth briefly considering our emotional response to them. Near misses, non-injury accidents and even very minor injury or minor damage accidents often appear comic and produce laughter. This applies particularly where the cause of the accident was a booby trap which the victim had created unwittingly. Major injury and major damage accidents are tragic or horrific and produce shock and numbness. The dividing line between these emotions is thin and fragile.

These emotive reactions probably predate man himself and appear to us to occur with chimpanzees and other primates. 'Don't sit there grinning at me like an ape'. These reactions must be part of nature's safety training. The comedy of clowns is based largely on such minor accidents. The strong emotional reaction and the ease with which it is communicated, ('Did I tell you the story of what happened to old Smithers last week?') confirm the importance of their role in nature's educational and survival programme. When our laughter subsides we may possibly reflect 'There but for the Grace of God go I'. This is healthy for it means the lesson is sinking in. Less healthy is the reaction 'Of course *I* would not be so stupid as to do a thing like that'. The Greeks had a word for that.

This natural training process confirms the axioms of Heinrich and Bird which follow from their observations of the ratio of the frequencies of accidents of various severities stemming from the same hazard. The comic-tragic message is used in many safety posters.

But the use of comedy in this connection needs care and moderation, for people and particularly children are easily intoxicated by comedy. 'Do that again Daddy, its so funny', can all to easily become 'I nearly died laughing'. But equally it is unwise to be constantly harping on the horror. This produces a dull petrification of the senses and morbid preoccupation with death and destruction. Ideally we should learn from our comedies that we may be spared the worst tragedies.

If we ignore the lessons of the minor 'comic' accident, it is likely that Karl Marx's dictum will need revision to read: 'History repeats itself first as farce, second as tragedy'.

1.2.1.3 Causes of accidents

It is an axiom of accident prevention that all accidents have causes and a further act of faith that the great majority, perhaps 99%, are preventable. But since the cause of an accident itself must have its own cause, causes are usually classed as direct or proximate causes, contributory causes and underlying or enabling causes. (This can be taken further, depending on how far back along the chain one needs to look.)

The direct cause of a preventable accident could be human error or some unsafe condition — mechanical, physical, chemical or environmental. The cause of the unsafe condition, i.e. the underlying cause of the accident is again usually human error or ignorance.

It is seldom however that a single cause on its own results in an accident. Usually there is a combination of causes. Where one is predominant it is usually referred to as the direct cause whilst others may be contributory causes. Where it is impossible to make this distinction the cause is a 'multiple one'. Precise distinctions are often difficult, though it usually pays to make the effort.

Accidents then have a hierachy of causes which may be simply presented as:

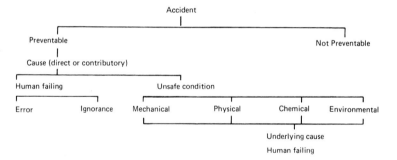

1.2.1.4 Proximate causes

The proximate cause of an accident is the nearest or most proximate cause to it. But in English law at least the proximity of the cause tends to be judged by its 'efficiency' rather than whether it was closest to it in time.[8] It assumes a special importance in insurance claims and lawsuits for compensation.

Before an insured can recover from his insurers for a loss, the proximate cause must be established. He can only recover if the proximate cause is covered by his policy. The legal maxim is 'causa proxima non remota spectatur' (the immediate, not the remote, cause is to be regarded). This can lead to endless legal arguments, which are discussed further in section 1.11.

A good illustration from the USA, where similar considerations apply, is given by Ralph Nader[9] in his famous exposé of unsafe American car design. A car just turned into a garage for service was being driven by the assistant manager to a service bay. The brakes were faulty and he was unable to prevent it hitting a mechanic, crushing his leg. The defective brakes resulted from an inadequate O-ring seal in the power braking system. This was a design or manufacturing fault common to that model of car. The manufacturer had failed to notify the car owner of this fault and the need to have it corrected. The owner had, however, mentioned to the assistant manager of the garage when he brought it in for service that his brakes had failed the previous day. The mechanic sued the car manufacturer for compensation for the loss of his leg. The judge said he believed General Motors had been negligent in not notifying the owner that the brakes were faulty and needed modification. But the mechanic lost his case because the action of the assistant manager in driving the car without brakes into the mechanics leg was a 'new and independent proximate cause of the injury'.

This, fortunately for the mechanic, was not the end of the story. He appealed to a higher court and as a result got compensation, while the question 'What was the proximate cause?' was left in mid air. The concept of 'proximate cause' seems to a layman to be unsound for the majority of accidents result from a combination of two or more causes or conditions, in which all had to b present before the accident could result. It is probably not wholly unfair to describe it as a fiction of the legal mind, to suit which facts have to be adjusted before judgement can be passed.

1.2.2 Hazards

The causes of accidents generally remain latent for some time before an accident occurs. These latent or potential causes are hazards. A hazard is a condition with the potential of causing injury or damage.

Hazards are sometimes referred to almost synonymously with accident causes, and with the same subdivisions. But a clear distinction should be made. (A hazard can exist without an accident whereas an accident cause without an accident is an absurdity.) Some American writers classify hazards as primary, initiating and contributory hazards.[10]

If an accident caused by one or more hazards sets in train a second accident, the first accident is a sequential hazard.

The gas explosion in a flat at Ronan Point, for example, not only blew out a wall panel of the flat but triggered a progressive collapse of the wall panels above and below it. The explosion which blew out the first wall panel was a sequential hazard. Some would identify the primary hazard as the method of tower block construction then in use, some as the gas leak, some as faulty plumbing and others as the use of gas in high rise flats. The primary hazard thus depends on our perception of causes and evaluation of effects.[12]

Hazard recognition, diagnosis and elimination are essential to any successful safety programme and form the major part of this book.

1.2.2.1 Hazard recognition

As with financial speculation, there are two approaches to hazard recognition, the 'fundamental' and the 'technical' approach, better known as Total Loss Control.[11]

The fundamental approach really amounts to a study of all possible hazards that could exist — first qualitatively, to try to be sure they are all recognised, then quantitatively to try to calculate the probability inherent in each one — and the overall probability of an accident.[9] This approach should be used when a new product is to be designed and sold, whether this be a car, aeroplane, moon rocket, child's nightie, sleeping tablet, contraceptive, chemical plant or electric blanket.

The technical or loss control[11] approach involves the careful recording and study of as many accidents as possible in order to identify and eliminate the hazards that led to them. This activity needs to be carried out in a methodical way for each department or operation within an organisation where losses are experienced. It is seldom if ever possible to record every accident whether loss causing or not, but the recording and analysis of property damage accidents as part of a safety programme was put onto an organised basis and became established mainly through the work of Frank Bird.[6] Causes are also generally easier to establish after an accident involving property damage than after one involving serious injury. There are several reasons for this, one being that those involved generally have less fear of incriminating themselves.

This approach has proved successful in many industries. Besides reducing the incidence of accidental injuries, it has paid off and shown a good profit on the reduced incidence of property damage. It could clearly not have been applied directly to the landing of the first man on the moon, but it may have paved the way to it.

A simple tree, for hazard recognition is thus as follows:

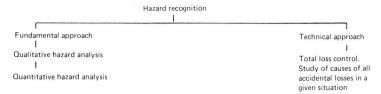

1.2.3 Non-accidental losses

Whilst we have seen that *some* accidents cause losses, i.e. injuries or damage, it does not follow that all losses are caused by accidents. They may be deliberate or planned, or they may be due to some slow and progressive detioration in human or machine function which is still faster than that expected from normal wear and tear or obsolescense.

The borderline between deliberate acts and unplanned events or accidents is a difficult one to draw, and fortunately the safety specialist or accident investigator is usually spared this responsibility. It is more the realm of lawyers, psychiatrists and the police.

There is another ill defined borderline *within* the class of accidents between avoidable and unavoidable accidents, or 'acts of God'.

We would not, as a rule, class a gradual impairment of hearing caused by excessive noise as an accidental injury, nor would we class the abnormally rapid failure of a piece of equipment, say by excessive vibration, through stress fatigue, as accidental damage. Such failure might however itself cause some other accident. The distinction is largely a matter of the time during which the damage occurred. An accident is usually thought of as something occurring within a short time spell, usually less than a second, and seldom more than a few minutes. It is a distinct, though brief, event.

The slower deterioration is less easy to define in words although it is as much the concern of those responsible for safety as are accidents themselves. 'Abnormal deterioration or impairment of function' seems the most general description. For humans it is often classified as an industrial or occupational disease, while for machines it may be fatigue, corrosion or *abnormal* wear and tear.

All losses, whether injury or property damage, whether caused deliberately, accidentally, or by abnormal deterioration, are likely to represent additional hazards in themselves, i.e. potential causes of further accidents and losses.

1.2.4 Other words

Danger expresses the degree of exposure to a hazard. By taking suitable precautions, the danger or degree of exposure to a given hazard is reduced. The guarding of machinery or the insulation of an electric cable reduces the danger or degree of exposure to the hazards.

Safety is the opposite of danger. Freedom from hazards represents absolute safety, but this is an ideal which is seldom realised. Safety is rather a matter of protection from hazards.

Safe and safety are indeed words which have many pitfalls and deserve a word of caution. First they are often misused for commercial, advertising and similar ends to manipulate a person into taking a decision desired by the person using the word. Second, something may be described as safe through genuine ignorance of a danger. Thalidomide and blue asbestos are examples in one way or another of both pitfalls. The most precise definition of safe known to the author is that 'A thing is provisionally categorised as safe if its risks are deemed known and in the light of that knowledge judged to be acceptable'.[12]

Chance, *probability*, *uncertainty* and *risk* are words which frequently recur and are most familiar to those concerned with insurance. Risk and uncertainty are ingredients of life, and we are constantly having to take risks merely to stay alive.

Probability is an objective mathematical term having a value between 0 and 1, where 0 represents complete impossibility and 1 represents absolute certainty. Statistics is the scientific study of probability.

Chance refers to the probability of an uncertain outcome of some event, and maybe spoken of qualitatively as 'a good chance' or 'a sporting chance'. *Uncertainty* exists only in our minds; and has much the same meaning as doubt. Its opposite is often regarded as faith, although Tennyson has pointed a paradox here:

"There lives more faith in honest doubt
Believe me, than in half the creeds."

Risk is used for uncertain eventualities where the probability can be expressed mathematically. As insurance contracts cover the consequences of undesired events, insurers refer to covering a risk.[8] Often the word is used loosely to refer to the subject insured, e.g. burglary, fire loss or having twins.

Risks may be classed as speculative or pure. A 'speculative' risk is one taken voluntarily where there is a possibility of gain or loss, such as betting on the pools. 'Pure' risks are ones where the only possibilities are loss or breaking even. Insurance allows us to exchange a small risk of a large loss for a small but certain loss — the premium. Every insurable risk should be measurable. The basic premium equation is:

$$\text{Premium} = (p \times \bar{c}) + E$$

where p = probability of a loss occurring;
$\bar{c}$ = average size of loss which occurs;
E = loading for expenses and profit.

REFERENCES

1. Carroll, L., *Through the Looking-Glass,* London (1872).
2. Crane, N.C., '*Just what is an accident*', Industrial Safety; **23** No. 3, 10 '1977).
3. Fife, I. and Machin, E.A., *Redgraves Health and Safety in factories*, 1st ed., Butterworth, London, 270 (1976).
4. Kafka, F., quoted by Flesh, *A dictionary of universal quotations*, 3rd ed., Cassell, London (1968).
5. Heinrich, H.W., *Industrial accident prevention*, 4th ed., McGraw Hill, New York, 16 (1959).
6. Bird, F.E., and Germain, G.L., *Damage Control*, 1st ed., American Management Association, New York, 23 (1966).
7. Ward, B., *Boolean algebra*, Howard W. Sams & Co. Inc., Indianapolis, USA (1971).
8. Chartered Insurance Institute, *Elements of insurance*, Chartered Insurance Institute, London, 2 (1974).
9. Nader, R., *Unsafe at any speed*, Grossman, New York, 43 (1965).
10. Hammer, W., *Handbook of system and product safety*, 1st ed., Prentice Hall, New York, 85 (1972).
11. Tye, J., *Management introduction to total loss control*, 9th ed., British Safety Council, London (1976).
12. The Council for Science and Society, *The acceptability of risks*, Barry Rose (Publishers) Ltd, London, 12 and 16 (1977).

1.3 THE COST OF ACCIDENTS

Contents

Can meaningful costs be given to accidents? If so, how will the knowledge help us?

For material loss accidents in which no injury occurs, accounting methods can give us reasonable figures. This is the essence of damage control which is discussed under Total Loss Control in section 1.5. A damage control programme works by pinpointing areas where unnecessary losses occur, by stimulating prevention measures and programmes, and thereby improving the firm's balance sheet. The measures taken to reduce material damage usually also remove hazards which cause human injury, and so improve safety generally. The damage costs obtained during the course of the programme are seldom complete, mainly because of the rather arbitrary dividing line between accidental damage and normal wear and tear. In a progressive firm, much of what was yesterday taken for granted as normal wear and tear will tomorrow be regarded as accidental damage and be subject to the scrutiny of a subsequent damage control programme.

For accidents leading to human injury and death, the costing becomes more difficult. How do we value a man's life? Can any meaningful price be assigned to it? Particularly since the abolition of slavery when human beings ceased to be bought and sold as commodities, many people have posed this question only to realise that human life just cannot be expressed in financial terms. Philosophically human life and money are on different planes. Compensation however generally has to be paid — usually by the insurers to the victim for serious injury or disablement or to his estate or dependants (or sometimes both) if he is killed.

The Law Courts, the insurers and the social services together have between them produced a large number of precedents which provide a broad spectrum of compensation costs for industrial injuries and fatalities. But the range between the highest and lowest compensation paid in similar circumstances is very wide, so wide in fact that the situation has been described by at least one writer as a lottery. Those who pursue their claims through the Courts as a rule do better than those who rely entirely on government insurance and social security, although the overall legal costs on average amount to 40% of the total payments and expenses.

Settlements in recent years rose very sharply, first in the USA. The same trend is apparent in the UK. A recent report of a Royal Commission on Civil Liability for Personal Injuries is discussed in section 1.10.4.

Most industrial risks are, however, insured by the employer so that the employer seldom pays compensation directly. The employer, as we see later, generally incurs a number of additional expenses incidental to the injury which are not covered by insurance, but his ability to pass the main risk on to the insurer does reduce his incentive to do as much as he otherwise might do to make his factory safer. The insurers for their part make a valuable contribution to fire safety (e.g. by charging lower premiums for premises with automatic sprinklers), but seem to be less effective in other areas of safety. They tend to hedge their own bets by spreading risks rather than discriminating much in their premiums between good and bad risks for the same type of operation (see section 1.10). This has proved frustrating to firms who have taken a leading part in studying hazards and reducing accidents. They naturally feel 'Why should we pay (through our insurance premiums) for the cost of all Joe Bloggs' accidents down the road? We know he only pays lip service to safety whilst we have a serious well organised programme on which we spend a lot of money.'

One approach[8] to this problem, described later in this section, is to treat the total costs of accidents and of preventative measures taken to avoid them as the total accident cost, and then set out to minimise this cost. Taken over the country as a whole (or over a large company which has a unified safety programme), in any particular field (such as agriculture or industry) the cost of accidents will fall as the expenditure on safety measures increases (see *Figure 1.3.1*). By plotting both accident and prevention costs and the sum of the two against some arbitrary 'risk reduction' scale (*Figure 1.3.2*) the sum of the two costs will show a minimum at a point not far on the risk reduction scale from the point where the prevention cost curve crosses the accident cost curve.

1.3.1 Industrial accident costs in the UK

In the Times newspaper a letter gave an estimate of the cost of industrial accidents in 1974 as £900 m. The precise arithmetic was not given, but the unit costs used in the calculation were as follows.

Fatalities	£3000
Group I injury	
(more than 4 weeks off work)	£1000
Group II and III injury	
(between 3 and 28 days off work)	£ 500
Non-reportable injury accidents	£ 10
Material damage-non injury accidents	£ 10

One interesting point of this estimate is that the cost of material damage caused by accidents at £660 m is higher that the costs of human death and

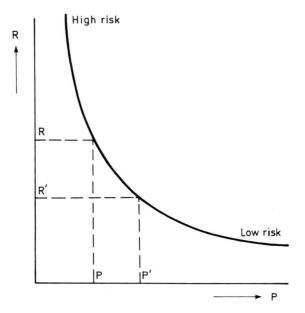

Figure 1.3.1. Relationship between risk level and prevention cost

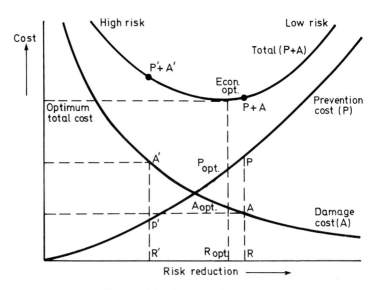

Figure 1.3.2 Costs vs. risk reduction

injury. The estimate was based on accidents reported under the Factory Act, 1961 and published in the 1974 report of H.M. Inspector of Factories. The scarcity of information on the subject in the UK is illustrated by the answer given by Roland Moyle, Minister of State, DHSS, on January 12th, 1977 to one of a series of detailed questions on industrial injuries put in the House of Commons by Jeffrey Rooker (Birmingham, Perry Bar).[2] *Mr Rooker* asked the Secretary of State for social services:

1. If he will publish an estimate of the national resource costs of occupational accidents and diseases, giving figures for:
   (a) fatalities
   (b) industrial accidents
   (c) the cost associated with prescribed industrial diseases
   (d) long term incapacity, and
   (e) other factors.
2. If he will publish an estimate of the subjective costs of occupational accidents and diseases;
3. If he will publish the resource and subjective costs of occupational accidents and diseases as a percentage of GNP!

In his reply Mr Moyle stated that 'No estimates are available of the resource and subjective costs of occupational accidents and diseases subsequent to those published in Appendix 9 to the 1972 report of the Robens Committee on Safety and Health at Work.[3] It was estimated in the report that the total of these costs amounted to 0.87% of GNP!

1.3.1.1 Data from the Robens report

The estimates made for the Robens Committee were prepared by the Research and Planning Division of the Department of Employment based on data for 1969. Other estimates of the annual cost of industrial accidents in the UK had been made by Beckingsale[4] in 1963 (£264 m) and Hanna[5] in 1970 (£220 m).

National resource costs are those born by the economy as a whole and need careful compilation to avoid the inclusion of transfers as between the employer and the Exchequer. Two sources of data were used, those of the Department of Health and Social Security and those of HM Factory Inspectorate. The DHSS data were more comprehensive, covering fatalities by statistics of industrial death benefit and non-fatal accidents by statistics of industrial injury benefit for the whole field of employment. The HMFI data gave fatalities and reportable accidents (see section 1.2) at premises covered only by the Factories Act.

The best estimates derived from the two sets of data are summarised in Table 1.3.1.

Table 1.3.1 ACCIDENT COST DATA

DHSS DATA	Number	Cost £m
National resource costs		
Fatalities	1 918	20.2
Industrial accidents	841 680	
lost output		84.5
medical and hospital costs		12.4
Damage and administration		42.1
Prescribed diseases		3.7
Long term incapacity		8.1
Non-reportable accidents		37.9
		208.9
Subjective costs		
Fatalities		9.59
Serious injury		42.08
Slight injury		77.75
Total subjective costs		127.42
Total resource + subjective costs		336.3

HMFI DATA (for premises within scope of Factories Act)	Number	Cost £m
Resource costs		
Fatalities	649	8.28
Industrial accidents	441 630	54.7
Under-reporting		19.8
Non-reportable accidents		14.5
Total resource costs		97.3
Subjective costs		36.6
Total resource + subjective costs		133.9

The Robens report mentions that it received many estimates of the total costs of industrial accidents and diseases ranging from £200 m to £900 m annually.

1.3.2 Accident Costs in the USA

Before leaving the subject of national accident costs, it may be useful to look at some readily available statistics for the USA in 1965 which cover accidents in the home and on the roads as well as at work. The total cost was estimated at $18 000 m, made up as follows:

	Million$
Accidental injuries	10 700
Property damage in motor vehicle accidents	3 100
Property destroyed by fire	1 455
Property destroyed and production lost in work injury accidents	2 800

The cost of accidental injuries, excluding material damage was $11 000 m, made up as shown in Table 1.3.2.

Table 1.3.2

Cost	Work	Home	Motor vehicle	Public non-motor vehicle	Total
Wages lost	1 400	850	2 400	850	5 500
Medical expenses	650	450	550	250	1 900
Overhead cost of insurance	750	10	2 850	10	3 600
Sub Total	2 800	1 300	5 800	1 100	11 000
Sub total as % *of grand total*	25.5	11.8	52.7	10.0	100.0

1.3.3 Hidden costs of industrial injury accidents

Whilst the following US analyses may not apply exactly in the UK because for example of the different medical systems, they provide a useful insight into the total cost of an industrial accident.

Heinrich[6] found an average ratio of 4:1 between the hidden costs of an injury causing accident in American industry and the 'direct costs' (meaning the costs of medical treatment and compensation paid to the employee).

The main factors in these hidden costs are:
1. Cost of lost time of injured employee.
2. Cost of time lost by other employees who stop work to assist injured employee, out of sympathy, curiosity, etc.
3. Cost of time lost by foremen, supervisors and other executives in the following:
   Assisting injured employee;

Investigating accident cause;
Arranging for continuation of injured employee's work by other
persons;
Selecting and training a replacement;
Preparing official reports and attending hearings.
4. Costs of time spent by first aiders, hospital staff not included in direct
costs.
5. Costs of damage to machines, tools or other property or material in
process.
6. Incidental costs of lost production (failure to fill orders on time, loss of
bonus or payment of penalties).
7. Costs under employee welfare and benefit systems.
8. Costs of full wages of employee on his return to work before his full
recovery.
9. Cost of loss of profit on productivity of injured employee and idle
machines.
10. Costs arising from excitement or lower morale of other employees.
11. Overhead costs of lost production causes by accident — heat, light,
rent, etc.
Several examples are given by Heinrich and in other safety literature.

1.3.4 Costs of non-injury accidents

No general costs of non-injury accidents over a range of industries could be
found either in the UK or elsewhere. The accounting methods used in
most companies are seldom designed to distinguish between 'accidental
damage' and 'normal wear and tear'. Accident damage to cars and road
vehicles which are insured is an exception. Otherwise it is only where a
policy of 'damage control' or 'total loss control' is applied and implemented
and backed by appropriate accounting methods that valid estimates of
accidental damage can be made. The well-known ratio of 1:100:500 for the
numbers of disabling injuries: minor injuries: property damage accidents
was found by Frank Bird and George Germain[7] through the intensive
investigation of one US Company only — Lukens Steel — over a six year
period. During this time the costs of property damage fell from $325 545 to
$137 832 per million hourly rated man hours worked. Clearly no firm is
likely to undertake studies of this kind if it did not intend, through their use,
to reduce the incidence and cost of property damage. Thus the very study of
meaningful property damage costs should lead to a reduction in these costs.

1.3.5 Accident cost optimisation

The work mentioned briefly in the first part of this section (1.3.1) and
illustrated in *Figures 1.3.1 and 1.3.2* was initiated and carried out by T.
Craig Sinclair on behalf of the Robens Committee on Health and Safety.
Referring to *Figure 1.3.2*, it is clear that as the costs of injuries and
damage are reduced as more money is spent on prevention, the point must

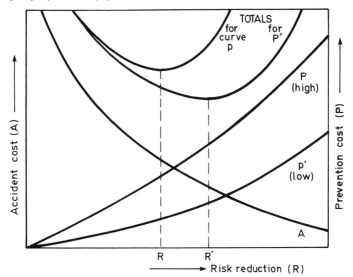

Figure 1.3.3   Cost vs. risk reduction. Uncertain prevention function

come when the incremental cost of preventative measures exceeds the incremental saving on accident costs. This is at the minimum total cost of accidents plus expenditure on prevention. The costs of accident prevention are relatively easy to assess but the costs of accidents contain many subjective elements, so that it is preferable to use minimum and maximum accident costs and draw both curves as in *Figure 1.3.3*, as well as both total cost curves. The minima are not far apart on the cost/risk reduction diagram. In both cases, the minima occur close to the point where the cost of accidents is equal to the cost of preventative measures.

Sinclair makes the simplifying assumption that the total costs of accidents (including preventative measures) are minimised when the annual cost of preventative measures has reached the annual cost of accidents. This assumption, as noted later, appears to create more questions than valid answers.

1.3.5.1 Estimating prevention costs

Craig Sinclair has grouped preventative costs under three headings:
   A. Design costs
   B. Operational costs
   C. Planning and consequence limiting costs.
These are summarised below but before examining them in detail, we should decide whether to be concerned only with personal injuries and their costs, or with accidental injuries and property damage as well. It is easier,

though perhaps less accurate, to start with personal injury costs on their own.

A. Design costs

The design costs of plant protection, which should be assessed for each plant item in turn, will then consist of:

1. All redundant control features installed for safety reasons.
2. All machine guards and protective devices installed on the machine to protect employees.
3. The part of the cost caused by additional metal thickness (called for under the appropriate design code) which is required to provide a margin of safety. (This should not include corrosion allowances since the part may reasonably be expected to corrode to that extent during its working life as a result of normal wear and tear.)
4. Systems installed to remove toxic and explosive materials produced by the process and maintain a safe and healthy working environment.
5. Fire protection measures — escape routes, fireproof materials and barriers.
6. Additional costs involved in layout for safety reasons.

B. Operational costs

1. Costs of the safety department which can be attributed to the plant under study;
   (a) salaries and overheads;
   (b) publicity material;
   (c) training;
   (d) protective clothing.
2. Cost of extra manning for safety reasons.
3. Costs of operating within restricted range of conditions (temperature, pressures, rate, etc) for safety reasons, as compared with the wider range which would otherwise be possible.
4. Cost of medical and first aid department.

C. Planning and consequence limiting costs

These, according to Sinclair, would include:
1. Cost of insurance.*
2. Cost of works fire brigade.
3. Cost of fault studies, hazard analyses, safety audits.
4. Costs of environmental sampling and analytical programmes; biological controls.
5. Costs of toxicity tests on drugs, food, etc.
6. Costs of testing for leaks of hazardous materials.
7. Costs of testing for electrical safety.

8. Costs of loading tests (structure) and pressure tests (pressure vessels).
9. Costs of flammability tests.
10. Related R and D costs.

Note: Sinclair does not explain this further, though it seems that insurance premiums should be included under costs of accidents rather than preventative measures. The payment of an insurance premium does nothing in itself to make the plant safer!

In all these costs, there is a great deal of arbitrariness as to where one places the zero or datum point. The law obliges employers and plant suppliers to adopt certain minimum safety features. Furthermore, there is no way of knowing what the accident rate or cost would be if no safety features or margins were allowed.

It therefore seems reasonable to modify Sinclair's approach (for an existing plant or factory) by taking the plant as designed and built as ones datum point, and only include those costs incurred by providing additional protection. In this case one can no longer postulate, as Sinclair did, that the optimum total cost is minimised when the costs of preventative measures equal the cost of accidents. Instead both prevention and accident cost curves must be plotted, and the minimum in the total cost curve determined.

1.3.5.2 Estimating accident costs

The costs of accidental injuries are made up of three parts, fatalities, serious injuries (over four weeks off work) and other injuries.

Thus we can write

$$C_A = R_D \times (A_{SD} + A_{OD}) + R_S(A_{SS} + A_{OS}) + R_O(A_{SO} + A_{OO})$$

where C_A = annual accident cost per worker;
R_D = annual risk of death per worker;
R_S = annual risk of serious injury per worker;
R_O = annual risk of other injury per worker;

average A_S = subjective element of cost
A_O = objective element of cost
with second subscript D, S or O for death, serious injury and other injury.

The preventative cost P is then compared with the accident cost C_A per worker.

Sinclair has applied this method with meaningful results to large groups of workers in entire industries, such as agriculture. The method clearly cannot be applied to very small groups where the results will have little statistical significance, but it can be applied to large groups working on identical plant and machinery.

Whilst there are many pitfalls in application, including the time lag needed before preventative measures become effective, the above method nevertheless provides a useful mental yardstick with which to approach the costs of accidents and prevention programmes. The reader who may occasionally wish to apply it quantitatively is advised to read Craig Sinclair's paper together with other recently published works on the subject.

1.3.6 Can there be an economically rational safety policy?

Craig Sinclair would be the first to admit that his analysis is based on the assumption of virtually full employment. Thus he does not take into account the saving to the nation in reduced unemployment benefits paid when a dead or disabled worker is replaced by one on the dole. If faced with the option of working at some risk of injury to themselves or not working at all, most workers would choose the former.

Taking the UK only, (although the situation is similar in other countries), many now think that job opportunities will become less favourable over the remainder of this century as our non-renewable energy resources gradually disappear. On this view it is fortunate that our population forecasts indicate a decline from 53 million odd now to about 50 million by the year 2000. In line with this plausible line of thought, voices have been heard, even at meetings on industrial safety, suggesting that the loss of a few more heads by the year 2000 might even make life a bit easier for the remainder.

Some further points should not be overlooked. The loss of a relatively young but experienced worker who has undergone long and expensive training represents the loss of a considerable investment, whereas the death of a worker in his sixties will save the expense of keeping him in retirement. Disabling injuries, particularly of young workers, are more expensive to the nation than fatalities. If we wanted a rational economic safety employment policy, we might consider reserving the safest jobs for the young and newly trained, gradually increasing their risk exposure in middle age and reserve the most hazardous occupations for the elderly, and probably more reliable, worker.

The other side of the coin should not be overlooked. Safety and accidents in their various aspects constitute an industry in themselves, with sales and export potential. They provide considerable employment to firemen, policemen, inspectors, doctors, nurses, lawyers, investigators, newsmen, publishers, insurers, expert witnesses, safety equipment suppliers and public relations men, besides forming an inspiration for script writers and the world of show business. This applies particularly to long and expensive public inquiries, paid for by the taxpayer who hopes for enlightenment but at the least expects his money's worth in entertainment. Also on the profit side of the balance sheet should appear the knowledge and experience gained. 'It's an ill wind that blows nobody good.'

It is clear then that the cost and economic consequences of industrial accidents and diseases are by no means clear or conclusive. One is forced in the end to recognise that economics is only one among many factors to be considered in conjunction with industrial accidents, injuries and disease. The intangible advantages of a whole and healthy population are overriding. Cost considerations and calculations of cost effectiveness can at best only be used as a guide to determining priorities and means of improvement.

Before leaving the subject, it is worth mentioning two other estimates made on the value (positive or negative) of human life.

The first made in connection with family planning studies assesses the gain to various societies through reduction in birth rate by various means.[9, 10] The second which was on somewhat similar lines to Craig Sinclair's exercise was also made to provide an economic yardstick for expenditure on various measures of accident prevention. The author, Melinek, concluded that in 1972 the value of a human life was about £50 000. He added, however, that there is a large difference between the value people place on their own lives and the value placed on them by society at large, which is usually much lower.[11]

REFERENCES

1. Boyle, A.G., Letter to the Editor, *The Times*, London, 8th December (1976).
2. Weekly Hansard, House of Commons Parliamentary Debates, HMSO, London, 516, Issue No. 1058 (10 to 13 Jan, 1977).
3. Lord Robens Committee, *Health and Safety at Work*, Appendix 9, HMSO, London (1972).
4. Beckingsale, A.A., *The cost of industrial accidents*, Alexander Redgrave Memorial Lecture, Institution of Industrial Safety Officers, Royal Society of Arts, London, 26 April, 1963.
5. Hanna, V., *Sunday Times*, London, 8 February, 1970.
6. Heinrich, H.W., *Industrial accident prevention*, 4th ed., McGraw Hill, New York, 50-61 (1959).
7. Bird, F.E. Jr., and Germain, G.L., *Damage control*, American Management Association.
8. Sinclair, T.C., *A cost effectiveness approach to industrial safety*, HMSO, London (1972).
9. Laing, W.A., *The costs and benefits of family planning*, PEP, London (1972).
10. Cowles, R.B., 'The Non-baby bonus', p. 339 in Hardin, G. *Population, Evolution and Birth Control*, 2nd Ed., Freeman & Co, San Francisco (1969).
11. Melinek, S.J., *A method of evaluating human life for economic purposes, Fire Research Note 950*, Fire Research Station, Boreham Wood, Herts, England (1972).

1.4 SAFETY RESPONSIBILITY AND ORGANISATION

Contents

This section deals with safety responsibilities and organisation within industrial and commercial firms, public companies and enterprises owned by the state and local authorities. They may vary in size from very small firms with only a handful of employees to large multi-national corporations. International corporations may present a special case with problems of their own when, as sometimes happens, the technical and managerial control (with all its implications for safety), is exercised from abroad. Safety responsibilities, particularly those of senior management, could be split under the headings: economics, human relations, and the law.

In economic terms, accidents cost money. Whilst many of the risks are insured (and some must be by law), the true costs nearly always exceed the agreed claims, since they include many incidental costs not covered by the policy. While a company with a good accident record and an effective safety organisation may secure substantial reductions in premiums, industrialists with good records do complain that insurers do not discriminate sufficiently between good and bad risks.[1]

Human relations generally transcend purely economic considerations and are governed by conscience, honour and ethics. Good human relations within a firm require that management pay proper attention to employee health and safety.

The law lays down minimum standards which have been increased considerably by the Health and Safety at Work etc, Act 1974. The Robens report[2] which led to this act found the previous position unsatisfactory and pointed to two essential needs:

explicit policy objectives
effective organisation in which individual responsibilities are clearly defined.

In the words of the report, 'In our investigations we formed the impression that undivided line management responsibility for safety and health matters more often than not stops at some point in the middle-management chain: further up the chain the responsibility tends to become diffused and uncertain. Safety and health should be treated like any other major management function, with a clear line of responsibility and command running up to an accountable individual at the very top. The other crucial level is that of first-line supervision. It is the supervisor who is on the spot and in a position to know whether or not safety arrangements are working in practice. His influence can be decisive. Both here and abroad, whenever we have seen outstanding safety and health arrangements it has been clear that a key role is played by well trained supervisors who are held accountable for what happens within their sphere of control. We are not at all satisfied that this key role in safety is sufficiently recognised throughout industry generally or that enough is done to equip supervisors for it.'

Responsibilities for safety in an industrial organisation were explicitly stated many years ago by Heinrich[3] who envisaged four categories of persons: directors and senior management, safety officers, supervisors and foremen, and employees. His clear and simple ideas are still valid and are summarised below.

1.4.1 General responsibilities for safety

Employees at all levels are involved in accident prevention, but the main responsibility rests with the directors and managenent who alone have the authority to issue orders and direct work. Provided, however, that the correct orders are given, the actual time required from top management may be quite small. More time and effort are required from the safety officer, the supervisors or foremen, the training department and the workers themselves, although once a hazard has been successfully eliminated it should not reappear provided normal caution is exercised.

The implementation of proper safety measures is sometimes resisted on the grounds that they slow down or interfere with production. Of course the introduction of any new and safer method or piece of machinery is likely at first to cause a temporary drop in production while the method is learnt and new habits are being formed. Soon, however, production reaches its previous level and then overtakes it as employees find they are able to work faster, knowing that certain risks have been eliminated and that they no longer have to worry over them. If this does not happen it is generally a sign that the incorrect remedy has been applied. A stop in production rarely happens when a properly organised safety programme is in operation, although it may occur where ill judged remedies are applied in a haphazard and disorganised way. In the general experience of industry the safe works is the efficient works.

1.4.1.1 Responsibilities of directors and senior management

Management is responsible for controlling the unsafe acts of employees chiefly because the unsafe acts occur in the course of employment which the management creates and directs. Management can control its employees by selection, training, instruction and supervision.

The responsibilities of the directors and senior management of a company for safety are basically twofold:

1. Management is responsible for the safe mechanical and physical conditions in the work places of which it has charge;
2. Management, because of its ability and opportunity, is responsible for preventing unsafe working practices by its employees.

Typical hazards in the first category are unguarded machines, worn or defective tools, inadequate light or ventilation. These are obviously within the control of management which is the sole authority in decisions about purchasing, positioning, operation, maintenance and guarding.

Hazards in the second category include oiling moving machinery, removing guards, placing material in gangways and working spaces, and riding on loads suspended by cranes.

1.4.1.2 Responsibilities of the safety specialist

The safety specialist or engineer is employed by management and is part of the managerial and supervisory staff directing the work of employees. The safety specialist thus shares the responsibilities of management as described above. Because of being specially qualified in safety work and often in direct charge of it, the safety specialist has the opportunity of advising others. He (or she) will periodically inspect the plant, machinery, tools and various work operations in order to determine mechanical and physical hazards and unsafe practices of employees. The specialist suggests and recommends improvements, takes part in the training and education of supervisors and employees, conducts or participates in safety meetings, acts as co-ordinator of safety work and as liaison agent with higher executives. In general, the task of the safety specialist is to supervise and promote the work of accident prevention.

1.4.1.3 Responsibilities of supervisors and foremen

The responsibilities of management also apply to supervisors and foremen. The person who gives direct instructions to employees is in a key position in reducing accidents, is closely associated with the work force, knows them personally, and is acquainted with their habits, grievances, attitudes and personal qualities. The foreman is generally at least as skilled as they are and may work beside them and controls the work force by influence and example. This control is a most important factor in safety performance. The sympathetic and intelligent support of able supervisors and foremen is essential to any safety programme. Proper training of supervisors in accident prevention methods is essential.

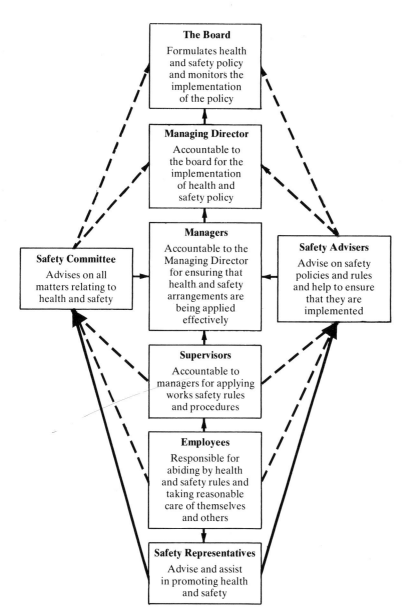

The Board
Formulates health
and safety policy
and monitors the
implementation
of the policy

Managing Director
Accountable to
the board for the
implementation
of health and
safety policy

Managers
Accountable to the
Managing Director
for ensuring that
health and safety
arrangements are
being applied
effectively

Safety Committee
Advises on all
matters relating to
health and safety

Safety Advisers
Advise on safety
policies and rules
and help to ensure
that they are
implemented

Supervisors
Accountable to
managers for applying
works safety rules
and procedures

Employees
Responsible for
abiding by health
and safety rules and
taking reasonable
care of themselves
and others

Safety Representatives
Advise and assist
in promoting health
and safety

*Figure 1.4.1   Company safety programme (Reproduced by permission of the
Controller of HMSO)*

1.4.1.4 Responsibilities of employees

The main responsibility of the non-supervisory employee is to oneself and ones dependants, to avoid being injured and thus impairing the ability to support them. Employees have no authority to instruct fellow employees nor are they responsible for inherent hazards in the equipment which they operate. But they have responsibilities to their employer for the safe conduct of their work and the maintenance of equipment:

1. Every employee should report unsafe working conditions which are outside his own control to his supervisor.
2. They should not interfere in the work of fellow employees in a way liable to create hazards.
3. They should not allow other employees to interfere with their own work in such a way.

Despite their lack of authority, employees may have opportunities to assist their fellows by passing on their knowledge and experience in safety matters and they should be encouraged to take active and helpful parts in safety meetings.

Employees who are members of a union or other organised labour group may participate directly or through their representative in safety inspections, in formulating safety rules and making recommendations. Although the opportunities and responsibilities of any individual employee are limited, their collective opportunities and responsibilities in a department or works are considerable.

Accident prevention is thus a co-operative task in which management has the main responsibility, employees must do their share, and the foreman or supervisor bears most of the burden of detail.

1.4.2 Responsibilities under the Health and Safety at Work etc Act 1974

The Act emphasises the all embracing responsibilities of employers in the following areas:

General responsibilities;
Directors' responsibilities;
Records;
Organisation, training and communication;
Duties of suppliers;
Control of contractors;
Role of inspectors;
Safety Programme and Implementation.

1.4.2.1 General responsibilities

'Employers must take account of circumstances in which . . . a reasonably foreseeable cause of injury to anybody acting in a way in which a human being may be reasonably expected to act in circumstances which may be reasonably expected to occur.' (Lord Reid in John Summers & Sons Ltd. v. Frost (1955))

The Act spells out the need to exercise foresight. This depends on having good information which will enable management and workers to identify potential hazards in good time and take appropriate action. Above all it requires competent scientific and engineering knowledge of all aspects of the processes carried out and the materials handled at every stage in them. The Act requires managements to disclose information in the following ways:

1. Publishing the company's safety policies and record, e.g. through the directors' annual report;
2. Advising employees of their safety policies and all other information on health and safety which may affect them;
3. Assisting inspectors in informing employees about unsatisfactory health and safety conditions which they have found which affect them;
4. Informing the local inhabitants on health and safety hazards caused by the company's activities;
5. Informing purchasers of the company's products of any risks to safety or health.

1.4.2.2 Directors' responsibilities

The Act makes directors responsible for any offence committed with their consent or connivance, or through their neglect. Thus a board executive should be responsible for health and safety within the company, for co-ordinating policies, ensuring that they are implemented, and reporting on these matters — eventually in the annual report. The report will require the agreement of line management, safety advisers and workers representatives.

Directors are required to publish a statement of the company's health and safety policy which is not just a declaration of intent but a detailed description of how its aims are to be achieved. The policy should take into account existing health and safety codes of practice and any new codes approved by the Health and Safety Commission.

The policy statement implies that the management should develop formal safety procedures and rules. These should cover the specification and maintenance of protective equipment, development of standards, methods of inspection, hazard identification, safety audits and accident reporting and investigation.

1.4.2.3 Records

The Act implies that management should keep comprehensive records of its safety procedures, audits and inspections, accidents and injuries and damage caused by them, as well as action taken as a result. Such records and statistics are also needed by management to monitor the effectiveness of its policies and procedures.

These reports and statistics should be accurate, relevant, well indexed, readily accessible and easy to understand. They should indicate how results compare with predetermined standards and should be made available to those whose job it is to take action on them.

1.4.2.4 Organisation, training and communications

The Act implies that certain employers may be compelled to appoint qualified safety advisers as well as instructing and supervising employees in safety matters. While not specifying any special form of organisation, it is recommended that safety and health be treated like any other major management function with a clear line of responsibility and command running up to an accountable individual at the very top.

The provision of safety training is the responsibility of industry itself. The normal job training of every operative should include instruction in the potential hazards of the job and general principles of safe working. Management is also responsible under the Act for keeping its work people informed and consulting their representatives on everything that concerns their health and safety.

1.4.2.5 Duties of suppliers

Manufacturers and suppliers of plant equipment and materials have a new obligation to ensure that their products are safe in the hands of their users, provided they are properly used. This means that designers, manufacturers and suppliers must:
1. Ensure that their products are designed and constructed so as to be safe in use;
2. Examine and test their products to ensure that this is so;
3. Establish what risks are involved in the use of their products, warn clearly of any dangers and how to avoid them;
4. Provide clear and detailed instructions on the safe use of their products.

1.4.2.6 Control of contractors

The Act obliges employers to provide for the safety of employees of contractors who work on their premises as well as their own employees. Employers must therefore ensure that contractors are aware of their shared responsibilities, and are advised to provide rule books for contractors, making it obligatory for them to abide by these in their contracts.

1.4.2.7 Role of inspectors.

Every workplace must be open to inspectors from the Health and Safety Executive which combines the functions of the former separate inspectorates discussed in section 1.1. Companies should, where necessary, seek the advice of inspectors in their health and safety problems and never try to conceal them. Experienced inspectors will frequently spot hazards missed by management and employees because of over-familiarity. The smaller industrial concerns have most to gain from the advice of Health and Safety Executive inspectors.

Quite apart from safety inspections by HSE inspectors, machines, equipment and buildings subject to wear and corrosion require to be inspected regularly by competent persons, the results recorded, and action taken to replace or suitably repair main items showing signs of deterioration long before failure occurs. This subject is dealt with in more detail in section 2.11.

1.4.2.8 Programme and implementation

The main steps required under a H & S programme conforming with the Act may be summarised as:
1. Appointing a director to be responsible for all H & S matters;
2. Establishing a working party to assist the director in planning and implementing the programme. This could be an existing safety committee or the nucleus of a future one;
3. Education of all concerned in the provisions and implications of the Act;
4. Study of existing and future approved codes and standards;
5. Implementation of regulations on health and safety requirements, safety representation and committees, accident notification and statistics;
6. Obtaining guidance as required from the HSE, private consultants or appropriate trade associations in interpreting the Act;
7. Reviewing present procedures and revising them as required;
8. Preparation of a policy statement and programme for Board approval;
9. Introduction of policy statement to employees;
10. Development of a detailed action plan covering assignment of responsibilities, ways and means of improving performance, training, accident recording and investigation, medical and first aid services, and a continuing programme to stimulate health and safety awareness;
11. Means of monitoring performance through safety audits and analyses of records and statistics.

Most industrial firms, or similar organisations, with a pay roll of 500 or more should employ at least one full time qualified safety or total loss control officer. Smaller firms will frequently manage with a part time adviser or consultant, while large firms and corporations will have their own safety departments with experts in various fields. Industries such as atomic energy, explosives and petrochemicals which have a high loss potential need an experienced core of engineers and techniciams with a purely monitoring function and divorced altogether from day to day production problems.

REFERENCES

1. Kletz, T.A., *Some of the wider questions raised by Flixborough'*, Institution of Chemical Engineers Symposium (The Technical Lessons of Flixborough), Nottingham (16 December 1975).
2. Lord Robens Committee, *Health and safety at work*, HMSO, London, 17 (1972).
3. Heinrich, H.W., *Industrial accident prevention*, 4th ed., McGraw Hill, New York, 43 (1959).
4. Fife, I., and Machin, E.A., *Redgraves health and safety in factories*, 1st ed., Butterworth, London (1976).

1.5 ACCIDENT REPORTS, RECORDS AND ANALYSES

Contents

1.5.1 Legal requirements

The legal requirements to report accidents at work are dealt with first. They are contained in section 80 of the Factories Act (1961)[1] the Social Security Act 1975[1] and the Social Security (Claims and Payments) Regulations 1975 (SI 1975, No. 560).[1] These are quoted in full below:

1.5.1.1 Factories Act (1961)

'80 Notification of Accidents

'1. Where an accident
(a) causes loss of life to a person employed in the factory; or
(b) disables any such person for more than three days from earning full wages at the work at which he was employed;
written notice of the accident in the prescribed form (c) and accompanied by the prescribed particulars, shall forthwith be sent to the inspector (d) for the district, unless it is an accident of which notice is sent in accordance with the requirements of the Explosives Act 1875, or the Petroleum (Consolidation) Act 1928.'
'2. Where an accident causing disablement is notified under this section, and after notification thereof results in the death of the person disabled, notice in writing of the death shall be sent to the inspector (d) for the district by the occupier (e) of the factory as soon as the death comes to his knowledge.'
'3. Where any accident to which this section applies occurs to a person employed and the occupier (e) of the factory is not the actual employer of the person killed or injured, the actual employer shall, if he fails to report the accident to the occupier immediately, be guilty of an offence and be liable on summary conviction to a fine not exceeding ten pounds (f).'

The Factories Act 1961 contains extensive interpretation of the terms (a) 'Accident', (b) 'factory', (d) 'inspector', (e) 'occupier' and details as to proceedings against offenders (f). All these have clearly been modified by HASAWA 1974, which in particular has greatly extended the definition of the term 'factory'.

The prescribed forms, however, remain in force and are shown in *Figures 1.5.1 and 1.5.2*. They are Form 43, prescribed by the Factories (Notice of Accident etc.) Order, 1965[2] and Form 43B, prescribed for building operations and works of engineering construction by the Construction (Notice of Accident etc.) Order 1964.[3]

Most employers are also obliged to record details of certain injuries in the Factory General Register (Part 3, Form F31 or 33) under section 80 of the Factories Act[1] or in the Accident Book (B.I. 510) under section 88 of the Social Security Act 1975.[1] Regulations made under section 15 of HASAWA 1974 will probably modify these existing reporting procedures.[1]

1.5.1.2 The Social Security Act 1975

'88 Notification of Accidents, etc –

"Regulations may provide
"(a) . . .
"(b) for requiring employers;
(i) To make reports, to such persons and in such form and within such times as may be prescribed, of accidents in respect of which industrial injury benefit may be payable.
(ii) To furnish to the prescribed person any information required for the determination of claims, or of questions arising in connection with claims or awards.
(iii) To take such other steps as may be prescribed to facilitate the giving notice of accidents, the making of claims and the determination of claims and of questions so arising.
"NOTES
"General note. This section contains provisions formerly in the National Insurance (Industrial Injuries) Act 1965, s 26.
"(a) Regulations. See the Social Security (Claims and Payments) Regulations 1975 (S.I. 1975, No. 560) set out below.

1.5.1.3 The Social Security (Claims and Payments) Regulations 1975 (SI 1975, No. 560)

'Special provisions relating to industrial injuries benefit only

'22 Notice of accidents

'1. Every employed earner who suffers personal injury by accident in respect of which benefit may be payable shall give notice of such accident either in writing or orally as soon as is practicable after the happening thereof:

'Provided that any such notice required to be given by an employed earner may be given by some other person acting on his behalf'

'2. Every such notice shall be given to the employer or (if there is more than one employer) to one of such employers, or to any foreman or other official under whose supervision the employed earner is employed at the time of the accident or to any person designated for the purpose by the employer, and shall give the appropriate particulars.'

'3. Any entry of the appropriate particulars made in a book kept for that purpose in accordance with the provisions of regulation 22 shall, if made as soon as practicable after the happening of an accident by the employed earner or by some other person acting on his behalf, be sufficient notice of the accident for the purpose of this regulation.

'4. In this regulation:

'employer' means, in relation to any person, the employer of that person at the time of the accident and 'employers' shall be construed accordingly; and

'employed earner' means a person who is or is treated as an employed earner for the purposes of industrial injuries benefit.

'5. In this regulation and regulation 23, the expression 'appropriate particulars' means the particulars indicated in Schedule 4 to these regulations.'

'23 Obligations of employers

'1. Every employer shall take reasonable steps to investigate the circumstances of every accident of which notice is given to him or his servant or agent in accordance with the provisions of regulation 22 and if there appears to him to be any discrepancy between the circumstances found by him as a result of his investigation and the circumstances appearing from the notice so given, he shall record the circumstances so found;

'2. Every employer who is required to do so by the Secretary of State shall furnish to an officer of the Department within such reasonable period as may be required, such informations and particulars as shall be required:

'(a) of any accident or alleged accident in respect of which benefit may be payable to, or in respect of the death of, a person employed by him at the time of the accident or alleged accident; or

(b) of the nature and other relevant circumstances relating to any occupation prescribed for the purposes of Chapter V of Part II of the Act in which any person to whom or in respect of whose death benefit may be payable under that Chapter was or is alleged to have been employed by him.'

'3. Every owner or occupier (being an employer) of any mine or quarry or of any premises to which any of the provisions of the Factories Act 1961 applies and every employer by whom ten or more persons are normally employed at the same time on or about the same premises in connection

DEPARTMENT OF EMPLOYMENT
HM FACTORY INSPECTORATE

F 43

A notice in this form should be sent to HM Inspector of Factories immediately the accident or dangerous occurrence becomes reportable. (See instruction overleaf.)
If the accident is fatal, HM Inspector should be informed immediately in writing.

FACTORIES ACT 1961, sections 80 and 81 and the
Dangerous Occurrences (Notification) Regulations 1947
Prescribed form of written notice of
ACCIDENT OR DANGEROUS OCCURRENCE

NOTE:

For accidents occurring:
(a) on building operations or works of engineering construction, use form F43B.
(b) in offices or shops, use form OSR2.

FOR OFFICIAL USE
District and date of receipt

No. of copies required

MR Group	
Ref. to	
M of T, etc.	
1 Serial No.	
2 MWBG	
3	
4 F., N.F., D.O.	
4(a)	
4(b)	
5 Process	
6 S.I.C.	
7(a) Causation	
7(b)	
7(c)	
7(d)	
7(e)	
7(f)	
7(g)	
7(h)	
7(j)	
7(k)	
7(l)	
8 Occupation	
9 Injury Nature Site	
10	
11	
12	
13	

1 (a) **OCCUPIER** of factory (or person carrying on processes at Docks, and certain other places).

   Name

   Address

   Industry

   (b) Actual employer of injured person if other than above

   Name

   Address

2 **PLACE** where accident or dangerous occurrence happened:

   (a) Address (if different from 1(a) above)

   (b) Exact location

   (c) Nature of work carried on there

3 **INJURED PERSON**

   (a) Full name (surname first)

   (b) Sex .... Age ....

   (c) Address

   (d) Occupation

4 **ACCIDENT or DANGEROUS OCCURRENCE**

   (a) Date .... Time ....

   (b) Full details of how the accident or dangerous occurrence happened and what injured person was doing. If a fall of a person or materials, plant, etc. state height of fall (if necessary continue on separate sheet).

   (c) If due to machinery, state:

      (i) Name and type of machine ....

      (ii) Part causing injury ....

      (iii) Whether in motion by mechanical power at the time ....

      (iv) If caused by crane or other lifting machine, specify type ....

5 **INJURIES AND DISABLEMENT**

   (a) Nature and extent of injury (e.g. fracture of leg, laceration of arm, scalded foot, scratch on hand followed by sepsis)

   ....

   ....

   (b) Whether fatal or non-fatal...................................

   (c) Was injured person disabled for more than three days from earning full wages at the work at which he was employed?...............................

6 Has accident (or dangerous occurrence) been entered in the General Register?............................

   Signature of Occupier, Employer, or Agent .... Date ....

Figure 1.5.1   Forms 43 used in the United Kingdom

FACTORIES ACTS, 1937 to 1959

FOR OFFICIAL USE

Form 43B.

Form prescribed by the Minister of Labour in pursuance of Sections 64 and 65 of the Factories Act, 1937, for

Notice of Accident or Dangerous Occurrence
ON A BUILDING OPERATION OR WORK OF ENGINEERING CONSTRUCTION

to be sent (immediately on the accident or dangerous occurrence becoming reportable) to H.M. Inspector of Factories.

(See Instructions overleaf.)

Note: If the accident is fatal it is desirable that H.M. Inspector be informed immediately by telephone.

District and date of receipt

M.R. Group
Ref. to
M. of T., etc.

		FOR OFFICIAL USE
1. (a) Person (or firm) undertaking building Operations or Works of Engineering Construction (see overleaf).	NAME .... Registered Office or Address .... Trade	1. Serial No.
(b) Actual employer of injured person if other than above.	Name .... Address .... Trade	2. M.W.B.G.
2 SITE where accident or dangerous occurrence happened		3. Age Group
(a) Address of site		4. F., N.F., D.O.
(b) Exact place on site		4(a).
3. NATURE OF WORK carried on at:—		
(a) **Building Operations** (*Tick items which apply*)	(iv) Industrial Building	4(b).
(i) Construction	(v) Commercial or Public Building	
(ii) Maintenance .... of		5. Process
(iii) Demolition	(vi) Block of Flats .... (vii) Dwelling House .... (viii) Other	6. S.I.C.
(b) **Work of Engineering Construction** (Specify type)		7(a). Causation
4. INJURED PERSON (a) Full Name (Surname first) .... Sex ...... Age		7(b).
(b) Address		
(c) Occupation (*Tick item which applies*) Bricklayer ... Carpenter/Joiner ... Painter ...		7(c).
Plasterer ... Plumber ... Scaffolder ...		7(d).
Steel Erector ... Demolition Worker ... Labourer ...		
Steeple Jack ... Slater/Tiler/Other Roofing Worker ...		7(e).
Other Occupation (Specify)		
Note:—Semi-skilled men or apprentices should be classified under the appropriate occupation.		7(f).
5. ACCIDENT (OR DANGEROUS OCCURRENCE) (a) Date .... Time		
(b) Full details of how the accident happened. (If necessary continue overleaf.) (If a fall of person or materials, plant, etc., state height of fall.)		7(g).
....		7(h).
....		7(j).
(c) State exactly what injured person was doing at the time		7(k).
(d) If due to machinery, state:—		7(l).
(i) Name of machine and part causing accident		
(ii) Whether in motion by mechanical power at the time		8. Occupation
6. INJURIES AND DISABLEMENT (a) Nature and extent of injury (e.g., fracture of leg, laceration of arm, scalded foot, scratch on hand followed by sepsis)		9. Injury
....		10.
(b) Whether fatal or non-fatal		
(c) If not fatal was injured person disabled for more than three days from earning full wages at the work at which he was employed?		11.
7. Has accident (or dangerous occurrence) been entered in the General Register?		12.
Signature of Contractor, Employer, or Agent		
Date		

Figure 1.5.1 (cont.)   Form 43B

with a trade or business carried on by the employer shall, subject to the following provisions of this paragraph:

'(a) Keep readily accessible a book or books in a form approved by the Secretary of State in which the appropriate particulars (as defined by regulation 22) of any accident causing personal injury to a person employed by the employer may be entered by that person or by some other person acting on his behalf; and

(b) preserve every such book, when it is filled, for the period of 3 years beginning with the date of the last entry therein.

'Schedule 4

(Regulations 22 and 23)

Particulars to be given of accidents.

'1. Full name, address and occupation of injured person;

2. Date and time of accident;

3. Place where accident happened;

4. Cause and nature of injury;

5. Name, address and occupation of person giving the notice, if other than the injured person.'

The accident report is made on National Insurance Form B.I. 76 which is sent to the injured person or his nominee on request by the DHSS after the injured person has received his 'first doctor's certificate'. Accidents reported in this way are, like those reported by the employer to the HSE Inspector, restricted to those causing absence from work for over three working days.

1.5.1.4 The legal reasons for accident reporting[4]

The report made by or on behalf of the accident victim to the DHSS serves several purposes:

1. It allows him to claim industrial injury insurance while off work; this, as noted in section 5.2, is significantly more than unemployment benefit, as well as being tax free;

2. It may form the basis for a claim under National Insurance for a disablement pension if the disability is assed at 20% or more;

3. It may form the basis for a lawsuit against the employer for compensation if the accident was caused by the employer's negligence. This might be significantly higher than the payments under National Insurance, although it would almost certainly be paid by an insurance company rather than by the employer. The report made by the employer to the H and SE inspector serves to inform the latter so that he can investigate the accident and its cause and advise the employer on any steps required to prevent any recurrence. Where the accident resulted from negligence or breaches of regulations or codes, the H and SE inspector may institute legal proceedings or apply for a restraint order.

There is no legal requirement to report minor accidental injuries such as cuts and bruises, and in many cases it is not clear at the time of the accident whether the victim will be absent from work for more than three days as a result (thus making the accident reportable) or not.

1.5.2 Reporting accidental injuries as they arise

Patterns of accident reporting vary widely. In some works and factories, a high level of reporting of injury-causing accidents is achieved, whilst in others only a few minor injury accidents are reported and even a number of legally notifiable accidents fail to get reported. Before discussing how accidents are, or should be, reported, and by whom, it is as well to note that there are serious psychological elements in any accident situation which tend to inhibit reporting or invalidate its objectivity.

The most important of these are legal liability and moral responsibility. This is really a question of self interest. Few people will volunteer information which appears to incriminate them or render them liable to penalties of any sort. Neither are insurance companies likely to encourage claims upon their pockets arising out of accidents. This fear of liability affects both sides of industry, management and employees alike. The limited and uncertain prospects of compensation are quite outweighed by the fears of adverse publicity and costly and unsuccessful lawsuits.

A second and sometimes related obstacle to accident reporting lies in the frequent polarisation between shop floor and office workers with a resulting communications gap between the two sides. Nearly all the accidents occur on the shop floor, but reports are kept 'by the other side' in an office.[5]

Unless safety specialists go out of their way to spend much of their time on the shop floor getting to know most of the workers there personally and discussing frankly their work and the associated hazards, this gap is likely to remain. Indeed it probably suits the short term interests of many managers that it should remain, if it reduces the number of accidents that are reported and the problems which these reports create.

A third common obstacle is that often nothing constructive appears to be done as a result of an accident report. Unless the report stimulates effective preventative action to avoid a recurrence of the accident, it will become merely another meaningless statistic, and lead to apathy on the part of the employees. Accidents, particularly minor ones, are generally regarded as 'part of the job'.

1.5.2.1 Whose job to report injury-causing accidents?

Heinrich, whose views on accident prevention carry great weight, recommended that the shop floor supervisor (chargehand or foreman) be primarily responsible for accident reporting.[6] But this system can clearly only work when:

Supervisors themselves are fully safety conscious and are effectively supported by their managements in reporting and investigating accidents and their causes; and

Supervisors have sufficient time among their many other responsibilities to do this properly.

A valuable report[5] by a team of engineers, sociologists and occupational psychologists who spent over a year as accident observers on the shop floor of four different workshops, suggests that the supervisors had neither the motivation nor the time to report and study accidents properly. They found that in factories with a first aid room, dispensary or surgery with a trained nurse in full-time attendance, the nurse was in a far better position to prepare a thorough, complete and objective report — based on the information given to her by the casualty during treatment, the injuries received, and the knowledge she has gleaned about the plant and machinery. The advantages of the nurse's report are threefold:

1. The casualty will generally have fewer inhibitions about discussing the accident with the nurse than with his (or her) supervisor;
2. Since the nurse is most unlikely to be held responsible in any way for the accident, her report will be free from the possible bias (caused by this fear) in a supervisor's report;
3. Even though she lacks the supervisor's knowledge of the plant, process or machine or the experience and skill of the injured employee, she will generally have enough time to draw out and report the facts. Where appropriate the nurse can be provided with a tape recorder to record her questions to the casualty and his or her reply, and prepare her written report from the tape later when, for example, he has been taken to hospital.

There can, of course, be difficulties in placing the entire onus of accident reporting on the nurse (assuming that she exists!). She may find her report at the centre of a legal battle between the firm and the employee or his trade union, and she may not have the confidence of the workers. One nurse reported by the authors of the study quoted insisted on treating all hand and finger injuries (of which there were a number), of machine operators with loose gauze bandages. The workers not unnaturally preferred to treat their injuries themselves with first aid dressings in the wash room!

Neither the works safety specialist nor the union safety representative have been mentioned so far in this question of accident reporting. Neither are, however, likely to have been present on the spot when the accident occurred, although both may be expected to visit the scene of the accident later and to make their own reports (possibly to find themselves in opposing legal camps afterwards).

While reports by eye witnesses may be valuable, one cannot rely on an eye witness being present at the scene of every accident, nor to have witnessed it as it developed. Often the eye witness report is very fragmentary and unreliable as a source of information.

The reporting of accidents becomes far more difficult in a factory where there is no full-time nurse in attendance, and casualties are dependent on treatment by one of several part-time first aiders — i.e. workers on another job who have undergone a recognised first aid course — and who are obliged to leave their work to assist any of their colleagues who are injured.

These people may give valuable first aid service, but they are not as a rule expected or encouraged to make out a proper accident report and often an uppermost thought in their minds is to get back to their normal job as soon as possible.

The greater use of time clerks with whom workers are in frequent contact to assist injured employees in reporting their own accidents has been recommended. Even in factories where the top management is actively interested in promoting safety, employees may find it difficult to report accidents without apparently wasting time or incurring the displeasure of their supervisor or workmates.

A 100% reporting factor of all injury-causing accidents is an ideal which is rarely likely to be achieved, and managers should recognise that the true number of accidents is likely to be far greater than the number reported.

A complete and effective accident reporting system requires the involvement of the following people:

(a) The industrial nurse or part time first aiders;
(b) The supervisors;
(c) The casualties;
(d) The time clerks.

The main reliance upon supervisors as recommended by Heinrich does not appear likely to be successful in British Industry today.

1.5.2.2 Accident Report Forms

In addition to the official Form 43 used for reporting notifiable accidents, most companies employ their own accidental injury forms or cards on which any and every accident can be reported, whether causing injury or not.

RoSPA provides accident record punch cards, incorporating standard accident classifications, to enable records and statistical analyses to be compiled quickly, together with a booklet describing their use. Another recording card system specially designed for the engineering industry is available from the Engineering Employers' Federation.

The use of standard forms can cause some important information about an accident to be omitted; this applies particularly when the form or card is small and the space available restricted. But the reporter, left with only a blank sheet on which to make his or her report may equally well forget to supply some standard information which will be required later.

One form specially designed to stimulate remedial action is the Supervisors Accident Report designed and supplied by the National Safety Council, Chicago.[7] This is shown in *Figure 1.5.2*. Though designed for use by supervisors, most entries could as well be completed by the works nurse attending the casualty. However a part-time first aid attendant would hardly have the time or knowledge to complete such a report, and would require a much simpler form.

SUPERVISOR'S ACCIDENT REPORT   Incident No. _____

(To be completed immediately after accident, even when there is no injury)

Company name and address _____

Plant or location address _____
(if different from above)

1. Name and address of injured_____ SSN_____ 2. Age_____
   (or ill) person
   _____ 3. Sex_____

4. Years of service_____ 5. Time on present job_____ 6. Title/occupation_____

7. Department_____ 8. Date of accident_____ 9. Time_____

10. Accident category (check) ☐ Motor Vehicle; ☐ Property Damage; ☐ Fire; ☐ Other_____

11. Severity of injury or illness   ☐ Non-disabling; ☐ Disabling; ☐ Medical Treatment; ☐ Fatality

12. Amount of damage $_____ 13. Location_____

14. Estimated number of days away from job _____

15. Nature of injury or illness? _____

16. Part of body affected?_____

17. Degree of disability?_____
                                    (Temporary total; permanent partial; permanent total)

18. Causative agent most directly related to accident? (Object, substance, material, machinery, equipment, conditions)

_____

_____

_____

Was weather a factor?_____

19. Unsafe mechanical/physical/environmental condition at time of accident? (Be specific)

_____

_____

_____

20. Unsafe act by injured and/or others contributing to the accident. (Be specific, must be answered)

_____

_____

_____

21. Personal factors (improper attitude, lack of knowledge or skill, slow reaction, fatigue)

_____

Figure 1.5.2   Supervisors accident report as used in the USA

22. Personal protective equipment required? (Protective glasses, safety shoes, safety hat, safety belt) _ _ _ _ _ _ _ _

Was injured using required equipment? _

23. What can be done to prevent a recurrence of this type of accident?
(Modification of machine; mechanical guards; correct environment; training)

_ _

_ _

24. Detailed narrative description (How did accident occur; why; objects, equipment, tools used, circumstance, assigned duties.

Be specific) _

_ _

_ _

_ _

(Use additional sheets, as required)

25. Witnesses to accident _

_ _

Date prepared _ _ _ _ _ _ _ _ _ _ _ _ Signature of Foreman / Supervisor _ _ _ _ _ _ _ _ _ _ _ _ _ _ _.

Department _

SUPERINTENDENT'S APPRAISAL AND RECOMMENDATION

a. In your opinion what action on the part of injured (or ill) person or others contributed to this accident?

_ _

_ _

b. Your recommendation _

_ _

Date _ _ _ _ _ _ _ _ _ _ _ _ Signature of Superintendent _ _ _ _ _ _ _ _ _ _ _ _ _ _ _ _ _ _

FOR SAFETY OFFICE USE ONLY

Temporary Total ☐   Permanent Partial ☐   Death or Permanent Total ☐

Started losing time_ _ _ _ _ _   Part of Body_ _ _ _ _ _ _ _

Returned to work_ _ _ _ _ _ _   Per cent loss or
                                      loss of use_ _ _ _ _ _ _ _

Time charge_ _ _ _ _ _ _ _   Time charge_ _ _ _ _ _ _ _ _   Time charge: 6,000 days

Compensation $_ _ _ _ _ _ _ _Medical $_ _ _ _ _ _ _ _ _ Other $_ _ _ _ _ _ _ _ _Total $_ _ _ _ _ _ _ _

Name and address   Name and address
  of hospital _ _ _ _ _ _ _ _ _ _ _ _ _ _   of physician_ _ _ _ _ _ _ _ _ _ _ _ _ _

1.5.2.3 Further accidental injury records

Many recorded statistics of industrial accidents are unreliable or distorted by variations in the levels of reporting, some of which are caused by reasons described earlier in section 1.5.2. Further suppression of data may result from well intentioned devices to reduce accidents such as competitions between departments for the best accident record. The safety manager or officer must nevertheless keep adequate records even though he may have some scepticism as to their accuracy or the value of obvious comparisons made between them.

Injury records of employees should be kept, on which every injury is recorded briefly with cross reference to the accident report form. An employee with a higher than average injury rate should not be branded automatically as an unsafe worker. His injury record could be caused by several other reasons, i.e.:

He may work harder than average;
He may report more minor injuries than average;
He may be engaged on tasks with higher than average risk.

A typical personal injury record card is shown in *Figure 1.5.3* and a record card as used in the USA is reproduced in *Figure 1.5.4*.

Injury Record of (name) (number) Occupation Department Date Employed						
Incident No.	*Date of injury*	*Type Fatal Notifiable Non-notifiable*	*Days off work*	*Claim No.*	*Cause*	*Remarks*

Figure 1.5.3   Personal injury record card — UK

Annual accident reports for the firm should be prepared with individual and cumulative monthly entries showing the number of man hours worked, the average number of employees, and the number of injuries divided between fatalities, notifiable injuries, and non-notifiable injuries, with severity rates and costs (where possible).

Accident Analysis Sheets should be kept on a monthly or quarterly basis, giving date and number of incident, name and department of casualties, nature of injury and part of body injured, injury classification, days lost, brief description of each accident and corrective action taken.

Injury rates in Britain are usually either given as frequency rates or incidence rates.[8] The injury frequency rate for a factory or department is usually expressed as the number of injuries per 100 000 manhours worked while the incidence rate is the number of injuries per 1000 persons employed in any year. These should be qualified according to whether they are total injuries, notifiable injuries, fatalities or injuries or various grades of severity.

'Accident Severity Groups' are classified by the Health and Safety Executive as follows:

Group 1

Code	Injury
1	Fractures (excluding 20 & 21).
2	Dislocations (excluding 21).
3	Amputations (excluding 23).
4	Crushing without fracture (excluding 25).
5	Concussions (excluding 26).
6	Deeply penetrating burns and scalds.
7	Burns and scalds covering more than 1 sq. ft.
8	Eye injuries involving loss of eye.
9	Eye injuries involving permanent impairment of vision.
10	Lacerations and wounds requiring at least 5 stitches.
11	Lacerations and wounds requiring skin grafting.
12	Wounds resulting in severed tendon.

Groups 2 and 3

Code	Injury
13	Abrasions and surface injuries.
14	Lacerations and open wounds.
15	Bruises.
16	Strains and sprains not resulting in 17 or 18.
17	Hernias.
18	Slipped discs.
19	Eye injuries (excluding 8 and 9).
20	Fractures of single bone of finger, thumb or toe.
21	Hairline cracks.

INVESTIGATOR'S COST DATA SHEET

Class 1 _____
(Permanent partial or temporary total disability)

Class 2 _____
(Temporary partial disability or medical treatment case requiring outside physician's care)

Class 3 __X_____
(Medical treatment case requiring local dispensary care)

Class 4 _____
(No injury)

Name _____Thomas Black_____

Date of injury __6-23____   Its nature __Abrasion on right leg__

Department __31_____   Operation __Trucking___   Hourly wage __$4.29____

Hourly wage of supervisor $__$6.75__

Average hourly wage of workers in department where injury occurred $__$4.10__

1. Wage cost of time lost by workers who were not injured, if paid by employer $__42.70__

    a. Number of workers who lost time because they were talking, watching, helping __1_.
    Average amount of time lost per worker __0___ hours ___20___ minutes.

    b. Number of workers who lost time because they lacked equipment damaged in accident or because they needed output or aid of injured worker __30_. Average amount of time lost per worker ___0___ hours ___20___ minutes.

2. Nature of damage to material or equipment __Hydraulic cylinder crushed by__ __fork on lift truck. Cylinder replaced with new unit._____

    Net cost to repair, replace, or put in order the above material or equipment $ 280.00

3. Wage cost of time lost by injured worker while being paid by employer $ 11.80
    (other than workmen's compensation payments)

    a. Time lost on day of injury for which worker was paid __2__ hrs. __45__ mins.

    b. Number of subsequent days' absence for which worker was paid ___----___ days
    (other than workmen's compensation payments) ___----___ hours per day.

    c. Number of additional trips for medical attention on employer's time on succeeding days after worker's return to work __----_.
    Average time per trip __----__ hrs. __----__ min. Total trip time __----__ hrs. _____ mins.

    d. Additional lost time by employee, for which he was paid by company __----__ hrs. __----__ mins.

Figure 1.5.4   Personal injury report card — USA

4. If lost production was made up by overtime work, how much more did the work cost than if it had been done in regular hours? (Cost items: wage rate difference, extra supervision, light, heat, cleaning for overtime.) $ __40.50__

5. Cost of supervisor's time required in connection with the accident $ __15.20__

    a. Supervisor's time shown on Dept. Supervisor's Report __1__ hrs. __30.__ mins.

    b. Additional supervisor's time required later ____ hrs. __45__ mins.

6. Wage cost due to decreased output of worker after injury if paid old rate $ __6.86__

    a. Total time on light work or at reduced output __1__ days __8__ hours per day.

    b. Worker's average percentage of normal output during this period __80__ %.

7. If injured worker was replaced by new worker, wage cost of learning period $ __0__

    a. Time new worker's output was below normal for his own wage __----__ days __----__ hours per day. His average percentage of normal output during time __----__ %. His hourly wage $ __----__ .

    b. Time of supervisor or others for training __----__ hrs. Cost per hour $ __----__ .

8. Medical cost to company (not covered by workmen's compensation insurance) $ __15.00__

9. Cost of time spent by higher supervision on investigation, including local processing of workmen's compensation application forms. (No safety or prevention activities should be included.) $ __22.50__

10. Other costs not covered above (e.g., public liability claims; cost of renting replacement equipment; loss of profit on contracts cancelled or orders lost if accident causes net reduction in total sales; loss of bonuses by company; cost of hiring new employee if the additional hiring expense is significant; cost of *excessive* spoilage by new employee; demurrage). $ __20.00__

    Explain fully.

    Total uninsured cost.. $ __454.56__

Name of company __Midwest Mfg.__

22	Dislocation of finger, thumb or toe.
23	Amputations involving loss of less than single joint of finger, thumb or toe.
24	Minor scalds and burns.
25	Crushing involving absence of seven calendar days or less.
26	Concussions resulting in loss of seven calendar days or less.
27	Gassing.
28	Poisoning.
29	Sepsis.
30	Other.

Cases coded 13 to 30 are Group 3 unless admitted to hospital for at least 24 hours or absent for more than 28 days. Group 2 include strains and sprains where there is legitimate doubt whether they were caused by accidents happening at work.

Different frequency rates are employed in the USA using ANSI Standard Z. 16.1, *'Standard methods of recording and measuring work injury experience'*, and A.S.I. Standard Z. 16.2, *'Standard method of recording basic facts relating to the nature and occurrence of work injuries'*.

American frequency rates are expressed per million manhours worked instead of per 100 000 manhours worked as in Britain. Other terms used in the USA are 'Severity Rate', which is the total days charged per million employee hours exposure, 'Average Days Charged' and 'Disabling Injury Index'.[7]

A term used quite frequently in accident statistics is the Fatal Accident Frequency Rate (FAFR). This is generally defined as the number of deaths per *hundred million exposed hours*.[5]

1.5.3 Reporting accidental damage

Accidents may cause damage to either or both the products of a works or factory and the plant or machinery used in their production. Damaged products are picked up by the inspection department, usually before despatch, and damaged plant and machinery are revealed by plant inspections and by examining the calls made on the engineering and maintenance departments, and by what appears in the scrap heap.

Damage control has developed into a discipline of its own, based historically on its use in military operations, especially air and naval warfare. It has been applied apparently with success, to damage caused by industrial accidents. It is now combined with control of industrial injuries into a combined system known as *Total Loss Control*. This involves reporting to control centres all property damage accidents and injury-causing accidents, supplemented by spot checks and safety audits.

Whilst the term *Total Loss Control* appears to have been overdone in some quarters, the study of total systems of work and of the working environment which it requires have been largely followed in this reference book. But the use of economic audits as essential features of damage control must not be overlooked.

1.5.4 Accident investigation analyses

Investigation and analysis go together. The facts revealed to the investigators by the reports of the accidents are first analysed. This analysis may be sufficient to reveal one hypothesis which explains all the facts adequately, and to eliminate all other possible hypotheses. But sometimes, especially where there has been a major accident or disaster, it is necessary to carry out further investigations to examine doubtful points or fill in gaps in the provisionally accepted explanation, to choose between rival explanations or events to recognise that none of the hypotheses so far examined are compatible with all the facts, and that a new hypothesis must be found. It must also be recognised that most accidents result from a combination of causes, in the absence of any one of which the accident would not have occurred. The single cause accident is quite a rarity, as a scientific study of over 2000 accidents has shown.[5]

Accidents resulting in minor injuries or damage or neither should be investigated if time and facilities permit, especially when certain types of accident are frequent. These types are usually easier to analyse than those causing serious injury or damage since there is less incentive for anyone to withold information or attempt to mislead the investigator.

The main purpose of an investigation should be fact finding and not fault finding. It is however impossible to disguise the fact that one consequence of the investigation *may* be that some persons are found responsible through lack of knowledge or failure to take some action required of them as part of their duties.

While it is important that they be given a full and fair opportunity to show that such allegations were unfounded, it is equally important that they should not be allowed to exert influence during the investigations or upset, distort or prolong the investigation by hiding essential facts or laying false trails.

As a model statement of the purposes of an accident investigation, that given by the (American) National Safety Council can hardly be bettered and is quoted below:

'1. To learn accident causes so that similar accidents can be prevented by mechanical improvements, better supervision or employee training;
2. To determine the 'change' or deviation that produced an 'error' that in turn resulted in an accident (systems safety analysis);
3. To publicise the particular hazard among employees and their supervisors, and to direct attention to accident prevention in general;
4. To determine facts bearing on legal liability. An investigation undertaken solely for this purpose, though, will seldom give enough information for accident prevention purposes. On the other hand, an investigation for preventative purposes may disclose facts which are important in determining liability.'

It should, however, be added that sometimes an investigation will reveal a serious unsuspected and socially unacceptable hazard in a product (or process) or in a material used in the manufacture or formed during the course of it. Examples which are mainly toxic hazards are the thalidomide tragedy, the Seveso disaster, the use of yellow phosphorous in matches and high beryllium alloys for general engineering. Such investigations generally lead to government intervention and sometimes the complete banning of the hazardous material or process.

1.5.4.1 Who should investigate?

When a serious accident occurs, which even though it has killed nobody, clearly lies well in the 'reportable' category, the company director responsible for safety should be informed at once. The local inspector of the Health and Safety Executive ought also to be informed promptly, through the director responsible for safety or his nominee (who would normally be the safety specialist) and advised what action is being taken by the firm itself to investigate the accident. The director with responsibility for safety should lay down the procedure to be followed in such cases in his absence. Speed is often vital in an investigation but if the Health and Safety Inspector is not advised of the accident until it becomes reportable after the statutory three working days, he might find himself presented with a 'fait accompli' if the investigation is already complete by the time he learns of the accident.

Whatever subsequent investigations are made, the foreman or shop floor supervisor should always be required to make his report (see *Figures 1.5.1 and 1.5.2*) whenever a serious accident causing a disabling injury or costly damage occurs. Completion of this standard report will itself require a minimum of investigation on his part, since he must be able to substantiate any conclusions which he draws.

The safety specialist should make his own investigation carefully checking the findings of the supervisor and the works nurse, against the evidence of the casualty and other witnesses and his own on the spot examination. The safety specialist should, by training, be able to search for key facts, whether hidden or apparent and produce an unbiassed report.

The works safety committee may also be called upon or feel bound to investigate, particularly in cases where the action of an employee has been thought to be responsible for the accident by the supervisor and/or the safety specialist. This finding will be more readily accepted if it comes from the Works Safety Committee, but it is not always appropriate to expect the Committee to investigate every accident, especially if few of its members have any real background knowledge of the particular circumstances.

Accidents involving specialised machinery and accidents with a strong technical content may require the assistance of technical experts, i.e. the company's own engineers and scientists, those from the firm that designed the machine involved in the accident, engineers acting on behalf of the insurance company, and sometimes an independent consultant. The decision as to who should be called in will rest largely with the company director who is responsible for safety.

Whilst speed is of importance in most investigations, nothing should be done to disturb evidence that may be required by officials from the Health and Safety Executive, or to usurp their right to make their own investigation. A frank and open attitude should always be adopted. Their experienced assistance in investigations of serious accidents will generally be found to be valuable.

No one should be assigned to investigations unless he has earned a reputation for fairness and has experience in gathering data. But once an investigator has been appointed, nothing should be done to try to influence him in the lines he decides to pursue or in the verdict he reaches. He must be truly independent. Where more than one investigator is appointed, and their views and approach differ, they should be afforded the same facilities and nothing must be done to favour one at the expense of the other. Unless they can reach agreement between themselves without outside interference, there is no alternative but to require both to prepare their separate reports with their own conclusions, and refer these to an impartial and expert outside body such as the Health and Safety Executive.

1.5.4.2 Determining and analysing the key facts in accidents

The use of a code which caters for the majority of industrial accidents can greatly facilitate the work of accident analysis. In the absence of any such authoritative code in the U.K., the use of A.S.I. standard 2.16.2 is suggested.[7]

This lists eight key facts and questions. Typical answers to six of these questions are selected from *Table 1.5.1.*

Fact	*Question*
1. Nature of injury.	What was the injury?
2. Part of body.	What part of the body was affected by the injury named in 1?
3. Source of injury.	What object, substance, exposure or bodily motion inflicted the injury named in 1?
4. Accident type.	How did the injured person come into contact with the object, etc. named in 3? Or during what personal movement did the bodily motion named in 3 occur?
5. Hazardous condition.	What hazardous or environmental condition or circumstance caused or permitted the occurrence of the event named in 4?
6. Agency of accident.	In what object, substance or part of the premises did the hazardous physical or environmental condition named in 5 exist?
7. Agency of accident part.	In what specific part of the agency of the accident named in 6 did the hazardous condition named in 5 exist?
8. Unsafe act.	What unsafe act of a person caused or permitted the occurrence of the event named in 4?

Contributory factors. These should be noted when they can be determined.

Table 1.5.1 Checklist for identifying key facts

1. NATURE OF INJURY

Foreign body	Strain and sprain	Amputation	Dermatitis
Cut	Fracture	Puncture wound	Ganglion
Bruises and contusions	Burns	Hernia	Abrasions
			Others........................

2. PART OF BODY

Head and Neck	**Upper Extremities**	**Body**	**Lower Extremities**
Scalp	Shoulder	Back	Hips
Eyes	Arms (Upper)	Chest	Thigh
Ears	Elbow	Abdomen	Legs
Mouth, teeth	Forearm	Groin	Knee
Neck	Wrist	Others........................	Ankle
Face	Hand		Feet
Skull	Fingers and thumb		Toes
Others........................	Others........................		Others........................

4. ACCIDENT TYPE

Struck against (rough or sharp objects, surfaces etc. exclusive of falls)	Struck by sliding, falling or other moving objects	Overexertion (resulting in strain, hernia, etc.)	Inhalation, absorption, ingestion, poisoning, etc.
Struck by flying objects	Caught in (on or between)	Slip (not a fall)	Contact with electric current
	Fall on same level	Contact with temperature extremes, burns	Others........................
	Fall to different level		

5. HAZARDOUS CONDITION

Improperly or inadequately guarded	Defective tools, equipment, substances	Hazardous arrangement	Poor housekeeping
Unguarded	Unsafe design or construction	Improper illumination	Congested area
		Improper ventilation	Others........................
		Improper dress	No unsafe condition

6. AGENCY OF ACCIDENT

Machine	Can and end conveyors (belt, cable, can dividers, chain, twisters, drops, can elevators, etc.)	Hoists and Cranes	Chemicals
Vehicles		Elevators (passenger and freight)	Ladders or scaffolds
Hand tools		Building (door, pillar, wall, window, etc.)	Electrical apparatus
Tin and black plate (sheet, stock, or scrap)	Conveyors (chutes, belt, gravity)	Floors or level surfaces	Boilers, pressure vessels
Material work handled (other than tin and black plate)		Stairs, steps, or platforms	Others........................

7. AGENCY OF ACCIDENT PART
8. UNSAFE ACT

Operating without authority	Using equipment, tools, materials or vehicles unsafely	Unsafe loading, placing and mixing	Adjusting, clearing jams, cleaning machinery in motion
Failure to warn or secure	Failure to use personal protective equipment	Unsafe lifting and carrying (including insecure grip)	Distracting, teasing
Operating at unsafe speed		Taking an unsafe position	Poor housekeeping
Making safety devices inoperative	Failure to use equipment provided, (except personal protective equipment)		Others........................
Using defective equipment, materials, tools or vehicles			No unsafe act

CONTRIBUTING FACTORS

Disregard of instructions	Lack of knowledge or skill	Failure to report to medical department	Others........................
Bodily defects	Act of other than injured		No contributing factor

The bases on which identification of key facts are made are as follows:

1. *Nature of injury.* If two or more injuries are incurred, one being obviously the most severe, that one should be stated. If there are several minor injuries of different nature the term 'multiple injury' should be used.

2. *Part of body.* If the injury was localised to one body part, that part should be named, but if it extended to several sections of a major body part, that major body part should be named. Thus an injury to fingers, hand, wrist and forearm should be given as *upper extremities, multiple*. If the injury was internal, the body system affected should be named.

3. *Source of injury.* The object which directly produced the injury should be identified. If the injury results from forcible contact with two or more objects, either simultaneously or in rapid sequence, the choice should normally be as follows:
   (a) If one object is moving and the other stationary, choose the moving one;
   (b) If both objects are moving or both are stationary, choose the last one; but
   (c) If the injury results solely from the stress or strain caused by a free movement of the body (e.g. in reaching, twisting or bending), state bodily motion as the source of injury.

4. *Accident type.* This is fairly straightforward. If the source of injury was bodily motion the personal action or movement during which the motion occurred should be stated as the accident type.

5. *Hazardous condition.* The hazardous condition is related both to the accident and the agency of accident, and so will generally determine the agency of accident named. The hazardous condition may be the sole accident cause or one of several causes when more than one cause is involved.

6. *Agency of accident.* This may or may not be identical with the source of injury. The distinguishing characteristics of the agency of accident are that it was significantly hazardous and for that reason contributed to the occurrence of the accident. Its selection is based on this fact whether or not it inflicted the injury.

7. *Agency of accident part.* If the agency of accident had a specific hazardous part that contributed to the accident, that part should be named, e.g. the defective rung of a ladder that broke causing the employee to fall to the ground.

8. *The unsafe act.* The unsafe act may or may not have been by the person injured. The unsafe act might have been 'negative', i.e. a failure to do something the person should have done. The unsafe act may have been

deliberate or committed in ignorance or through forgetfulness. The unsafe act may be the sole accident cause or one of several causes.

The contributory factor may also be an accident cause.

The accident investigator clearly has not done his job unless he selects a hazardous condition, an unsafe act or a contributory factor as an accident cause, since all causes lie within these categories, and without a cause there would have been no accident.

Two examples illustrate this analysis:

Example 1. A fork lift truck went out of control when one wheel hit a block of wood which projected into the aisle. The truck left the aisle, entering a working area where it struck a machine operator, breaking his leg.
Analysis
1. Nature of injury — fracture.
2. Part of body — lower leg.
3. Source of injury — lift truck.
4. Accident type — struck by.
5. Hazardous condition — improperly placed wooden block.
6. Agency of accident — wooden block.
7. Agency of accident part — none.
8. Unsafe act — unsafe placement of material.
   Contributory factor — none.

Example 2. A labourer working in a trench was suffocated under a mass of earth when an unshored wall caved in.
Analysis
1. Nature of injury — suffocation (fatal).
2. Part of body — respiratory system.
3. Source of injury — earth.
4. Accident type — buried.
5. Hazardous condition — lack of shoring.
6. Agency of accident — trench.
7. Agency of accident part — none.
8. Unsafe act — needs further investigation (i.e. failure to shore, failure to warn, working without authorisation).

1.5.4.3 Classifying key facts

Rarely do two accidents occur in exactly the same way, although they tend to follow general patterns. Grouping accidents according to pattern is necessary for analysis.

For each key fact a general classification should be established in which similar data may be grouped. More specific classifications should then be set up within each general one to preserve as many details as possible. For example, ANSI standard Z 16.2 lists under 'Defects of Agencies' the following sub-headings.

1. Composed of unsuitable materials.

2. Dull.
3. Improperly constructed, assembled, etc.
4. Improperly designed.
5. Rough.
6. Sharp.
7. Slippery.
8. Worn, broken, etc.
9. Other.

Classifications are often best set up and developed as reports are reviewed. To be of most value the classification must cover situations which are relevant to the factory or company. The sub classifications should each be given a code number when analysing a number of accidents of the same type.

Thus, a group of ladder accidents in the same factory could be analysed as follows:

Code No.	Unsafe Conditions	Total number of accidents
1	Slippery rungs	6
2	Broken rungs	3
5	Weak, worn or cracked rails	2
	Unsafe Acts	
10	Failure to secure ladder at top	29
20	Failure to secure ladder at bottom	12
30	Placing ladder improperly (total)	22
31	On boxes etc	14
34	On inclined or other equipment	2
35	At too great an angle	3
40	Working in unsafe position (overreaching, improper stance, straddling between ladder and window ledge, etc.)	10
50	Ascending or descending improperly (insecure grip, jumping to ground, carrying heavy load in hand, etc.)	35
60	Using ladder known to be defective	8

Single classifications of each fact were sufficient to reveal most of the information needed to assess ladder accidents and take remedial steps. Where the agency of accident is more complicated, cross classification are required, e.g. between hazardous and conditions and accident type.

When tabulating accidents, hand sorting and tallying is generally preferred to the use of mechanical punched card equipment, since it forces the analyser to look at and think about each accident; mechanical sorting should only be used when large numbers of accidents are concerned which have already been analysed in smaller groups and appropriate action taken.

Merely tabulating accident analyses and filing them away without taking preventative action is a complete waste of time. The action which the above analysis of ladder accidents should have lead to would have been:

1. Retraining programme of all employees using ladders, with emphasis on accident causes revealed—backed by posters illustrating these specific points.
2. Insistence on regular ladder inspections.
3. Insistence on removing defective ladders and destroying them or repairing them promptly.

REFERENCES

1. Fife, I., and Machin, E.A., *Redgrave's Health and Safety in Factories*, Butterworth, London (1976).
2. *The Factories (Notice of Accident) Order* (1965) HMSO.
3. *The Construction (Notice of Accident) Order* (1964) HMSO.
4. Atiyah, P.S., *Accidents, compensation and the Law*, 2nd ed., Weidenfeld and Nicolson, (1975).
5. Powell, P.I., Hale, M., Martin, J., and Simon, M., *2000 Accidents*, National Institute of Industrial Psychology, London (1971).
6. Heinrich, H.W., *Industrial accident prevention*, 4th ed., McGraw Hill, New York (1959).
7. National Safety Council, *Accident prevention manual for industrial operations*, 7th ed., National Safety Council, Chicago (1977).
8. Arscott, P., and Armstrong, M., *An employers guide to health and safety managenent*, Kogan Page, London (1976).

1.6 WORKERS OR MACHINES?

Contents

The appropriate degree of mechanisation and the capital employed per industrial workplace were questions first decided in a rough and ready way by economic forces. This is still largely the case today.

The replacement of the worker by machinery, whatever its ultimate benefits, has the immediate effect of throwing men and women out of work, with consequent reduction in (or total loss of) income. The industrial revolution in Britain during the early nineteenth century was accompanied by fierce strife between bands of manual workers, particularly in the textile industry, and factory owners over the introduction of newly invented power looms, stocking frames and other machinery. This is strikingly described[1] by Charlotte Bronte in 'Shirley', which includes an account of an attack on a textile mill. The fact that the work done by the manual worker was often monotonous, unhealthy and dangerous was of little consolation to those thrown out of work by machinery and left to starve.

At that time the introduction of machinery probably increased the risks to the reduced labour force, and so had few short term merits in the eyes of the workers. The process is described in detail by Marx in his chapter on Machinery and Modern Industry in 'Capital'.[2]

Figure 1.6.1(a)   Workers in an old factory (Wills Collection of Tobacco Antiquities. Photo by W. D. & H. O. Wills, Photographic Dept.)

Today mechanisation and especially automation is often considered as a solution to a job involving risks to health and safety, but in so far as it reduces employment, it is bound to cause conflicts. The worker in an unsafe or unhealthy industry is almost always in a dilemma. If he complains or organises complaints about the health or safety hazards, he runs the risk not merely of incurring his employer's disapproval, but more importantly, of causing the entire manufacture to be discontinued. This fear is undoubtedly one of the main causes for the slowness in the recognition of the health hazards of many industries, e.g. those involving the use of asbestos, lead, cadmium, benzene, chromium compounds and mineral oil. It also lies at the root of many manual worker's ambition to achieve an office or 'white collar' job. Even when there is no improvement in status or pay there is usually a lower exposure to hazards and potentially hazardous materials.

There is today a very wide range in the degree of mechanisation and capital: labour ratios of different industries and countries. Some industries such as oil refining have a high capital:labour ratio while others like jobbing joinery and prison industries such as trawl net making are mainly manual with a low capital:labour ratio. The degree of mechanisation also varies greatly from highly developed industrial countries to primitive societies.

The social problems resulting from mechanisation, together with the rundown in most of the world's accessible mineral resources, especially petroleum, have caused many serious writers to question the wisdom of high degrees of mechanisation and high capital:labour ratios in industry. One of the most persuasive of these was the late E.F. Schumacher[3] in 'Small is Beautiful' although his enthusiasm often tended to obscure the difficulties of getting back to low capital:labour ratio industries.

In discussions of the appropriate degree of mechanisation, most attention is given to economic and social factors. The question of safety is usually overlooked or taken for granted, although it clearly needs to be considered in this connection.

1.6.1 Safety as a factor in determining degree of mechanisation

Some hazardous industrial processes where there are risks of fires and explosions, or exposure of personnel to radiation, high temperatures or dangerous chemicals now tend to be highly mechanised and automated. Control is then from a remote location in which a safe environment is maintained. Accidents and injuries are limited because the number of workers is low, although the injury rate may nevertheless be above average. Injury and fatal accident frequency rates are discussed in section 1.5.

In other less hazardous industries, where capital is short, or areas where unemployment is high, health and safety tend to be thought of more in terms of injury rates (e.g. number of injuries per 1000 persons employed) rather than in terms of total numbers of injuries.

Another view of this question can be formed by considering the differing abilities of workers and machines to perform various functions. Such a comparison, adapted from Willie Hammer's Handbook,[4] is given in Table 1.6.1 parts A, B and C.

Figure 1.6.1(b)   A highly mechanised modern factory (W. D. & H. O Wills)

Part A gives properties for which humans are generally preferable to machines, part B gives those for which machines are generally preferred, and Part C gives properties where the choice is less clear and needs detailed study.

Table 1.6.1 COMPARISON OF HUMAN AND MACHINE CAPABILITIES
A. *Properties for which Humans are Preferred*

Property	Human	Machine
1. Availability and cost	Good. Available at short notice. Often cheap.	Limited by cost and production time.
2. Ease of maintenance and repair.	Good. Can maintain and often repair himself without special care.	External maintenance and special care generally needed.
3. Fuel and power requirements.	Low. Single fuel source serves all power functions.	Generally higher. May need several special power supplies.
4. Cooling requirements.	Not normally required.	External cooling often required.
5. Overload tolerance.	High for short periods.	Limited.
6. Performance with damaged internal communication system.	Good. Nervous system has in-built redundancy.	Poor. Complete breakdown often follows failure of single circuit.
7. Self protection.	Good.	Normally none, unless specially built in.

People, profits and safety

Table 1.6.1 *(cont)*

Property	Human	Machine
8. Performance flexibility.	Excellent.	Limited.
9. Reasoned decision making.	Good.	None outside limits of built in programme.
10. Interpretation of unclear input signals.	Good. Can often correctly interpret vague statements.	Poor. Needs clear and precise signals.
11. Adaptability to new programmes.	Good. Can reprogramme himself.	Poor. Needs thorough reprogramming.
12. Improvisation in coping with unexpected events.	Good, based on memory and experience.	Poor.
13. Learning from experience.	Good.	Poor.

B. *Properties for which Machines are Preferred*

Property	Human	Machine
1. Obedience.	Variable. Often sensitive to abuse.	Good. Does only what it is told.
2. Dependence on proper oxygen supply.	Complete.	Oxygen seldom required.
3. Resistance to radiation.	Poor. Needs special protection.	Much improved.
4. Temperature range for efficient working.	Narrow.	Wider.
5. Resistance to toxic hazards.	Poor. Needs special protection.	Improved.
6. Mechanical power.	Low and limited for prolonged periods.	Limited only by design. Can function continuously.
7. Performance of repetitive functions.	Fair. Impaired by boredom, fatigue, illness and cycling.	Good. Impaired only by lack of maintenance or calibration.
8. High speed information searching.	Fair.	High performance possible with computers.
9. Disposibility	Often fraught with problems.	Easy.

Table 1.6.1 *(cont.)*

C. *Properties where choice needs detailed study*

Property	Human	Machine
1. Liability to error.	Sometimes makes errors but often detects and corrects them.	Seldom makes errors, but once made cannot correct them.
2. Sensing abilities.	Has variety of senses but cannot detect some important signals.	Improved abilities possible for specialised routine duties.

The comparison is necessarily general and may soon become outdated as more sophisticated machines are invented. Even so it seems worth summarising and re-iterating the lessons to be learnt.

1. As a rule the deciding factors will be cost and availability.
2. Hazardous processes should be mechanised and automated as far as possible. Personnel required to control the processes should be able to do so remotely in a controlled micro climate and be protected against the accidents which the hazards could create. The design of protective buildings and windows needs special advice and attention, since humans are better able to withstand blast than most buildings. In a large explosion men are more likely to be killed by the collapse of a building than by the blast itself.[5]
3. If there is a serious safety risk through human disobedience, machines should be chosen.
4. Prolonged arduous mechanical work should be performed by machinery unless there is evidence that it can be performed by willing men without risk to health.
5. Monotonous repetitive tasks which impair human functions should be performed by machines.
6. Tasks which require flexibility, decision making, adaptability and improvisation should be reserved for humans.
7. Mutual 'back' up between humans and machines should be used to check for faults and errors which may cause accidents. Thus an alert human being can generally detect a fire (by sight, smell, heat or noise) more efficiently than a fire detection instrument, but the latter has the advantage that it can always be on the spot and needs no sleep.

1.6.2 The worker-machine interface

In practice, few industrial processes are entirely manual or entirely automated. Most involve a complex blend of manual and machine operations, with workers, particularly where labour is cheap and abundant, filling gaps in a series of mechanical operations. Men also direct and control the operation of the machines.

Accidents frequently occur at these man-machine interfaces. Their causes are usually complex and involve both operator error in misjudgement and hazards arising from the machine itself. The most fruitful approach to the problem lies in the study of the relation of the machine to the worker and is taken up further in section 4.5 under Ergonomics.

REFERENCES

1. Bronte, C., *Shirley*.
2. Marx, K., *Capital*, Volume 2, Chapter XV, Machinery and Modern Industry, Allen and Unwin, Woking (1946 reprint).
3. Schumacher, E.F., *Small is Beautiful*, Abacus, London (1974).
4. Hammer, W., *Handbook of system and product safety*, Prentice Hall, New Jersey (1972).
5. Kletz, T.A., "Some of the wider questions raised by Flixborough", Paper in Symposium *The Technical Lessons of Flixborough*, The Institution of Chemical Engineers, Nottingham, (December 1975).
6. Powell, P.I., Hale, M., Martin, J., and Simon, M., *2000 accidents*, National Institute of Industrial Psychology, London (1971).

1.7 HUMAN FACTORS, SELECTION AND TRAINING

Contents

Since a considerable proportion of industrial accidents is attributed to human error, it is only natural to expect that human factors and personality profiles would have a marked influence on the frequency of industrial accidents, and that by careful selection and training, i.e. finding or fitting the worker for the job, such accidents would be appreciably reduced.

The idea of the 'accident prone personality' was at one time much in vogue. Accident proneness is greatly influenced by the mental attitude of the subject. According to Hunter[1]

'The accident prone are apt to be insubordinate, temperamentally excitable and to show a tendency to get flustered in an emergency. Those and other defects of personality indicate a lack of aptitude on the part of the subjects for their occupation.'

A good account of several theories of accident proneness is given by Tiffin and McCormick in *Industrial Psychology*.[2] Some obvious factors making for accident proneness include poor vision, unpopularity among ones fellows, and manic depressive personalities. A more subtle correlation (discussed later under personnel selection) is found between the relative level of spatial perception and muscular activity, and accident rates. Individuals showing levels of muscular activity higher than their levels of spacial perception tend to have higher than average accident rates.

Nowadays more attention is paid to people's attitudes, motives, mental conflicts and frustrations. Reactions to frustration can take varied forms — aggression, regression, fixation, resignation, negativism, repression and withdrawal. These are sometimes accompanied by strong emotions which can create the stresses discussed in the last section. Management should try to avoid allowing these frustrations in the first place, but since this is not always possible, it should also provide a few safety valves. A 'permissive listener' will very often relieve tensions by helping a person get things off his chest.

While there can be little doubt that training is important for accident prevention there is now far less agreement among those who have studied the subject about what qualities make a safe or unsafe worker, and the extent to which personality factors affect job safety. Perhaps one reason for this lies in the difficulty in making truly scientific and objective studies of the relative safety performance of different workers, and in eliminating extraneous and incidental factors from such comparisons.

It has also to be recognised that most 'misfits' and those with obvious psychopathic tendencies have already been excluded by present methods of personnel selection, and that the working population in any factory is more homogeneous than the population as a whole.

'Fitting the task to the man (or woman)'[3] is ultimately likely to prove more effective in terms of both production efficiency and accident prevention than 'fitting or finding the worker for the job'. This however involves capital expenditure and the time for it is usually when a new production unit is being planned and designed. It is discussed in section 4.5 Ergonomics.

Even when 'the task has been fitted to the man (or woman)', this is seldom anyone in particular, but an 'average man or woman' — usually of some particular race or region. Workers still have to be found and fitted to the job, and that is what concerns us here. Ideally a worker should fit the job both physically and mentally, but fortunately man is adaptable, and the man-job relationship can survive even a considerable degree of misfit with its resulting stress.

1.7.1 Human factors in accident causation

Here one has to make many distinctions, i.e. between inherent factors and temporary factors resulting from stresses, training, etc., between physical and mental or psychological factors and between factors of age and sex.

In a very exhaustive shop floor study lasting over a year in three different factories, over 2000 accidents were analysed and the influence of personal factors was studied.[4] Some physical factors were found to give expected correlations with accident frequency in particular situations; thus men less than 5ft 9in tall, working in a dispatch department, where large packages were handled, had a significantly higher accident rate than the big fellows.

Older workers had a significantly lower accident rate than younger workers in a machine shop, but in the assembly shop the reverse was true. Whilst it would seem that there is an optimum age for safe performance, it is very difficult to study age in isolation from other factors such as experience. Neither marital status, nor number of children gave any significant correlation with accident frequency, nor did nationality provided the different nationalities present could communicate readily with each other without inter-racial tension.

Talkative people were found to have a higher accident rate than the less talkative. The investigators felt this may have been simply because they reported more of their accidents than the others, but their actual observation seems to ring true. One would expect talkative people to have only half their mind on their job, and even to cause others to have accidents by distracting them. Compulsive talkers can be a menace in many jobs, and even deafness may sometimes have blessings. But totally withdrawn and uncommunicative people may also be injurious to safety, particularly when they fail to report accidents or warn others of hazards. So somewhere there is probably an optimum in talkativeness.

Similarly, extraverts showed a higher accident frequency rate than intraverts in most situations, but since extroverts were generally more talkative the same explanations may well apply. Neither smoking, drinking moderately, nor differences in intelligence showed any correlation with accident frequency, nor did eyesight, education, medical history or domestic stress. The jobs themselves did not, however, require very high standards of intelligence, education or eyesight. But the lack of positive results in all these comparisons is no proof that correlations did not exist which might have been revealed if the effects of extraneous factors could have been eliminated.

Those with longer service in a particular shop and those who had repeated a task most frequently had fewer accidents than those with shorter service or those who changed tasks frequently. On a quickly repetitive task, the highest accident rate occurred towards the beginning of the period on the task. Training which includes techniques of risk avoidance to the tasks in hand should reduce accidents, especially among recruits. All too often this training arises simply through the experience of having a minor accident.

Boredom and monotony in a repetitive task were thought to influence accident frequency, but this was not firmly established.

The above study was related to activities where the workers studied worked in groups rather than in isolation, although their tasks were in most cases done individually.

1.7.1.1 Effects of stress

The effects of various stresses on working performance have been studied carefully by several psychologists[5], and although statistical proof is not easy to find, it seems fair to conclude that many accidents attributed to personal failure occur simply because the person involved was overstressed.

Many accidents involving women are caused by pre-menstrual stress; this unfortunately is insufficiently recognised by employers, supervisors and safety experts[6]. Undoubtedly it affects some women more than others, and at some periods of their lives more than others. It requires sympathetic treatment and consideration by employers, and re-arrangement of work schedules where necessary to reduce the exposure of women workers at these critical times to accident hazards.

However it is wrong to imagine that all stresses are harmful. There is generally an optimum stress level for any individual in any particular task. As there are several different kinds of stresses, the term 'degree of arousal' is used by psychologists to give a common quantitative measure for stresses of different kinds. Typical degree of arousal-performance curves are shown in *Figure 1.7.1*. The most common kinds of stress include:

A feeling of being personally threatened.
Too much to do at once, (e.g. talking and/or eating while driving).
Noise.
Excessive body temperature.
Loss of sleep.
Influence of drugs.

Tea and coffee contain caffeine which increases arousal, as do the amphetamine drugs benzidrene and dexadrine. If a person is feeling sleepy

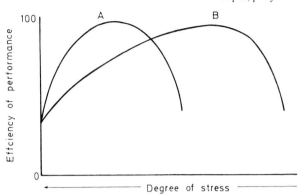

Figure 1.7.1   Performance efficiency vs. stress Curve A. Intellectual problem solving
Curve B. Heavy manual work

these stimulants increase his arousal to normal level. Sleeping tablets, tranquillisers and anti-histamines reduce arousal and produce behaviour similar to that of loss of sleep.

A stress may either increase or decrease arousal to or from the optimum. An overstressed person may behave similarly to an overloaded computer. A person may attempt to compensate for the fall in his efficiency by concentrating on the more important aspects of his task at the expense of the less important. It is not possible to make two decisions at the same time, but a routine well-practised skill can be performed while doing something else as well. Experiments have been done to determine the optimum level of arousal for various activities. Continuous noise for instance will increase a person's alertness or degree of arousal and increase his speed or efficiency for simple operations, though decreasing his accuracy for complex work. Problems which require deep thought can sometimes only be solved by a person who is isolated from noise and all ordinary stresses: the bath seems a favourite place for doing this.

A typical experiment on the influence of stress was described by Brown, Tickner and Simmonds.[7] Each of twenty-four drivers had to answer questions over a radiotelephone while deciding whether or not a gap between two obstacles was wide enough to drive the car through. Drivers made more wrong decisions while driving and answering questions than in a control situation while driving without questions. If the driver delayed his decision for too long, he had to drive through the gap whether he liked it or not because he was not allowed to stop or swerve sharply. If the driver delayed dealing with the question until he had decided whether or not to accept the gap, he sometimes had partly forgotten the question, so that his answer was wrong as well as late. But once he had decided to accept the narrow gap, his skill in driving the car through it was not degraded by having to answer questions simultaneously.

Training and experience of the job, particularly the latter, are important in enabling the worker to carry it out under stress; thus an unfamiliar task is performed less satisfactorily while under stress.

The stresses under which workers operate should be seriously studied by all those responsible for industrial safety.

The dangers of compulsive, irrational and accident-causing behaviour in overstressed people were well illustrated in a study of a number of major railway accidents in the UK.[8] The men concerned were signalmen and drivers working in isolation. In several instances, overstress (resulting in one case from criticism of the man's previous actions by his superiors), resulted in an altered mental condition which produced abnormal and psychotic behaviour such as setting the points wrongly or moving them while a train was passing over them. The phenomenon is termed by the authors of the study 'coping failure'. Early signs are said to be difficulty in getting off to sleep, early waking, and anxiety. Other symptoms are changes in weight and sleeping habits, alteration in sexual habits and compulsive behaviour. Three groups of coping failure were quoted:[9]

1. Inadequate coping capacity — found in people who appear unable to cope with challenges which others surmount without difficulty.
2. Excessive challenges either in terms of degree or frequency of challenge.
3. Excessive internal drive, deriving from innate ambition or social pressure.

Coping failure seems to have something in common with the legendary death wish said to be evidenced in the careers of famous commanders and leaders such as General Wolfe and Adolph Hitler.[10]

The phenomenon is mentioned here as a warning that it should be born in mind. The works doctor should be best placed to advise on it. Its consequences are most serious when the failure occurs in personnel upon whose behaviour under stress the safety of themselves and many others depends. It would not seem unreasonable to prepare a short list of these positions in any works or factory, and then examine the performance of their occupants under various stresses. Cases of earlier mental in-stability and any drug habits of these people should also be studied, conclusions drawn and acted upon.

1.7.2 Selection

It should not be necessary in this section to discuss the work of personnel selection in detail except in relation to safety at work. The need to avoid square pegs in round holes should, however, always be born in mind.

The use of written tests[11] involving choosing one of several possible answers to a number of questions such as those used for vocational guidance allows personalities to be graphed or tabulated in terms of percentiles for such qualities as:

Emotional stability.
Self-sufficiency.
Extroversion.
Dominance.
Confidence.
Sociability.

Similar tests for occupational interests and aptitudes can build up a very useful picture of a person's real innate abilities which are not always revealed by examination results. The skilled application of such tests and results can do much to ensure that personnel are correctly placed. Records of previous mental instability are of obvious importance for the placing of personnel who have the safety of others in their hands, although for equally obvious reasons these are not always forthcoming.

Medical examination should be insisted upon for recruits to be employed in operations where physical defects such as deafness, poor eyesight and colour blindness would prove hazardous. The same rule should be applied to sub-contractors and their employees.

Selection of people for particular tasks can only be done rationally when the task itself is properly defined. This is mainly a matter of *Job Analysis*. It is discussed lucidly by Boydell in *A guide to job analysis*[12]; training and selection are discussed in the same context.

Tests designed to test the fitness of applicants for particular jobs are of

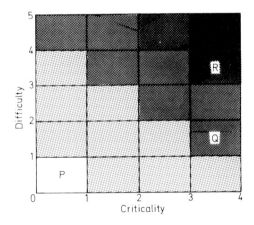

Figure 1.7.2   Difficulty vs. criticality (From 'A guide to job analysis (Boydell))

two types, aptitude tests and achievement tests. The aptitude test is intended to show how well the individual will be able to perform the job after appropriate training. The achievement test is intended to show how well he is actually performing in a job he is already doing or has been trained to do. Only aptitude tests are discussed further here. The characteristics to be 'measured' form three groups:

A. Mental ability
B. Spacial perception
C. Motor ability

Tests of mental ability of different kinds are used to determine whether an individual comes up to the standard required for a particular job. Some jobs require special alertness, or social intelligence, (e.g. especially supervisory jobs) in addition to general intelligence.

Spatial perception is not just good vision, but includes the ability to perceive spatial patterns accurately and compare them rapidly with each other so as to be able to recognise small differences in otherwise identical objects.

Motor abilities are generally measured by psychomotor tests which are really combined tests of sensory and muscular abilities. They may be grouped as follows:

1. Control precision.
2. Multi-limb coordination.
3. Response orientation.
4. Reaction time.
5. Speed of arm movement.
6. Rate control.
7. Manual dexterity.
8. Finger dexterity.
9. Arm hand steadiness.
10. Wrist finger speed.
11. Aiming.

Different individuals show quite specific differences in their performance in these psychomotor tests which should reflect their aptitudes for various skilled manual occupations. It is, however, of little value to apply such tests unless the jobs themselves have been properly evaluated and the required psychomotor skills identified.

When such tests are used in selecting applicants for various jobs, the relative levels of spatial perception and muscular activity should be examined, and only those with a high ratio of spatial perception to muscular activity should be employed in delicate and potentially hazardous skilled manual jobs (*Figure 1.7.2*).

Senses are more complex than commonly thought. In addition to the five commonly recognised sensory channels — vision, hearing, touch, taste and smell, there are several others of which vibration, pressure and the kinaesthetic sense are most important.[13] The last means the ability to recognise and coordinate the relative position of various parts of the body without looking at them — such as touching or scratching a particular spot on one's back.

1.7.3 Education and training

Education is taken to mean learning to think and apply one's mind to new situations, whereas training in the narrower sense is associated with moulding habits. Training can be used in a broader sense to include education, and this, unless stated otherwise, is the sense in which the word training is used here. Training is needed at all levels and in several areas in which various aspects of safety are important. Managers need education rather than training in the concepts, philosophy and economics of safety.

Safety directors, professionals and advisors should have had training which includes knowledge of the following:

1. Relevant legal requirements.
2. Methods and requirements of inspection, accident investigation and safety audits.
3. Principles and practice of accident prevention.
4. Specific hazards met in the work of the organisation.
5. Emergency procedures and situations.

Line managers and supervisors need training in safe working methods, accident prevention, human relations, and how to train and supervise those working under them in safe working methods.

It is important, (as mentioned earlier) that the safety specialist is closely involved with training. In some firms, the functions of training and safety are combined within one department. This is sound provided it is recognised that training is but one of many activities with which the safety specialist is concerned. An organised though flexible programme is needed for safety training within any organisation.

The main areas in which safety training is needed are given in the following paragraphs.

1. General safety

In well-managed factories general safety is usually covered by a short induction course for all new entrants in which subjects such as housekeeping, correct methods of lifting and handling, machine hazards, the use of protective clothing, etc, are dealt with. Particular hazards of the works and its departments, materials and machines should be brought to light and safety rules and responsibilities explained.

Procedures to be followed in various emergencies will be discussed, but since this is an induction course for entrants who are unfamiliar with the works, it is more important at this stage to make the new employee aware that these things exist, of the need to take them seriously, and how they will be brought to his attention, rather than smother him in too much detail at the outset of his employment.

General safety training must not stop after a brief initial induction course. The safety adviser must take steps continuously to maintain workers' interest in the subject and to involve them in it. He will only succeed in this if the managers, senior staff and supervisors themselves have a positive attitude towards safety. Further information and training sessions for all employees backed by films, talks, demonstrations, a newsletter and posters should be held to meet the particular needs of the works. It is also suggested that every works should maintain a small lending library, accessible to all employees, containing selected books on industrial safety and hazards appropriate to the materials and the operations carried out.

2. Fire fighting, escape and rescue

All workers need to be instructed on what to do in the event of a fire, and should be examined from time to time to ensure that they fully understand. Fire drills should be held to consolidate this training in factories where there is a significant fire hazard. All workers should be trained when and how to raise the fire alarm and how to use fire fighting equipment provided on the spot, such as hose reels and extinguishers. In addition, many works have their own fire brigades manned by a skeleton full-time crew, and part time firemen recruited from the firm's employees, and specially trained. Fire training is discussed in section 6.6.

3. First aid

First aid is discussed in Appendix B. In all factories where more than fifty people are employed, it is a statutory requirement that the person in charge of the first aid box must possess a current first aid certificate issued by one of the voluntary first aid societies, St. John's Ambulance Association and Brigade, the British Red Cross Society, and the St. Andrew's Ambulance Association. First aid training is not only valuable in itself. It is also good safety training in bringing home the consequences of unsafe acts to the trainee.

4. Specific job training

Many workers will have had specific and relevant job training or experience before being engaged, e.g. in a technical college, an industrial training school or rehabilitation centre or with a previous employer. This former training and experience may or may not have paid adequate attention to safety. Even so, it is unlikely that it will correspond exactly to the materials, machinery and equipment which the worker will now be required to handle.

Job training in works and factories differs widely in depth and scope, and it is difficult to generalise about a subject whose needs vary so widely according to the operations carried out. A widely noted criticism of present job training is that it is haphazard and consists of picking up the required skills from a more experienced worker, assisted by a little intermittent instruction from a chargehand or foreman.

The most serious training problem is that of training raw recruits with no comparable experience of machines, equipment and materials. Traditional training methods of apprenticeship whereby the trainee picks up his skills from an experienced worker are recognised as inefficient and time wasting. Even where the experienced worker is thoroughly efficient at his job, he may not have any aptitude for training and will often have performed an operation so many times with apparent ease that he has entirely forgotten the problems he encountered as a beginner. Persons in charge of training, whether they be experienced workers on the job, who are training in addition to their ordinary work, or full time trainers, must themselves be taught to train.

It is often debated whether job training should be done on the spot in the working environment, or in a simulated artificial environment. The advantage of training on the job is that it is carried out in realistic working conditions. The advantage of training in a simulated environment is that it helps to enable the various components of the task to be isolated and learnt separately. This is not always possible. A balance between the two methods is needed.

An essential part of training is rapid feedback to trainees of data about their own performance, showing how and why they may have succeeded or fallen short. The good trainer will not hesitate to show trainees evidence of their own progress and success and to help them to build up confidence. The trainer will also stop the trainee when a mistake is made and ask the trainee to think out and explain what went wrong before explaining in so many words. In this way the trainee is more likely to remember the lessons, as well as learning how to correct mistakes. The trainer should teach trainees to criticise their own performance.

Sometimes it will prove helpful to select those workers who are to train recruits working alongside them and give them a short basic course in training. The components of a task usually need to be studied, separated and broken down into a number of separate skills which are taught individually so that each skill is mastered before they are combined.[13] When learning to drive a car, the actions of changing gears and steering are taught and practised separately before they are combined. The important thing is not to increase the stress or degree of arousal in the trainee beyond the point where his performance falls off until he has become so used to the task that the stresses are reduced to acceptable levels.

Training in a new task is usually best given in short spells, a little at a time, rather than over protracted periods. In this way the trainee has time to adjust to the new stresses between sessions.

Examples of training programmes are given in section 5.8 for drivers of overhead travelling cranes, and in section 5.9 for fork lift truck drivers. The method of training used must be appropriate to the job. A wide variety of artificial training aids are available — programmed-learning courses, visual aids, simulators, teaching machines and films. The film with sound-track which conveys audio and visual information together and commands undivided attention is a most useful medium of instruction.

Sound training of workers in their various tasks results in the formation of habits which enable the job to be performed without excessive stress, which itself is beneficial for safety.[14] Safety specialists should at the same time have studied the hazards of the tasks and the lessons to be drawn from them, and incorporate practical lessons on avoiding these specific job risks into the training programme. Safety specialists should also study carefully whether the applications of these lessons will cause conflicts, i.e. between safety and productivity or bonus packets. These conflicts must be resolved within the firm. Failure to resolve them is the cause of many accidents.

One corollary of the need for training is the need to ensure that the training is properly used. It is an utter waste of time to train a worker for one job and then post him to an entirely different one, and employ another untrained worker for the job for which the first one had been trained. Workers in skilled and specialised tasks where a slight error could cause a serious accident (e.g. crane drivers) should be licensed and periodically re-examined. While some mobility of labour is always essential in any factory it is in the unfamiliar task that most accidents occur.

Assistance for safety training may be obtained from:

1. *Voluntary safety organisations who arrange courses at different levels in various topics i.e.*
The British Safety Council, 62 Chancellors Road, London, W.6.
RoSPA, Cannon House, The Priory, Queensway, Birmingham, B4 6BS.

2. *Colleges of further education who run safety courses at various times.*

3. *Courses mainly for supervisors which include 'job safety'.*
Training Services Agency, TWI Section, Ebury Bridge House, Ebury Bridge Road, London, SW1.

4. *Industrial Training Boards responsible for the firm's industry.*

5. *General information on safety.*
The Industrial Health and Safety Centre, 97 Horseferry Road, London, SW1P

REFERENCES

1. Hunter, D., *The disease of occupations,* 2nd ed., The English Universities Press, London, 977 (1957).
2. Tiffin, J., and McCormick, E.J., *Industrial Psychology*, 3rd ed., Allen & Unwin, London (1966).
3. Grandjean, E., *Fitting the task to the man*, 2nd ed., Taylor and Francis, London (1975).
4. Powell, P.I., Hale, M., Martin, J., and Simon, M., *2000 accidents*, National Institute of Industrial Psychology, London (1971).

5. Poulton, E.C., 'Skilled performance and stress' in *Psychology at Work* edited Peter B. Warr, Penguin, London (1971).
6. Dalton, K., *Premenstrual syndrome and progesterons therapy*, William Heinemann Medical Books (1977).
7. Brown, I.D., Tickner, A.H., and Simmonds, D.C.V., 'Interference between concurrent tasks of driving and telephoning', *Journal of Applied Psychology*, 53, 419-24 (1969).
8. Atherley, G.R.C., 'People and safety—stress in its role in serious accidents', Symposium Paper Eurochem Conference, *Chemical Engineering in a Hostile World*, Birmingham, June (1977).
9. Mills, I.H., 'The disease of failure and coping', *Practitioner*, 217, 529-38 (1976).
10. Waite, R., *The psychopathic God, Adolph Hitler*, Harper and Row, London (1977).
11. *Private Communication*. The Vocational Guidance Association, London.
12. Boydell, T.H., *A guide to job analysis*, British Association for Commercial and Industrial Education, London (1973).
13. Edholm, O.G., *The biology of work*, Weidenfeld and Nicolson, London, see Chapter 11, (1967).
14. Health and Safety Executive, *Basic rules for Safety and Health at Work*, HMSO, London (1975).

1.8 WORKING HOURS, INCENTIVES AND OUTSIDE INFLUENCES

Contents

Shift work, unusually long working hours, and piece-rate systems are all possible contributory causes of personal errors which can lead to accidents. But firm data linking these to accident incidence is scanty, although a great deal more has been learnt about their effects on human performance in various fields, including production. There is generally an inverse correlation between human performance in common industrial tasks and accident frequency, so it seems fair to fill in the gaps in our knowledge of the relation between these factors and accidents with information derived from studies of the effects of these factors on human performance, especially in the field of production.

The influences discussed here are all related to the effects of personal stress levels or degrees of arousal on performance. The first question to ask is whether they result in the worker performing at stress levels above or below the optimum for the job. One must be careful not to paint too simplified a picture, for the stress level can be mental or physical, and different kinds of stresses affect people differently. But some simplification is essential if we are to obtain a clear overall picture.

Two extremes have to be distinguished, fatigue and boredom, even though a worker can be both bored and fatigued in some circumstances at the same time. Boredom implies insufficient stress or arousal and fatigue implies excessive stress.

1.8.1 Rest pauses, fatigue and boredom

Industrial psychologists agree that rest pauses during the working day are essential, and the evidence suggests that short frequent pauses of say five minutes every hour are more effective than longer and less frequent pauses.[1] The need for rest pauses when hard physical work is done is usually evident. But they are equally important in sedentary work and mental effort. The effects of fatigue have been well documented.

All tasks include the receipt of signals, usually visible or audible, before some movement is made. As fatigue develops, the delay between receipt of the signal and the beginning of the movement increases and becomes erratic. In a repetitive task, some of the cycles become longer with an

Figure 1.8.1   Typical modern control room (Foxboro-Yoxall Ltd. Photographer, Harold White)

unusually long delay every now and then, following which the response time quickens. These delays or 'blocks' represent involuntary rest pauses within the central nervous system which indicate a gradual accumulation of inhibitory substances within it which finally blocks the transmission of nerve impulses. The inhibitory state disappears during the block and recovery takes place. An inhibitory state can be partly overcome by stronger stimuli, as when the person makes a greater effort. The effects of fatigue are specific as though they are local to one part only of the central nervous system. They can be combatted by a change of activity, so long as the alternate task is not too similar to the fatiguing one.

While fatigue appears to be due to overloading of part of the nervous system, boredom appears to be due to underloading. This may happen in so-called vigilance tasks, where somebody has to observe a situation (e.g. a group of instruments) for long hours to detect an occasional and usually fleeting signal which is evidence that something abnormal has happened which will quickly lead to trouble.

The nervous system appears to suffer from a low degree of arousal so that boredom sets in after a time, and the person fails to notice the signal. But fatigue and boredom may have more in common than is first apparent. In both, the person's central nervous system has been placed in a state of alert to expect the signal. To maintain this sense of alertness can be tiring in the same way that muscles tire when held in tension without doing any actual work. This happens in many disciplinary punishments, e.g. children being made to stand in the corner, or soldiers kept immobile for long periods with shouldered arms.

Sometimes the fatigued or bored person, after apparently reaching the limit of exhaustion at an early stage, develop's a 'second wind', perhaps through discovering some new and interesting facet of the subject of his fatigue or boredom, which now occupies the forefront of his attention, allowing him to continue performing the fatiguing or monotonous task 'automatically' at a lower level of consciousness.

Whilst Man's ability to adapt to most situations is truly amazing, clearly we should not expect too much. There is always a time when Man fails to adapt, and this is where the danger lies.

1.8.2 Working hours and overtime

Industrial conditions in the nineteenth century were unsafe and unhygienic and very long working hours were common. Workers were treated rather as machines on the assumption that their hourly productivity was the same however many hours a day they worked. This is reflected in the notion then popular among industrialists that all their profit was obtained during the last hour of their employee's work, first put forward in a pamphlet[2] written by Professor Nassau W. Senior, an Oxford economist in 1837. This was written in support of the cotton manufacturers in their fight against the newly passed Factory Act and the still more menacing 'Ten-hours agitation'.

The dangers of long working hours were demonstrated in the munitions crisis of World War I when very long hours (up to 100 per week) were worked. This resulted in a drop of output.

The lesson was, however, soon forgotten, for the same story was repeated in Britain in 1940. A paper by H.M. Vernon published that year showed that during a 12-hour working day women experienced two and a half times as many accidents as during a 10-hour day.[3]

These points were brought out in the report of the National Board for Prices and Incomes on 'Hours of Work, Overtime and Shift Working' in 1970.[4] They found that both overtime and shift work are favoured by workers and managements. Workers get an opportunity of augmenting their pay packets, managements make greater use of existing manpower, while shift working allows fuller implementation of plant. Even today it is not uncommon to find an actual working week of 60 hours including overtime, and men working for periods of two weeks without a rest day. Efficiency in these circumstances must inevitably fall and accident rates will increase unless productivity is also very low. A vicious circle is reached from which it is difficult to escape. *(Figure 1.8.2)*.

The moral to be drawn is that managers who are concerned with safety and accident prevention need to look closely at overtime working, and would be well advised to prohibit actual working weeks (including overtime) in excess of 50 hours other than in exceptional circumstances.

1.8.2.1 Accident distribution over the days of the week

The study[5] of 2000 accidents discussed in section 1.7 showed interesting differences in accident rates over the days of a working week. The length of the study and the large number of accidents studied enabled statistically significant conclusions to be drawn. The highest accident rate was nearly always found on Mondays and the lowest rates on Thursdays and Fridays.

On closer examination it was found that the underlying factors which

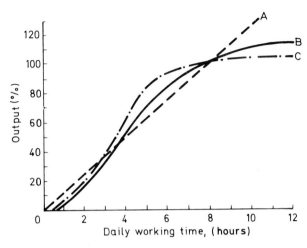

*Figure 1.8.2   Relationship between working times and performance. Performance in
an 8-hour day equals 100%
Curve A. Proportionality between working time and performance.
Curve B. Relationship between performance and working time with an average
workload.
Curve C. With a high workload (Lehmann)*

caused these differences were work load and absenteeism. Absenteeism
was always higher on Mondays than on other days of the week, which
resulted in several workers having to undertake unfamiliar jobs that day.
Work load in most departments was lower on Thursdays and Fridays than
on other days. Pay day was Thursday in some departments and Friday in
others. Pay days were always well attended and the atmosphere on these
days was more relaxed.

1.8.2.2   Distribution of accidents over the hours of the day

The same study showed three characteristics in the distribution of accidents
over the hours of the day:

1. The accident rate was higher in the mornings than in the afternoons,
   the peak time for accidents occurring after mid morning. This pattern
   was persistent and appeared to reflect the work load.
2. Local accident peaks occurred just before breaks. These could have
   been the result of fatigue, but more probably appeared due to a
   speed-up of production at these times to meet targets before the
   break.
3. At the end of the afternoon, the accident peak was less pronounced

and preceded a decline as people stopped work to tidy up at the end of the day.

These studies showed that the incidence of accidents was very much a function of the work done, and they also teach us to be on our guard against picking on certain individuals as accident prone. An adaptable hard-working man who performs a variety of tasks is unfortunately likely to have more accidents than one who sticks to a single job and works at a more leisurely pace. If he is talkative as well and reports all his accidents he is almost certain to have a higher accident record than average!

1.8.2.3 Shift work

Shift work is essential in many industries, e.g. chemical manufacture, oil refineries, steel plants and power stations, and is widespread in others despite its disadvantages.

Shift workers are generally under greater stress than day workers for two reasons:

1. Their natural rhythm of rest and activity has been interrupted and they are constantly in a process of readapting to a different rhythm.
2. Night shift workers will suffer fatigue through loss and interruption of sleep in the daytime in most domestic circumstances.

In spite of this, shift work has compensations for many workers such as:

A. The ability to enjoy shopping and recreational facilities — pubs, swimming pools, tennis courts, golf courses and the countryside at off-peak periods without having to queue.
B. The absence, especially on night shift, of disturbing and complicating factors caused by maintenance work and changes in production schedules, operating conditions, etc. ordained by the managers who mostly work on days only.

However, long spells on night shift are unattractive to most workers. As a worker quaintly put it, 'a week is quite long enough for the wife to be left alone at night'.[6]

Various studies of man's performance on night shift have shown a distinct deterioration in output and in the time taken to perform simple tasks such as answering the telephone.

1.8.2.4 The rhythm of rest and activity[7]

Man has a natural rhythm of rest and activity, commonly monitored by his deep body temperature which in normal life rises from a minimum at about 07.00 hours to a plateau between 11.00 and 14.00, thereafter continuing to rise slowly to a peak at about 22.00 hours. The deep body temperature then

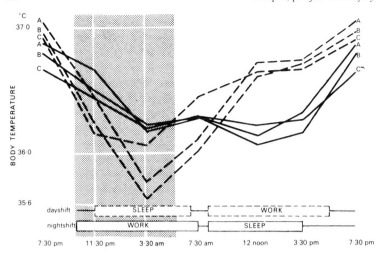

Figure 1.8.3   Body temperature curves of three subjects, (A.B.C) during dayshift and nightshift). The night is indicated by the shaded portion. The diurnal range of body temperature averages 1.1 °C in dayshift and 0.6 °C in nightshift. (From 'Biology of Work' by O.G. Edholm, published by Weidenfeld & Nicolson)

falls rapidly to its minimum at about 03.00 hours. The overall range of variation in a healthy person is about 0.6°C (*see Figure 1.8.3*).

This cycle of body temperature is accompanied by changes in various physiological and biochemical activities such as activity of the adrenal cortex, and urine flow, and there is little doubt that Man's physical and mental performance are impaired during the low body temperature part of the cycle. There appears to be some built-in timing mechanism which controls the cycle.

The bodies of people on shift work attempt to adjust their cycles to the new hours of work, some succeeding better than others. Some shift workers starting a spell of night shifts manage to adapt or invert their cycles by the second or third day, others may take a week or more to adapt, and some cannot adapt at all. Adaptation is partly related to the type of work done and partly to their domestic and outside activities. Workers engaged in strenuous manual work adapt more easily than those engaged in control, monitoring and vigilance activities.

Men working beyond the artic circle in total winter darkness adapt more easily than those working in lower latitudes. Examples are, however, found of men and wives with no children working permanent night shifts. Their diurnal cycles have completely inverted and adapted entirely to their lives.

The daytime sleep of night workers is seldom as deep or refreshing as normal sleep at night time, and matters are made worse if the worker lives in poor housing conditions and is exposed to noise from the street or his family. In a typical situation, father may return from work at 07.30. The children leave for school at 08.00. Father has a meal at 08.30 and gets to bed at 09.30, when mother starts the housework, which keeps father awake for some time. At 15.30 the children return from school, waking father prematurely.

1.8.2.5 Shift systems[1 6 7]

A great many shift systems have been tried, with varying degrees of success. Medical experts who have studied shift work, almost universally condemn the rotating three shift system so commonly worked, wherein morning afternoon and night shift workers rotate in weekly intervals. They recommend instead one of two alternatives:

1. A system whereby not more than two consecutive night shifts are worked, so that the diurnal cycle is unaffected, and does not have to readjust every week that nights are worked.
2. Selection and recruitment of a permanent night shift.

A limited number of workers are generally prepared to work nights indefinitely. Their diurnal cycles then invert permanently and the problem of constant changes in cycle is avoided. This can work well with man-wife combinations. But it is important that their domestic circumstances allow such workers to sleep adequately during the day. Other workers then rotate alternately between morning and afternoon shifts, without serious adjustment to their diurnal cycles. The main obstacles to this system are perhaps:

(a) Not enough workers prepared to work on permanent night shift;
(b) Problems of communication between day managers and a night shift whom they seldom see.

Managers concerned with the health and safety of shift workers should consider these suggestions and examine sympathetically the very real problems of shift workers, especially their sleep when on night shift.

1.8.3 Incentives

The incentives considered in this sub-section are of two kinds, those intended to stimulate production and those intended to promote safety. The first take the form of piecework and production bonuses, the second take the form of good housekeeping and safety competitions. Both have acquired an unfortunate reputation among authors of books on safety. Piecework and the self-employed worker, especially in building trades, have a reputation for high accident rates through cutting corners to secure maximum earnings. Self-employed building workers (the *lump*) seem also to be on the whole rather ill-informed about the risks they are running, especially in regard to back injuries through faulty habits of lifting and carrying.

Schemes which give workers bonuses for increased production appear to do less harm, mainly because they are so often a lottery and have little to do with the workers' efforts. As the authors of *2000 accidents* observed,[5]

'in the general accident situation, bonus pay is unlikely to correlate with accidents because there are few cases where it properly reflects the human work content of the task'.

Nonetheless, the same authors found that the number of accidents generally correlated with the amount of work actually done rather than the time spent on the job. Incentive schemes to higher productivity should probably only be condemned when they lead to increases in accident rates per unit of production.

Again the authors of *2000 accidents* found a correlation between accidents and bonus payments in at least one department where the bonus system was such that the bonus rate increased for each extra ton handled in a day after a certain tonnage had been cleared. This led to a rush to finish loads between 17.00 and 17.30 and an accident peak at that time.

The effect of incentive schemes on workers with limited coping capacity must also be watched carefully; when such workers have been identified, it is wisest that they should not be exposed to piecework or incentive schemes for faster work and greater output.

Incentive schemes for greater output should also be carefully examined to see how far incentives to safer performance can be incorporated in them.

Safety incentives generally take the form of competitions between departments and firms for good housekeeping and lowest accident rates, or the longest accident-free period. Incentive schemes are especially common in the USA. Unfortunately we have probably heard more of their abuse than of the good they have achieved. It is almost inevitable that such competitions will increase safety awareness and lead to more thought and better co-operation between workers and managements over safety. The obvious abuse they can lead to is failure to report and record accidents, which led to the remark of an American union official[8] about his company's receipt of an award for 2 million hours without a lost time accident: 'It's all walking wounded. That's what we call it.'

The general conclusion reached is that incentives towards greater output and greater safety should not be discouraged, but should be carefully examined both as schemes on paper and in operation to ensure that they do not contain features which are undesirable for genuine safety and accident prevention.

1.8.4 Domestic and outside influences

Domestic and outside influences can affect the stresses of workers themselves to a considerable degree. This has been noted particularly in the case of shift workers.

Although there appear to be no general correlations between domestic circumstances and accidents at work, a wise employer will look carefully into the welfare of his work force and do what lies within his power to obviate the stressful influences in their lives.

REFERENCES

1. Edholm, O.G., *The biology of work*, Weidenfeld and Nicolson, Chapter 12 and 13 (1967).
2. Senior, N.W., *Letters on The Factory Act as it affects the cotton manufacture*, London (1837) (quoted in "Capital", Part III, Chapter 9, Section 3 by Karl Marx.)
3. Vernon, H.M., "An experience of munition factories during the Great War", *Occupational Psychology*, 14, 1-14 (1940).
4. National Board for Prices and Incomes, *Hours of work, overtime and shift working*, Report No. 161, HMSO, (1970).
5. Powell, P.I., Hale, M., Martin, J., and Simon, M., *2000 accidents*, National Institute of Industrial Psychology (1971).
6. Murrell, K.F.H., *Ergonomics*, Chapman and Hall, London, Chapter 19 (1965).
7. Wilkinson, R., 'Hours of work and the twenty-four hour cycle of rest and activity', in *Psychology at work* (edited by Peter B. Warr), Penguin, London (1971).
8. Page, J.A., and O'Brien, M.W., *Bitter wages*, Grossman, New York (1971).

1.9 SAFETY INSPECTIONS AND AUDITS

Contents

Inspections and audits are necessary for virtually all forms of organised human activity. They are familiar features of factory life and industrial production.

Their objectives generally include some element of safety, although it is not always the main objective of the inspection. Examples of such inspections are product quality control tests, personnel interviews and tests for employment selection, building inspection, and regular inspection of plant and machinery in use for wear, corrosion and metal fatigue.

There are some inspections however where safety is recognised to be the prime objective. These include statutory inspection of boilers, pressure vessels and machinery, fire safety inspections, monitoring and analysis of the industrial atmosphere for harmful dusts and vapours, and medical examination of persons who may have suffered exposure to them.

As organised industrial safety has come to be a recognised part of industrial operations, safety inspections, led by safety specialists have become necessary as part of this activity and as a means of checking its effectiveness. These inspections are of two types:

1. Inspections to check the safety of the working environment, machinery, tools and plant.
2. Inspections to check the safety of the working practices of employees.

These may have to be extended further to consider the organisation and line of control in greater depth, the procedures themselves and how they come to be established.

It is clear that there are some types of inspection which will be primarily the concern of the safety specialists, while there will be others usually of a specialised, technical nature, with an important safety content, for which other departments are responsible. To these, the safety specialists may only be able to make a limited contribution, depending on his education, previous experience, time available and terms of reference.

To assess the overall standard of safety performance in an industrial organisation requires a wider examination than can be given by safety inspections alone. Thus the concept of the safety audit has come into being. It has been developed by the Chemical Industries Association mainly for use within the Chemical Industry. Since it can be applied generally in practically all industrial operations the safety audit is described later in this section. Where the system of Total Loss Control is followed, the safety inspection and auditing systems discussed should dovetail into the Total Loss Control scheme.

In all works and factories, whether safety audits or Total Loss Control systems are employed or not, it is recommended that senior management responsible for safety in a works or factory make a systematic list of all regular inspections and control tests carried out for all purposes within the organisation, and evaluate the safety implications of the inspections and tests in collaboration with the safety man. Where these fail to satisfy safety requirements, additional tests will need to be introduced to fill the gaps, and clear directions given as to who, when and why.

1.9.1 Regular inspections with safety as secondary objective

It is impossible to draw up a full list of such inspections or point out all their implications for safety. However, it is important that their safety content be evaluated and properly recognised. Nearly all inspections fall within the wider audit of a Total Loss Control system, since they have as their objective the reduction of losses resulting from faulty raw materials, production or worn, faulty or inefficient machinery, buildings or equipment.

Roads, car parks and pedestrian paths should be regularly inspected and maintained to keep them in good order. Cracks, holes, breaks, subsidence, tripping hazards and drainage are causes of accidents and call for inspection on grounds of safety.

Works rail sidings require regular track inspection and maintenance and additions of ballast to avoid dangers of derailment.

All exposed steelwork requires regular inspection, protection and replacement or reinforcement of corroded parts for functional reasons. Neglect leads to risks of collapse and injuries.

Stairways require regular inspection to check the condition of treads and risers, handrails, lighting and obstructions.

Walls and foundations require inspection for cracks, settlement, spalling, frost damage, and loose cladding which may be dislodged and fall causing damage or injury.

Roofs need inspection, particularly for security of anchorage.

Chimney stacks need regular specialist inspection.

Floors need regular inspection to check for overloading, shrinkage, slipperyness, holes, unguarded openings, cracks, sagging and warping.

Raw materials, materials in process and finished products are subject to regular quality control and inspection mainly to secure efficient production meeting customer requirements. Inspection and testing should also include such tests as are needed to ensure that raw materials and materials in process are safe at all times with the methods of handling and processing employed, and that the products themselves are safe. Sometimes minor changes in raw material shape, size or composition can cause hazards to personnel handling it, and minor changes in the process or method of finishing may cause hazards to customers.

1.9.2 Inspections with Safety as main objective

A list of all such inspections should be prepared giving details of what should be inspected, where it is located, when it should be inspected and by whom. The list will form the basis for individual checklists by each inspector.

A typical list of groups of things to be inspected for reasons of safety is given in *Table 1.9.1*. This is by no means comprehensive or detailed. Ready-made lists such as this serve as a reminder of things to be included and are no substitute for lists appropriate to the particular factory.

In works with a fire department separate from the safety specialist or department, inspection of fire protection equipment would fall under the fire department. It is, however, best that fire protection and safety be combined in a single department. First aid inspection may come within the orbit of the safety department or of the works medical officer.

Monitoring of toxic gases, vapours and dusts requires special apparatus and training, and will normally be carried out by works laboratory staff, reporting to the works medical officer, safety specialist and works management. Pressure equipment inspection will generally be carried out by a specialist insurance inspector. Clear responsibility for these varied types of inspections should be laid down after discussion with those most concerned.

1.9.2.1 Inspections of working conditions

Although the safety specialist will be primarily responsible, it is important that personnel at all levels be involved and encouraged to inspect things themselves. At times of a shutdown and general inspection when large areas have to be inspected within a short time, it is unlikely that the safety specialist alone will be able to cope, and many inspection responsibilities will have to be delegated. Plant supervisors and foremen in particular should make their own housekeeping inspections. Safety inspections bring the safety specialist or other inspector into touch with individual employees thereby stimulating a two-way flow of ideas. If the inspections succeed in revealing unsafe conditions which are then speedily corrected, the employees' confidence in management will be greatly increased. The absence of effective inspections will lead to apathy.

General inspections should include places where 'no one ever visits' and where 'no one ever gets hurt', such as overhead places and roof beams which cannot be seen from the floor. These may have loose metal objects which could become dislodged and fall or accumulations of a combustible dust which could contribute to an explosion. Inspection of working conditions may be periodic, intermittent, continuous or special. Most of those so far discussed are periodic.

Table 1.9.1. SAFETY INSPECTION LIST. TYPICAL GROUPS OF ITEMS TO BE INSPECTED

1. *Fire Protection*
Portable extinguishers
Standpipes, hoses
Sprinkler heads, valves
Exits, stairs and signs
Storage of flammables

2. *Housekeeping*
Aisles, stairs and floors
Storage and piling of materials
Wash and locker rooms
Waste disposal
Car parks and yards

3. *Tools*
Power tools, wiring
Hand tools
Storage of tools

4. *Personal Protective Equipment*
Face and Eye Protection
Shoes
Gloves
Gas masks and respirators

5. *Working Environment*
Lighting
Ventillation
Heating
Noise
Other environmental hazards

6. *Material Handling*
Trucks, hand and power
Elevators
Cranes and hoists
Conveyors
Slings, ropes, chains

7. *Bulletin Boards*
Content, up-to-date and relevant
Neatness
Lighting

8. *Machinery*
Point of operation guards
Belts, pulleys, gears, shafts
Oiling, cleaning, adjusting
Maintenance, oil leaks

9. *Unsafe Practices*
Excessive speed
Improper lifting
Running, horseplay
Removing guards
Working on unguarded machines

10. *First Aid*
First aid kits and rooms
Stretchers, blankets
Emergency showers
Injury reporting

11. *Toxic Condition Monitoring*
Automatic detectors
Air sampling and analysis

12. *Pressure Equipment*
Boilers and pressure vessels
Gas Cylinders

13. *Ladders, scaffolds and stairs*

Tools and appliances which are kept in a store, being borrowed as required against a requisition and returned when the work is complete may be inspected in the store on return, and maintained where needed.

Intermittent inspections made unannounced help to keep employees alert, and may be needed where accident rates are above normal. Intermittent inspections may also be made by safety representatives and committees, and by individual employees. Continuous inspections are needed in some factories with highly-automated machinery and production

Figure 1.9.1   Sampling and testing atmosphere for toxic contamination (Rotheroe and Mitchell Ltd)

lines. Here selected employees have the sole job of roaming about observing operations and making adjustments and minor repairs.

Special inspections are needed when a modification is made or new equipment is installed. Accident investigations require special inspections by the investigator and safety personnel.

Health surveys are necessary when a health hazard, e.g. the use of asbestos or vinyl chloride has been identified. These will involve sampling and analysis of the atmosphere by experts and medical examination of all employees who have been exposed (*Figure 1.9.1*).

1.9.2.2 Inspections of working practices

Safe working procedures must be established for jobs where accidents occur. In practice this means most jobs. This requires observation and analysis of the job being performed by various workers, and discussion between supervisors, workers and safety specialists. Safe working procedures will usually be established for most jobs. If accidents still occur when the 'safe' procedure is being followed, the procedure must be reviewed.

Safety specialists can assist supervisors by observing workers in order to eliminate unsafe procedures.

Some unsafe practices to be looked for are listed below.

1. Operating machinery or using equipment without authority.
2. Working at unsafe speeds.
3. Working without guards or with ineffective guards.
4. Using defective tools or equipment.
5. Using tools or equipment in unsafe ways.
6. Using hands or other parts of body instead of tools.
7. Handling objects or materials unsafely, e.g. overloading.
8. Standing or working under suspended loads or open hatches.
9. Mounting or dismounting moving equipment or vehicles.
10. Repairing or adjusting equipment while moving, pressurised, electrified, or with hazardous materials inside.
11. Acting so as to distract or frighten other workers.
12. Failure to use protective equipment or safety devices when needed.
13. Use of inadequate, incorrect or faulty protective equipment.
14. Poor housekeeping.
15. Horseplay.

Observation of work practices requires planning between safety specialists and the supervisor concerned. They must be selective, know what to look for and practice observation to improve their own powers. They should be on guard against familiarity, record observations systematically and prepare a checklist. Whilst the work procedures of all employees should be checked occasionally, it is the inexperienced workers and accident repeaters who need special attention.

When close observation of workers is necessary, they should be approached and asked for their consent to the observation and its purpose explained. When a worker is observed, working unsafely, the supervisor should correct him tactfully and in private, making a point of finding the man's reasons for acting unsafely, and securing his agreement on future performance.

1.9.2.3 Night inspections

When a firm works a two or three shift system, inspectors should occasionally work with all shifts. Safety conditions may change considerably after dark, mainly as a result of artificial lighting. Discharge tube lighting may cause stroboscopic effects, causing misjudgement of machine speeds. Colour contrasts may disappear and shadows may cause hazards.

The inspector should have a good working knowledge of illumination and have suitable light meters available (see section 4.1). An occasional visit to the night watchmen to observe their working conditions should also be made.

1.9.2.4 Photographic aids

Safety personnel should be encouraged to use photographic aids —
particularly an instant developing camera for recording unsafe conditions
and training supervisors and employees. Sometimes the use of a lightweight
video recorder is justified.

1.9.3 Checking plans, specifications and new materials

Checking plans and specifications of new factories, factory extensions,
plant and machinery are of prime importance to safety. This subject is dealt
with in chapter 2.

New materials used in a factory or process frequently bring hazards and
should only be introduced when their hazard potential has been fully
checked. Chemical hazards are discussed in section 6.2.

1.9.4 Reporting inspections

Every inspection should be followed promptly by a written report. This may
be an emergency report — when urgent action is needed — a routine report
covering observations of unsatisfactory conditions, or a periodic report
which is usually a summary of several routine reports, the action called for
and the results achieved.

Reports must be detailed and positively identify the dates, times,
departments and machines, etc where the faults were found, and they
should give the name, position and signature of the inspector. Reports
should be directed to the head of the department or area inspected with
copies to higher management. Specific recommendations should be made
in the reports for review and approval by the appropriate manager or
committee. These should be listed in order of importance. Once a
recommendation has been approved it should go through the normal
procedure for changes in design or working methods, and be fully
integrated into the system.

1.9.5 Condemning equipment or vetoing practices

Depending on the hazards involved, it may be necessary to give the
inspector authority to condemn equipment, shut down plant or machinery
on the spot, or veto certain current practices. This is largely a matter of
policy and judgement.

Such powers are given to inspectors in explosives and munitions
factories, and they are fully justified with highly hazardous processes and
machines. There should be recognised procedures for the exercise of such
powers. Special tags should be used and attached prominently on
condemned machinery and equipment, and steps taken to remove
condemned items without delay from the workplace.

1.9.6 Safety audits

A safety audit is designed to measure the effectiveness of a company's safety programme in every aspect. The objectives should be clearly defined, an example being:

1. To carry out a systematic and critical appraisal of all potential hazards involving personnel, plant, services and methods of operation.
2. To ensure that the occupational health and safety standards fully satisfy the legal requirements and those of the company's written safety policies, objectives and programmes.

The subject is dealt with in detail in the booklet *Safety audits* prepared and published by the Chemical Industries Association Ltd.[2] A safety audit requires preparation and the formation and assignment of a team including managers, company safety specialists and impartial outside consultants to carry it out.

REFERENCES

1. National Safety Council, *Accident prevention manual for industrial operations*, 7th ed., National Safety Council, Chicago (1976).
2. Chemical Industry Safety and Health Council of the Chemical Industries Association Ltd, *Safety audits – a guide for the chemical industry*, Chemical Industries Association, London (1975).

1.10 INSURANCE AND COMPENSATION

Contents

The situation in the UK regarding insurance for accidents at work and industrial diseases is in a process of change and anything written today is almost certain to be out of date in a year or two. It, therefore, seems better to try and look at the subject broadly and in perspective rather than to attempt to treat the present and likely future position in detail. How far does insurance affect safety and hazards in industry and how far should safety personnel and those responsible for safety concern themselves with insurance matters? The answers to the first question are rather ambiguous.

In some respects, insurance works towards improving safety standards. In others it seems to encourage people to take greater risks. This is not surprising considering the origins and function of insurance. To quote a well-known authority[1] — 'Insurance not only spreads risk; it also encourages enterprise'. 'Spreading risk' can be taken in two ways, both of which apply to insurance. It pools the losses of the unfortunate few risk takers who suffer over the larger number of risk takers who avoid losses. In doing this it tries to disseminate its knowledge and experience of accidents and accident prevention so that the risks are reduced or 'improved'. It can hardly be fortuitious that H.W. Heinrich, pioneer and architect of scientific accident prevention, was for many years superintendent of the Engineering and Loss Control Division of an American insurance company.

At the same time insurance encourages the more cautious and less adventurous spirits to take risks. Insurance has probably been an essential factor in the development of western industrial society; modern industry can no more function without insurance than the housewife without her 'cuppa'. The sly remark, 'I hope you've got that car/barn/mother-in-law or whatever well insured' between friends voices a common feeling that insurance allows people to assume risks with greater equanimity than they would in its absence. It is also probable that the remark was first made by an insurance salesman! We must be clear that insurance is *not* accident prevention, though it may be a substitute for it; money spent on insurance cannot legitimately form part of an accident prevention budget.

This partly answers our second question, which is dealt with firmly by the authors of *2000 accidents*:[2]

'Insurance is not prevention

'Some safety officers appear to spend time dealing with insurance claims. A firm line must be drawn between these activities and those of training and safety. The objectives of the two are disparate.

The insurance claims man is concerned with minimising his payments out. He wants to blame the injured party for any loss. He wants reports and statements from witnesses on which legal arguments about blame can be based. The idea of blame inhibits constructive thought about accident prevention, especially if the blame is imagined to rest on the (absent) injured party.

On the other hand, the training and safety specialist is concerned with minimising injuries. Whilst no one would quarrel with him if he wanted to know exactly how an accident happened in his factory, his concern must be for the changes in design, layout, working arrangements and training which will reduce or eliminate injury. This is a strictly objective activity and it needs to be seen as such if all concerned are to co-operate willingly.

Most firms have someone who specialises in insurance; this person should handle all claims under employer's liability and national insurance.'

1.10.1 Types of industrial accident insurance

This covers both insurance of personnel against accidental injury death and occupational diseases, and insurance of factory premises, plant, machinery and materials in store or process against damage from almost any cause (fire, explosion, weather perils, riots, flooding, theft, impact, leakage and contamination, and most other types of accident). Further consequential loss risks (business interruption) resulting from any of these risks are also insured.

But while property and business risks are handled almost exclusively by the commercial market, state or national insurance has largely but not yet entirely replaced commercial insurance for employers' liability for industrial injuries and diseases. This change has been steadily taking place over the last 100 years, and its present culmination may be seen in New Zealand following the Accident Compensation Act of 1972[3], which provides universal coverage for accidental injury. Common law liability for personal injuries and death has been completely abolished, and with it the system of third party liability insurance.

The scheme is funded from three principal sources, levies on employers, levies on the self employed and levies in respect of motor vehicles, with some small contribution from general revenue. This now seems the direction in which Britain is moving, judged by press comment[4] on the work of the Royal Commission on Civil Liability and Compensation for Personal Injuries under Lord Pearson, set up in 1973.

Insurance and law have been closely interlinked. First a legal liability, either under common law or statute was established, e.g. for employers to their employees for injuries received at work. The employer then had little alternative but to insure against the risk. Then, following an accident causing serious injury, if no satisfactory settlement could be reached, employees had no alternative but to sue their employer for damages, although it was the employer's insurer who was in fact potentially liable.

The insurer would pay for the legal defence, initiate appeals, etc. and finally perhaps pay damages, but all on behalf of the insured employer. Thus legislation created business for insurance companies, and insurance companies in resisting claims, created business for the courts, judges, barristers, solicitors and clerks. According to Professor Atiyah[5], some 80% of the cases brought before the High Court are for damages in respect to personal injuries. If a scheme similar to that now operating in New Zealand were introduced in Britain, a considerable amount of commercial insurance business would disappear.

1.10.2 Industrial injury insurance and compensation in the UK

This brief review is based largely on Professor Atiyah's objective and scholarly book, *Accidents, Compensation and the Law*, which is recommended to serious readers. It must be one of the most remarkable exposés of the shortcomings of our legal system as regards industrial injuries ever to be written by a professor of law.

Victims of industrial accidents in the 19th century were usually left with nothing except poor or charitable relief to live on. Their only hope of compensation for injuries received at work was to prove in a law court that the injury was caused by the employer's negligence. This rarely happened, but when it did, the employer had three ready-made defences:

The doctrine of common employment, which denied liability for the negligence of a fellow worker.

The doctrine of contributory negligence denying liability where the workman was partly responsible for his injuries; and

The doctrine of *volenti non fit injuria* which denied liability for injuries occurring from a known and obvious risk.

The Employers Liability Act of 1880 reduced the first of these defences, but the first serious reform was Joseph Chamberlain's Workman's Compensation Act of 1897. This broke away entirely from the common law principle that liability must be based on fault, and gave a workman and his dependents a right to compensation for any accident 'arising out of and in the course of employment'. The statutory liability for compensation fell on the employer, as a result of which a brisk insurance business sprung up, although employers were not compelled to insure this liability.

The Act did not attempt to provide full compensation, but was based on the assumption that the employer was to bear half the loss and the employee the other half. This was subsequently altered leaving the employer liable for 80% or more of the loss (i.e. of earnings).

Although the Act intended that the victims should receive compensation by weekly payments, the insurance companies preferred to dispose of claims by lump sum payments. Both the victim and the insurance company were more interested in a 'favourable settlement' than an objectively fair one, but the odds were against the victim. While the Act watered down in practice had made progress in accidental injury compensation,

compensation for industrial diseases proved a trickier hurdle. Eventually only those industrial diseases (e.g. pneumoconiosis) which could definitely be shown by medical science to be caused by certain types of work were included.

The next major statute affecting industrial injury compensation was Lloyd George's National Insurance Act of 1911. This was a national insurance system, established and administered by the State. Commercial insurance companies were not involved at all, and the 'adversary procedure' played no part in the system.

The Act provided income in cases of disease and unemployment rather than of accidents. But since many diseases were industrial in origin and much unemployment the result of accidents, the Act helped victims indirectly. The insurance was compulsory and based on flat rate contributions from the employer, employee and the state. Benefits were also flat rates. They were not intended to support anybody in themselves but to alleviate distress.

The next milestone was the Beveridge Report of 1942 and the National Insurance Act 1946 and the National Insurance (Industrial Injuries) Act 1946, which were consolidated in 1965. These replaced the workman's compensation scheme and the national insurance system.

Beveridge retained a separate scheme for industrial injuries, awarding the victims higher compensation than provided for sickness, but brought them under a state national insurance system, hoping that private insurance schemes for employers liability and the adversary process of settlement and litigation would disappear. But he did not entirely abolish employers liability, trying instead to limit the employers responsibility at common law to actions for which he was 'morally and in fact' responsible.

The result of Beveridge's work was not, however, to abolish litigation for industrial injuries, but, because of the 'premium' he set on industrial injuries (compared with injuries received at home), to greatly increase the liabilities for industrial injury under common law. While Beveridge wanted to limit special compensation for industrial injuries to cases of death or prolonged disability, the government gave preferential treatment to all industrial injuries.

It also decided on disability pensions and gratuities for the disability 'as such' irrespective of lost earnings. Compensation was to be paid weekly throughout the period of disablement, and lump sum payments were rejected except in cases of very minor disabilities. The benefits are mainly administered by civil servants — non-lawyers — in social security tribunals.

The main features of the two Acts of 1946 (consolidated in 1965) have remained intact but their benefits have increased faster than wages and prices. Since 1975 however, contributions to the scheme have been wholly earnings related.

One result of inflation is that private employer's industrial accident liability insurance becomes less and less viable or attractive compared with National Insurance. Professor Atiyah shows a number of comparisons between the awards of Courts under Common Law for industrial injuries and the industrial accident benefits provided by State Insurance, which have the great advantage of being free of income tax. The comparisons are somewhat in favour of National Insurance, and for this and other reasons it seems probable that employers liability insurance and the resulting litigation will eventually disappear. However the Employers Liability (Compulsory Insurance) Act 1963, made insurance compulsory (in the UK) by employers of their liability to employees.

The Royal Pearson Commission reported in 1978 (section 1.10.4). A prime reason for this Commission was public dissatisfaction with the long legal delays (up to ten years in some cases) before compensation was available for parents and victims of the thalidomide tragedy.

Compensation for occupational diseases has tended to lag behind compensation for industrial injuries, due partly to the greater difficulties in diagnosing both the diseases and their causes. At present there is still only a very limited number of recognised occupational diseases caused by exposure to toxic chemicals, dusts and vapours. These however are liable to strike residents in areas adjacent to the factories where they are produced as well as workers engaged in their production, and reasons for compensating employees more favourably than local residents under a National Insurance scheme are quite obscure.

1.10.3 Property damage insurance

Insurance of industrial plant is largely in the hands of private insurance companies. The subject is too large and complicated to be dealt with here. The positive aspects of industrial insurance as regards accident prevention are to be found in the various codes for building construction and engineering and fire protection and in the engineering inspection services which have come into existence as a result of insurance requirements.

1.10.4 The Pearson Report[6]

The general summary given in the report of its own conclusions is reproduced below:

'Our compensation system should be looked at as a whole. Tort should be retained and, while the two systems of tort and social security should continue side by side, the relationship between them should be significantly altered. Social security should be recognised as the principal means of compensation. Double compensation should be avoided by offsetting social security benefits in the assessment of tort damages. Money available should be spent on the more serious injuries rather than minor injuries. The range of those receiving compensation should be extended.

No-fault compensation should be introduced for motor vehicle injuries. The no-fault provision for work injuries should be improved. A new benefit for all severely handicapped children should be introduced.

The range of tort should be extended by introducing strict liability in some areas. Under tort, provision should be made for periodic payments for pecuniary loss.

In administering compensation existing systems and institutions should be used, but considerable simplification of the highly complicated social security system is desirable.

Our terms of reference do not cover all injuries, and at least one million injuries every year, mostly those occurring in the home, would remain outside the scope of our proposals.

On its own reckoning, the Commission estimates the total compensation for personal injury and death resulting from injury in the UK to be £800 million a year at 1 January 1977 prices. The Commission anticipates that its recommendations will result in an increase in the cost of social security compensation of £130 millions a year and a reduction in the cost of tort compensation and other payments of £89 millions a year.

With regard to work injuries, the Commission summarised its recommendations as follows:

'We concluded that the industrial injuries scheme administered by DHSS should provide the basis for improved provision for those injured at work. Higher benefits should be paid for the first six months of incapacity or widowhood, followed by pensions calculated in the same way as in the new state pensions scheme, but in all cases at the improved levels that would accrue if contributions to the scheme had been paid for 20 years. Widowers should be treated in the same way as widows. The scheme should be extended to cover the self employed and to include commuting accidents. The conditions for compensation for occupational disease should be less restrictive. The increased compensation costs should be met by employers.'

Any further comment on the report at this stage, and before the relevant legislation is passed, would be premature. The success of the Commission can only be judged in retrospect — perhaps in the light of the length of time before another commission has to be appointed to review the practical effects of its recommendations.

REFERENCES

1. Dinsdale, W.A., *Guide to Insurance,* David and Charles, Newton Abbott, 11 (1970).
2. Powell, P.I., Hale, M., Martin, J., and Simon, M., *2000 accidents*, National Institute of Industrial Psychology, London, 44 (1971).
3. Hodgkinson, N., 'Giving accident victims the compensation they need as soon as they need it', *The Times*, 29th June (1977).
4. Hodgkinson, N., 'Automatic Compensation urged for people injured accidentally', *The Times*, 17th June (1977).
5. Atiyah, P.S., *Accidents, compensation and the Law*, 2nd ed., Weidenfeld and Nicolson, 575 (1975).
6. The Royal Commission on Civil Liability and Compensation for Personal Injury, Chairman: Lord Pearson, 3 volumes HMSO (1978).

1.11 LEGAL ASPECTS AND INQUIRIES

Contents

Since the Health and Safety at Work, etc Act, 1974 came into force, many books and texts on it have been published.[1][2][3][4][5][6] These explain it in detail and give advice to all parties on their duties and rights, and how to avoid trouble. Most are by legal experts.

In this section a different approach has been adopted in which the UK health and safety laws are viewed in their geographical and historical perspective. The reasons for the sweeping changes introduced by the Act, and which are still going on, are discussed. An example of the working of the Law in a typical case involving personal injury is given and the methods of conducting inquiries into accidents and major disasters are outlined.

The mutable and often ambiguous laws made by governments and their agents on health and safety at work are often in marked contrast to the more permanent natural laws, discovered by men learning through the ages, to which all matter, both animate and inanimate is subject. Such laws are those of gravity, conservation of energy and matter (modified by relativity), the Phase Rule, and Murphy's Law.[7]

The bearing of these natural laws on the causes of accidents and disasters must be frankly and competently examined by our inquisitors before the accusers pass judgement.

1.11.1 Different legal systems

Legal aspects of industrial health and safety can only be discussed in the context of a particular legal system, in this case English parliamentary statutes and common law. This is technically limited to England and Wales and does not automatically apply in Scotland, Northern Ireland, the Isle of Man or the Channel Islands. Although respected and taken into consideration in other countries, it is but one of several major legal systems.[8]

Common law has its origins in the laws made by the King's judges in the days when Parliament had little power. It is basically 'law by precedent'. Once a judgement was made in a case with certain circumstances, it was highly likely that a similar verdict would be reached when these circumstances were repeated. Even today most Parliamentary statutes leave whole areas open to interpretation by the courts, and common law by precedent continues to be made and applied to fill these gaps.

North American Law which like English Law is basically Common Law has many important differences. In particular it has both State and Federal Laws, with a double hierachy of courts, in some cases the State Court being the first court and in others the Federal Court.

Most of Western Europe, Scandinavia and Latin America has an entirely different legal system — the Romano Germanic system. This is derived from the Institutes of Justinian which distinguished between 'positive law' decided by men for their practical needs, and 'natural law' based on idealised concepts. The Romano Germanic legal system is heavily codified, e.g. Code Napoleon which has strongly influenced the legislation of Western Europe.

The USSR and East European countries now have an entirely different legal system from those operating in Western Europe and English speaking countries. It is based on the political and philosophical ideas of Marxism-Leninism. The more important legislation is said to be initiated by the Politburo of the Communist Party or the Council of Ministers and ratified by the Supreme Soviet.

Chinese law is different again. The Confucian system has largely been replaced by laws reflecting the recent phase of revolutionary communism in China.

Japanese law reflects strict rules of traditional behaviour and personal relations (the 'giri') and is said to be closely associated with ideas of punishment and prison.

Large areas of the world are subject to Muslim, Hindu and Afro Malagasay legal systems.

1.11.2 English Legislation

English law appears to outsiders to be overburdened with legal definitions (often unintelligible to laymen) and detailed solutions which easily inflate it. Of historical interest is the fact that legislation on industrial health and safety developed earlier in England than most other countries, coinciding with the earlier industrial development in England.

Industrial health and safety legislation in England has hitherto had the habit of growing in piecemeal fashion over the years, rather like creeper until it threatens the fabric of the house it was intended to adorn. Thus every twenty five years or so a Royal Commission is set up with wide powers to prune it, repealing dead statutes and streamlining and consolidating live ones. But history shows it is easier to beget new laws than do away with old ones, and no sooner has the old wood been cut away than new shoots sprout forth, leading in a short time to as thick and tangled a jungle as before. In spite of this, progress is made in fits and starts, and hope springs eternal.

In 1802 the first industrial health act was passed to improve conditions in cotton mills, covering the working hours of apprentices, cleansing and ventilation of factories, and the provision of working clothing. After a gap of seventeen years, a succession of statutes were passed covering the conditions of women and adolescents in textile and allied industries, and later in specified non-textile factories. By 1875 the law consisted of a number of uncoordinated statutes and regulations. These were reviewed by a Royal Commission whose report resulted in the Factory and Workshop Act 1878, which for the first time attempted comprehensive legislation. Further statutes were again passed piecemeal until, by 1901, further rationalisation had to be carried out with the passing of the Factory and Workshop Act 1901. This act was frequently amended and extended until repealed by the Factories Act 1937, which provided a more comprehensive code for health and safety in all factories. This contained new safety provisions affecting standards and testing of lifting tackle and cranes, steam and air receivers and contained requirements for floors, stairs and work places. Ship repair, power generation and engineering construction were also included.

The 1937 Act after amendment in 1948 and 1959 was replaced along with other Acts (including one to protect against lead poisoning) by the Factories Act 1961. Although this Act contained powers to make new regulations for dangerous processes and plant, all existing regulations affecting them were to continue in force even when the statutes under which they were made had been repealed.

The Factories Act 1961 covered about 11 million workers and the subsequent Offices, Shops and Railway Premises Act of 1963 covered about 8 million employees. Other legislation extended protection to a further 3 million workers.

1.11.3 The Factories Act 1961[1]

The Factories Act 1961 is a complex document containing provisions for General Health, General Safety, Fire Precautions and Prevention, First Aid, Welfare, Notification and Investigation of Accidents and Industrial Diseases, Employment of Women and Young Persons, and Notices and Returns, Five groups of people may be liable under the Act:

1. The Occupier — both civil and criminal liability
2. The Owner — for matters within his control
3. Owners, sellers or hirers of dangerous machinery
4. Employees who wilfully prejudice the health, safety or welfare of their fellows.
5. Officers of the company who contribute to offences under the Act by connivance, consent or negligence.

Four separate bodies were responsible for the administration under the Act — The Factory Inspectorate, The Fire Authorities, Factory Doctors and Local Authorities. Liability under these various Acts is generally irrespective of breaches of Common Law.

In spite of the fairly detailed provisions of the Act, liability was not always clear cut, and many injured workers and relatives of diseased employees who sought compensation under the Act found themselves in a situation not unlike a lottery.

1.11.4 Some case history

The following examples are quoted in order to show the wide variation in judgements which can occur when compensation is sought by recourse to the Courts.

1. Cummings (or McWilliams) v. Sir William Arrol & Co. Ltd. (1962)
The widow of a steel erector who had been killed by a fall at a shipbuilding yard, sued, *inter alia*, the occupiers of the yard alleging a breach of s.26(2) of the Factories Act 1937 in that the occupiers had failed to provide the deceased with a safety belt. The occupiers admitted that no belt had been provided but contended that, since there was evidence that the deceased would not have worn a belt had one been provided, their failure to provide a belt did not cause the death. This contention was upheld by the House of Lords.[1]

2. Pearce v Stanley Bridges (1965)
The Plaintiff was injured when his arm was caught between the rising platform of a lifting machine and a conveyor belt.
Held, the Act did not impose an obligation to fence a 'gap' between two machines. Moreover, the alleged danger was not reasonably forseeable, and for both these reasons the claim must fail.[9]

3. Summers (John) & Sons Ltd., v Frost (1955)
A workman injured his thumb while using a grinding stone. To fence the machine would be to make it unusable.
Held, the fact that it was commercially impractical to fence the machine was no defence.[9]

4. Pengelly v Bell Punch Co. (1964)
In a factory, reels of paper were stored in racks and on the floor. While trying to take one down, the plaintiff caught his foot in between two reels and injured himself.
Held, the reels were not an obstruction, for they were properly there. In order to be an obstruction the article must be something which has no business to be where it was.[9]

5. Dorman Long v Bell (1964)
'The respondent slipped on a greasy metal plate which was an obstruction.'
Held, the employers were liable.[9]

These examples illustrate how Judges' Law or Common Law has grown to fill the gaps in Parliamentary Statutes, through most of which the proverbial 'Horse and Cart' might otherwise be driven. It also shows that the law on industrial accidents is complex and specialised and cannot as a rule be successfully grasped or approached by anyone who is not prepared to devote a good deal of time to studying it carefully. Even then, the verdict in any situation can not always be forecast with confidence, and may be annulled or reversed on appeal.

1.11.5 Injury compensation in the future

An enlightened attempt to ensure fair compensation for those injured by accidents at work and on the road has been made in New Zealand following the Accident Compensation Bill of 1971.[10] In place of remedies at common law, all employees and self employed persons who suffer injury through accident are entitled without proof of negligence to compensation equivalent to 80% of normal earnings, plus lump sum payment for loss of faculties. Administration of the scheme is in the hands of the Accident Compensation Commission which inter alia fixes levies paid under the Act by employers. Similar legislation in England might remove the apparent anomalies noted from the cases quoted above.

While it is only possible in this section to skim the surface of English Industrial Health and Safety legislation, an attempt has been made to refer to the appropriate current regulations in sections dealing with particular hazards and industries.

1.11.6 The Robens Committee Report[10] and HASAWA 1974[1]

The Health and Safety at Work etc Act 1974 which resulted largely from the Robens report is a much shorter Act than the Factories Act 1961. Yet its proposals are more revolutionary and its consequences therefore likely to be more far reaching. Its powers to repeal or modify existing legislation have already been exercised over a wide field. The extent of the pre-1974 legislation affected by the 1974 Act is shown in *Table 1.11.1* HASAWA 1974 has already been referred to in section 1.1 in connection with the role of the Health and Safety Executive and Commission and further in section 1.4 in connection with the responsibilities it places on Industry. Some of the further changes in legislative attitudes which the Act is hoped to achieve are discussed next.

Chapter 9 of the Robens Report dealing with sanctions and enforcement illuminates this new approach. Speaking of infringement of statutory regulations, paragraphs 258 onwards are quoted from below:

'258. The conventional sanction in this field is prosecution in the criminal courts followed on conviction by the imposition of fines . . . In the evidence submitted to us the maximum levels of fine . . . were widely criticised as being derisory . . . leading to . . . fines that appear to be little more than nominal.

'259. But what are the chances of an offender ever being prosecuted? . . . It is . . . evident that despite the existence of voluminous legal requirements, only a very small proportion of offences ever lead to prosecution . . . according to factory inspectors it is rare for any inspection visit not to reveal a number of breaches of the law for which criminal proceedings could be instituted. Nevertheless, some 300 000 visits made by factory inspectors in 1970 resulted in the prosecution of less than 3000 offences.

'260. ... we do not believe that the traditional sanction commands any widespread degree of respect or confidence.

'261. ... the typical infringement arises rather through carelessness, oversight, lack of knowledge or means, inadequate supervision or sheer inefficiency. The real need is for a constructive means of ensuring that practical improvements are made and preventative measures adopted ... the laborious work of preparing prosecutions — and in the case of the Factory Inspectorate, of actually conducting them — consumes much valuable time which the inspectorates are naturally reluctant to devote to such little purpose.

'262. The sanctions available should provide scope for distinguishing between situations where the accent should be on punishment and the more frequent situations where the accent should be on constructive remedial action. The sort of thing we have in mind is to be found in the system followed in Ontario where substantial financial penalties imposed on employers by the Workman's Compensation Board are frequently suspended on condition that a specified programme of remedial measures is undertaken.

'*The future use of criminal proceedings*

'263. We recommend that criminal proceedings should ... be instituted only for infringements of a type where the imposition of punishment would be ... supported by the public ... the maximum permissible fines should be considerably increased.

'264. ... provision should be made for the imposition of higher penalties in the case of repeated offences ... the fact that not only corporate bodies but also individuals such as directors, managers and operatives are liable to prosecution should be spelt out very clearly.

'*Administrative sanctions*

'265. ... in future much greater reliance should be placed on non-judicial administrative techniques for ensuring minimum standards of safety and health at work. Where advice and persuasion fails and pressure is necessary, the pressure should be exerted in a form that is positive and constructive as well as quick and effective.

'268. ... the existing powers concerning enforcement orders should be reorganised and strengthened so that inspectors are able ... to exert effective pressure to ensure the prompt rectification of unsatisfactory conditions and circumstances.

'*Improvement Notices*

'269. We recommend that inspectors should have the power without reference to the Courts, to issue a formal Improvement Notice to an employer requiring him to remedy particular faults or to institute a specified programme of work within a stated time limit.

'*Prohibition Notices*

'276. The Improvement Notice would be the inspector's main sanction. In addition an alternative and stronger power would be available to the inspector for use where he considers the case for remedial action to be particularly serious.

'*Licensing*

'280. Licensing systems provide enforcing authorities with a powerful sanction. Conditions of licence can be imposed with various penalties for non-observance.

'282. ... whilst licensing provides a tight means of control and a powerful sanction against abuse, licensing systems should be used very selectively. We have in mind that the licensing approach should be adopted mainly for the control of high-hazard installations such as bulk storages of intrinsically dangerous chemicals, or for particularly hazardous activities such as demolition work.'

In chapter 10 dealing with public safety, the report deals with the problems facing local authorities and states the need for consultation between local planning authorities and safety inspectorates when required. It refers particularly to the Department of the Environment Circular 1/72 *Development involving the use or storage in bulk of hazardous material*:

'Local planning authorities are advised that the Factory Inspectorate should be consulted before granting planning permission for development with might involve the use or storage of such materials in excess of the quantities stated.'

The Flixborough disaster which occurred as HASAWA was being put into effect, added weight to these concepts which have been reinforced by the formation of the Major Hazards Committee and the Risk Appraisal Group of the Health and Safety Executive.

The history of the influence of legislation on the health and safety record in other fields suggests that firm legislation backed by effective policing and prosecution of offenders is essential to ensure the changes in habits necessary to secure a worthwhile improvement. In nearly every case the legislation aroused strong opposition initially, which gave way to grudging respect and finally approval when the measures were shown to be effective and former unsafe habits had been broken.

Some typical examples are:

1. International acceptance of Samual Plimsoll's proposals which reduced the hazards of overloaded ships (see section 1.1).
2. The introduction of tests for drunken drivers and the imposition of stiffer penalties which led to a significant improvement in road accidents late in the evening.
3. Quarantine regulations which have kept this country relatively free from rabies.

The effort of policing such legislation though heavy initially subsequently fell off as its benefits were more generally appreciated.

The formulation and enforcement of effective safety measures by Government in industry appears to be a far more difficult problem than, say, enforcing rules for safe driving. The main difficulties are:

1. Industrial premises are not on view to the public as our highways are, and the processes are often secret.

2. Industrial processes are far more varied and complicated than motor cars, and standards that may be safe for one process may be totally unsuitable for another.
3. The people responsible for industrial processes are often the heads of very large and wealthy companies, who are far more adept at fighting prosecution than the average car owner.

Another aspect of the Robens Report and HASAWA 1974 is the greater importance placed on voluntary standards and codes of practice. The Report has this to say:

'142. We have advocated that statutory regulations should be simpler in style . . . We go further than this. We recommend that in future no statutory regulation should be made before detailed consideration has been given to whether the objective might adequately be met by a non-statutory code of practice or standard.
'156. At general policy level the Authority should be advised and assisted in the forming and revision of regulations by a standing Advisory Committee on Regulations and Codes.
'Technical working parties
'159. The Advisory Committee would have neither the time nor the expert capacity to become closely concerned with the technical detail of individual regulations, codes and standards. We envisage that technical working parties be appointed as the need arose to undertake the detailed work on new proposals. These working parties would be serviced by the Authority and the aim would be to include the best available expertise from independent organisations and from industry.'

It is as yet far too early to say whether the approach to health and safety at work advocated by Lord Robens Committee and implemented by HASAWA 1974 will succeed better than its predecessors. The new approach which is in many ways revolutionary, is far more discriminating in its penalties than its predecessors, and it has caused far more discussion.

The effect of employee safety and health representatives nominated and trained by Trade Unions has yet to make itself felt on these attitudes.

1.11.7 Public and consumer protection

Responsibilities of manufacturers to warn the public of dangers (e.g. emissions) which their operations could accidentally cause and to warn users and consumers of hazards in their products were discussed in 1.4. These are defined in certain general duties (sections 3, 4, 5 and 6) of HASAWA 1974, which have the following headings:

3. General duties of employers and self-employed to persons other than their employees.
4. General duties of persons concerned with premises to persons other than their employees.

Table 1.11.1. EARLIER LEGISLATION WITHIN THE SCOPE OF THE HEALTH AND SAFETY AT WORK ETC., ACT 1974

Chapter	Abbreviated Title	Relevant Provisions
1875 c. 17	Explosives Act 1875	Whole Act except sections 30-32, 80 and 116-121
1882 c. 22	Boiler Explosions Act 1882	Whole Act
1890 c. 35	Boiler Explosions Act 1890	Whole Act
1906 c. 14	Alkali, etc. Act 1906	Whole Act
1909 c. 43	Revenue Act 1909	Section 11
1919 c. 23	Anthrax Prevention Act 1919	Whole Act
1920 c. 65	Employment of Women, Etc. Act 1920	Whole Act
1922 c. 35	Celluloid, etc. Film Act 1922	Whole Act
1923 c. 17	Explosives Act 1923	Whole Act
1926 c. 43	Public Health (Smoke Abatement) Act 1926	Whole Act
1928 c. 32	Petroleum Consolidation Act 1928	Whole Act
1936 c. 22	Hours of Employment Act 1936	Whole Act except section 5
1936 c. 27	Petroleum Licences Act 1936	Whole Act
1937 c. 45	Hydrogen Cyanide (Fumigation) Act 1937	Whole Act
1945 c. 19	Min. of Fuel & Power Act 1945	Section 1 so far as it affects health, safety and welfare

1946 c. 59	Coal Nationalisation Act 1946	Section 42(1) and (2)
1948 c. 37	Radio Active Substances Act 1948	Section 5(1)(a)
1951 c. 21	Alkali, etc. (Scotland) Act 1951	Whole Act
1951 c. 58	Fireworks Act 1951	Sections 4 and 7
1952 c. 60	Agriculture (Poisonous Substances) Act 1952	Whole Act
1953 c. 47	Emergency Laws Act 1953	Section 3
1954 c. 57	Baking Industry (Hours of Work) Act 1954	Whole Act
1954 c. 70	Mines and Quarries Act 1954	Whole Act except Section 151
1956 c. 49	Agriculture (Safety, etc.) Act 1956	Whole Act
1961 c. 34	Factories Act 1961	Whole Act except Section 135
1961 c. 64	Public Health Act 1961	Section 73
1962 c. 58	Pipelines Act 1962	Sections 20-26, 33, 34 and 42, Schedule 5
1963 c. 41	Offices, Shops & Railways Premises Act 1963	Whole Act
1965 c. 57	Nuclear Installations Act 1965	Sections 1, 3-6, 22 and 24, Schedule 2
1969 c. 10	Mines & Quarries (Tips) Act 1969	Sections 1-10
1971 c. 20	Mines Management Act 1971	Whole Act
1972 c. 28	Employment Medical Advisory Service Act 1972	Whole Act except sections 1 and 6 and Schedule 1

5. General duties of persons in control of certain premises in relation to harmful emissions into atmosphere.
6. General duties of manufacturers etc. as regards articles and substances for use at work.

Broadly speaking, consumers and members of the public can prosecute successfully for injuries or damage which they receive through accidents caused by dangerous acts of firms employees, explosions and emissions, and through the use of hazardous articles and materials which they purchased if no clear warning of the hazard was given.

Injuries and damage sustained by third parties and caused by the negligence of employers and their employees also form the basis of claims for compensation under Common Law. But the onus of proof lies with the injured party. The employer is liable for his employee's negligence if it occurred while he was engaged in the normal duties of his employment, but not if he was clearly doing something else. In that case the employee would be liable. Employers are obliged to insure against these and other claims under the Employers Liability (Compulsory Insurance) Act, 1969, but not so their employees. So usually it is the Insurance Company who has to fight such claims.

The usual tactic of an employer (or his insurer) on receiving a claim for injury compensation from a member of the public seems to be to ignore it. Not until a court action is filed and on the point of being heard, when the injured party has already spent a good deal of time and money in securing proof and building up a firm case, will the defendant show his hand. Then, if he decides that the claimant really has a case, he will usually make an offer out of court, and after some bargaining the case will be settled in this way.

The writer can speak from personal experience, having once been injured by a blow on the head from a scaffolding clip thrown by a labourer in the back of a lorry. Both the employee (of a firm of scaffolding contractors) and the writer were invisible to each other, their line of vision being obscured by the load of scaffolding clips between them. It took over two years to obtain compensation which when received barely covered the expenditure of time, effort and nervous energy, although the successful exercise gave some emotional satisfaction.

A certain amount has already been said about establishing the cause of the accident under 'proximate cause' in section 1.2. The whole subject is quite complex, as is that of 'contributory negligence', but further information may be found in the references given.

1.11.8 Investigations and inquiries

When a claim for compensation for industrial injury or damages is contested, the parties may take the matter to expert arbitration, or it may result in a court case. Often highly complex technical causes are decided in this way by judges, hearing evidence from the parties to the claim backed by expert witnesses.

The Health and Safety Commission has powers under section 14 of the

Figure 1.11.1 An inquiry in progress

Health and Safety at Work, etc Act to direct investigations and inquiries into the causes of any accident, occurrence or situation where this is felt necessary.

The Commission may direct the Executive or authorise someone else to investigate and report on the matter, or it may, with the consent of the Secretary of State for Employment, direct an inquiry to be held on the matter. Such inquiries are normally held in public, except where a Minister of the Crown directs otherwise, e.g. on grounds of national security. The Commission may then confer powers on someone to hold the inquiry and provides him with the necessary assistance and funds.

Although such inquiries are held in public, representation at them is usually restricted to the following persons:

(a) The Commission.
(b) Any enforcing authority concerned.
(c) Any employers associations and/or trade unions representing employees concerned.
(d) Anyone who was injured or suffered damage.
(e) The owner or occupier of the premises where the subject of the inquiry occurred.
(f) Anyone who carried out activities which may have given rise to the accident.

Such inquiries usually have a dual object:

1. To discover the cause or causes of the accident;
2. To decide who was responsible for it and in what measure.

In inquiries after a major disaster, the parties to it are generally represented legally by counsel, and obliged to pay their own costs as well as those of their scientific and technical advisers. The costs of such inquiries can be considerable, especially if they are protracted, which is rather unfair on the innocent parties.

Unfortunately, the second object of the inquiry can produce a situation in which the full cause tends to be obscured. This can be particularly serious if the hazard not brought to light is one which may remain latent in other similar plants or situations, and there lead to a repetition of the disaster.

While lawyers are clearly the appropriate persons to argue and decide responsibilities once the cause or causes have been clearly diagnosed and agreed, it is only those with the appropriate knowledge, training and experience who can do the latter. Where, as can happen, those persons with these necessary qualities are prevented from performing this diagnostic function because the results might conflict with the short term interests of the party with whom they are involved, a difficult and unsatisfactory situation can arise, wherein serious hazards are perpetuated as a result of the inquiry. An example of this is given in a recent paper.[11] Of the dual objectives, the discovery of the full and true causes is clearly more important from the point of view of future accident prevention than apportioning blame or responsibility.

REFERENCES

1. Fife, I., and Machin, E.A., *Redgrave's Health and Safety in Factories*, 1st ed., Butterworth, London (1976).
2. Jones, W.T., *The Health and Safety at Work Act – a practical handbook* Graham and Trotman, London (1975).
3. Mitchell, E., *The employers' guide to the Law on Health, Safety and Welfare at Work*, Business Books, London (1975).
4. Cusworth, G.R.N., *Health and Safety at Work etc Act 1974*, Butterworths, London.
5. Handley, W., *Industrial Safety Handbook*, McGraw Hill, London (1977). See Chapter 39, W.D. Gold 'Legal aspects of industrial safety'.
6. Health and Safety Commission (leaflets): HSC 2 *The Act outlined* (1976); HSC 3 *Advice to employers* (1976); HSC 4 *Advice to the self employed* (1976); HSC 5 *Advice to employees* (1976); HSC 6 *Guidance on employers policy statements for Health and Safety at Work* (1975); HSC 7 *Regulations, Approved Codes of Practice and Guidance Literature.*
7. Klipstein, D.L., *Edsel Murphy. The history of a great mind*, Magazine of Circuit Design (13 Aug. 1967).
8. Rene, D., and Brierley, E.C., *Major legal systems in the world today*, 1st ed., Stevens and Sons, London (1968).
9. Selwyn, N.M., *Industrial Law Notebook*, 1st ed., Butterworths, London (1969).
10. Lord Robens Committee, *Safety and Health at Work*, HMSO, London (1972).
11. King, R.W., 'Post accident investigations', Industrial Safety, 12 (October, 1977).

1.12 SECRECY

Contents

'This story has no moral. If it points out an evil, at any rate it suggests no remedy'. — *from The Unbearable Bassington by 'Saki' (H.H. Munro)*

Those who have been involved in major accident inquiries are generally aware of the harmful effects of secrecy in masking causes and thereby perpetuating hazards. Yet rarely is the subject discussed by speakers and writers on industrial safety. One noteworthy exception is the American crusader, Ralph Nader, who put the matter aptly at a conference organised by RoSPA in Birmingham in 1976:[1]

'But one has to ask in whatever country one is working, to what extent is secrecy itself an occupational hazard. In our country there have been some judicial decisions involving the Free Information Act which in effect say that the very suppression of this information in itself is leading to a continuance of avoidable death and injury and disease in various workplaces and other environments. I suspect from my conversations with people here that there is a considerable amount of information that is not either readily available or is held secret under the Official Secrets Act, for example the guidelines that the inspectors are given to determine what and how information is to be given to workers under the '74 Act in your country. I am told these guidelines are being withheld from public request.'

The last sentence was, according to the editor of the conference proceedings, 'promptly and comprehensively denied by the Health and Safety Executive in the UK.' However, there does, according to an unofficial spokesman of the H & SE, appear to be real conflict between some clauses in HASAWA '74 and the Trade Secrets Act or the Statistics of Trade Act, 1947, without even bringing our Official Secrets Act into the picture. Moreover, we have no Freedom of Information Act such as that which exists in the USA.

To see the problem in perspective, one should not only consider the Official Secrets Act, but also the commercial secrecy which normally surrounds many industrial processes — particularly new ones which involve advanced technology. This secrecy is considered not only legitimate but necessary by many industrial organisations in order to safeguard their competitive position. Firms can hardly be blamed for withholding 'know how' about novel processes which has been acquired at considerable cost. Their employees with access to such information are generally required to sign an undertaking not to disclose it to anyone outside the firm as a condition of employment. Even safety information is a marketable commodity, and as such is sometimes 'guarded'.

It is easiest to describe the problem by way of an example, albeit a hypothetical one, which allows motives to be examined without fear of libel proceedings.

Company XYZ develops a new industrial process for making poly-wopperlene, a resilient non-stick germicidal material which is ideal for lavatory seats. But at one stage in the process there is danger of a serious explosion if three possible conditions inherent in the process should occur simultaneously. The hazard is not easily foreseen, but company XYZ has an experienced team of engineers and chemists who spot it. They take the necessary precautions, which are expensive. These add 20% to the cost of the product, but there is now no danger that the hazard will eventuate. For several years company XYZ has a monopoly of the manufacture of polywopperlene. It has patent protection and prospers accordingly.

This position, highly desirable as far as company XYZ is concerned, is rudely shaken when a rival company, UVW, enters the scene. UVW was once a very solid respectable family business, but now, unbeknown to the general public, a whiz-kid has gained control, sacked the former managers, and introduced a cowboy management in his own style. Bemused by the success of XYZ Co. with its polywopperlene he learns the secret of its manufacture by commercial espionage, while failing to learn about the hazard (which he might not understand anyway).

UVW then starts making polywopperlene and marketing it at a lower price than XYZ. They claim that they have invented a cheaper and better process. XYZ are fairly sure that their own patents are being infringed, but cannot get enough information to prove this because of the tight secrecy within UVW where no one knows what anyone else there is doing.

Meanwhile XYZ's profit margin is reduced almost to zero, and they are in danger of going out of business. They have but one prospect of salvation, that UVW are unaware of the hazard and that their plant will blow up. This raises the delicate moral question, should XYZ publicise the hazard, whereupon UVW might take the necessary precautions, or should they keep quiet about it?

If they warn of the hazard they will probably go out of business, and if they keep their mouths shut UVW's plant stands a 50:50 chance of blowing up within two years with considerable loss of life.

Faced with such a dilemma, human choice is not always altruistic. Let us suppose the plant does blow up as expected, within three years, killing all those present and destroying the control room and all records. Let us then suppose there is a public inquiry. Many explanations are put forward, but none really fits the facts. The court of inquiry finally has to admit that it really does not know what caused the disaster.

Should company XYZ then publicise the hazard and explain what happened?

If they did would UVW be grateful?

Would anyone in fact have thanked them?

This hypothetical situation is by no means as far fetched as it sounds. Similar things do happen in real life; sometimes it is a rival company from whose prying eyes the secrets are concealed, sometimes it is its own workers. It is even possible, where a multi-national company is concerned, whose head office is centred in some distant land whose inhabitants speak a strange tongue, that nobody in the British subsidiary is aware of the hazard. But it is always the workers (and often the neighbouring inhabitants) who suffer.

1.12.1 Robens on secrecy[2]

The report of the Robens committee has little to say on secrecy, although implicitly recognising the existence of the problem in several passages.

1.12.2 HASAWA (1974) on secrecy[3]

The subject appears in various forms in several places in this report. The most important for this discussion are:

Section 2. *'General duties of employers to their employees'. These include:*
    Subsection (2)c. 'The provision of such information, instruction, training and supervision as is necessary to ensure, so far as is reasonably practicable, the health and safety at work of his employees.'

Section 3. *'General duties of employers and self employed to persons other than their employees.'*
    Subsection (3). 'In such cases as may be prescribed (g), it shall be the duty of every employer and every self-employed person, in the prescribed circumstances and in the prescribed manner, to give to persons (not being his employees) who may be affected by the way in which he conducts his undertaking the prescribed information about such aspects of the way in which he conducts his undertaking as might affect their health or safety.'
    Note (g). 'Prescribed. For definition see s. 53(1). No regulations have yet been made under this provision.'

Section 27. *Obtaining of information by the Commission, the Executive, enforcing authorities, etc.*
    Subsection (1) authorises the Commission to serve a notice on any person requiring him to furnish to the Commission or the enforcing authority in question such information as may be specified in the notice and to do so in such form and manner and within such time as may be so specified.
    Subsection (2) states that 'nothing in section 9 of the Statistics of Trade Act 1947 (which restricts the disclosure of information obtained under that Act) shall prevent or penalise:
    '(a) the disclosure by a Minister of the Crown to the Commission or the Executive of information obtained under that Act about any undertaking (f) within the meaning of that Act . . .
    '(b) the disclosure by the Manpower Services Commission, the

Enployment Service Agency or the Training Service Agency to the Commission or the Executive of information . . .'

Section 28. *Restrictions on disclosure of information*

This is a lengthy section, but the meat is contained in subsection (2) which states, subject to a number of clauses that 'no relevant information shall be disclosed without the consent of the person by whom it was furnished'. This of course excludes disclosures to the Health and Safety Commission.

So in the hypothetical case discussed earlier, the technical director of XYZ Co. could well disclose the hazard to the Health and Safety Executive, but refuse consent to further disclosure by the latter — perhaps on the reasonable grounds that the hazard only applied to the patented process which XYZ were operating, and that, as nobody had obtained a licence to operate it, the hazard would exist nowhere else.

Section 14. *Power of the Commission to direct investigations and inquiries*

This section contains The Health and Safety Inquiries (Procedure) Regulations 1975 which were discussed in section 1.11.8 of this book.
Regulation 8, clause (3) is important for secrecy: 'The appointed person:

(a) shall, to the extent to which he has been so directed in writing by a Minister of the Crown, hold the inquiry otherwise than in public for the purpose of hearing evidence relating to matters specified in the direction, being matters of such a nature that it would, in the opinion of the Minister, be against the interests of national security to allow the evidence to be given in public; and

(b) may, on application made to him in that behalf, hold the inquiry otherwise than in public to such extent as he considers necessary for the purpose of hearing evidence, the giving of which is in his opinion likely to disclose information relating to a trade secret.'

This raises the possibility that a Court of Inquiry may hold part of the Inquiry in public and part in private, without anyone not privy to the private hearings being aware that they have taken place. Moreover, it would be unlikely that the findings of the private hearings would ever be published! One might presume that the Court would then prepare two reports, one for public consumption and one for private!

All things considered, Ralph Nader's suggestion that secrecy may itself be an occupational hazard seems only too well founded. One has only to follow the results of a few inquiries into industrial diseases and disasters — not to mention revelations which sometimes follow in their train — to appreciate this.

Secrecy about hazards can all too readily become a pathological condition among some directors and managers who succeed in convincing themselves that a particular hazard does not exist. After all, they have to convince their employees and insurers that it does not exist, and it is hard to do this unless they are convinced themselves. The next step is that any employee daring to suggest that the hazard could be real becomes dubbed 'unfriendly to the company' and is 'drummed out of the brownies'. A process of natural selection takes place whereby all independent thinkers are removed and the principal is surrounded only by yes-men.

It is Hans Anderson's old story *The Emperor's New Clothes* which is being re-enacted. Tragedy is seldom slow in following. Even then, the Emperor will still not have learned.

REFERENCES

1. Nader, R., 'Health and Safety priorities for US industry', in *World Safety Brief World Safe 1976*, RoSPA, Birmingham, 26 (1977).
2. Lord Robens Committee, *Safety and Health at Work. Report of the Committee 1970-72*, HMSO (1972).
3. Fife, I., and Machin, E.A., *Redgrave's Health and Safety in Factories*, Butterworths, London (1976).

1.13 PLANNING FOR MAJOR EMERGENCIES

Contents

A major emergency is defined[1] as one which may affect several departments in a works and possibly extend beyond the works itself. It may cause serious injuries, loss of life and/or extensive property damage. It will require the use of outside resources as well as those of the organisation itself to handle it effectively. The type of emergency considered here is one involving a very large fire or explosion, and/or the sudden release of large quantities of flammable or toxic materials to the atmosphere, capable of causing many human casualties. These are essentially emergencies envisaged by the Advisory Committee on Major Hazards (set up after the Flixborough Disaster), in its first report[2] (1976).

This is, of course, only one kind of major emergency. Others can and do arise through plane crashes, in rail and road transport, building failures, abnormal climatic conditions and through enemy or terrorist action. These are not discussed here either because they are not primarily industrial hazards or because, when they are, their chances of happening are too remote to allow effective planning. Major hazards caused by atomic explosion or the release of large quantities of radioactive materials are also excluded from this discussion.

1.13.1 Does planning help in emergencies?

If one examines the emergency plans that have been made before a disaster, and then examines what happened in practice, some sobering and disagreeable facts sometimes emerge.

To take the Flixborough disaster as an example, the emergency plan called for plant and laboratory personnel to assemble in the control room. Everyone in the control room was killed, and one man who had run from a place where others survived was killed just before reaching it. But those in the adjacent laboratory who ignored the plan and wisely decided to

124

get out of the building and run as fast as they could away from the imminent vapour cloud, all survived, although they were caught by the blast in the open and hurled some distance by it. One lesson from this is that rigid plans worked out on paper in advance are often inappropriate for survival in a particular and largely unforeseen emergency. Intelligent human beings can and often do decide on their best survival course as an on-the-spot decision, especially when it comes to 'getting the hell out of here fast'.

Sometimes rigid plans and orders *are* necessary for action in a major emergency, but planners must think very carefully when making them to ensure that they have considered all likely consequences, and that they cannot reasonably aggravate a disaster or add to the death roll.

Plans should, as far as possible, be simple and flexible, and allow those affected by a major emergency to use their knowledge and initiative to best advantage.

One trouble with emergency planning is that it is often for the last disaster, not the next one.

Planning, however, is needed not only for survival but for rescue and to prevent spread and escalation of the disaster. The operator who after the Flixborough disaster went straight to the valve on the base of an ammonia storage sphere and closed it, may have prevented many casualties.

1.13.2 Special features of major emergency planning

Planning for major emergencies has several common features with normal emergency planning which are discussed in section 1.12. These include:
1. Identification of possible emergencies.
2. Assessment of potential losses arising from them.
3. Selection of measures to protect personnel, minimise losses and prevent their spread.
4. Implementation of these measures.
5. Checking their effectiveness in real emergencies, and adapting them as necessary.

Planning for major emergencies has several differences:
1. It requires more help and cooperation with outside organisations, police, fire, ambulance services, news media, voluntary services and other organisations in the neighbourhood.
2. It requires one or more effective control posts with first class communications, and personnel who are closely familiar with all aspects of the situation to take charge or advise. Communications are, perhaps, the key to major emergency planning, as many of the normal channels of communication may have been destroyed or rendered inoperative, with those that remain being mostly overloaded.
3. Accurate information of what has happened, is happening and is likely to happen in an affected area will be at a premium. The plan must take this into account, and ensure that people with specialised knowledge of the plant, processes and materials are available to advise on the likelihood and consequences of any spread or escalation of the emergency.

1.13.3 Identification of major hazard situations

In the case of flammable materials, the greatest threat lies in the massive escape of volatile liquids present in a process or in storage under pressure at temperatures above their normal boiling point, (i.e. at atmospheric pressure). During such an escape the liquid partly vapourises and the mixture entrains air to form a large and persistent combustible cloud which is likely to explode when ignited.

Liquefied hydrocarbon gases stored in bulk under pressure, and light petroleum fractions and other light hydrocarbons in process plants under pressure at elevated temperatures are the most potent sources of hazard. Less volatile flammable liquids as well as gases under pressure on the whole present less serious threats.

The sudden release of large quantities of toxic materials can, in certain weather conditions, produce lethal concentrations many miles from the point of release.

There have been a number of disasters caused by accidental vapour cloud explosions throughout the world — one or more seems to happen somewhere every year. At the time of writing there has been one large one and three smaller ones in Britain.

There have been fewer known disasters caused by the sudden massive release of toxic materials, although a number of creeping disasters have occurred caused by the gradual release of toxic materials, the consequences of which were not immediately appreciated. These include asbestos, arsenic, lead and mercury compounds, organo phosphates and other highly toxic organic compounds.

The Advisory Committee on Major Hazards took a somewhat pragmatic approach to the problem, by proposing a list of 'Notifiable Installations' where these hazards are plainly present. The list[2] is as follows:

(a) Installations storing or processing toxic material where, if containment is lost, there can be an emission of toxic gases or vapours equivalent in effect to more than 10 tonnes of chlorine.

(b) Installations storing or processing flammable materials where, if containment is lost, there can be a rapid emission of flammable gases or vapours of more than 15 tonnes.

(c) Installations storing or processing materials which are intrinsically unstable or of very high exothermic reactivity, where the total inventory is more than 5 tonnes. Examples are ethylene oxide, acetylenes, organic peroxides.

(d) Installations with a large inventory of stored pressure energy, typically process operations at 100 bars or above using gas phase reactions.

(e) Installations storing or processing flammable materials which have a flash point of less than 22.8°C, where the total inventory is more than 10 000 tonnes.

(f) Installations storing or processing liquid oxygen where the total inventory is more than 135 tonnes.

(g) Installations storing or processing ammonium nitrate where the total inventory is more than 5 000 tonnes.

(h) Installations storing or processing materials which in a fire can cause an emission of toxic gases or vapours equivalent in effect to more than 10 tonnes of chlorine.

Such installations while including many chemical and oil refining plants, are also found in depots (especially for LPG), stores and other process plants (e.g. blending fertilizers) which are rather peripheral to the chemical industry. Since many rail and road tank cars handle at least the minimum quantities of many of the listed materials, it is clear that many notifiable installations are in fact mobile and travelling daily on our roads! Such hazards cannot be dealt with while they are moving, but they must be included in any works emergency plan where they may be present (while loading, unloading or waiting to unload).

A major emergency plan should be developed where a major hazard exists which falls into one of the categories of this list. When assessing the quantities of toxic materials other than chlorine needed to qualify as a notifiable installation, three things should be considered:

1. The toxicity of the gas or vapour, e.g. as given by its TLV (see section 3.2.4). Most toxic gas and vapours have higher TLV's (i.e. are less toxic) although some such as arsine have much lower TLV's.
2. Whether the gas or vapour is readily emitted or dispersed. This depends partly on its boiling point and whether it is present under pressure and/or at elevated temperature.
3. Whether the vapour is heavier or lighter than air. Heavy vapours persist longer at low level and in pits and depressions, although once the vapour is well mixed into the atmosphere its density has little effect on its further dispersion.

1.13.4 Assessment of risk[1]

When a major hazard has been identified, the possible consequences in terms of risk to people and spread of damage must be assessed. Account needs to be taken of:

1. The type of incident expected (fire, explosion, toxic release).
2. The area or location of the incident in relation to neighbouring plants, storage areas and built-up areas.
3. Prevailing winds.
4. Areas most likely to be affected (known to insurers as the target areas).
5. Population densities in the possible areas affected.
6. Possible damage or contamination of drains, crops, water supplies.
7. Possible 'domino' effects, i.e. an explosion in one area causing the release of flammable or toxic materials elsewhere.
8. Possible effects of collapse of buildings and structures.
9. Presence of radioactive sources.

Next one needs to examine how adequate the existing resources (first, works and second, local authority fire, ambulance and hospital services) are to handle the most serious foreseeable emergency, and then to decide in collaboration with outside services what further provision or action is needed.

1.13.5 Objectives of plan

The plan should make best possible use of works and outside services, firstly to
Rescue victims and treat them;
Safeguard others, (evacuating them where necessary);
Contain the incident and control it with minimum damage.
Secondly the combined services should be organised to:
Identify the dead;
Inform relatives of casualties;
Provide authoritative information to the news media;
Preserve relevant records and equipment needed in evidence in any subsequent inquiry;
Rehabilitate affected areas.

1.3.6 Liaison with outside authorities

Available resources include Police, Fire, Ambulance and Hospital Services, District Inspectors of the Health and Safety Executive, and sometimes the Local Authority. In the event of a major emergency the Police will coordinate the activities of the Emergency Services, and a senior officer will be designated as the overall Incident Controller.

The Fire Service is duty bound to maintain a brigade sufficient to meet normal requirements, but may not be able to cope adequately with incidents in isolated high risk situations. This will require discussion between the works managements and the Chief Officer.

When called to a fire, the senior Fire Brigade Officer will take charge of fire fighting, rescue and salvage operations. Fire Brigades are also authorised to assist in handling the escape of toxic materials.

The Health and Safety Executive will wish to see proper plans for dealing with major emergencies (as well as lesser ones), and satisfy themselves that these are practical and sufficiently detailed and rehearsed. The Local Inspector of the Health and Safety Executive should be advised promptly by telephone of a major emergency once the outside services and key personnel have been informed.

Liaison with the outside services is needed to ensure that:
The plan is properly coordinated and its operation effectively controlled.
Works procedures and existing plans developed by an outside authority are in harmony:
Outside services are clearly aware of the nature of the risks and have

appropriate knowledge, equipment, materials and apparatus to deal with it;

Equipment of the works and outside services is compatible;

Works and outside personnel who are likely to be involved during an emergency know each other;

The appropriate type and number of outside services arrive on the scene promptly.

1.13.7 Works organisation for major emergency

Whilst no detailed procedure suitable to all works and situations can be given here, certain basic common features are given.

1.13.7.1 Emergency control centre

A pre-arranged emergency control centre with adequate communications, maps, site plans and all relevant information is necessary (*Figure 1.13.1*). Whilst this needs to be as close as possible to the scene of the emergency, it should be far enough away and well enough protected so that it is operable

Figure 1.13.1   An emergency control centre (Robert Nagelberke)

in a major emergency, and not put out of action by it. It may have to be blast resistant and be provided with an independent clean air supply.

The siting, construction and provision of services for an emergency control centre need a great deal of thought. It has got to withstand the worst that can happen in a major emergency. It must not collapse as a result of an explosion, or worse, be buried by the collapse of floors above it. It must not have windows that can be blown in — even wire-reinforced glass can be fractured in an explosion and the pieces cause wounds worse than shrapnel to its occupants. It is probably best that it has no windows at all. The air inside it must be clean and uncontaminated while that outside may contain dangerous concentrations of toxic gases. It must be effectively lit when the power supply has been disrupted. It must not be subject to flooding. The temperature and other conditions inside it should enable men inside to work coolly and calmly when surrounded by chaos.

In some cases it may be necessary to have two alternate emergency control centres. The centre should contain:

1. An adequate number of direct dialing PO telephones, at least one of which should be ex-directory or capable of use for transmitting external calls only, together with a list of telephone numbers and addresses of key personnel.
2. An adequate number of internal works telephones, with list of numbers.
3. Battery operated radio equipment, which will be the only effective means of communications if telephones are rendered inoperable by the emergency.
4. Adequate lighting with emergency supply (batteries).
5. An adequate number of tables, chairs and light refreshments.
6. Toilets (chemical) and washing facilities with an adequate water supply for several hours in case the mains are put out of action.
7. Plans of the works suitably mounted on boards showing:
   (a) areas where hazardous materials are present in quantity, as well as gas cylinders and radioactive materials.
   (b) Locations of safety equipment, fire water systems and alternate water sources, as well as stocks of foam and other fire extinguishing media.
   (c) Works entrances and road system, constantly up-dated to show any impassable roads.
   (d) Assembly points and casualty treatment centres.
   (e) Map showing works in relation to local community.
   (f) Location of all utilities and their isolation valves.
8. Additional works plans which may be marked up during the emergency to show areas affected and at risk, deployment of emergency equipment and personnel, special problem areas (e.g. fractured pipes), evacuated areas and other relevant information. These can be covered in glass or transparent plastic which may be marked with coloured erasible markers.
9. Note pads, pens, pencils.
10. Roll of employees, showing shifts rotas where applicable.

11. Suitable personnel protective equipment for persons visiting affected areas.
12. A long playing tape recorder, to record the sequence of events.

1.13.7.2 Assembly points

Several key assembly points should be chosen to which employees should proceed in case evacuation is necessary. These should be in safe places, well away from the areas at risk. They should be located so that employees can choose which one to proceed to without having to approach an affected area to reach it or pass down wind of any fire or likely emission of toxic materials.

Each assembly point should be clearly marked and provided with a means of communication (radio or telephone) with the emergency centre. Unless put out of action by the emergency, every assembly point will be manned by someone to record names and details of those reporting there, call for an ambulance if needed, and give or relay any further instructions to those reporting on what to do or where to go and report to the controller.

1.13.7.3 Effect of shift working

Most notifiable installations involve shift work, and there is a higher chance of a major emergency when day supervisors are absent than when they are present. Once a major emergency starts, no procedure will cause it to wait for some key day man to arrive and take charge before developing or 'blowing its top'.

It follows that the shift workers on the spot are those most likely to have to cope with a major emergency during its most critical and destructive phases. A most important task of those not involved in rescue and fire fighting is to shut down and make safe operating plant, close valves and isolate flammable and toxic materials in plants, tanks and pipelines, and cut off the flow of material to any burst pipe or point of escape. This they may have to do while making their escape from the scene of the emergency.

1.13.7.4 Key personnel

Two key positions should be filled in following a major emergency by nominated persons:

1. Works incident controller.
2. Works main controller.

Both will normally be day staff (i.e. plant managers) but their work may have to be done by deputies, particularly shift staff until they arrive.

Works incident controller. The works incident controller will proceed to the scene, assess the scale of the emergency, and take responsibility, where necessary, activating the major emergency procedure if this has not already been done. He should wear a distinctive hat and/or jacket and have a portable two-way radio and a runner in attendance. (The runner will normally carry the radio for him, but if he drops it, he will have to run!) His tasks include:

Securing the safety of personnel.
Minimising damage to plant, property and surroundings.
Minimising loss of material.
Directing fire fighting and rescue operations until the arrival of the outside brigade when he will hand this over to the senior brigade officer.
Ensure that the affected area is searched for casualties.
Evacuate non-essential workers to the appropriate assembly point.
Set up a communications point and establish contact with the Emergency Control Centre, and report developments to the Works Main Controller stationed there.
Inform and advise the senior fire officer present.
As far as possible avoid destroying evidence needed in any subsequent inquiry.

Further, in the absence of the Works main controller or pending his arrival, the incident controller will direct shutting down and evacuation of plant that may be affected and ensure that outside services and key personnel have been called in.

Works main controller. The works main controller will normally be the senior manager at the works. He will go to the Emergency control Centre as soon as he is aware of the emergency and take over from whoever has been deputising for him. His duties will include the following:

Ensure that outside services and key personnel are called in and that neighbouring firms are informed.
Establish communications with the Works Incident Controller, liaise with him and exercise operational control of those parts of the works outside the affected area, directing the shutting down of plants and evacuation of personnel where necessary.
Ensure that casualties are adequately attended to, calling more help if needed, and see that relatives are informed.
Liaise with Police and Fire Services and the Health and Safety Executive, and advise on possible effects of the emergency on areas outside the works.
Control traffic movements in the works.
Record or arrange for a chronological record to be made of the emergency.
Arrange for the relief of personnel and the provision of food and drinks.
Obtain early warning of changes in weather conditions from the local Meteorological office.

Issue statements as required to the news media and inform and liaise with his company's head office.

Ensure that evidence is preserved and, finally

Control the clean-up and rehabilitation of affected areas after the emergency.

(*Note:* If he is ever to sleep he must deputise.)

1.13.8 Raising the alarm and declaring a major emergency

The alarm should be audible everywhere in the works (with more than one alarm where necessary) and there should be enough points spread over the works for anyone to raise it without going far. (Often the emergency will make its presence felt to most people without the formality of sounding an alarm, but even so the alarm should be sounded.) Everyone should be able to raise the alarm thereby giving an early warning signal.

The major emergency signal will be different and distinguishable from the early warning signal. A limited number of senior personnel on any shift should be authorised to give it, as well as key day personnel. If several different types of major emergency are possible on the same site, i.e. release of flammable or toxic gases, consideration should be given to providing distinctive alarm signals for each, although the number should be strictly limited and all employees carefully instructed in their meaning in order to avoid confusion.

Everyone in the works should recognise and understand the major emergency signal, and know what his duties are. (Some will be part-time firemen or first aiders, others will have the responsibility of shutting critical valves to isolate tanks, pipework and plants, and others will have to take charge of particular assembly points.)

1.13.9 Key personnel and essential workers

Besides the works incident controller and the works main controller, the activities of many other key personnel need to be planned to meet a major emergency; these include senior managers responsible for production, engineering, technical services, laboratory, personnel, medical services, transport and safety and security. Special attention needs to be given to the carrying out of atmospheric tests, although many other needs will arise calling for the specialised work of various departments.

All key personnel should on declaration of a major emergency report to the Emergency Control Centre unless otherwise instructed.

Various workers will have essential duties in shutting down and isolating plant as detailed above; others with special training will be needed for first aid, emergency engineering work, transporting equipment, moving tankers from areas of risk, atmospheric testing, and acting as runners if other forms of communication break down.

This work is best organised by the creation of a task force of men with various skills and specialities located at a suitable centre which is in telephone contact with the Main Controller.

1.13.10 Evacuation and accounting for personnel

Evacuation may be spontaneous in case of a sudden major disaster, but it is more likely to be controlled, especially in areas adjacent to the one primarily affected. Plans must be made to ensure that employees in these areas are quickly warned when they should evacuate.

By arranging for all employees to evacuate through the manned assembly points, most of those present can be accounted for. Even so, it can be difficult to know how many and which people were present on the site at the time of the emergency, especially if this happened at a shift change. The following action should be taken to assist in establishing the whereabouts of personnel:

1. The incident controller should arrange for a search to be made to locate any casualties. A further search may be made by the Local Authority Fire Brigade.
2. Nominated personnel should record the names and numbers of casualties taken to the respective reception areas and the addresses of the hospitals, mortuaries, etc. to which they are taken. The names of fatal casualties should be obtained and provided to the police.
3. Nominated works personnel should record the names and departments of people reporting at assembly points.
4. A responsible person at the Emergency Control Centre should collate the lists, check them against the nominal roll of those believed to be on site, and the police informed of any thought to be missing. Where missing persons could reasonably have been in the affected area, the incident controller should be informed and arrangements made for further search.

1.13.11 Other post-emergency duties

Many other duties which lie outside the scope of this book follow any major disaster. These include comforting relatives, ministering to the injured and dying, public relations and rehabilitation of affected areas.

1.13.12 Action outside works

The most serious risk is that caused by a large release of toxic vapours. Charts and plans can be drawn up in advance relating the likely spread of the vapour cloud taking its buoyancy and local topography into account for a number of possible weather conditions at the time of the release. Expert assistance, including that of meteorologists is needed to plan for the spread of such clouds. The police will be closely involved, and roads may have to be closed.

1.13.13 Training

Major emergency plans should be drawn up clearly in writing and explained to all site employees. The component parts of the plans — communications, mobilisation of emergency teams, search and rescue of casualties, and emergency shut down and isolation of plant — should be tested in exercises and the exercises monitored by independent observers.

The results should be reviewed and discussed by senior works personnel and officers of the Emergency Services, and the plans where necessary revised to take account of any deficiencies found.

REFERENCES

1. Chemical Industry Safety and Health Council of the Chemical Industries Association Ltd., *Major emergencies,* 2nd ed., Chemical Industries Association, London (1976).
2. Advisory Committee on Major Hazards, Health and Safety Commission, *First Report,* HMSO (1976).
3. Symposium Papers by Bruce, D.J., and Diggle, W.M., Duff, G.M.S., and Husband, P., also Maas, W., 'Emergency Planning' in *Proceedings of the 1st International Loss Prevention Symposium, the Hague/Delft, 28-30 May, 1974.* Elsevier (Amsterdam-Oxford-New York) (1974).
4. Diggle, W.M., 'Major emergencies in a petrochemical complex. Planning action by emergency services'. Paper in Symposium *Process Industry Hazards, Accidental Release, Assessment, Containment and Control,* The Institution of Chemical Engineers, Symposium Series No. 47 (1976).

2 DESIGN, CONSTRUCTION, INSPECTION AND MAINTENANCE

2

Design, construction, inspection and maintenance

2.1 INTRODUCTION

Contents

The title of this chapter may appear presumptuous, since designers in all engineering disciplines have always had to design for safety and would justifiably be offended if anyone were to suggest otherwise. In spite of this we all know that the risks and accident records in some works are very much higher than in others, and that the difference is often to be found in the degree of attention given to safety aspects in the design.

The subject can perhaps be split into two parts:
1. Compliance with legal and statutory requirements and codes of practice.
2. Other safety considerations not covered in 1. This is the most significant part.

Regulations and codes of practice will be discussed in section 2.2. These cover every engineering profession or discipline: civil, structural, mechanical, electrical, chemical and illuminating, etc. Some of these take into account the inherent hazards of the site or process. Thus the civil and structural engineer will consider the seismic factor, wind and snow loadings of the location where the works are to be built.

The electrical engineer will consider *inter alia* the possible ignition of flammable vapours which may be accidentally present, by the switchgear or other electrical apparatus. The fire protection engineer will consider the flammability and fire loads of various parts of a factory and their contents before planning means of escape, sprinkler and hydrant systems. These aspects are fairly well codified.

2.1.1 Fires and explosions

The main aspects of fire protection to be considered in building design are discussed in section 2.6. The hazards to which they relate are also discussed

139

in chapters 4, 5 and 6. But in other engineering fields, especially process and chemical engineering, it has so far proved very much more difficult to quantify the risks on any sort of a hazard scale, so that far more is left to the skill, judgement and experience of the designer or design team.

Section 2.8 provides some useful guidelines from sources in the USA on the safe design and layout of oil and chemical plants and processes. The Dow Chemical Company have made a notable contribution to quantifying fire and explosion hazards. The Hazard Index[1] is useful in two ways, firstly in making a rational judgement on the level of protective features needed, including standards of layout and spacing, and secondly in focussing attention on high risk areas and examining the specific hazards in those areas in detail.

Safety is concerned with people, and a cardinal object of the designer is to create a safe and healthy working environment. This involves a study of the various requirements discussed in chapter 3, illumination, noise, ventilation and air conditioning, heating and ergonomics.

But even when all these matters have been taken into account, there are often gaps in the safety net where unacceptable hazards can arise. Sometimes these result from interactions between hazards of different engineering and other disciplines. The hazards may appear acceptable to professionals in each discipline viewed from their own standpoint but, when combined, the hazards interact producing a far more serious overall hazard than they would individually. An example of this is the Ronan Point building disaster[2] of 1968 where one wing of a multi-storey block of flats collapsed as a result of a gas explosion in the kitchen of one of the flats (*Figure 2.1.1*). The form of construction employed the minimum of steel structure, and made use of factory-built units stacked like matchboxes, the load being carried by the outside walls which incorporated concrete panels. The method had been approved and complied with local by-laws and the relevant codes of practice, which were mainly concerned with wind loads.

For cooking and heating town gas was provided. This had caused a number of explosions in domestic dwellings with destruction of property and sometimes fatalities among the occupants. This limited gas explosion risk is accepted by the public for houses and similar dwellings as the probability is low and injuries and property damage are generally limited to the occupants of the dwelling where the leak and explosion occurs.

It should have been possible to foresee that a gas explosion in a high rise block of flats of this construction could destroy a load bearing wall and lead to the progressive collapse of a complete wing of the building. However nobody in authority appeared to have seriously considered this possibility until it actually occurred at Ronan Point.

2.1.2 Safety in design

Another aspect of safety which is still insufficiently appreciated by designers is in allowing for human foibles or psychological limitations. This view[3] was stated by Charles Critchfield, group safety adviser to Unilever Ltd. in 1974 in these words:

Figure 2.1.1   Collapsed flats at Ronan Point (London Express News and Feature Services)

'. . . we are compelled to deduce that past generations of designers have made these mistakes because their technical training has led them to be more concerned with the safety of things or artifacts than with the safety of the people using those things.'

From this, Critchfield argued strongly for the safety specialist to be a member of and adviser to the design team, since he is in Critchfield's view 'the one person who can look at a layout design and point out from his training and experience the unsuspected design features which, whilst not technically unsafe, may tempt people to take short cuts and place themselves at risk.'

A further advantage of this is that by bringing the safety specialist into close contact with designers, their own appreciation of safety aspects of design is increased. Some of the common design failings noted by Critchfield are included in the other sections of this chapter.

In addition to the safety specialist the design team should also include people who have been involved with the running of similar factories or plants and who will be responsible for the operation of the one now being designed. It is very easy for designers to become too divorced from the plants or machines they are designing and to fail to appreciate some of the pitfalls they can introduce. Positive feedback of this kind of information from the factory to the designer is absolutely essential.

Other useful check lists for safe design of various plants have been made, notably one[4] prepared for process plants by a Dutch committee under the chairmanship of Ir. A. W. M. Balemans, Directorate General of Labour, Voorburg. This source has been particularly useful in preparing sections 2.3 and 2.4.

Hazards to personnel during building construction are dealt with in chapters 5 and 6.

2.1.3 Storage of equipment

A point frequently missed in the design of factory buildings and plant is the need for local storage of safety apparatus (e.g. breathing apparatus), and essential ancilliaries and tools, e.g. ladders and steps, solvents and cleaning materials, lighting fittings, as well as waste materials (until they can be disposed of). This can be a serious problem if the building is some distance from the main store, or if shifts are worked in the building while the main store is manned on days only. Unless proper storage space is provided these

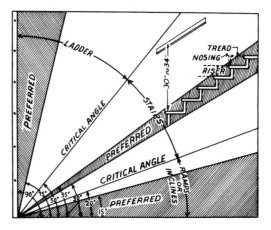

Figure 2.1.2   Preferred angles for fixed ladders, stairs and ramps. Dimensional details on the right (National Safety Council, U.S.A.)

things tend to be left lying about and cause hazards of their own. Another point often missed is the need to control the spillage of dusty, toxic, flammable or corrosive materials.

2.1.4 Ramps, stairs and fixed ladders

Ramps and slopes both inside and outside buildings should be avoided wherever possible, and traffic must be so routed that lift trucks do not have to traverse slopes. The angles of fixed ladders *(Figure 2.1.2)*, stairs and ramps should be restricted to the following ranges (see also section 5.4).

Ramps   0 to 15° to horizontal
Stairs   30 to 35° to horizontal
Fixed ladders   75 to 90° to horizontal

Truck routes and pedestrian routes should be separated and free of blind corners, and all truck routes should have hard surfaces.

REFERENCES

1. Dow Chemical Co., *Fire and Explosion Index*, Dow Chemical Co., Michigan, USA (1976).
2. Griffiths, H., Pugsley, A., and Saunders, O., *Report of the Inquiry into the Collapse of Flats at Ronan Point, Canning Town.* HMSO (1968).
3. Critchfield, C.E.M., 'Engineering For Human Safety', Paper in Symposium: *Loss Prevention and Safety Promotion in the Process Industries,* Elsevier, Amsterdam (1974).
4. Balemans, A.W.M., 'Check-List, Guide lines for safe design of process plants', ibid.

2.2 REGULATIONS AND CODES OF PRACTICE

Contents

2.2.1 Regulations

Statutory regulations were discussed generally in section 1.11 'Legal Aspects and Inquiries'. Under the Health and Safety at Work, etc. Act 1974, much of the former legislation (see Table 1.11.1) is to be replaced by a system of new regulations, supplemented by Codes of Practice. A leaflet[1] setting out these procedures has been issued by the Health and Safety Commission. The scope of the regulations may be:

General, e.g. concerning notification of accidents,
Limited, e.g. on the use of abrasive wheels,
Functional, e.g. electricity,
Industry related, e.g. construction or docks.

Regulations may be detailed setting out just what needs to be done, and they may refer for further details to codes of practice or standards which may be updated without amending the regulations. Codes of practice when so referred to directly become part of the statutory requirement and must be complied with. However, codes of practice not referred to in the regulations have not the same legal force as ones referred to in regulations.

Regulations proposed by the Commission, accepted by the Secretary of State and laid before Parliament become law and are enforceable through the Courts. Breaches of duties imposed by these regulations generally become actionable in civil proceedings when they cause damage.

2.2.2 Codes of Practice

There are now several types of codes of practice, the most important legally being codes of practice approved and generally issued by the Commission. These are termed 'Approved codes of practice'.

2.2.2.1 Approved Codes of Practice

Approved codes of practice will either be underpinned by a set of regulations or else will state precisely the limits of their application. An approved code of practice has special status even when not underpinned by

Figure 2.2.1 Selection of Codes of Practice Regulations, etc., governing factories and other premises

a regulation. For although failure to observe a provision of an approved code does not itself render a person liable to criminal or civil proceedings, any provisions of an approved code which appear to be relevant may be quoted in evidence in criminal proceedings against a person alleged to have contravened a statutory requirement or prohibition.

Approved codes will usually be generated by the Commission, the Executive, industry advisory committees set up by the Commission, or possibly in conjunction with such a body as the British Standards Institution.

Codes of practice published by the British Standards Institution and other professional and industrial bodies are not approved codes of practice under the provisions of the Health and Safety at Work etc. Act.

Only a limited number of approved codes have been issued or are in preparation.

2.2.2.2 Non-approved Codes of Practice and Standards[3]

These are divided into four main groups:

1. *National (i.e. British) Codes and Standards.* BS Codes of practice give general guidance on a variety of subjects including design. A list of British Codes and Standards is published yearly in the British Standard Yearbook[2]. The standards specify detailed requirements and meet a variety of needs:

(i) *Application Standards.* These specify the principal characteristics of equipment and provide a basis for the scope of supply, e.g. BS 132 *Steam Turbines.*

(ii) *Design Standards.* These specify calculation methods, manufacturing and testing requirements, e.g. BS 5500 *Pressure vessels.*

(iii) *Material Standards.* These define the chemical composition and properties (mechanical, electrical, optical, etc) of various materials.

(iv) *Manufacturing Standards.* These standardise dimensions and other basic requirements thereby reducing costs.

2. *Industrial Codes and Standards.* These are set up by professional institutes and trade associations and are intended to supplement the National Standards by setting out the requirements of particular industries.

3. *Company Standards.* These also have a complementary role although they are sometimes forerunners of National Standards, e.g. *Electrical installations in flammable atmospheres* (ICI/RoSPA).

4. *International Standards.* There are at present a limited number of International Standards, which have been prepared by the International Standards Organisation. Although their use is being encouraged by the EEC, progress is slow, and many manufacturers are reluctant to abandon their national codes which act as barriers to imports.

2.2.2.3 Overlapping of Codes

In some fields there seems to be a proliferation of overlapping standards produced by different organisations. The storage and handling of liquefied petroleum gases which is covered by several different codes, etc as follows:
Statutory Instrument
1. *The Highly Inflammable Liquids and liquefied Petroleum Gases Regulations 1972.* Much of this is couched in legal language which the layman cannot be expected to understand. The DOE have issued an *Explanatory Booklet*
2. *Highly flammable liquids and liquefied petroleum gases: guide to the Regulations 1972.*
National Codes
3. *Codes of Practice for the storage of liquefied petroleum gas at fixed installations 1971.* Published by the Home Office.

4. *Storage of LPG.* H & SE Booklet No. 30.
Industrial Codes
5. *Codes of Practice 1 to 12.* Published by the liquid Petroleum Gas Industry Technical Association. These Codes of Practice cover all aspects of the storage and handling of LPG.
6. *Liquefied Petroleum Gas Safety Code 1967.* Published by the Institute of Petroleum (This is now under revision).
7. *Storage and Handling of LPG.* Published by the Fire Protection Association.
Company Code
8. ICI/RoSPA. *Liquefied Petroleum Gases, Storage and Handling, Engineering Codes and Regulations 1971.* This is considerably more demanding than Nos 1 to 7 and in many quarters is recognised as a national code.

Perhaps we can justify the existence of so many codes, but it would seem better if the various interested parties could join forces as they do in the USA under the National Fire Protection Association and form a working party to produce a single and more detailed code, which could be brought up to date at frequent intervals, say every two years.

There also seems to be a need for more Standing Committees to interpret and review proposed amendments to Codes and Standards.

2.2.2.4 The usefulness of Codes

Codes and standards have their advantages and disadvantages.
Their advantages lie in:
1. Helping the purchaser of an item by defining what is accepted good practice.
2. Limiting the liberties that can be taken by an unscrupulous designer or manufacturer.
3. Spreading information on new techniques and encouraging their use.
4. Eliminating the need for each company to produce its own standards.
Their disadvantages are:
1. Working to the letter rather than the principle of the code can lead to expensive and sometimes unsafe designs.
2. Unless regularly and frequently amended in the light of best current information, they can perpetuate out of date design features, and result in uneconomic and uncompetitive designs.

REFERENCES

1. Health and Safety Commission, *Regulations, Approved Codes of Practice and Guidance Literature*, Health and Safety Commission (1976).
2. British Standards Institution, *British Standards Yearbook*, BSI.
3. Stead, W. R., 'Do We Need More Codes of Practice?', Symposium paper, *Institution of Chemical Engineers, N.W. Branch*, Manchester (November 1976).

2.3 SITE SITUATION

Contents

Before a new factory or plant can be built, the approval of the local authority has to be obtained. This will be concerned, among other things, with possible hazards created within the site which could 'spill over' into the local community, and with discharges from the proposed site which could adversely affect the environment.

The local authority will consult the local Health and Safety Inspector on hazards. He in turn will, where appropriate, seek guidance from the Risk Appraisals Group of the Health and Safety Executive. Many of the routine precautions to be observed are given in the Department of Employment leaflet *So you want to start a factory?*.[1] If the works is liable to contain a 'major hazard' the first report of the Advisory Committee on Major Hazards should be studied.[2]

The main onus for selecting a site for a new works or for rejecting a site offered by the government or a local authority, rests however with the directors and owners of the company. The following check points are intended only as guidelines on safety aspects with a view to avoiding some of the pitfalls which often arise in site selection.

When considering a possible site for a new factory or plant, the following requirements must be studied and satisfied:[3]

1. Size of site.
2. Nature of site.
3. Climate.
4. Accessibility.
5. General provisions.
6. Environment.

2.3.1 Size of site

The site must be large enough for a safe layout, with sufficient room for parking facilities, roads, loading and unloading bays, service buildings, stores, storage tanks, effluent treatment and possible future extensions.

Where dangerous fluids have to be stored in bulk, the possibility of underground storage should be considered.

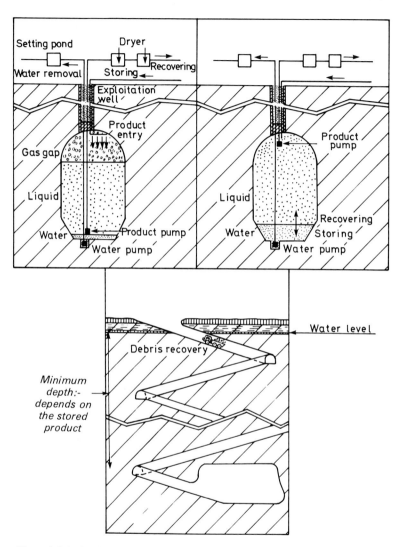

Figure 2.3.1   Underground storage of LPG. The operation is shown in the top two diagrams.
(left) Storage with fixed water level. In this the level of ground water entering the cavern is regulated by automatically controlled pumps. The water removed is checked and if necessary, treated. There is normally a vapour space above the stored product.
(right) Storage with variable water level. Here, the cavern remains full, and the quantity of product stored is regulated by varying the water level below the product.
The lower diagram shows tunnel access. (Geostock)

2.3.2 Nature of site

The following geophysical factors should be studied:

1. *Water table.* The site must be made safe against floods and high water table, the cost of which may be considerable.
2. *Soil settlement.* Settlement is likely in areas which require drainage, with back filled soil near former refuse dumps, mines, tunnels and watercourses.
3. *Ground disturbance.* Existing or expected ground disturbances arising from mines, geological and other reasons should be considered.
4. *Acidity and sulphate reducing bacteria.* The acidity of the soil and the presence of sulphate reducing bacteria should be investigated to anticipate possible attack on foundations, cables, pipes, tanks.
5. *Relief and slope.* If the proposed works involves handling hazardous liquids or heavier-than-air gases, and the site is sloping, the possible hazards of their escaping from higher to lower levels must be considered.

2.3.3 Climate

If flammable or toxic fumes or vapours are liable to be released in an emergency, wind direction and velocity frequency tables should be studied. Calm conditions greatly increase the hazards of such discharges, as well as the probability of fog formation in the neighbourhood of cooling towers, stacks and steam discharges.

Precipitation and temperature should be considered in relation to drainage, fire fighting and other safety aspects.

2.3.4 Accessibility

The adequacy of road access should be studied. If hazardous materials are to be supplied to or delivered from the works, the routes should be checked to ensure that they are safe and free from restrictions on carrying certain materials.

If rail traffic is required, the possibility of making a safe connection with the existing system must be checked. If waterborn transport is required, the possibilities for safe mooring, loading, unloading and manoeuvring of vessels should be investigated. Access roads for outside fire brigades and ambulances should be free from obstacles such as level crossings and movable bridges.

2.3.5 General provisions

An adequate supply of water for fire fighting should be available. Drainage of rain water, fire water and process water must be technically feasible and any environmental problems so caused must be considered.

The safe disposal of waste products must be considered, e.g. temporary storage and collection, processing or destruction.

The reliability of power and fuel supplies must be considered.

2.3.6 Environment

The sensitivity of the processes employed to impurities in the air or water should be considered, e.g. traces of hydrocarbons in the air intake to air separation plants have caused serious explosions.

The interactions between emissions (normal and accidental) and the occupiers of adjacent sites need to be considered both for the proposed works and its neighbours. Where such emissions could ignite or explode, sources of ignition both within and adjacent to the site should be taken into account.

Sources of noise and vibration from neighbouring industries, transport vehicles, etc should be considered and their possible effects on personnel, equipment and buildings examined. If the site is near an airfield, the hazards of low flying aircraft particularly to stacks and high structures required on the site, should be considered.

REFERENCES

1. Department of Employment. *So you want to start a factory?*, Leaflet HSW 8, HMSO.
2. The Advisory Committee on Major Hazards, *First report*, HMSO (1976).
3. Balemans, A.W.M., and Others, 'Check-list. Guidelines for safe design of process plants', Paper in Symposium *Loss Prevention and Safety Promotion in the Process Industries*, Elsivier, Amsterdam (1974).

2.4 SITE LAYOUT AND PLANNING[1][2]

Contents

Site layout is considered under the following headings:
1. Administration and service buildings and car parks.
2. Roads and internal railways.
3. Production units.
4. Fire fighting.
5. Storage, loading and off-loading.
6. Noise.

The layout and planning of hazardous oil and chemical plant are dealt with in section 2.5.

2.4.1 Administration and service buildings and car parks

All buildings (including general stores, administration buildings and canteens) and car parks should be located at the periphery of the site to limit internal traffic. Buildings where concentrations of people occur should be well separated from hazardous units or areas.

Parking lots should be planned as far as possible on the outside of the fenced site area.

2.4.2 Roads and internal railways[3]

The entrances to the factory should allow vehicles to turn off and enter the public highways safely. The site road system should be designed for one way traffic. Weighbridges should be positioned so as not to cause obstructions.

The road and rail plan should minimise traffic in the operation (i.e. production) area. The roads should provide good multiple access to production buildings, process units and stores; these have roads between them. There should be adequate hard standing correctly placed for fire service appliances, especially turntable ladders. Roads and railways should not present an ignition risk to emissions from production units. Where the possibility of such emissions exists, means of rapidly closing the road should be provided. There should be adequate room for vehicles to manoeuvre, wait and turn safely near loading and unloading bays.

152

If internal railway lines have to be provided, they should be planned to avoid crossing a production area and to exclude frequent blocking of road traffic.

2.4.3 Production units

Vulnerable and potentially hazardous units and equipment should be specially located to localise the consequences of any incidents and disruptions.

Large installations should be built in an elongated form with roads on either side or else split into units divided by roads to allow access for fire fighting, maintenance vehicles and cranes. The layout should allow adequate room for maintenance and such operations as removing heat exchanger bundles without interfering with roads, passageways and entrances.

The flow of materials through the works between production units and within production units should be carefully studied to minimise the distances travelled and the amount of handling. Where fluids are handled in pipes they should be carried on pipe racks, preferably located on one side only of the production unit.

Cooling towers where necessary should be so located to minimise mist and ice hazards.

2.4.4 Fire fighting

Multiple access to the site for outside services from several directions is desirable. Mobile equipment kept on site should be stationed in a safe place but with easy access to anywhere on the site.

2.4.5 Storage, loading and off-loading

Loading and offloading facilities should be positioned to minimise disturbance to traffic. Storage of explosive, unstable, flammable and toxic materials should be carefully designed to give ample distances within the storage area as well as between it and other areas.

Maximum limits of hazardous materials in storage in any area should be set, and if larger quantities have to be stored, additional storage areas, well separated from the first, should be provided for them.

2.4.6 Noise

Noise needs to be considered in at least two contexts when planning the layout of an industrial site.

1. Buildings housing occupations which require a very low noise level which include most managerial and office work and skilled precision

Figure 2.4.1 An aerial view of the Beecham Pharmaceuticals antibiotics plant at Irvine, Scotland. This example of a well laid out pharmaceutical factory in which potential hazards are greatly reduced was engineered and constructed by Humphreys and Glasgow Ltd. It won the Irvine Development Corporation's Good Design Award. (Beecham Group Ltd.)

work, should be sited well away from noise sources, both inside and outside the site perimeter.

2. Operations which are unavoidably noisy and which cannot be adequately screened should be sited as far as possible from areas where there are high concentrations of personnel, and where possible separated from them by areas or buildings (vehicle parks, stores) where there are few people in attendance.

REFERENCES

1. Balemans, A.W.M., and others, 'Check-list. Guidelines for safe design of process plants', Paper in Symposium, *Loss Prevention and Safety Promotion in the Process Industries*, Elsivier, Amsterdam (1974).
2. National Safety Council, *Accident prevention manual for industrial operations*, 7th ed., National Safety Council, Chicago (1976).
3. Health and Safety Executive, *Booklet 44, Road transport in factories*, HMSO, London.

2.5 LAYOUT AND DESIGN OF BUILDINGS AND PLANT[1][2][3]

Contents

Many engineering disciplines are involved in the layout and design of buildings and plant. In all these safety plays a prominent part, and indeed the skill of the engineer lies largely in designing something which is safe without being impractical or uneconomic. There are also problems of safety at the interfaces between different disciplines. Sometimes these are not fully appreciated in the design stage, as was illustrated by the Ronan Point disaster (see section 2.1). This section can only give a selection of general points which are sometimes missed; engineers responsible for layout and design of particular types of buildings and plant should build up their own safety checklists.

2.5.1 Layout

2.5.1.1 Walking distances

These should be studied and reduced to a minimum consistent with good accessibility and other factors.

2.5.1.2 Hot parts

All parts normally hot (e.g. small unlagged steam and hot oil pipes) should have protection to avoid burns to employees. No hot exposed parts whose temperature is within 50°C of the ignition temperature of the flammable material should be permitted (see section 5.7) in hazardous areas (zones 0, 1, and 2) where flammable gases, vapours or dusts may be present.

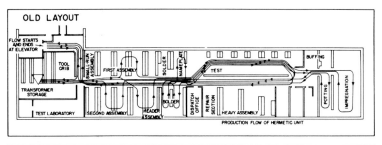

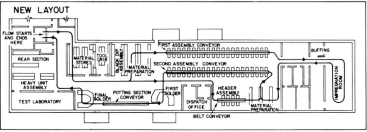

Figure 2.5.1 Minimising cross flow of hazards incidental to materials handling in transformer manufacturing operation (National Safety Council, U.S.A.)

2.5.1.3 Cross flow

Layout should be studied to minimise the cross flow of materials and/or pedestrians (*Figure 2.5.1*).

2.5.1.4 Adjacency of hazards

Operations involving ignition hazards (e.g. a welding bay) should not be located next to combustible materials (e.g. a carpenters shop). Similarly a toxic hazard should not be located next to a populated area.

2.5.1.5 Maintenance operations

The space requirement for maintenance operations must be taken into account, and checked that it does not obstruct fire exits or encroach on busy passageways.

2.5.1.6 Weather hazards

These hazards, e.g. ice and heavy rain, must be considered when planning access to plant built in the open.

2.5.2 Design

2.5.2.1 Technical safety

The technical safety of all equipment must be checked in the design stage, e.g. stresses in centrifuge bowls and shells of pressure vessels.

2.5.2.2 Water full load capacity

Since nearly all pressure vessels and columns will have to be hydraulically tested when full of water, the load bearing capacities of their foundations and supports must be adequate for the weight of the equipment when full of water.

2.5.2.3 Ergonomics and workers environment

The ergonomics of plant design and the maintenance of a satisfactory working environment must be considered when designing or purchasing plant (see chapter 3).

2.5.2.4 Noise

The noise level produced by the plant in all modes of operation, particularly bought plant, must be considered, and equipment specifications should include a maximum noise emission (see section 3.4).

2.5.2.5 Siting and separation

Separation between plant items must be adequate in all respects.

2.5.2.6 Passages and gangways

These must be adequate, taking into account the possible use of lift trucks and other wheeled equipment and their turning radii.

2.5.2.7 Access to work points

All work points must have safe means of access. Blind corners should be avoided but, if not, mirrors should be placed at suitable points (*Figure 2.5.2*) so that works traffic can be seen easily.

2.5.2.8 Booby traps

Headroom must be adequate everywhere, tripping hazards avoided, and temptations to take dangerous short cuts (e.g. under moving conveyors) must be eliminated.

2.5.2.9 Access to overhead work points

Fixed ladders or stairs should be provided to give access to overhead work points which require the presence of an operator.

2.5.2.10 Machine guarding

All moving machinery must be adequately guarded, and in specially hazardous situations, interlocking or similar guarding should be provided.

2.5.2.11 Electrical safety

The electrical safety of the installation must be thoroughly checked during the design stage by a qualified electrical engineer.

2.5.2.12 Protection of electrical cables and apparatus

All electrical cables and apparatus must be protected against mechanical damage, heat, water, chemical attack and unauthorised access (fences, locks, etc).

Figure 2.5.2 Mirror sited in factory yard (H. & S.W. Booklet 43)

2.5.2.13 Securing parts of overhead equipment

Locking washers and similar devices should be used on bolts, etc of overhead equipment (cranes, hoists) to ensure that parts cannot work loose and fall on people below.

2.5.2.14 Rescue of injured persons inside vessels, etc.

A suitable attachment point (beam, eye-bolt, etc) should be provided above the top manhole or other opening in vessels, hoppers, etc to enable a chain block to be used to winch out an injured man.

2.5.2.15 Isolation of process and storage vessels

Vessels, heat exchangers, etc must be provided with means of positive and complete isolation (e.g. blind flanges) for complete isolation of services during internal inspection and maintenance.

2.5.2.16 Electrical bonding

All plant containing flammable liquids must be electrically bonded and earthed. This includes steel framed buildings and reinforcing rods in concrete.

2.5.2.17 Product spillage

The design should ensure that product spillages are controlled by bunds, troughs, etc to avoid risks of slippery floors and corrosive splashes.

2.5.2.18 Misuse of plant and equipment

The designer should anticipate misuse of the plant and equipment and ensure by interlocks and similar devices that it remains safe even if misused.

2.5.3 Materials in process

2.5.3.1 Dust control

Hazards of inhalation and explosion of dusts must be eliminated by suitable methods of dust control.

2.5.3.2 Handling hazardous materials

Flammable, toxic and corrosive materials should, where possible, be handled mechanically and not carried by hand.

2.5.3.3 Monitoring flammable and toxic vapours

Where materials giving rise to flammable and toxic vapours are used regularly, the use of permanent monitoring instruments should be considered.

2.5.3.4 Overhead conveyor design

Care is needed in selecting and designing overhead conveyors to ensure that goods cannot jam, build up and fall off.

2.5.3.5 Wet conveyors

Troughs must be provided with wet conveyors to take spillages to appropriate drains.

2.5.4 Stores

2.5.4.1 Handling

Mechanical handling should be used where possible to avoid hazards of manual handling.

2.5.4.2 Bund drainage

Care must be taken to ensure that water can be drained from all bunds liable to contain flammable liquids.

2.5.4.3 Local stores and stocks

Local stores should be planned to ensure that they do not cause obstruction, that movement to and from them does not interfere with traffic and that access is adequate. If the stored material is flammable or combustible a limit should be placed on its inventory and care should be taken in its siting.

2.5.4.4 Product stores

Separate buildings should be provided for the storage of particularly flammable products and combustible stocks should be separated from sources of ignition.

2.5.5 Fire water damage

Stocks which could be damaged by fire or water should be stored on non-combustible pallets to prevent water absorption from the floor. Electrical motors and machinery liable to water damage should be mounted on plinths rather than at floor level. Racks and shelves should not extend to ceilings or be built directly on to walls. They should not retain water and be non-combustible if possible.

Vital parts of plant and machinery should be provided with waterproof covers for use in an emergency. Stacks should allow all-round access and be of such proportions that they can quickly be covered with waterproof sheets. Electrical switchgear should be mounted on battens and not directly on a wall so that it will not be affected by water running down. Valuable records should be stored in fire-resistant cabinets or strongrooms.

2.5.6 Engineering workshops

2.5.6.1 Layout

Adequate clear gangways must be provided in the layout, with adequate clearance around dangerous machines (e.g. automatic lathes and planing machines).

2.5.6.2 Welding

Welding should be carried out only in permanent welding bays, with adequate shielding against radiation, with good ventilation and proper fume extraction.

2.5.6.3 Gas cylinders

Racks for cylinders containing flammable, toxic or asphyxiating gases should be on the outside wall (preferably the north wall) of the building where the gas is used. The manifold and high pressure valves and reducer should be outside the building, with only low or medium pressure piping or tubing entering the building. The cylinder rack and valves should be protected against sun and rain.

2.5.6.4 Lifting gear

Adequate permanent lifting gear should be provided in the correct places for all foreseeable loads.

2.5.7 Works laboratories

2.5.7.1 Fume cupboards

Adequate fume cupboards fitted with shatter-proof glass, must be provided, and consideration given to the disposal of the fumes extracted.

2.5.7.2 Floors

Floors should be non-slip and consideration should be given to the effects of likely spillages on them when choosing floor materials.

2.5.7.3 Storage of flammable and toxic chemicals

Laboratory solvents should be stored in an external fire resistant locker, while working quantities of flammable liquids for daily use should be stored in a smaller fire resistant locker in the laboratory. Dangerous chemicals should be kept in a locked store.

2.5.7.4 Other safety precautions

Adequate mechanical ventilation should be provided where toxic hazards exist; where necessary, safety showers should be provided. Methods of detecting fire, flammable and specific toxic vapours should be considered.

2.5.8 Offices

2.5.8.1 Storage shelves

Proper access must be provided, and people should not have to reach down to floor level or up above eye level for books, files and other articles in frequent use.

2.5.8.2 Electric cables

Adequate socket-outlets should be provided adjacent to or as part of desks and tables on which electrically operated machines are used, to avoid tripping hazards.

2.5.8.3 Strong rooms and archives

These should be provided with a method whereby anyone who gets shut in can give an alarm which will be noticed by a responsible person (e.g. the security guard or the police) if the office is empty.

2.5.8.4 Office furniture and environment

This should be carefully selected with a view to avoidance of back troubles (see section 5.2).

Office environment needs careful consideration in order to benefit the office staff. This subject is fully discussed in chapter 3.

2.5.9 Canteens and kitchens

2.5.9.1 Kitchen equipment

Only safe and approved equipment should be used.

2.5.9.2 Steam and condensation

Special attention must be given to the removal of steam and products of combustion from kitchens by suitable extraction hoods, fans and ducting.

2.5.9.3 Environment

Environmental conditions, particularly in kitchens, must receive careful attention (see chapter 3).

2.5.9.4 Cold stores

Cold stores where persons could get shut in should be provided with a means of opening from inside and/or an alarm as considered in para. 2.5.8.3.

REFERENCES

1. Private communications.
2. Balemans, A. W. M., 'Checklist. Guidelines for safe design of process plants', Paper in Symposium *Loss Prevention and Safety Promotion in the Process Industries*, Elsevier, Amsterdam (1974).
3. National Safety Council, *Accident prevention manual for industrial operations*, 7th ed, National Safety Council, Chicago (1976).

2.6 BUILDING DESIGN AND FIRE PROTECTION

Contents

The Law[1] imposes a number of requirements on owners and occupiers of industrial premises on various aspects of fire protection, both of premises and persons employed. The agency at present responsible for inspecting such premises and checking that the legal requirements are met is the county fire authority which is represented by the local fire brigade.

The main Acts and Regulations which affect the design of industrial premises as regards fire protection are:

The Thermal Insulation (Industrial Buildings) Act, 1957, Section 3. This empowers the local fire authority to ban the use of thermal insulation materials which do not conform to recognised standards of resistance against the spread of flame.

The Factories Act, 1961 (Sections 40 to 52 inclusive). These sections cover in considerable detail the means of escape in case of fire. All occupiers of industrial premises must apply for and obtain from the local fire authority a certificate which shows that the premises have proper means of escape, complying with various requirements which are covered in the text that follows. The Act also requires apparatus for giving an alarm in case of fire and means of calling the fire brigade and also that employees be instructed in their use, and that they be tested at least once every three months. It also requires fire fighting equipment to be kept on the premises and for employees to be instructed in its use.

Section 148 of the same Act empowers officers of the local fire authority to enter industrial premises to inspect fire risks and the precautions taken to control them.

The Fire Precautions Act, 1971, covers all these points in greater detail.

The Highly Flammable Liquids and Liquefied Petroleum Gases Regulations, 1972 regulate the design of installations where such materials are stored and used, as well as the means of fire fighting in such installations.

The Health and Safety at Work, etc. Act, 1974, Section 78. This section amends the Fire Precautions Act, 1971. The main effect of these amendments is to give wider scope to the fire authority to designate premises which require a fire certificate concerning their means of escape in case of fire.

The risks of fire spreading are to a great extent determined by the care and consideration given when the factory or building was planned and designed. If a room in which a fire has started has fire-resistant walls, floor and ceiling, and if all openings in the room, both vertical and horizontal can be closed with tight fitting fire resisting doors and hatches, it should be possible quickly to confine and even extinguish the fire by closing the doors and hatches, since this shuts off the oxygen supply.

The intensity of a fire in a room or building depends on its size and what is in it. Rooms and buildings with an *undivided capacity* of more than 1800 m³ are dangerous, as this is the largest size in which a fire can be dealt with by conventional fire appliances. Where such rooms are necessary their contents should be protected by automatic sprinklers, particularly if they contain flammable materials.

Special care is needed when buildings contain materials which may spread fire by flowing, such as flammable liquids and rubber, waxes and tars which melt easily. Stores where combustible materials such as paper reels are stacked high present a particular problem as normal sprinklers cannot cope with such high fire loads.

Heat radiated from intense fires can ignite combustible materials at over 30 m, and burning brands often spread fires over even greater distances.

Figure 2.6.1(a)   Spread of fire in building level. This shows the destruction of a hairdressing saloon caused by the fire.

Figure 2.6.1 (b) and (c)   These show how easily the fire was able to spread along the roof space. (Photographs by courtesy of the Fire Research Station)

In single storey buildings, heat from fires spreads most readily at roof level and combustible roofs and roof linings can spread fire rapidly. Unprotected columns, trusses and other roof supports are exposed to heat at an early stage of a fire and may collapse quickly. Heat also spreads through undivided areas, open doorways and other openings. In multi-storey buildings, hot gases from fires may rise through openings in floors for staircases, lifts and ducting and spread fire to other floors. Combustible floors are consumed and floors and walls which are not fire resisting may collapse, thus spreading the fire and making fire fighting more difficult *(see Figure 2.6.1)*.

The most positive and effective way of limiting the spread of fire within a building is by subdividing it into the smallest possible compartments of fire resisting construction, each of which would contain any fire likely to break out in it and at the same time ensure the stability of the building if a fire did break out.

As well as fire damage, the damage that may be caused by water from fire hoses and sprinklers must also be considered in building design, since this may be even more costly than that caused by the fire itself. Basically the building must be designed to collect and drain freely the maximum likely water load from hoses and sprinklers, and the floors must be designed to carry the weight of their contents when completely water laden, not merely in their normal dry state.

2.6.1 General requirements of building design in fire prevention

In order to withstand fire and prevent its spread, buildings should comply with the following requirements.

1. All load bearing elements such as walls, floors, columns, beams and partitions forming protected enclosures (including staircase, lift and hoist enclosures) should be of fire-resisting construction.
2. Structural steelwork should be protected against the effects of fire and heat by being encased in concrete, brickwork or other thermal insulating material.
3. Subdividing walls should rise to the full height of the building and should project above the roof if the roof is not of fire resisting construction.

The degree of fire resistance required for the structure and fire partitions of a building depends on its function. Stores and buildings containing large quantities of combustible materials require a fire resistance of four hours whereas one hour is sufficient for most office premises.

The value of fire resisting walls is lost if they contain openings which are not effectively closed when a fire occurs.

2.6.2 Walls and openings

In general, the external walls of buildings should be non-combustible throughout and have a fire resistance of at least two hours. For single storey buildings, more than 10 m from the site boundary, the external walls need have a fire resistance of only one hour. The risk of fire spreading is increased by the presence of windows and other openings in outside walls.

To prevent the spread of fire through the openings of fire resisting walls, the following are necessary:

1. Doorways and other openings in fire-separating walls should be fitted with fire-resisting doors or shutters which are no less fire resistant than the walls themselves.
2. Chutes, hoists and lift shafts passing through floors should be totally enclosed in fire-resisting walls which have at least half the fire resistance of the floor. All openings into these enclosures should be fitted with doors or shutters whose fire resistance is equal to that of the enclosure.
3. If fire-resisting doors or shutters are normally liable to be left open, it is advisable to fit types which close automatically in the event of fire.
4. Where ducts and pipes penetrate fire-resisting walls and floors, the openings round them should be sealed with non-combustible material. Large ducts should have internal fire traps at these points.

The use of combustible material in finishing a building should be limited as far as possible. Wall, ceiling and roof linings should be of material which is preferably non-combustible and at least fireproofed. Air spaces between linings and walls, ceilings or roofs should be fire-stopped at intervals.

2.6.3 Roofs

A roof may spread fire in a number of ways:

1. By transmitting fire internally at roof level from one part of the building to another and by dropping burning particles ahead of the fire onto the floor;
2. By radiating heat from the underside, thereby assisting the ignition of materials ahead of the fire;
3. By allowing fire to penetrate it from below and ignite combustible materials on top of it;
4. By igniting other buildings or materials as a result of flying brands or radiated fire.

Roofs of buildings should be constructed in such a manner that they will not contribute to the spread of fire. This is particularly important where the roof surface cannot be broken at intervals by fire resisting walls rising above roof level. Roofs should incorporate the following features:

1. The roof covering should be of non-combustible material;
2. The supporting columns and beams should be encased in concrete or other heat-resisting material to protect them against collapse. This is less essential for roof trusses which only cause localised damage if they collapse;
3. The insulation material or linings should preferably be non-combustible or should at least be fireproofed;
4. All voids or openings in a roof should be broken by fire stops.

2.6.4 Floors

Floors must be designed to minimise the upward spread of fire from rooms below, i.e. they should be solid and free from openings, and they must also be designed with a view to preventing water damage. The main points needed to meet the second requirement are:

1. Floors above valuable equipment and goods should be watertight.
2. Floor levels should be designed and sloped to direct the flow of water to the drains and out of the building.
3. Adequate drains should be provided to take the maximum flow of water from hoses, even when the drains are partially blocked by floating debris washed down from the fire by hoses. Drains must be clearly marked and provided with manholes and scuppers, and should be located in areas of the floor designated as aisles or passageways, not in storage or working areas.
4. Where drains cannot be provided, sumps with automatic pumps to collect and remove water are recommended.
5. Sills or ramps should be provided at doors and other openings.
6. Drainage channels should be provided across large floor areas to prevent the lateral spread of water.
7. Floors should be designed to carry the goods and equipment for which they are intended when these are water laden.
8. Concrete floors on which lubricating or other oil may be spilled should be finished with a screed which is non-oil-absorbent.

2.6.5 Basements

Basement rooms require special care. Two alternate escape routes at opposite ends should be provided. Openings in the floor above them for access stairs and ventilation must be arranged to prevent cigarette ends, sparks or hot metal falling into them.

Special care must be taken over drainage to ensure that basement rooms cannot be flooded with fire water. Means of releasing smoke and heat from a basement fire must be provided in the design to enable firemen to reach the seat of a fire. Foam inlets to hydrants and associated pipework to enable firemen to deal with basement fires should also be considered.

2.6.6 Electrical installations

Faulty electrical installations are a frequent cause of fire, while failure of emergency electrical equipment such as sump drainage pumps as a result of a fire can add to the losses. Electrical cables and equipment should be well shielded and protected against vibration, mechanical damage and heat exposure. Non-combustible insulation should be employed where heat exposure is unavoidable or would have serious consequences as a result of a fire.

The electrical installation should be designed to facilitate maintenance and overhaul.

2.6.7 Adapting buildings for other purposes

When a building is used for a purpose different from that for which it was built, a careful examination should be made to ensure that it has adequate fire resistance for its new function. A check must also be kept on all structural alterations and changes in services to ensure that the fire resistance of walls and floors is not destroyed.

Holes left by the removal of pipes and ducts should be filled in with non-combustible material to the full thickness of the wall or floor. Doorways and other openings no longer required should be bricked up.

2.6.8 Access for fire appliances

Entrances to all areas and large buildings in a works should be large enough to allow a fire engine to pass through. This generally requires a minimum width of 3.1 m and a minimum height of 3.7 m.

Distances between buildings should not be less than 3.7 m in order to limit the spread of fire between buildings and to allow fire engines to pass between them. These spaces should be kept clear, and no equipment or materials should be stored in them. Passages between buildings are best not covered.

The design of a building should not make it difficult for firemen to pitch ladders and escapes against it.

2.6.9 Control of smoke and hot gases

In large single storey buildings, means should be provided for controlling the spread of smoke and hot gases and for enabling them to be released to the atmosphere as an aid to fire fighting. This may be done by providing special fire vents in the roof and non-combustible curtains extending from the roof ridges down to the level of the eaves and thus subdividing the roof space.

Light glazing may be provided in the roof at the head of lift shafts and staircase enclosures which can be opened or broken to assist the release of smoke and heat. Smoke extractor fans should be provided in basements.

2.6.10 Escape from buildings

Adequate means of escape from fire must be provided for all personnel in a works, factory or office building, and the following four requirements should be satisfied:

1. No one should need to go towards the fire in order to escape.
2. All escape routes should be as short as possible, of adequate capacity, and should lead to open air at ground level, either direct or by way of a fire resisting enclosure.
3. Protected parts of escape routes should not be exposed at any point to penetration by smoke or fire.
4. Escape routes should be clearly marked with arrows at every corner and intersection.

These conditions may be met by making the following provisions:

1. At least two escape routes leading in opposite directions should be available to everyone.
2. Escape routes should be planned so that no one has to go more than 30 m to reach open air or a smoke-free fire-resisting stairway, corridor or lobby. Single-storey buildings such as steel mill buildings may be exceptions to this when the amount of flammable material present is small. Lifts should not be considered as escape routes because they may not be operable in the event of a fire.
3. Escape corridors, stairways and exits should be wide enough and numerous enough for everyone in the building to leave quickly and without confusion or panic. The minimum width for an escape exit should be 750 mm and sixty normal people can be reckoned to pass through it in one minute. The aim in designing an escape route is to permit everyone using it to pass through and reach a safe place in less than four minutes. It must be borne in mind that some of the escape routes may not be usable because they have been penetrated by smoke or fire, and it should therefore be assumed that only half the escape routes may be usable. The width of escape routes is calculated from these considerations.
4. Stairways, corridors and lobbies forming part of escape routes should be enclosed by walls, floors and ceilings capable of withstanding a fire for 30 minutes, with self-closing fire resisting doors at every entrance to the enclosure. Two sets of self-closing doors are required at a lobby, one leading into it and the other leading from it to the staircase or corridor. It may also be necessary to provide doors to stop smoke at intermediate points in long enclosed corridors.
5. Exits from escape routes should be sited so that people leaving the building can move well clear of it without exposure to danger. For example, escape routes should not lead into confined spaces where congestion and panic may arise or where people may be trapped and overcome by heat radiated from the fire.
6. Escape stairways and corridors should be well ventilated to open air

so that smoke entering them is cleared at once, or they should be entered through well ventilated lobbies provided for this purpose.

7. External escape stairs should only be provided as a means of escape from the upper floors of a building if it is not possible to provide enclosed stairways. All windows adjacent to external staircases should be fitted with wired glass and should be fixed shut.

8. Where space is restricted, ladders and spiral stairs may be provided as external means of escape for limited numbers of fit and able bodied people.

9. Stairways which are not part of an escape route, lift shafts and hoists should, where practicable, also be enclosed in walls which have a fire resistance of at least half an hour and have all openings protected by fire resisting doors or shutters, to prevent fire and smoke from spreading through. Other openings in floors should where possible, be protected by fire resisting shutters or filled in.

Every fire escape exit, other than normal exits, must be clearly marked as such. Stairs and corridors which do not lead directly to a place of safety should be marked 'No Thoroughfare'. If confusion is likely to be caused, some additional indication should be given about the nature of the premises to which the staircase or corridors give access.

The colour of fire escape doors which is usually red frequently causes them to be confused with doors (also painted red) which must on no account be opened in the event of fire. Some large firms paint their fire escape doors green, and the ones to be kept permanently closed, red. This is to be recommended, provided everyone is trained to recognise the difference.

Doors intended as fire exits or which may be used as such must be designed so that they can always be easily and quickly opened from the inside. Any doors opening on to any staircase or corridor from any room liable to contain more than ten people (with the exception of sliding doors) should be constructed to open outwards.

2.6.11 Stores and other buildings

Bulk quantities of goods and raw materials are best stored either in the open or in a separate building set apart for the purpose. If the goods are combustible, the building should be protected by an automatic sprinkler system. When flammable materials are stored in a building, it should be subdivided by fire resistant walls into compartments each with a maximum capacity of 8000 m³. Service stores which are supplied from the bulk stores may be sited in the same building as the manufacturing process, but separated from the rest of the building by a fire resistant wall when the goods are flammable *(Figure 2.6.2)*.

A special storeroom should be built for flammable liquids in cans or drums. This should be at least 6 m from any other building and from the nearest road or boundary or from stacks of flammable materials. The building should carry notices showing that it contains flammable materials

Figure 2.6.2   Store for containers of flammable gases (ICI Ltd)

and forbidding smoking or the use of naked lights in it. All electrical fittings in it should be flameproof.

Stores for liquified and compressed gas cylinders should be well ventilated, free from direct sunlight, and well removed from every possible source of ignition (see section 6.4).

Offices, computer rooms, workshops, laboratories, changing rooms, canteens and stores should generally not be attached to factory buildings, particularly where the factory building has a significant fire hazard.

Any building which is lightly manned and which contains valuable equipment such as computers or large quantities of combustible stores should be provided with an automatic fire detection and alarm system. It is usually justified to integrate this with an automatic water sprinkler or carbon dioxide or halon smothering system.

2.6.12 Hydrants, appliances, detectors, sprinklers and alarms

Fire protection equipment, both fixed and mobile, is discussed in chapter 4. Provision must, however, be made for the installation of fixed equipment when the building is designed, and thought must be given to the location of hose reels and other fixed and portable appliances, and to ensure that sufficient space has been allocated to them.

REFERENCES

1. Fife, I., and Machin, E.A., *Redgrave's Health and Safety in Factories*, Butterworths, London (1976).
2. Fire Protection Association, *Planning Fire Safety in Industry,* and other publications.

2.7 SOME CAUSES OF BUILDING FAILURE[1]

Contents

(This section is based on 'Building Disasters and Failures' by the late Geoff Scott. See also 'Catastrophic Failures' by Bignell, Peters and Pym[3]).

A steadily increasing number of buildings using modern methods of construction have been failing over the last twenty-five years. In this section the direct causes are first discussed, with examples of failures resulting from them, followed by a short discussion of some of the underlying causes.

2.7.1 Direct causes and examples

Cause 1. The use of high alumina cement without proper control of the water: cement ratio and the temperature of curing. High alumina cement needs far more critical control, both in mixing and curing, than conventional Portland cement. This material was actually banned in France as far back as 1943, but continued to be used here in spite of a paper given in 1963 by Professor Neville to the Institution of Structural Engineers which highlighted the risks.
Examples. The most spectacular failure in Britain was the collapse of the assembly hall roof of Camden School for Girls on the evening of June 13th, 1973, twenty-four hours after a meeting in the hall which 500 parents had attended. Over a thousand school buildings built over the previous twenty years were found to be suspect a year later when several other failures had occurred.

Cause 2. The extensive use of calcium chloride in in-situ *concrete works.* Calcium chloride is added to concrete mixes for frost protection and to facilitate stripping of shuttering. However, it causes accelerated corrosion of pre-stressed wires within the concrete and its effects are comparable to a time bomb.
Example. One of the early failures was that of a pre-stressed roof beam in a Cadbury Schweppes factory in North London in August 1974. Since then the roofs of Felton Lower School, the floor of Glebe Primary School at Ickenham, and a 15-floor block of council flats in Salford, were found to be in danger of collapse, due mainly to corrosion of stressed steel wires in the concrete. Many other recent buildings are suspect, but it is almost impossible to detect the corrosion without cutting away the concrete to expose the reinforcement.

176

Cause 3. Box girder design of bridges causing failure during construction. This form of bridge construction came into regular use in the 1950s and 1960s without adequate basic design criteria and based on BS 153, a Code of Practice for simply supported steel girder bridges. This code according to the Milford Haven Bridge Disaster Inquiry Report, was not wholly applicable to box-girder structures nor sufficiently comprehensive.

Examples. The Milford Haven Bridge collapsed during construction with the loss of five lives. A bridge of similar design at the Yarra river, Australia collapsed while being built prior to this with the loss of thirty-five lives. A similar bridge under construction over the Lieser river in Austria collapsed in May, 1975 with the loss of ten lives. Two box girder bridges under construction in New Zealand also collapsed.

Cause 4. Failure of plywood box beams in roofs, particularly of swimming pools. The main problem appears to be periodic wetting of the beams by condensation which causes decay in the wood and progressive creep deflection of the beams. Decay is hastened by volatile compounds of chlorine and ammonia which are formed in the swimming pool by reaction between the chlorine added for disinfection and urine from some of the pool users. The urea formaldehyde adhesive used in the beams decays under such conditions. The construction is inherently lightweight and is only intended to support the roof covering.

Example. A third of the roof of the swimming pool at Ilford County High School for boys collapsed on the night of 2nd October 1974.

Cause 5. Use of wood wool slabs as formwork for in-situ *concrete trough flooring.* The problem here is that defects within the floors due to air pockets causing inadequate bonding between the reinforcement bars and the concrete cannot be detected without opening up the structure.

Examples. The Youth Employment Office in Belfast, a city centre office block in Newcastle-upon-Tyne, and the 11-storey Inland Revenue building at Malden, Surrey, were all found suffering from this defect, more or less accidentally. Expensive repairs became necessary.

Cause 6. Foundation settlement. This is sometimes found when constructing buildings over old mine workings or tunnels.

Examples. The language laboratory and social science rooms of Kent University subsided 18 in into the 150-year old tunnel of the Canterbury-Whitstable railway in mid July 1974. Conversely, some damage was caused to an 18th century building, the Royal Overseas League on the edge of the Green Park, by recent tunnel construction for the Fleet Line underground.

Cause 7. Wind and inadequate knowledge of its effects on buildings.
Examples. Collapse of three Ferry Bridge cooling towers on 1st November 1965 and a high cooling tower at ICI's Ardeer Nylon Works on 27th September 1973. Several other high cooling towers and chimneys could be suspect.

Figure 2.7.1 Collapse of swimming pool roof (From 'Building Disasters and Failures' published by the Construction Press Ltd.)

Cause 8. Excess floor loading in offices caused by heavy modern equipment (computers, filing cabinets, photocopying machines and safes).
Example. Defects found in the floor of the Inland Revenue building at Malden were probably partly due to this cause. Many office floors are probably overloaded.

Cause 9. Inadequate bearings for beam seatings resulting mainly from design error and inadequate inspection during construction.
Example. This was a partial cause of the roof failure at Camden School in 1973 (see Cause 1). Failure of a concrete staircase under construction at high rise flats at Mogden, Middlesex was also partly due to this cause.

Cause 10. Excessive roof voids and vertical wall cavities in buildings leading to rapid fire spread.
Examples. Fairfields old people's home, Notts, in December 1974; Usworth comprehensive school, Co. Durham, in October 1974. There have been several others. Methods of construction were amended in 1971 to incorporate fire checks within the roof voids, but about 1000 buildings escaped the net.

Cause 11. Use of flammable material for building cladding.
Example. Summerland Leisure Centre fire, Isle of Man, in August 1973, when fifty lives were lost. Voids in the wall through which fire spread rapidly were a contributing factor.

Cause 12. Gas explosions. The danger of these has increased since changing to natural gas because of its higher pressure and higher calorific value.
Examples. Explosions in Bristol, Brentford and Bradford, December 1976 and January 1977.

2.7.2 Some contributory and underlying causes

Most of the failures and disasters listed above occurred through the use of new or modified systems and techniques which had been introduced on a large scale with little experience of their behaviour or possible hazards in use.

Whilst there generally appears to have been an element of compulsion in the decisions to press ahead rapidly with such developments, it would appear that technical monitoring and professional criticism at the conception stage has been inadequate.[2]

In addition, the training of professional architects and engineers has become more divorced from the practical experience of building, while traditional craftsmen in the industry are a dying breed. The size and complexity of modern buildings have increased. Bye-laws and codes of practice are often incomprehensible, even to architects and engineers, let alone the everyday builder or site operator. In these circumstances, the possibility of building failure is more likely, unless stringent precautions are taken and strictly adhered to.

REFERENCES

1. Scott, G., *Building Disasters and Failures*. Construction Press, Lancaster (1976).
2. Council for Science and Society, *Superstar technologies*, The Council for Science and Society, London (1976).
3. Bignell, V., Peters, G. and Pym, C., *Catastrophic Failures*, The Open University Press (1977).

2.8 OIL AND CHEMICAL PLANT

Contents

The hazards of some oil and chemical plants are at least on a par with those of explosives factories. The latter are under tight government control and need not be discussed further. In the UK all plans for oil refineries and potentially hazardous chemical plants are referred by the local authority to the area inspector of the Health and Safety Executive who in turn will usually refer them for review by the Risk Appraisal Group of the Executive before approval is granted.

The hazards fall largely into two types:

1. Toxic and radiochemical hazards
2. Fire and explosion hazards.

The principal guidelines in the hands of the designer of plants with toxic hazards is Technical Data Note[1] 'Threshold Limit Values' published by the Health and Safety Executive, based largely on the list drawn up by the American Conference of Government Industrial Hygienists. Toxic hazards are discussed in sections 3.2, 3.9 and 6.2 and radio-chemical hazards in sections 6.9 and 6.10.

Several guides to layout and design, particularly for fire and explosion hazards in Chemical plants have been published in the UK, e.g. by the Chemical Industries Association,[2] the Institute of Petroleum,[3] the Institution of Chemical Engineers[4] and the British Standards Institution.[5] The recommendations of many of these guides are in the main qualitative, and leave a great deal to the judgement of the designer.

In the USA the home of the oil and petrochemical industries which has had these problems for longer and on a larger scale than the UK, several more detailed and quantitative guides are available, notably those published by the Oil Insurance Association,[6]* Dow Chemical,[7] Factory Mutual Engineering Corporation,[8] and the National Fire Protection Association.[9] Those faced with difficult design decisions regarding plant safety which cannot be resolved easily by British published guides would do well to check whether the point has been covered in any of these American guides.

---

*The 'Oil Insurance Association' has now been incorporated with the 'Factory Insurance Association' in 'Industrial Risk Insurers'.

2.8.1 Oil and petrochemical plant

One of the most useful guides is the Oil Insurance Association's *Plant construction recommendations*[6] which applies to oil refining and petrochemical plants. Although the standards laid down are in many ways stricter than those commonly found in the UK, major incidents such as the Flixborough disaster are compelling our insurers and designers to move towards similar standards in the UK.

The following points are summarised from *Plant construction recommendations* published by the Oil Insurance Association.

Plant layout

1. As a minimum requirement, there should be at least 30 m of clear and open space between the battery limits of various processing units.
2. Pipe racks of process units may run down the middle of those units, thus splitting them into two areas. Main pipe racks into which these unit pipe racks feed should be separated by 7.5 m from all operating equipment.
3. No storage tanks for flammable fluids should be allowed within the battery limits of any process unit. The number and size of vessels containing flammable fluids within process units should be reduced to an absolute minimum, and they should be as close to the ground as possible. If they have to be elevated they should be in as open a position as possible and near to the battery limits.
4. Spacing between equipment within battery limits should be adequate for all maintenance needs. Furnaces in any units should be at least 15 m from any processing equipment and individual furnaces should be at least 7.5 m apart. Compressors should be at least 7.5 m from exposed equipment. Oil pumps should not be located directly below overhead pipe racks or any equipment.
5. Each furnace should ideally have its individual stack. Common breachwork whereby the flue gases from several furnaces pass into a common stack should be avoided, to prevent a fire on one furnace spreading to the others. In practice this is often impossible to achieve as a single very tall stack may be necessary to disperse flue gases with high sulphur dioxide contents. In this case the possibility of subdividing the flue into separate channels for each furnace should be considered.
6. Drainage must be adequate to handle fire water as well as rain and cooling water. All sewers must be properly trapped and vented, and all spilled oils must be separated in an open area.

Buildings and enclosures

1. Control rooms should be located in relatively unexposed conditions and should be at least 30 m from processing equipment. They should

Figure 2.8.1   Blastproof control room and safety store (ICI Ltd.)

be designed to be blast proof to resist a minimum of 0.2 bar (3 psi) over-pressure and should have an absolute minimum of windows — preferably no windows. They should be pressurised, with at least ten air changes per hour and flammable gas detectors and alarms fitted in the air intakes, with automatic shut off of the air supply if the flammable gas concentration in it reaches a predetermined level. A typical blastproof control room is shown in *Figure 2.8.1.*

2. Pumps and compressors handling flammable fluids should be located in the open. The only concession is a possible weatherproof shield, but this should be as small as possible.
3. Administration buildings, workshops and stores should be located well away from process equipment handling flammable fluids.
4. Plant buildings should be of non-combustible materials and no combustibles should be used in the interior construction for partition walls or insulation.

Fireproofing

1. Structural steel, vessels, columns and skirts should be properly fireproofed up to a height of 9 m or higher depending on loading.
2. Both vertical and horizontal supports of major piperacks in the heart of processing areas should be properly fireproofed for 3-hour ratings. The fireproofing materials should withstand direct fire hose streams.
3. Supports for fan coolers located over piperacks should be fireproofed for at least 1-hour rating when water spray is used and for 2-hour rating when no water spray is used.

Fire protection equipment

1. Refineries with capacities of 100 000 Bbl per day and more should have a minimum of 10 000 g.p.m. fire water pumping capacity at a pressure of at least 8.5 bar (125 psi), with at least half the capacity being driven by diesel engine with independent fuel tanks, the rest being either electric or steam driven.
2. Fire water pumps should be well away from process areas and be well housed in contrast to process pumps. The waterline from the fire pumps to the hydrant grid mains should be duplicated.
3. There should be a minimum of 4-hours water supply at full fire pump capacity.
4. There should be at least two hydrants within 75 m of all buildings, tanks and structures and an adequate number of monitor nozzles in and around all processing units. Foam nozzles must also be provided.
5. Water spray protection is required above hot oil pumps, above large elevated vessels and below fin-fan coolers. Liquefied gas pressure storage vessels should be adequately protected by water sprays to prevent weakening of the shell and failure by fire exposure at normal relief valve pressures. The water spray systems and lines supplying them should be protected as far as possible from damage by explosion. If the water spray system is manually controlled there should be two or more alternate and well protected control stations so that one at least is always accessible in a fire.
6. Instrument control lines should be protected against fire and flying objects, and should be enclosed in fire resistant trays.

Process equipment

1. All process equipment must be properly relieved and must be capable of being blown down, following API 520 and 521. Non-toxic but flammable vapours may in extreme emergency be vented to atmosphere at the highest possible point (if the quantity is in excess of what the flare system will handle), but liquids should not be vented near process equipment.
2. Remotely-operated emergency control valves for isolating, dumping and venting flammable fluids are needed in the base of all large vessels containing flammable fluids. It may be necessary to provide a second control station from which they can be operated in case the control room is involved in an explosion or fire.
3. Furnaces and boilers must have flame failure devices.
4. Process equipment should be designed with a view to ease of metal inspection as well as maintenance, and may incorporate elements for condition monitoring and non-destructive testing (see section 2.10).
5. The best grades of steel and other metals appropriate for the duties should be used. Cast iron and threaded fittings should be avoided. No brass or other low melting alloys should be used for valves in plants processing flammable fluids. Small threaded nipples should be back welded. Drain openings should be plugged.

Tanks

1. All tanks for flammable liquids should be at the very least 60 m from process units, generally more.
2. Hot hydrocarbon tanks should be avoided if at all possible — but where this cannot be done, their water contents and heating arrangements need to be carefully designed as well as supervised (*vide* Pernis explosion 1968).
3. LPG tanks should be located as far from processing areas as possible and should be protected with copious water sprays.
4. Crude oil tanks should be located in individual bunds capable of holding total tank content. Common bunds for several tanks should be kept to a minimum.
5. Floating roof tanks are advisable for all flammable liquids with flash points up to 50. Fixed roof tanks should only be used for low flash oils, e.g. residual fuel oils. Means of applying foam to fixed roof tanks should be provided.
6. Oil tank farms should never be installed at levels above those of process units.

2.8.2 Process and chemical plant

The Dow Fire and Explosion Index[7] has been referred to earlier (see section 2.1) in connection with chemical plant. One main problem with such plant is to know how hazardous the plant is and how far special precautions such as extra spacing between units, blast proof control rooms, fire proofing of structures and special safety instrumentation are required.

The cost of the special features needed for highly hazardous plant is very considerable, quite apart from the number of skilled man hours needed to install them correctly. The indiscriminate installation of such features whether they are needed or not can lead to scepticism about the whole safety programme, with the result that on the next occasion they may be omitted on a plant where they really are needed, and people get hurt or killed in consequence.

The Dow Fire and Explosion Index goes some way towards filling this need. It serves only as a guide however. Every case must be studied on its own merits.

The following discussion is based on the fourth edition of Dow's guide,[7] which was published in 1976. Several changes have been made in this edition, particularly in the numerical values of the index for different degrees of hazard. Care must therefore be taken when using the guide to ensure that the same edition is used and referred to throughout.

The fire and explosion index can be calculated objectively for any separate unit of a chemical and process plant from four factors:

1. Materials used in process;
2. Special material hazards (e.g. oxidizing materials, materials subject to explosive decomposition),

3. General process hazards (whether reactions are involved or not, and whether a batch or continuous process is employed),
4. Special process hazards (temperature, pressure, ease or difficulty of control, and quantity of combustible materials present, etc.)

These factors should be calculated during the process design of the plant before detailed engineering design starts. The index is calculated by combining the factors. It is a number which normally lies in the range of 1 to 134. Hazard indices higher than 134 are theoretically possible. Such processes are usually so hazardous that every effort should be made to reduce the hazard index by modifications to the process design or plant layout before proceeding with the detailed design.

Table 2.8.1 DOW FIRE AND EXPLOSION HAZARD INDEX

Degree of hazard	Index range
Light	0-50
Moderate	51-81
Intermediate	52-107
Heavy	108-133
Severe	134 and over

In the fourth edition of the guide the index is used not only to assess the preventative and protective features needed on the plant, but also to estimate the maximum probable property damage (MPPD) which could be caused by fires and explosions. A base 'MPPD factor' which is applied to the replacement cost of the unit to give the base MPPD is first obtained from a graphical correlation of the Dow Index and the material factor for the unit. The addition of protective features then enables some rebate to be made to the base MPPD to give a corrected MPPD. The cost of the preventative and protective features can then be compared with the reduced losses likely to occur, and in this way an attempt can be made to strike an economic balance between the probable cost of fires and explosions and the cost of protection.

The Dow guide gives two lists of preventative and protective features, a list of basic features which should always be applied regardless of the fire and explosion hazard, and a list of recommended features to be applied depending on the nature and degree of the hazard.

Another new feature introduced into the fourth edition of the Dow guide is a Toxicity Index which is intended to quantify the hazard of a release of toxic material from the plant or unit to the atmosphere which could arise from an upset in the process.

The Fire and Explosion Index of the plant or process has a wide bearing on many other things such as the detail with which operational procedures are drawn up, the selection and training of operating and maintenance personnel, the need or otherwise for an independent inspection organisation with responsibility for safety monitoring, and its degree of authority.

There are frequent and rather pointless arguments about these questions as applying to the oil and chemical industries in general, when in fact the answer really depends on the hazardousness of the plant or process in question.

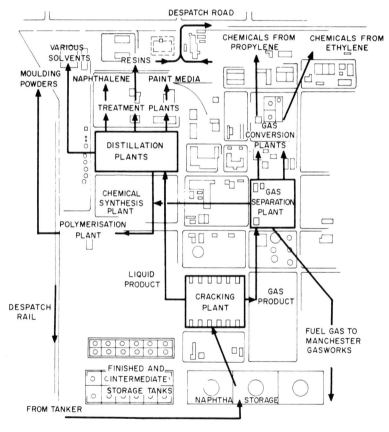

Figure 2.8.2   *Diagram of flow of products through petrochemical site superimposed on site layout (by courtesy of H. W. Charlton, Manchester Association of Engineers) (From 'Plant Layout' by J. C. Mecklenburgh published by International Textbook Co. Ltd.)*

2.8.3 General oil and chemical plant design and layout

The references given at the end of this section deal in more detail with these questions which are rather peripheral to the subject of this book. Oil and chemical plants are mostly 'tailor made', that is designed to give particular production rates of certain products, each with clearly defined properties and specifications, from equally well defined raw materials.

The design starts with a flow scheme and material balance which gives the flow quantities, compositions, temperatures and pressures of the materials being processed as they pass through pipes, pumps, vessels, reactors, columns etc. From this the duties, sizes, materials of construction, metal thicknesses etc. of the individual equipment items are calculated, and drawings and specifications made. The instruments and control systems are specified and having decided on the sizes of each different piece of equipment, trial general arrangement drawings are made showing everything in plan and elevation. Often a scale model is made at this point to assist in reaching the best layout.

Such tailor-designed plants, when designed by an experienced and competent team can, in the final stages of design, be subjected to a

Figure 2.8.3   The system for the protection of this large sphere in fire includes special provision for the vulnerable leg to shell joints. Concrete has been applied over the metal legs to protect them (H. & S. W. Booklet 30)

Figure 2.8.4   Sprinkler system for LPG road tank carloading station (H. & S. W. Booklet 30)

qualitative hazard analysis or operability study which aims at discovering all possible types of malfunction and examining their consequences. Many of these arise from the failure of some specified valve, instrument or piece of equipment. Some of the consequences may be serious and others trivial. All serious consequences (which could result in an accident with property loss or even a disaster) are then listed and a fault tree built up to show the various ways in which the serious accident could arise, and due to which initial malfunction (*see* Section 2.9).

The probabilities of each of these initial malfunctions are then estimated from the best data available, and a quantitative hazard analysis is made to show the probability of the accident occurring. If this is too high, the design must be changed or modified to reduce the probability to an acceptable risk.

Such exercises are generally only followed through for complex costly and hazardous plant. They are discussed in more detail in section 2.9. There is no doubt that they can do a great deal to reduce hazards and improve the reliability of the plant. Operating procedures for such a plant would be clearly defined and laid down in detail.

Some plants, however, may start as 'general purpose plants' where the operating company is unclear of the process requirements, and so asks for so many columns, pumps, vessels, etc to be erected in certain places, with the idea of working out later and in practice how they will be operated and how they will be used to make the required products. They tend to start out as batch plants and to be modified to continuous ones later (growing like Topsy all the time). They are, in effect, large scale laboratory experiments. Such plants are inherently more hazardous than ones which have been well laid out and tailor made for any given Hazard Index. One reason that they are more dangerous is because it is difficult to know precisely what the plants contain at any given time and what their limitations are under those conditions. Fortunately such plants are generally found with processes of the less hazardous type, with a relatively low fire and explosion index. But wherever possible proper plans and process engineering flow schemes should be made for the modified plant.

These enable the operation of the plant to be worked out in detail. Standard procedures can then be laid down, operability studies made and any hazard analyses found necessary can be carried out.

REFERENCES

1. Health and Safety Executive, *Guidance Note EH/79, Threshold Limit Values for 1979*, Health and Safety Executive.
2. Chemical Industry Safety and Health Council of the Chemical Industries Association Ltd., *Major Hazards — Memorandum of Guidance on Extensions to Existing Chemical Plant introducing a Major Hazard*, Chemical Industries Association, London, 1975.
3. The Institute of Petroleum, *Refinery safety code*, The Institute of Petroleum, London (1965).
4. Mechlenburgh, J. C., *Plant layout*, Leonard Hill/The Institution of Chemical Engineers (1973).
5. British Standards Institution, *Code of Practice for Fire Prevention in the Chemical Industry*, British Standards Institution (1975).
6. Oil Insurance Association, *Plant construction recommendations*, Oil Insurance Association, (Industrial Risk Insurers), Chicago (1969).
7. (See Ref. 1, section 2.1).
8. Factory Mutual Engineering Corporation, *Loss prevention in chemical plants*, Factory Mutual Engineering Corporation (1974).
9. National Fire Protection Association, Standards (6 volumes), Boston, Massachussetts.

2.9 HAZARD ANALYSIS AND PLANT AND EQUIPMENT DESIGN[1]

Contents

2.9.1 Reliability and safety

We all know or believe that some machines, cars, forms of construction and even people are more reliable than others. Sometimes we think of this as being due to 'higher factors of safety', greater conservatism, and the use of well tried methods and systems rather than 'risky innovations'. Reliability is generally associated with experience and 'good engineering practice' which usually goes beyond strict adherence to regulations, national standards and codes of practice. We are particularly concerned with reliability when our own safety depends on it.

We try to arrange that when critical components in a machine fail they 'fail safe' although the expression is only relative. In a process controlled automatically by a number of pneumatically-operated valves, we would arrange that in case of air failure, the valves assumed the safest mode (open, closed or even perhaps partly open), so that the process automatically shut itself down. A 'fail safe' system ensures that failure of the air supply does not lead to damage to the plant and equipment or escape of flammable or toxic materials.

However this cannot disguise the fact that failure has occurred. The process has shut down, so that production is lost and probably some material in process has been spoiled. A 'fail safe' system therefore merely minimises the loss in the event of failure.

Fail safe systems are only possible in limited fields. If the rope of a crane, a bridge, or a railway line breaks suddenly under load, there is probably no way in which it can fail safe, and disaster seems inevitable. These things must be constructed and maintained to high standards of reliability, which in practice means high safety factors and frequent inspections.

Several new and highly innovative technologies have been developed over the last few decades which have called for the ultimate in reliability and yet have not permitted the traditional 'trial and error' approach to be adopted. Perhaps the most remarkable of these has been manned space travel. The success of these missions has been largely due to the development and application of 'Systems Safety' based on painstaking analysis of every possible failure or malfunction, its probability and consequences. This is known as hazard analysis.

2.9.2 Accident analysis

Hazard analysis has evolved from accident analyses which were primarily attempts to reconstruct causes of accidents which had specially serious consequences.

By going in detail through all the steps which led up to the accident, and examining the effects of everything that happened, the cause should ultimately reveal itself to the trained observer. A good example is the two BOAC de Havilland Comet disasters at Rome airport in 1953. On January 10th and April 8th similar accidents occurred shortly after take-off as the aircraft reached their cruising altitude of about 30 000 ft, when the aircraft exploded violently, killing all on board. The violence of the first explosion led to the belief that a bomb had been placed on board the plane and very special precautions were taken to rule out this possibility. When the accident was repeated on April 8th, sabotage was quickly ruled out, and it was found eventually that metal fatigue had started a crack in the top of the fuselage where an aerial for an automatic direction finder protruded. Cabin pressure then caused the crack to extend rearward, tearing out a section of the aircraft.

On a much lower plane, a reconstruction of the events leading up to the Flixborough disaster provides a similar though less familiar example[2]. Technical calculations based on the data collated by the official inquiry[3] provided strong circumstantial evidence that a sudden pressure surge was to be expected within the plant at about the time of the disaster. On this theory, some 200 kg or more of water which could not be drained was present in the plant in start-up. It passed into an unstirred reactor next to the bridge pipe and settled in the base below circulating cyclohexane. As the temperature rose, there came a point where boiling at the water-hydrocarbon interface would be expected. This would have caused the two phases to mix, with a sudden evolution of vapour and a rise in pressure. This is illustrated in *Figure 2.9.1*.

Such reconstructions are only possible when a great deal of reliable information about the events leading up to the accident has been collected. They require a combination of relevant theoretical knowledge, practical experience, patience, curiosity, logic and an atmosphere free from pressures and prejudices in which to reflect.

Valuable as the results of such post accident investigations may be, an accident followed by an investigation is an inefficient and costly method of determining the hazards present, and no organisation can sustain such losses for long. In the wisdom of hindsight one might suppose that there is nothing special about the logical steps of the post-accident reconstruction which pointed to a particular cause which could not have allowed this to be done before the disaster occurred.

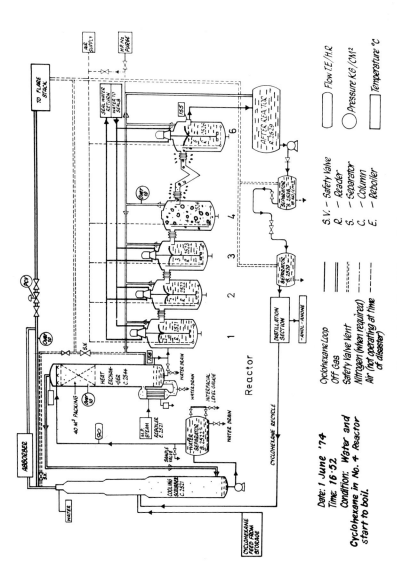

Figure 2.9.1 The Flixborough disaster

(The diagram shows Ralph King's explanation for the bursting of the by-pass).

Although this is true in theory and may even be possible in a perfect world, perhaps one may be excused for taking a somewhat sceptical view of the prospects of discovering and removing all hazards which could have disastrous consequences from a system in the design stage. Every system of human checking and counter checking will have its Achilles heel. Once a hazard has slipped through the net and the system has been rubber stamped 'safe', an air of authoratitive approval surrounds it, and reputations of large companies as well as individuals are at stake. The refutation of any criticism becomes a public relations exercise, which in extreme cases takes the form of the biblical adage 'If thine eye offend thee, cast it out'. In so many recent large-scale man-made disasters, the danger not only could have been foreseen, it actually was, yet no one took any notice.

The R.101 disaster, the thalidomide tragedy, Ronan Point, the use of high-alumina cement in building, and now the excessive use of calcium chloride in building are all examples of this. The dangers of asbestos fibres to the respiratory system were known for many years before action was taken.

If nothing else, these considerations only emphasise the need for hazard analysis, which is now vital in the design of nuclear installations, hazardous oil and petrochemical plant, aeroplanes and advanced and complex machines, where failure to spot all the hazards in advance can have disastrous consequences.

2.9.3 Types of hazard analysis

Hazard analysis falls basically into two compartments, *Qualitative Analysis,* to determine what hazards are present, and *Quantitative Analysis,* to determine the probability of certain serious consequences arising as a result of these hazards.

As in chemical analysis there is a sort of half-way house termed here 'semi quantitative hazard analysis' which gives a rough idea of the probability of certain accidental events. In fact virtually all hazard analysis is semi-quantitative. However precisely the exercise is carried out, the results can seldom give much more than an order of magnitude of the probability of a particular accidental event.

Among chemical engineers and plant operators in the UK, the term 'operability study' has come to be used in place of 'qualitative hazard analysis', while 'hazard analysis' refers to quantitative hazard analysis only. American specialists in systems safety use the terms qualitative and quantitative as described here.

Analyses may also be made in the *pre-design* stage and in the *post-design* stage. Some organisations carry out hazard studies at a number of different stages from the time of the inception of the project, during the design stage, and after the commissioning of the plant.

A pre-design analysis discovers and evaluates the hazards likely to be present in the system to be developed, and should also reveal all relevant legislation, codes of practice and specifications. It should provide the basis for preparing specifications and standards and criteria to be followed in the design and indicate conditions and practices to be avoided. It may reveal specific precautions to be observed, and determine the suitability or otherwise of particular hardware, materials and procedures. The pre-design analysis is very important because it is at this stage that problems can be revealed and resolved with least friction. Designers can be most upset when members of other organisations (e.g. safety specialists) point out after the equipment has been designed or ordered, that safety requirements exist of which they, the designers, were ignorant.

A post-design analysis determines whether the selected equipment and procedures meet the standards and criteria established as a result of the pre-design analysis. The post-design analysis will be more detailed and quantitative than the pre-design analysis. It will not only determine potential hazards of normal operation under normal conditions and procedures, but also examine what dangerous effects could be produced by abnormal environments, operating conditions, and procedural errors. It should also examine modes of failure. The post-design analysis should determine whether a design which does not meet the highest safety standards should be modified or redone or whether it can be accepted. Both pre-design and post-design hazard analyses are required.

2.9.4 Qualitative hazard analysis

Qualitative hazard analysis is required to identify hazards which cannot readily be quantified as well as those which can, in order to prepare the way for a quantitative analysis of the latter.

Qualitative analyses are carried out by one major chemical company on all projects whose capital value exceeds £100 000[4]. They consist of an in depth investigation of 'models' — mainly the piping and instrument diagram, by a team of four or five — the process engineer responsible for chemical engineering design, the project engineer responsible for mechanical engineering design, the commissioning engineer, the hazard analyst and another safety professional. A comprehensive check list of possible hazards compiled from literature and experience will be prepared in advance.

The analysis will usually be carried out in conjunction with the preparation of the plant operating manual. The analysis itself is said to require about 200 man hours per £1m capital, but it is so demanding that studies are restricted to 2½ hours per session and at a rate not exceeding three sessions per week. Each pipe and vessel is considered in turn using a series of guide words to stimulate creative thinking about what would happen if the flow deviated from the design intention, during start up, normal operation and shut down. Guide phrases used for continuous plants include high flow, low flow, reverse flow, high and low temperature and pressure and other important parameters. Sometimes a major hazard is revealed which necessitates modification or redesign, after which the qualitative analysis must be repeated.

2.9.5 Hazard rating[1]

The qualitative analysis has to consider all possible conditions and events to determine whether they could or could not cause or contribute to injury or damage, regardless of probability. But since cost will not allow the very unlikely hazards, such as a plane crashing on the plant, to be guarded against, some hazard rating is needed at quite an early stage. Such rating methods are of two kinds, relativistic and probabilistic.

2.9.5.1 Relativistic Methods of Rating

One method of evaluating hazardous conditions is to assign to them a numerical rating based on past experience of accidents caused by them, i.e.

1. Remote — no record of past accidents in similar systems.
2. Random — accident has happened once in the histories of the items reviewed.
3. Occasional — has occurred two or three times.
4. Chronic — has occurred more than three times.

Another method is by using safety factors or safety margins. The safety factor of a container is the breaking stress or pressure divided by the maximum working stress or pressure. The matter is not, however, quite as simple as this, since a thick wall welded vessel of mild steel will be more subject to post-weld stress and low temperature embrittlement than a thin wall vessel.

A third is to rate the hazard on a numerical scale 1 to 10, based on the opinions of various raters. This can lead to wildly discordant ratings depending on the rater and his role; designers tend to give safer ratings to their equipment than the men who have to operate and maintain them.

2.9.5.2 Probabilistic methods of rating

Quantitative hazard analysis is generally based on the probability ratings that an event (e.g. the failure or fault of a component) will occur. These methods are used by insurance companies, in statistical quality control and maintenance. The probability may be expressed as the number of times the fault occurs in a large number of trials or in a year of continuous use. A good deal of such statistical data has been collected for instruments and control systems.[5]

A low probability of a fault or failure guarantees nothing. It may tell us that failure is likely once in ten years, but it cannot say when this will happen. The probability of failure is also a function of the service life of the component and its environment and how well it has been maintained.

2.9.6 Reliability

Reliability (R) is the probability that a piece of equipment or component will perform as intended for a given length of time (usually the working life) in a stipulated environment. The working life may be twenty years for continuously operating plant. For machines used discontinuously or on day shift only, it may be 10 000 hours, but for expendable military hardware such as rockets, it may be a matter of only minutes or hours. If the probability of malfunction or unreliability is Q, and both Q and R are expressed as decimals, then
$R+Q = 1$ and $R = 1-Q$
Equipment failures are of three types:
   1. Early failures, which occur during the 'running in' period — due often to substandard components or poor assembly.
   2. Random failures, which occur at a more or less constant rate from the end of the 'running in' period throughout the useful life of the equipment.
   3. Wear out failures which begin when the components are past their useful life — due to fatigue, corrosion or similar causes.
The failure rate for a large number of identical components or pieces of equipment thus gives a 'bathtub curve' when the failure rate is plotted against time as in *Figure 2.9.2*.
The ratio t/T, between the required operating time t and the mean time between failures T is very important. When t equals T, whether this is one minute, one hour, or one year, reliability is only 0.368. To increase reliability, the ratio t/T must be decreased.
The reliability of a complex system depends on the individual reliabilities of its components. If the components operate in series so that the failure of any one would cause failure of the system, the system reliability equals the product of the reliability of each component.
System Reliability $R_s = R_1 \times R_2 \times R_3 \times R_4 \ldots$
This is the Product Law of Reliability.
If each component of a 6 component system has a reliability of 0.9 (i.e. 90%) the overall reliability is $R_s = 0.9^6 = 0.53$.

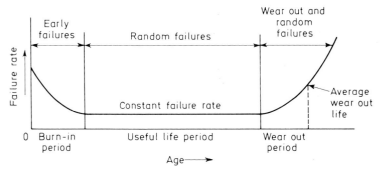

Figure 2.9.2   Bathtub curve: accident rate vs time

To keep the reliability as high as possible and minimise the effect of the product law, the following means are employed:

1. Use of components of very high reliability.
2. Use of parallel redundant systems.
3. Use of standby systems.
4. Operation under optimum environmental conditions.

Parallel redundant components perform the same function although only one is required for successful operation. Both components must fail for the system to fail. The probability that both will fail (within the time specified) is the product of the individual failure probabilities (Q) of each component.

Example

Two parallel components each have a reliability of 0.81.

Failure probability of each = 0.19

Probability that both will fail = $(0.19)^2$
$$= 0.036$$

∴ system reliability = 0.964

A doubly redundant system using three similar components in parallel would have an overall reliability of 0.993.

Redundant systems have some disadvantages. They increase cost, complexity and maintenance. Also, once one of two parallel redundant components has failed, the reliability of the system then falls back to that of the single component. Thus to retain high reliability, there must be a means of detecting failed components and replacing them promptly after failure.

Standby components which take over only when the operating one fails require failure detection and switchover devices. Although the standby should wear out at a lower rate than the operating unit, the failure detection and switchover devices will not be 100% reliable. Standby units are used in critical situations such as power supplies to hospitals.

Improvement of the environmental conditions is achieved by controlling temperature, humidity, shock, vibration, corrosion, erosion, radiation and friction. This can greatly increase the reliability of components, e.g. in a computer.

A large number of tests are required to establish the reliability of complex components and systems. Many systems do not have constant failure rates, but follow a Gaussian or other distribution.

2.9.7 Models, fault trees and quantitative analysis

Modelling involves the presentation of all parameters in a system, etc, in such a way as to indicate their interrelationships. There are three principal types of model, iconic, analog and symbolic.

An iconic model is a visual representation which closely resembles the system that it describes. It may be a three-dimensional scale replica such as a model aeroplane or a model of a molecule, or it may be two dimensional, i.e. sketch, photograph or drawing.

An analogue model represents one property or set of properties by another, e.g. the use of electric currents and voltages to represent the flow and pressure of a fluid in a system.

Analogue computers are widely used to simulate and predict the performance of industrial processes under different conditions.

Symbolic models have the least physical resemblance to the system they describe. They may be a series of mathematical equations, block diagrams, computer programs or sets of symbols.

The value of a good model is that it includes the elements and factors of the system it represents and obeys the same rules of behaviour. Models permit experimentation at minimum cost and without risk of costly equipment loss. In this way the optimum design can be selected and errors corrected.

A main use of models in hazard analysis is in the development of logic or fault trees. A fault tree is built up by considering a particular failure of a system with serious consequences and examining how this could arise from the failure of various components in the system.

Logic trees are used in the design and simplifications of computer circuits, and involve some knowledge of Boolean algebra. The knowledge required is elementary and not difficult to grasp since only certain basic elements are required to construct and build up the kind of fault tree used in hazard analysis.

These are the OR and AND gates, represented as shown in *Figure 2.9.3.*

Both gates take in certain inputs and produce a single output. In quantitative hazard analysis the gates are used somewhat differently from their use in Boolean algebra, but the essential features are the same. An OR gate is used for the inputs of components in series, the failure of any one of which would cause failure of the system. An AND gate is used for the inputs of components and subsystems in parallel, where all components or subsystems must fail before the system fails.

The use of an iconic model and the construction of a fault tree from it is represented by a simple example of a physical process.

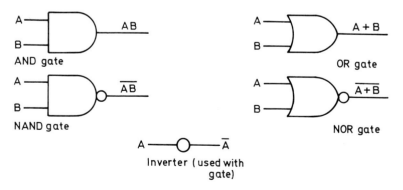

Figure 2.9.3   Logic element symbols

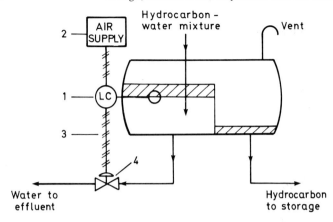

Figure 2.9.4   Flow scheme of separator

2.9.7.1   Example. A hydrocarbon separator

Figure 2.9.4 is a flow scheme of a simple settler for separating water from a light hydrocarbon liquid forming part of a process. It is elevated and operates at atmospheric pressure and a few degrees above ambient temperature. It consists of a baffled and vented horizontal vessel, pipes, a float operated interfacial level controller which is supplied with compressed air at 1 bar gauge and which transmits a compressed air signal to a control valve on the water exit line. Separated hydrocarbon flows to storage and separated water into the effluent system.

Failure of the interfacial controller which allowed hydrocarbon to pass into the water effluent would be dangerous and could lead to a fire. It is necessary to know at the design stage what the probability of this would be, and whether additional safeguards are needed.

Figure 2.9.5 shows a fault tree for the control system. The undesirable event — opening of the control valve and escape of hydrocarbon is shown at the top.

Four things could cause this to happen:
1. Failure of the level controller itself.
2. Failure of the air supply.
3. Failure of the air signal line from the controller to the control valve.
4. Failure of the control valve in the open position.

Any one of these events alone would be sufficient to cause the control valve to fail in the open position. The four events are therefore joined by an 'OR' gate. This means that their failure frequency rates must be added to give the frequency of overall failure.

Statistical data on the reliability of the four parts of the system shows the following failure rates:

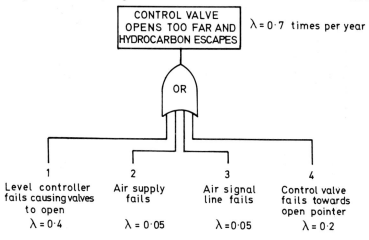

Figure 2.9.5   Fault tree for separator

Event	Failure rate, times per year
1. Controller failure — causing valve to open	0.4
2. Air supply failure	0.05
3. Air signal line failure	0.05
4. Control valve failure towards open position	0.2
Total	0.70

This is unacceptably high. An acceptable criterion might be 0.1 failures per year, although for a working life of twenty years this would give a low reliability of 0.135 or a probability of failure of 0.865 (86.5%) sometime during the working life of the system.

Some slight improvement would be achieved if the control valve were specified to close on air failure. This would eliminate events (2) and (3), but the overall failure rate would still be high — 0.6 failures per year.

The most acceptable way of improving the performance of the system would be to fit a low interfacial level switch to the separator and connect this to an audio-visual alarm in the control room. The flow scheme is shown in *Figure 2.9.6*. This would alert the operator to the failure of the control system some time before hydrocarbon had started to leave the separator, and in time for remedial action to be taken.

Now the level alarm system could itself be inoperational without anyone noticing it. To cover this situation the level alarm system would be tested say once per week. The overall result is assumed to be that the alarm would

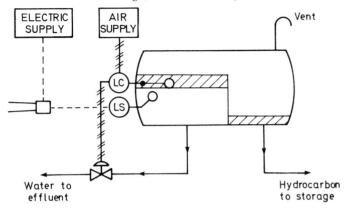

Figure 2.9.6   Flow scheme for improved separator

be out of action for a certain time, say two weeks per year. This gives a fractional dead time (FDT) of 0.04.

But before we draw our new fault tree, have we not forgotten something? What about the operator? Will he always respond to the alarm? Experience says he will do so ninety-seven times in a hundred. Hence we assign a fractional dead time to the operator of 0.03. To be exact one should multiply this by the fractional live time of the alarm (1.0-0.03 = 0.97) but the data does not justify such precision.

The fault tree is shown in *Figure 2.9.7*. The failure rates on this side are quoted as frequency rates, with the dimension T^{-1}. But on the right-hand side the data for the alarm and the operator are quoted as 'fractional dead time' which is dimensionless. These are additive, so they are joined by an OR gate.

The outputs from the two OR gates are fed into an AND gate at the top of the tree which multiplies them to give the overall frequency failure rate of 0.042 times per year.

Over twenty years working life, this gives a reliability of 0.43 or a probability of failure of 0.57 (57%), which would only be acceptable if the risk of fire resulting from escape of hydrocarbon from the separator were low.

2.9.7.2 Consistent units for fault tree

Although the fault trees shown here conform to those published in many papers on hazard analysis, the use of two different units, one with the units (time)$^{-1}$ and one dimensionless can be confusing, since like units can only be added through an OR gate and unlike units can only be multiplied through an AND gate.

While it is possible to avoid this confusion by the use of a dimensionless number, R for reliability, and Q for unreliability instead of a failure

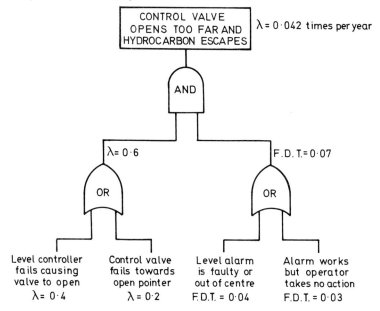

Figure 2.9.7   Fault tree of improved separator

frequency rate, this in turn causes additional complications which in this case tend to make matters worse. So it is probably best to stick with the units shown in hazard analysis of this type, taking care to remember that one is using two different units, one with the dimension T^{-1} and the other dimensionless.

2.9.7.3 Limitations of quantitative hazard analysis

The limitations on the type of analysis shown lie firstly, in the scarcity of good reliability or failure rate data for individual components (particularly for closely specified components) and secondly, in the time taken to draw fault trees for even quite simple systems.

Thus the fault trees drawn relate only to a single mode of failure of the system (loss of hydrocarbon to the water effluent) caused by failure of the control system.

There are other possible causes of hydrocarbon escape:
1. Blockage of the liquid outlet lines, and overflow through the vent pipe.
2. Failure of a pipe-joint of a pipe itself, or even the settler vessel.

As more reliability data are collected, and the work of detailing hazard trees is systematised and speeded up with the aid of computers the technique will become more widely used in industry and probably also in insurance. Even when not pursued in such depth as is now common with

Identification: _____ Mark I Flight System
Subsystem: _____Wings   Designer: ___ Daedalus

Hazard	Cause	Effect	Hazard Category	Corrective or Preventive measures
Thermal radiation from sun	Flying too high in presence of strong solar radiation	Heat may melt beeswax holding feathers together. Separation and loss of feathers will cause loss of aerodynamic lift. Aeronaut may then plunge to his death in the sea.	IV	Provide warning against flying too high and too close to sun Maintain close supervision over aeronauts. Use buddy system. Provide leash of flax between the two aeronauts to prevent young impetuous one from flying too high. Restrict area of aerodynamic surface to prevent flying too high.
Moisture	Flying close to water surface	Feathers may absorb moisture, causing them to increase in weight and to flag. Limited propulsive power may not be adequaₜ to compensate for increased weight and drag so that aeronaut will gradually sink into the sea. Result: loss of function and flight system. Possible drowning of aeronaut if survival gear is not provided.	IV	Caution aeronaut to fly through middle air where sun will keep wings dry or where accumulation rate of moisture is acceptable for time of mission.

Figure 2.9.8   Safety analysis table (From 'Handbook of System and Product Safety'
by W. Hammer published by Prentice Hall International)

military systems and spacecraft, it sharpens the appreciation of those employing it for the reliability and hazards of different machines and processes.

2.9.8 Aspects of detailed hazard analysis

It has only been possible in this section to scratch the surface of a technique which has been developed in great depth for specific advanced technological projects. Most of these are so complicated that it is necessary to break them down into systems, sub-systems and their components and carry out hazard analyses on the components, the sub-systems and the interfaces between the sub-systems, as well as on the system as a whole.

In reporting such analyses, it is usual to supplement the fault trees with safety analysis tables and narrative summaries.

2.9.8.1 Sub-systems

Every piece of equipment, plant or machine can be broken down into sub-systems which can be analysed. The following headings form a basis for sub-system classification:

Power. Nearly every system requires power for its operation. The source may be electrical, human, hydraulic, compressed air, etc.

Structural. The structure supports, unites and often protects the other sub-systems. The skeleton of a body and the frame of a building are examples.

Control. Every sub-system needs to be controlled. The word here refers to the direct means of control applied to or built into the sub-system, rather than the overall control system which includes sensors, operators and communications.

Sensor. A sensor senses the environment, outputs, inputs, and other factors which the operator needs to know for successful operation. Measuring elements of instruments are sensors as are human eyes, ears and noses.

Communications. The communications network of a human or animal is the nervous system; that of machinery and equipment is composed of electrical wiring and circuits, hydraulic and pneumatic lines and mechanical linkages.

There are boundaries or interfaces between most combinations of sub-systems to which hazard analysis is also applied to explore reliability and potential loss.

2.9.8.2 Safety analysis tables

These supplement the information given in models and fault trees, and usually have the following main headings:

Hazard.

Cause. (Factors which generate or contribute to the hazard).

Effect. (Particularly the results of hazards or their causes).

Corrective and preventative measures.

Other headings frequently found useful are:

Critical times when the hazard may occur.

Time available for corrective action.

Relative frequency of occurrence.

Hazard level.

Office responsible for improving design to reduce hazard level.

An entertaining safety analysis table prepared by Willie Hammer for the wings designed by Daedalus for his son Icarus is reproduced with permission of the analyst in *Figure 2.9.8*.

2.9.8.3 Narrative summaries

These are produced last and describe each portion of the analysis shown in the models (especially the fault tree) and safety analysis table, supplementing the information already given.

Their main purpose is to highlight possible problem areas and indicate where further design and safety effort should be directed.

2.9.8.4 Official and legal uses for hazard analyses

Records of hazard analyses are useful in supporting plans of hazardous installations submitted to planning authorities, and their use may be insisted upon when major hazards are concerned. They may also be used in support of evidence given in official inquiries and liability cases.

REFERENCES

1. Hammer, W., *Handbook of System and Product Safety*, Prentice Hall, New Jersey (1972).
2. King, R. W., 'Flixborough — the role of water re-examined', *Process Engineering* (London), September (1976).
3. Department of Employment, 'The Flixborough disaster — Report of the Court of Inquiry', HMSO, London (1975).
4. Gibson, S. B., 'The design of new chemical plants using hazard analysis', symposium paper in *Process Industry Hazards,* The Institution of Chemical Engineers, Rugby, Symposium series No. 47 (1976).
5. Lees, F.P., 'A review of instrument failure data', symposium paper in *Process Industry Hazards*, The Institution of Chemical Engineers, Rugby, Symposium series No. 47 (1976).

2.10 CONDITION MONITORING, NON-DESTRUCTIVE TESTING AND OTHER INSPECTION AIDS

Contents

Traditionally the prevention of failure of machines and equipment has depended on regular inspection and maintenance when the item is taken out of service. It is then stripped down, all parts thoroughly examined for wear, corrosion, fatigue, etc, worn parts built up or replaced, and the unit rebuilt and put back into service after final inspection. Whilst this method is still often used, sometimes as a back-up to the methods of inspection described here, it has two disadvantages which are particularly serious when wear rates are unknown or unpredictable:

1. The item may be taken out of service, stripped down and inspected too soon before any significant wear has occurred. This involves considerable expense as well as the risks of faulty re-assembly.
2. The item may fail prematurely between inspections causing injury, damage or even disaster.

The experienced driver of a vehicle or user of a machine has always developed an awareness of symptoms of trouble — unusual noise, excessive vibration, high temperatures and lowered performance. All of these tell him that a thorough check-up and maintenance is due. Condition monitoring has developed this subjective awareness of trouble into scientific techniques which will detect and often pinpoint the incipient trouble without human intervention or before it can be sensed by the human eye, ear, nose or touch.

Non-destructive testing covers a range of techniques for measuring, e.g. wall thicknesses of pressure vessels, the soundness of welds and castings, etc without having to drill holes in the item or test it to destruction. Some non-destructive methods of testing may be applied while the equipment is in service. The dividing line between condition monitoring and non-destructive testing is then difficult to draw.

There are also a number of special aids to inspection such as stethoscopes, endoscopes and instruments which employ fibre optics which enable the inspector to direct beams and see round corners into the most inaccessible of spaces. The dividing line between some of these aids and

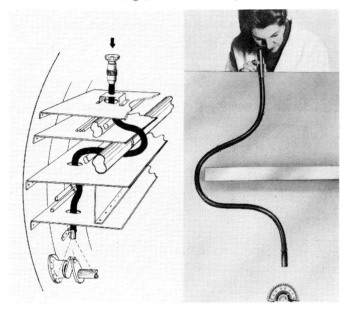

Figure 2.10.1   A flexible periscope (P. W. Allen & Co.)

'non-destructible testing methods' is again difficult to draw. Thus the reader should not be surprised to find methods which are here described under one heading being elsewhere described under a different one.

2.10.1 Condition monitoring

Condition monitoring embraces all techniques which may be used to diagnose the condition of plant and machinery whilst in operation, either by observation or inference. The techniques were first developed in high technology industries, e.g. aerospace, nuclear energy and computers, with considerable success.[1] Their wider application is described in a comprehensive book by Collacott.[2] The main techniques used for machinery are:
1. Contaminant analysis.
2. Vibration analysis.
3. Sound monitoring.
4. Performance trend analysis.
5. Static testing techniques.

2.10.1.1 Contaminant analysis[2]

Where a fluid is flowing through a system which is subject to deterioration or wear, the products of this process may be entrained in the fluid flow (particularly if the fluid is a liquid) and/or may react with the fluid producing observable changes. By monitoring the materials entrained, or the changes in the fluid, it is often possible to assess the nature and the extent of the deterioration within the system. Probably the best known use of this technique is in the analysis of engine lubricating oils.[3]

Other applications include dissolved gas monitoring, particularly in electrical systems where arcing or overheating may cause insulation breakdown with the evolution of gaseous breakdown products. Applications involving gas-in-liquid, gas-in-solid, and gas-in-gas analysis are all known.

2.10.1.2 Vibration analysis

Mechanical systems which involve moving components — so-called active systems — generate characteristic vibration signatures. Such active systems have been divided into three groups:[4]

1. Cyclic machinery (engines and transmissions).
2. Flow noise generators (pumps and boilers).
3. Single transient generators (switches and punches).

Of these, cyclic machinery lends itself most easily to this technique, although the methods used to separate the significant vibrations from the background noise are often complex.

2.10.1.3 Sound monitoring

In many ways this can be regarded as very similar to vibration analysis, with the characteristic vibrations falling within a specific range. However, the difference is important as it relates to the added difficulties of sound over vibration analysis.

In CM terminology sound monitoring refers to the monitoring of air-borne pressure waves and is primarily a scalar quantity with no directional parameter, whereas vibration is essentially a vectorial quantity with both magnitude and direction.

The main development of sound monitoring, to date, has been with cyclic machinery.

2.10.1.4 Performance trend analysis

This technique compares the value of a parameter or set of parameters with a base value at regular time intervals. Deviations from the norm can be gradual (i.e. drifting) or sudden. The main areas where this technique has been used in a CM programme are in aircraft and marine applications, although it has been used for many years in other fields including chemical engineering and even financial speculation.

Examples in the process industries include automatic recording of pressure drop over a catalyst bed, and of the temperatures of fluids flowing into and out of a heat exchanger.

2.10.1.5 Static testing techniques

There are many types of defect which can be detected without affecting the usefulness of the component being tested. Whilst the techniques are properly in the field of non-destructive testing, some of them can usefully be included in a CM programme.

From the safety point of view, some of the most interesting techniques are those used to monitor corrosion. Here, two methods predominate — radiography and ultrasonics — although eddy currents, magnetic particle and dye penetrant techniques can be very useful for certain types of corrosion investigation. A useful table on techniques for specific applications is given by Collacott, (Ref. 2, Table 11.5 p. 361).

In addition, recent developments in thermography show that this technique has considerable potential in the process industries.

2.10.1.6 CM techniques summary

These five techniques (p.208) all relate to ways of detecting incipient failure. Its objectives and benefits are partly in increased safety and partly economic by enabling mechanical equipment to be operated longer before it has to be shut down for maintenance. Great care is needed when the latter objective is desired to ensure that all possible modes of failure can be detected at an early stage by condition monitoring. The main benefits to safety from condition monitoring lie in the process and chemical industries where accidents are especially common during start up following a shut-down for maintenance.

The protection of recording devices such as those used in condition monitoring and of their records also has a bearing on safety, through assisting those engaged in disaster and accident investigations. This is illustrated by the following extract from the Report of the Flixborough disaster inquiry:[5]

'We would have been greatly assisted in the inquiry if essential records such as those relating to the pressure and temperature of the cyclohexane in circulation in the plant during the fatal shift had been preserved. It is recommended that consideration be given to installing devices or systems for recording vital plant information in a form which would survive the effects of fire or explosion. An example of such a device is the 'black box' used in aircraft.'

2.10.2 Non-destructive testing

Most physical properties and forms of energy have been utilised in different techniques of non-destructive testing. However, non-destructive testing is used mainly for detecting flaws or determining the thickness or the quality of metal parts, and the methods used may be grouped under the following headings:

1. Magnetic particle.
2. Penetrant.
3. Radiographic.
4. Ultrasonic.
5. Other methods.

Detailed information on the theory and practice of non-destructive testing methods is given in McMasters' *Handbook of non-destructive testing*[6] and other references.[7,8]

2.10.2.1 Magnetic particle

These methods are used primarily for detecting cracks and flaws at or near the surface of ferromagnetic materials. The object is magnetised, either by passing a current through it or by placing coils round it, and a finely divided magnetic powder is applied to the surface. The powder may be applied dry or as a wet suspension, and is available in various colours as well as fluorescent. Defects in the object interrupt the magnetic field which is shown by the pattern of the magnetic particles. The electric circuits generally employ low voltage sources with high currents, and require the usual precautions against electric shock as well as arcing (*Figure 2.10.2*).

The particles themselves usually present a dust hazard requiring local exhaust ventilation or respiratory protection. Goggles or other forms of eye protection should also be worn.

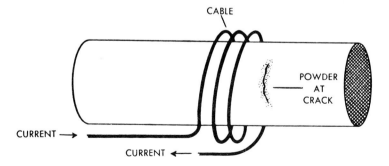

Figure 2.10.2   Detection of transverse crack by magnetic particle technique (Magnaplus Ltd.)

The method requires considerable skill and experience on the part of operators for reliable indication of cracks and particularly of flaws below the surface.

2.10.2.2 Penetrant

These techniques are used for detecting surface cracks and pores in metal and other objects. After cleaning the object, a special penetrating liquid is applied to the surface for long enough to soak into the crack or pores. Excess penetrant is then removed from the surface, leaving penetrant in the pores.

A developer is applied to the surface which produces a colour reaction with the penetrant in the cracks or pores which makes them clearly visible. The object is then examined with a suitable light source, which in the case of a fluorescent penetrant system may be a 'dark' ultra violet source.

A wide range of penetrants and developers is available. Most penetrants are organic compounds which can cause dermatitis and should not be allowed in contact with the skin. Suitable eye protection is needed when ultra violet light sources are used.

2.10.2.3 Radiographic

X-rays and gamma rays from radio-active sources are used for the examination of metal and other objects for flaws, based on differences in intensity in the transmitted radiation. These differences are transformed into visual images by exposure of photographic film, fluorescent screens, Geiger counters and ionisation gauges. X-rays have some advantages in sensitivity and contrast, but gamma ray sources have the advantage of portability and independence of electric power and expensive apparatus.

Beta ray sources which are less penetrating than gamma rays are used for measuring sheet metal thicknesses.

All forms of ionising radiation have hazards which are discussed in sections 6.9 and 6.10.

2.10.2.4 Ultrasonic

Three ultrasonic methods are available based on reflection, transmission and resonant frequency. The ultrasonic waves are in every case generated by a piezo-electric crystal transducer; the reflected or transmitted waves are detected and measured by similar transducers coupled to electric amplifiers.

Any material which transmits vibrational energy may be tested, and these methods are used for plastics, ceramics, glass, concrete and rubbers as well as metals. Metal objects may be tested in dimensions up to 10 m. Ultrasonic methods are specially useful for flaw detection and thickness measurement of the walls of pipes, pressure vessels, etc while the

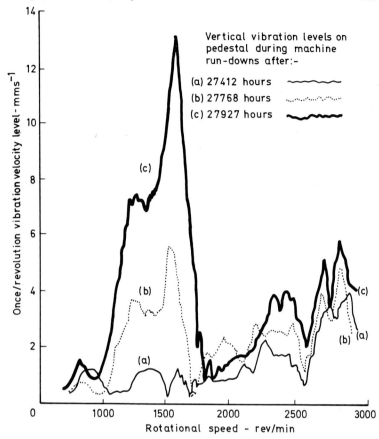

Figure 2.10.3   Detection of cracked rotor shaft from vibration signatures during run-downs (National Engineering Laboratory; reproduced by permission of the Controller of HMSO)

equipment is in use. Most ultrasonic methods require that the transducer be pressed against the surface of the part to be tested.

When equipment subject to wear and corrosion is being designed, the location and connection of suitable positions for the application of ultrasonic transducers should be considered to ensure that measurements can be made at critical points.

2.10.2.5   Other methods

Among other useful, though less commonly used, methods of non-destructive testing are tribolectric, magneto-inductive, eddy current, microwave and infra red.

2.10.3 Other inspection aids[9]

A number of other useful aids to inspecting plant and machinery used by inspectors, maintenance men and operators, are available. Most of these fall under the following headings:
Temperature indication and measurement.
Leak detection.
Stethoscopes.
Stroboscopes.
Optical inspection devices.

2.10.3.1 Temperature indication and measurement

While most plant and machinery have permanent temperature measuring instruments built in at critical locations, temperature checks at other points are often required to diagnose possible troubles. The following are among the devices available:

1. Temperature indicating 'stickers'. These are self adhesive strips for sticking on surfaces; they change colour (usually turning black) when the surface reaches a particular temperature. They are used to warn or prove that an excessive temperature has been reached, and to check temperature gradients or anomalies.
2. Portable electronic thermistor thermometers. These have a small probe with a temperature sensitive thermistor at the end of a thin rod which gives a rapid response to the temperature of any surface, gas, liquid, etc. with which it is in contact. It is connected by a cable to a battery powered pocket size indicator, and has many uses in 'trouble shooting'.
3. Portable optical pyrometers. Small pistol-shaped battery operated pyrometers measure the temperature of any surface at which they are pointed, so long as it fills the field of view of the instrument. They are available with various fixed temperature ranges between 20°C and 1650°C. They are most useful in the higher temperature range for checking the insulation of furnaces, furnace and tube temperatures, balancing heating elements and 'trouble shooting'.

2.10.3.2 Gas leak detectors for pressure and vacuum systems

Many different types of gas-leak detector have been developed, marketed and abandoned in favour of the old fashioned soap and water technique. Nowadays sensitive portable ultrasonic leak detectors with headphones and probes for detecting internal as well as external leaks are available (*see Figure 2.10.4*). For many purposes they are said to be both faster and more reliable than other methods. Their principle of operation is that the smallest leak of gas within a closed system or between the system and the atmosphere signals its presence by a high pitched whistle at a frequency well beyond the range of the human ear. This is converted electronically into audible noise.

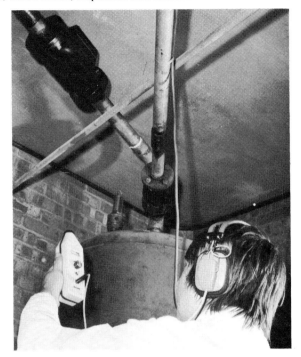

Figure 2.10.4   Leak testing with Dawe ultrasonic leak detector (Dawe Instruments Ltd.)

2.10.3.3 Tunable stethoscopes

Sensitive tunable stethoscopes with amplifiers are only a short step removed from ultrasonic leak detectors. They enable unwanted noise to be reduced by a tuning control and in skilled hands enable wear and other defects to be detected in mechanical and electrical equipment.

2.10.3.4 Stroboscopes

Portable battery-operated transistorised stroboscopes allow moving parts of rotating or oscillating machinery to be observed as though they were stationary. Faults may thus be observed which are not apparent when the machinery is stationary.

2.10.3.5 Optical inspection devices

Many devices whose origins probably lay in surgery and dentistry are

Figure 2.10.5   Small fibroscope being used for examination of pipework in chemical plant (P. W. Allen & Co.)

available for viewing the inside of tubes, tanks and equipment with tortuous internal passages. These provide both for illumination and vision, and include endoscopes — slim tubular optical instruments in which the light beam is carried along a Perspex tube which can be straight or curved — and fibre optic periscopes which consist of a flexible tube containing two fibre glass bundles, one for lighting and one for image transmission (*Figure 2.10.5*). These can give very high quality images when properly used and cared for.

A variety of mirrors are also available, including telescopic inspection mirrors for viewing 'places no one ever visits' *(see Figure 2.11.1).*

REFERENCES

1. Davies, A. E., 'Principles & practice of aircraft powerplant maintenance', *Trans. I. Mar. E.* 84, 441-447 (1972).
2. Collacott, R.A., *Mechanical fault diagnosis and condition monitoring*, Chapman & Hall, London (1977).
3. Scott, D., 'Examination of debris and lubricant contaminants, *Proc. I. Mech. E.*, 187 Pt 3G, 83-86 (1968).

4. Weichbrodt, B., *Mechanical signature analysis: a new tool for product assurance and early fault detection.* G.E.C., R & D Report, Schenactady, USA, 68-c-197 (1968).
5. Parker, J. C., Pope, J. A., Davidson, J. F., and Simpson, W. J., *The Flixborough Disaster, Report of the Court of Inquiry*, HMSO, 36, para 216 (1975).
6. McMaster, R., *Non-destructive testing handbook*, Ronald Press, New York (1959).
7. National Safety Council, *Accident prevention manual for industrial operations*, 7th Ed., N.S.C., Chicago, 925-928 (1974).
8. Dodge, D.D., *Non-destructive testing*, pp. 5, 106-113 in Marks Standard Handbook for Mechanical Engineers, 7th ed., McGraw Hill, New York (1967).
9. Clifton, R. H., *Principles of planned maintenance*, Edward Arnold, London, 109-121 (1974).

2.11 MAINTENANCE AND INSPECTION

Contents

Maintenance and inspection are defined in BS 3811:1964[1] as follows:

'Maintenance is work undertaken in order to keep or restore every facility, i.e. every part of a site, building and contents to an acceptable standard.'

'Inspection is the process of ensuring by assessment that a facility reaches the necessary standard of quality or performance and that the level is maintained.'

Maintenance thus clearly implies some measure of inspection as part of its function to determine

1. what maintenance is needed and when, and
2. whether the results are acceptable.

The word maintenance is therefore understood to include this minimum level of inspection. An example of a device which can be used in this type of inspection is shown in *Figure 2.11.1*. In addition however many organisations, depending on size and degree of hazard, have a separate inspection department independent of maintenance and production and responsible only to top management. Such inspection is closely related to safety. Maintenance is necessary to ensure the efficient, economic and safe operation of the things maintained.

Hazards arise in three ways connected with maintenance:

1. Through lack of maintenance, which allows buildings, plant or machinery to fall into a dangerous condition.

218

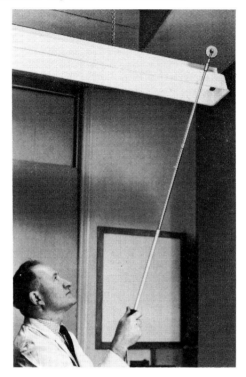

Figure 2.11.1   Telescopic inspection mirror (P. W. Allen & Co.)

2. During the course of maintenance, e.g. through accidental falls, falling objects, gassing and material handling.
3. As a result of faulty maintenance, e.g. nuts not tightened, motors wired to run in reverse; these create new hazards not present before maintenance was carried out.

Maintenance workers thus require a greater safety awareness than most production workers.

Whilst a product may be well maintained to the standards to which it was designed and constructed, it does not follow that these standards are always adequate.[2]

Maintenance may be planned or carried out on an 'ad hoc' basis, with various degrees of planning in between. Preventative maintenance is but one form of planned maintenance, although planned maintenance and preventative maintenance are often thought to be the same.

2.11.1 Forms of maintenance[3]

Completely unplanned and *ad hoc* maintenance is probably the most common form, and it is often the cheapest and most flexible. It means waiting until either a fault appears in an item, or the item totally breaks down, and then deciding in the light of prevailing circumstances whether to scrap it, sell it, or repair it, and whether to replace it with a similar item or by something else. This policy assumes the ready availability of replacements or repair facilities so that spares need not be carried. All forms of planned maintenance involve some loss of flexibility (assuming that the plan is followed), and some outlay in making the plan and implementing it (e.g. through ordering and carrying spares).

The following forms of planned maintenance are recognised.[3]

2.11.1.1 Replacement and scrapping of worn items

This applies particularly to cheap easily replaceable items such as drill bits, hacksaw blades, as well as to ropes, slings, and electric luminaires. It applies to many parts (subject to wear) of larger items to which planned preventative maintenance is applied.

If the item to be scrapped is liable to reach a dangerous condition (e.g. a rope or hand tool) before it wears out completely, it is important that this be recognised and a suitable form of preventative inspection employed so that arrangements are made to scrap and replace the item before it becomes dangerous.

2.11.1.2 Replacement and sale of worn items

This policy is applied particularly by operators of car fleets. Only basic maintenance — lubrication, servicing and adjustment — is carried out. The vehicles are sold when their operating efficiency begins to fall when they are likely to require costly and lengthy overhaul.

2.11.1.3 Breakdown maintenance

Items which operate as individual units or are separate from the actual manufacturing process whose breakdown does not constitute a safety hazard are often best run until they breakdown and then replaced by a reconditioned machine whilst the broken down machine is sent for maintenance. A sewing machine is an example. Where this system is employed it is important to establish that the item is not operated in a dangerous condition before it finally breaks down.

Where this can arise the method should be modified to some form of preventative maintenance to ensure that the limit of safe operation is not passed.

2.11.1.4 Preventative maintenance

Preventative maintenance is usually justified economically on continuous and/or highly automated processes, where the cost of lost production is high and where the failure of one piece of equipment brings the entire process to a halt. On safety grounds, preventative maintenance is necessary on other types of equipment, e.g. boilers and pressure vessels, whose failure may become dangerous.

This latter necessity is recognised in the Factories Act 1961[4] for many types of plant and machinery, so that a limited amount of preventative maintenance has become virtually obligatory.

The efficient organisation of preventative maintenance is a complex and skilled activity which requires systematic record keeping, item numbering, spare parts stock control and stock keeping, inspection procedures, job analysis, work programming and planning. Whilst safety specialists cannot be expected to be experts in all these activities, they should study the subject in sufficient depth to appreciate them, and to evaluate how well they are organised and carried out in their own factory, works or company. Preventative maintenance also requires a thorough analysis of the hazards that may arise:

1. Through failing to carry it out at the right time;
2. In the maintenance operations themselves;
3. Through faulty maintenance, which leaves the equipment in a more dangerous condition than it was originally.

2.11.2 Preventative Maintenance and Inspection required under the Factories Act 1961 and other Regulations[3,4]

Whilst statutory inspection of boilers, air receivers, cranes, and power presses is generally carried out by an independent firm specialising in these activities and having close links with insurers, there is no reason why this need be so, since the various regulations merely state that the inspection be carried out by a competent person. Many large organisations have their own inspection departments with trained personnel who undertake their own statutory inspections. Provided these inspectors are experienced and well trained and not subject to pressure from production departments, this has several advantages:

1. They are likely to be more familiar with the equipment inspected and its previous history than the outside inspector.
2. Their inspection activities can easily be extended to cover other equipment in use, the failure of which could be hazardous or economically serious but for which there is no statutory requirement to inspect.

In other words, the inspections required by law form a base on which a more thoroughgoing system of preventative maintenance may be built. The statutory requirements include, of course, that the items inspected be clearly marked with a securely fixed plate or tag which will not come off and that a detailed record of the equipment and of the inspections be kept — both indispensible features in any preventative maintenance programme.

Table 2.11.1 STATUTORY REQUIREMENTS RELATED TO PREVENTATIVE MAINTENANCE IN UK

Facility concerned	Act or Regulation	Section No.	Max. interval between attentions	Attention required	Details of facility and attention required
Factory cleanliness	Factories Act 1961	Sec. 1	Weekly	Cleaning	Floors of work rooms to be washed or cleaned by other method if effective.
	"		14 months	Cleaning	Inside walls and ceilings with smooth impervious surface to be washed with hot water and detergent or cleaned by other approved method.
	"		7 years	Painting	Inside walls and ceilings if painted or varnished to be repainted or revarnished.
	"		14 months	Whitewash or colour wash	Inside walls and ceilings if not painted or varnished.
Hoists and lifts	"	Sec. 22	6 months	Examination	Competent person to thoroughly examine.
Chains, ropes and lifting tackle	"	Sec. 26	6 months	Examination	Competent person to thoroughly examine.
	"	"	14 months	Annealing	All chains and lifting tackle except rope slings and others unsuitable for treatment to be annealed.
	"	"	6 months	Annealing	All chains and lifting tackle made of 12 mm bar and smaller, except as above, and all used in connection with molten metal or slag to be annealed.
Cranes and other lifting tackle	"	Sec. 27	14 months	Examination	Competent person to examine all parts and gear.
Breathing apparatus, safety belts and ropes and reviving apparatus	"	Sec. 30	Monthly	Examination and maintenance	Breathing apparatus, safety belts and ropes and revival apparatus in factories where work has to be done in confined spaces where dangerous fumes may be present.

Facility concerned	Act or Regulation	Section No.	Max. interval between attentions	Attention required	Details of facility and attention required
Steam boilers	Factories Act 1961	Sec. 33	14 months	Examination	All boilers except those detailed below:
	(including Examination of Steam Boiler Regs.)	Reg. 4	26 months	Examination	(a) Water tube boilers with fusion welded or forged drums and headers with capacity of at least 50 000 pounds steam per hour. (b) As (a) with capacities of at least 25 000 pounds steam per hour when forming part of group of boilers with capacity of at least 100 000 pounds steam per hour. (c) Waste heat boilers with fusion welded seams and fusion welded superheaters forming part of a continuous flow installation in a chemical or oil refining process plant.
Steam receivers and containers	,,	Sec. 35	26 months	Examination	All steam receivers with certain exceptions.
Air receivers	,,	Sec. 36	26 months	Examination and cleaning	All air receivers.
Water sealed gas holders	,,	Sec. 39	2 years	Examination	All water-sealed gas holders with storage capacity of at least 5000 cu ft to be externally examined by competent person.
Water-sealed gas holders	,,	Sec. 39	10 years	Examination	Gas holders of which any lift has been in use more than 20 years shall have internal state of sheeting examined by cutting samples from crown and sides.
Fire warning devices	,,	Sec. 51	3 months	Testing and Examination	To be tested and examined at least once every three months and whenever an inspector requires.

Facility concerned	Act or Regulation	Regulation No.	Max. interval between attentions	Attention required	Details of facility and attention required
Electricity insulating stands	The Electricity Regulations 1908	Reg. 23	As necessary	Maintained in sound condition	Insulating stands and screens to be provided and kept permanently in position where necessary.
		Reg. 24	As necessary	Periodically examined	Portable insulating stands, screens, gloves or other suitable means to be provided and periodically examined by a competent person.
Ventilation for metal grinding	The Grinding of Metals (Miscellaneous Industries) Regs. 1925	Reg. 17(a)	6 months	Examination and testing	All ventilating plant for dust control to be examined and tested and any defect revealed to be repaired.
		Reg. 17(b)		Recording results	Particulars of examination and tests to be recorded in a register, Form 89.
Power presses	The Power Presses Regulations 1965	Reg. 52	12 months	Examination and testing	Power presses on which tools are fenced exclusively by fixed fencing within immediately preceding 12 months or in any other case.
			6 months	Examination and testing	— within the immediately preceding 6 months.
		Reg. 53	6 months	Examination and testing	Protective devices on power presses other than fixed fencing.
Radiation sources	The Ionizing Radiations Sealed Sources Regs 1969	Reg. 15(3)	26 months	Test for leak of radioactive substance	Prescribed test for leakage of radioactive substance to be made by a qualified person.

The statutory requirements for inspection, cleaning, lubricating, painting, etc which would be included in a system of preventative maintenance are given in *Table 2.11.1*.

Preventative maintenance is not always part of a planned overall scheme, but is sometimes dependent on *ad hoc* checks of the condition of the plant and equipment, or based on methods of condition monitoring explained in section 2.10.

A unified system of planned preventative maintenance within a company requires a great deal of organisation and enforcement, and frequently arouses resentment among maintenance workers who feel that their skills are being depersonalised and that they are being turned into robots. This is not true of a good system.

As an example, a lubricator working to a planned programme might have a separate routine for every day of the week, so that every day he was lubricating only those machines which required particular grades of lubricant. The lubricator would also be required to make a rudimentory inspection of all machinery which he encounters. His instructions and tasks for each day would be clearly spelt out on a work order card which he would be required to sign for each lubrication task completed, and the card would be returned to records and filed at the end of his working day. He would also be required to initiate a 'Squark Sheet' for every defective or leaking item of machinery found on his rounds, with an indication of the urgency required for repair, and to channel this through the maintenance planner who is responsible for arranging corrective action.

In large and complex works whose plants must operate with a minimum of downtime, the organisation of preventative maintenance may require critical path planning methods and even the use of a computer. More commonly, punch card systems which can be sorted by a simple office machine are used and found to be adequate (*see Figures 2.11.2 and 2.11.3*). Wall charts and peg boards are also widely used to indicate the state of a maintenance programme and highlight outstanding work.

Whilst the need for a system is inherent in the concept of planned preventative maintenance, it should not be allowed to become over complicated or to generate more paperwork than is absolutely necessary. The magnitude of effort spent on the system and its paperwork should bear a sensible relationship to the cost of maintenance itself. The tail must not be allowed to wag the dog.

The best form of maintenance is to design everything so that no maintenance is needed. Unfortunately this is seldom economically possible.

2.11.2.1 Pressurised systems[5]

Current UK legislation on the inspection and maintenance of pressurised systems is piecemeal and applies only to particular types of pressure vessels — steam boilers, air gas and steam containers, and receivers (*see Table 2.11.1*). The result is that while on a certain works there may be legal obligations to open up and inspect an innocuous low pressure compressed air receiver every two years, there may in the close proximity be a

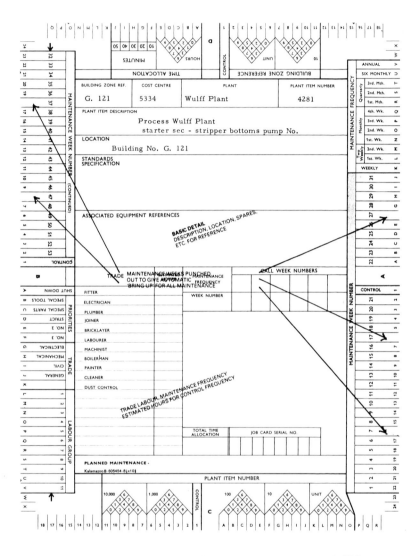

Figure 2.11.2   Kalamazoo punch card for recording maintenance work (Kalamazoo Ltd.)

BUILDING ZONE REF. NO.	PLANT DESCRIPTION	PLANT ITEM NO.	MAINTENANCE	PRIORITY	TRADE	TIME ALLOCATED	AVAILABLE
G. 121	Wulff Plant	4281	Routine	A	Mech.	$2\frac{3}{4}$ hrs.	

TASK CARD INSTRUCTIONS

MAINTENANCE WORK REQUEST FOR
ELECTRICAL CHECK LIST 216
Blg. No. G. 121 Item No. 4281
Process Wulff Plant
Item Starter-Sec. Stripper Btms Pump No. 2
Mech. Item and No.
Matce Routine Cost Code 5334
NATURE OF WORK - STARTER INSPECTIONS

		Item No.	Std. Time Per Man Hrs.	Std. Time Per Man Mins.	Time Taken On Hrs.	Time Taken On Mins.	Time Taken Off Hrs.	Time Taken Off Mins.	Total Hrs.	Total Mins.
1	Isolate unit in accordance with regulations	1	-	25						
2	Check condition of the main and auxiliary contacts and that the spring pressures are satisfactory	2	-	15						
3	Ensure that all connections are electrically and mechanically sound	3		10						
4	Check level and appearance of the switch tank oil, and oil of the immersed potential transformers. If oil is discoloured, or if black deposit is detected on mechanism, contacts or arc suppression devices, change oil	4	1	5						
5	Check that operating mechanism works smoothly and positively and that contacts close in correct sequence	5		15						
6	Check that phase barriers are clean and undamaged	6		10						
7	For contactor operated starters, check that air gap between closing magnet poles is within specified limits	7		5						
8	Ensure that shading rings and pull off springs, where provided, are in order	8		5						
9	Check for looseness of contactor coils and rectify	9		15						

MAINTENANCE INSTRUCTIONS NEED ONLY BE TYPED ONCE YET ISSUED AT ANY TIME WITHOUT EXTRA CLERICAL WORK
SAVING WRITING AND COPYING TIME

REFERENCE FOR AGREEMENT ON AVAILABILITY FOR MAINTENANCE ON PLANT

STANDARD MANNING (NO. OF TASK CARDS TO BE ISSUED) TOTAL

THIS SHEET MUST BE RETURNED TO THE ISSUING OFFICE WITHIN ____ DAYS OF ISSUE
DETAILS OF REPAIRS UNDERTAKEN OR REQUIRED SHOULD BE ENTERED →

REASON FOR DIVERGENCE FROM STD. TIME & REPAIRS UNDERTAKEN OR REQUIRED

SPECIAL TOOLS REQUIRED

OPERATOR SUGGESTIONS TO IMPROVE JOB EFFICIENCY

OPERATIVE'S NAME	CLOCK NO.	PAYMENT CODE	JOB CARD SERIAL NO.

Figure 2.11.2 (cont.) Reverse side of card

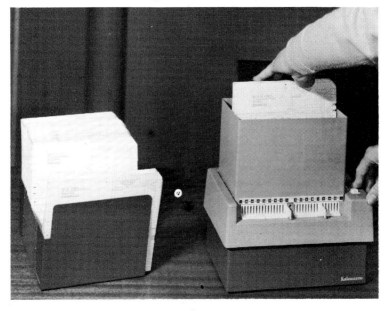

Figure 2.11.3   Punch card filing system (Kalamazoo Ltd.)

high-pressure chemical process reactor handling a corrosive and explosive mixture of chemicals. For this there is no legal obligation to inspect. This situation has long been recognised to be ridiculous.

Proposals for new legislation for pressurised systems are contained in a consultative document issued by the Health and Safety Commission.[5] The intention is to reform the existing legislation to cover nearly all types of static and transportable pressure vessels and pressurised systems. Since the document is at this stage purely consultative, and since its concepts may be altered substantially by the time legislation is passed, it is not discussed in detail here. Central to the proposals is that the user of the pressure vessels should appoint a responsible person to oversee a written scheme of examination and testing to ensure proper maintenance of the plant, and 'competent persons' to carry out the examinations and tests.

These 'competent persons' will usually be members of qualified and approved independent inspection organisations. They may, however, be employees of the company owning and using the pressure vessel provided they are able to demonstrate an adequate degree of independence from immediate production pressures. They will be required to examine any repaired or modified parts of a pressurised system before they are brought back into use, have their own design office staff backed by appropriately qualified specialists, have access to all necessary types of current inspection equipment and be experienced in their use.

2.11.3 Maintenance personnel and training[6]

Maintenance is so important to safety that maintenance men should be specially selected for their experience, ability and alertness and specially trained in accident prevention. Maintenance workers are faced with a complex and changing set of hazards instead of the regular pattern of activity usual for most production workers. They should be trained in the safe use of ladders, slings, ropes and protective equipment of various kinds as well as the tools of their trade, and also in first aid and rescue and fire fighting. In works and factories where they are not continuously engaged in maintenance work, they often form the core of the emergency first aid and fire fighting teams.

Where hazardous materials are present, maintenance workers should be trained in their properties and how to control the hazards. They should also be notified of any new hazard or hazardous material introduced into the areas they cover. They should be trained to inspect ladders, ropes and lifting gear, to recognise excessive wear and to remove worn equipment from service.

At the start of any non-routine maintenance job, the maintenance crew should meet to discuss the hazards and plan methods of safe working. For specially hazardous jobs, the safety manager should be present at the meeting, and after a safe procedure has been agreed (sometimes involving several trials with models), a detailed record of the various steps should be made and copies distributed to the workers involved. Maintenance workers should make a habit of checking all tools used for wear and defects.

The inspection duties of maintenance workers should be clearly defined and special training given where required, e.g. to inspect portable power tools.

Appropriate clothing should be worn for particular jobs. It should fit closely, with the minimum of pockets. Breast pockets should be removed to prevent items falling into machinery, etc when the wearer bends over. Neckties and loose clothing should not be worn and, for most work, rings, jewellery and wristwatches should not be worn.

When the number of tools carried does not warrant a tool bag or box, a belt should be worn with tool carriers at the side of the body, not at the back, to minimise injuries in case of a fall. The kit should generally include a flameproof handlamp. Persons working in high places or inside tanks, silos, etc should wear a lifeline, properly secured, and be trained in its use. Gloves and eye protection appropriate to the work are essential parts of the kit.

Special isolation procedures must be established for the maintenance of any machinery or equipment which might become energised during maintenance.

Maintenance workers should make a regular habit of making tool checks before and after a job to make sure nothing has been left inside a repaired machine or piece of equipment. Machines which have been repaired should where possible be turned over by hand before power is switched on. Guards must always be replaced and properly adjusted at the end of a job.

Maintenance managers should keep up to date with new methods, products and equipment, and arrange for personnel to attend special courses as required. Departments should carry a full stock of service manuals and engineering books which their workers should be familiar with and know how to use.

Maintenance procedures should be periodically reviewed for safety, and a suggestion scheme set up. Permanent ladders, catwalks, and hoists should be provided for frequent maintenance jobs.

2.11.4 Building maintenance

Buildings, as well as equipment, require maintenance to prevent them reaching a dangerous condition. Prompt checking and correction of faults on a regular basis is less costly and far safer than waiting for damage to reach serious proportions.

The safety specialist cannot be expected to be an expert on building maintenance, but should become sufficiently knowledgable at least to be able to distinguish between good and bad practice.

2.11.4.1 Foundations, footings and column bases

Foundations are difficult to inspect, but a careful watch should be kept for settlement, cracks and seepage of water. Settlement is checked by placing level marks on columns at a known elevation (say 1.5 m) above the basement floor; excessive settlement will affect the stability of the building and its equipment, and must be reported at once to the structural engineer for action.

Cracks in foundation walls, particularly below ground level, may allow water to enter and corrode steel columns or rot wooden ones. All cracks should be repaired promptly, rust at the base of steel columns be removed, and an anti-rust coating applied.

2.11.4.2 Structural members

Horizontal members — joists, beams and girders — are often covered by ceilings and inaccessible. The only evidence that something is wrong is found from sagging of the floor. A deflection greater than 1 in 360 of the span length indicates overloading.

Building columns should be checked for unusual distortion or buckling and for holes cut or drilled in them. Stell members should be checked yearly for rusting. Concrete members should be checked for cracks, spalling and chipping.

Wood members should be checked for dry rot, splits, shrinkage and slippage. Wooden beams, joists and girders should have full bearing on their end supports.

2.11.4.3 Walls

Exterior walls should be inspected for cracks. Mortar joints, especially round windows, often loosen and open, and need raking and pointing preferably with a gun which forces the caulking compound into the opening. Interior walls and partitions need similar checking for cracks, holes, loose mortar and worn areas where vehicles may have scraped them.

Walls and columns which are liable to be hit or scraped by vehicles should be guarded by substantial steel railing near floor level. They should be fastened to the floor.

2.11.4.4 Ceilings and floors

Ceilings should be checked for unusual sag (which may be due to excessive loading on the floor above) and for the need for cleaning, repair and painting.

Floors should be checked for holes and irregularities, rot, wear and excessive sag. Traffic must be kept off floors during and after repair until any cements or other compounds have set. Careful attention must be given to floor cleaning and waxes which make wooden floors slippery should not be used. Floors should be protected from leakage of liquid by pans or absorbent material. Allowable floor loadings should be obtained from a reliable source and signs showing these maximum loadings painted or fastened on walls or columns with warnings of the maximum quantities or heights of material that may be stored. The common practice of overloading floors with heavy equipment, trucks and stored materials is extremely hazardous.

2.11.4.5 Roofs, gutters and roof-mounted items

Some flat roofs are damaged by people walking on them and steps should be taken to restrict access to this type of roof. Gutters must be kept clean and checked for cracks.

All roofs need regular inspection. Flashings must be checked to see that the metal is tight with the roof as well as with the drain. Leaking roofs must be repaired as soon as possible after the leak appears; tracing the source of a leak is often difficult. All roofs must be securely anchored.

Water tanks, cyclone separators, stacks, vents and penthouses are often mounted on roofs, and their supports flashed at roof level. These should be checked during roof inspection.

2.11.4.6 Tanks, tank towers and stacks

Elevated tanks and their supporting towers must be carefully examined and maintained to guard against fire and structural failure.

Stacks are subject to deterioration, both inside and out caused by wind, weather, corrosive flue gases, lightning and settlement. They consequently require frequent examination. The lightning protectors of brick and concrete stacks should be checked to ensure that they are continuous from the top of the stack and well grounded.

2.11.4.7 Loading platforms

The edges of loading platforms should be protected against damage from vehicles by angle or channel iron. The surfaces should be kept in good repair and free from ruts.

2.11.4.8 Underground maintenance

Before entering sewers, tunnels and pits, the atmosphere should be tested for oxygen deficiency, methane, hydrogen sulphide, carbon monoxide and carbon dioxide and proper ventilation and/or respiratory equipment used. At least two men should work on sewer maintenance (see section 6.1 'Enclosed Spaces').

Blowers are preferable to suction fans for ventilating sewers, pits, tunnels and trenches because they supply air from a source above ground. All trenches over 1.5 m deep in which work is being carried out should be shored.

Waste disposal channels and trenches should be kept in good repair and checked for settlement which may occur as a result of leaks or seepage which carry earth with them.

The position of underground pipelines and cables must be carefully checked before excavation is begun. Underground pipelines should not be cut into or opened until they have been completely isolated, depressurised, emptied and, where necessary, purged. Excavations should not be made immediately below buried pipelines in use.

At night, open trenches must be protected by barricades, signs and lanterns.

2.11.5 Maintenance of lighting systems and interior decoration[6,7]

Various aspects of this are discussed in more detail in section 3.3.

Precautions must be taken when replacing lamps, whether incandescent, fluorescent or mercury vapour. Gloves should be worn when handling lamps. Replacement should, where possible, be done at weekends or at times when personnel are not present, to reduce exposure to flying glass and dust.

Spent lamps may need to be broken before they can be disposed of. This should be done out of doors after wrapping the lamp in newspaper to contain the broken glass and dust (which may be toxic).

Lighting conditions should be checked with a meter. Fluorescent fixtures should be of the locking type. Reflectors, walls and ceilings must be cleaned regularly to maintain illumination levels. Care is required in selecting means of reaching lamp fixtures for replacement. Ladders, step-ladders or maintenance platforms should be provided with built-in compartments, trays or fixtures for lamps, cleaning buckets and materials.

Attention should be given to the spectral distribution and composition of both the primary light source and of the factory interiors. There are many different fluorescent and mercury lamps, some of which appear to have a spectrum close to natural daylight. In fact this is deceptive. The spectral distribution of an 80 W colour matching fluorescent lamp (*Figure 2.11.4*) has four characteristic humps representing the spectrum of the mercury vapour discharge superimposed on a smoother curve given by the phosphors. One problem and potential hazard with such lighting is colour metamerism.

A metameric pair of colours will match under one light source but not under another. This presents the hazard that two objects or parts of an object which can be readily distinguished by the colour difference in natural daylight or in one type of artificial light, appear to have identical colours in another artificial light. Then if one object is superimposed on the other as a background it may be invisible.

Once a satisfactory lighting scheme which avoids this problem has been developed it is important that the precise types of all lamps used be recorded with their positions so that they are always replaced with similar

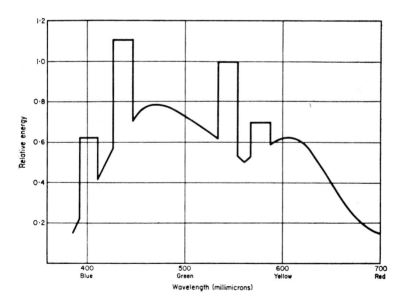

Figure 2.11.4 Spectral energy distribution of an 80-watt colour-matching fluorescent lamp. The vertical lines in the curve represent mercury vapour discharge. Phosphor output in fluorescent lamps is shown on the curved line (From 'Physical Working Conditions', ed. W. McCullough. Published by The Gower Press)

lamps. If the lighting as originally planned proves unsatisfactory and needs to be improved, it is important to check that the improvement does not introduce hazards of this kind.

The colour of interior surfaces brings two problems. In the past such surfaces have tended to absorb far too much light making adequate illumination levels difficult. The second is a psychological one and depends on achieving a good colour balance between the cool colours — blue, green and yellow green — which have shorter wavelengths and the warm colours — yellow orange, orange and red — which have longer wavelengths. The short waves produce reactions of coolness and recession (and increase heating bills!) whereas the long waves produce the reaction of warmth, nearness and dominance. Colour and colour balance is critical for a factory environment; lack of balance leads to monotony and feelings of contempt for the environment and equipment. A scheme that is too warm will produce irritation, discomfort and bad optical conditions.

When light-coloured interior surfaces are employed, it is of course important to have a regular cleaning programme to maintain effective illumination. If surfaces become dirty quickly — in, say, two weeks or less — this is a clear indication that the concentrations of particulate matter or toxic gases (e.g. hydrogen sulphide) in the atmosphere are excessive and probably a hazard to health as well as illumination.

2.11.6 Pipework maintenance

Identification of pipework by colour schemes, tags and stencils is important to prevent accidents[8]. Maintenance personnel need special protection and warning against opening wrong valves, disconnecting or reconnecting wrong pipes. Stories abound of such mishaps, such as connecting steam to handrails or linking toilet drains into oil-water separators in refineries.

Before any work is done on a pipeline it must be isolated completely by closing the valves; the valves must be locked and tagged and the section of the line depressurised and drained. The maintenance supervisor must determine what protective gloves, goggles and clothing are needed to protect workers against material which can not be drained. Clearly marked emergency showers must be provided for plants handling chemicals, and maintenance workers should know their position. Workers should have cotton waste or other absorbent materials at hand to clean up spillages. They should also have copies of piping drawings showing the location of equipment, valves and safety features.

When much pipework is involved, care must be taken that floors are not overloaded. Warning flags should be attached to the projecting ends of long lengths of pipe in transport. Where much overhead pipework is being joined and installed, it should be lifted and supported by suitable rigging while work on it is being done.

2.11.7 Maintenance shops and stores

These must be adequate for the work in hand, and have all the machine tools and other necessary equipment suitably arranged to handle the work. If welding, paint spraying or solvent cleaning is carried out, the shops should be well ventilated.

2.11.8 Lubrication

A full survey of lubrication requirements should be made, and the information entered on the machinery records. The lubrication engineer should be able to oil every part needed without danger from moving parts, and special features and fixtures provided where necessary to achieve this. The lubrication engineer should be provided with a diagram of every machine on which all the parts requiring lubrication are clearly marked showing the type of lubricant required at each point and how frequently it must be applied. Leakage and excessive use of lubricant can cause hazards.

Lubrication should be carried out according to a carefully planned

STANDARD CODING FOR LUBRICATION SYMBOLS

FREQUENCY CLASSIFICATION		LUBRICANT CLASSIFICATION		LUBRICANT CLASSIFICATION GRADE NUMBERS										
				VISCOSITY RANGE — REDWOOD SECS 140°F										
SHAPE-FREQUENCY		COLOUR	OIL	35-40	41-45	46-55	56-75	76-100	101-130	131-170	171-300	301-450	451-600	601-1000
Circle	Daily	Red	Lubricating—Light	1	2	3	4							
		Black	Lubricating—Medium						5	6	7			
		Orange	Lubricating—Heavy								8	9	10	11
		Blue	Hydraulic	1	2	3	4	5	6					
Triangle	Weekly	White	Slideway					4/5		6/7		8/9		
		Red stripes on white ground	Special Purpose This applies to any oil outside the above classification.	Grade numbers do not apply. Sequence numbers are applied to differentiate between types of special purpose oils.										

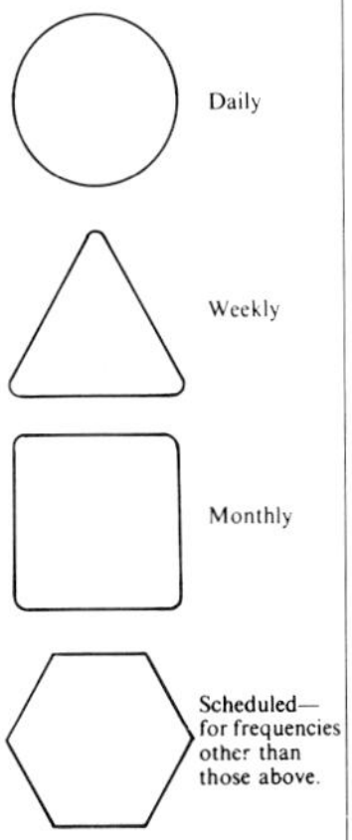

				N.L.G.I. No.			
Square	Monthly	COLOUR	GREASE	1	2	3	4
		Yellow	Lithium	1	2	3	4
Hexagon	Scheduled— for frequencies other than those above.	Pink	Soda	1	2	3	4
		Green	Lime	1	2	3	4
		Yellow stripes on black ground	Special Purpose This applies to any grease outside the above classification	Grade numbers do not apply. Sequence numbers are applied to differentiate between types of special purpose greases.			

Grade numbers are black or white dependent on background colour.

Figure 2.11.5   Standard coding for lubrication symbols (PERA)

programme, so that each worker responsible for lubrications has a supply of all oils, greases, guns, etc before starting work. A standard code for lubrication symbols has been prepared by the Production Engineering Research Association (*Figure 2.11.5*). The programme should make it clear whether it is necessary to stop or switch off any particular machine or motor before starting work, and whether it is necessary to remove any guards. In the latter case, the worker should be trained and instructed to remove and replace the guards correctly.

2.11.9 Equipment inspection

Nowadays, inspection of equipment is carried out by non-destructive testing and by condition monitoring while the equipment is operating. These subjects are discussed in section 2.10. Nevertheless, a great deal of inspection of plant and parts while the plant is shut down by special inspectors is still necessary.

Inspectors should be experts in the equipment under their care; this applies particularly to electrical equipment, which may stop or fail without any obvious warning. Inspectors must be equipped with suitable instruments for making their tests and measurements. Some of these are discussed in section 2.10.

REFERENCES

1. BS 3611:1964, *Glossary of general terms used in maintenance organisation.* British Standards Institution.
2. Bigell, V., Peters, G., and Pym, C., *Catastrophic failures*, The Open University Press, Milton Keynes, 67 (1977).
3. Clifton, R. H., *Principles of planned maintenance*, Edward Arnold (1974).
4. Fife, I., and Machin, E. A., *Redgrave's Health and Safety in Factories*, Butterworth, 26 *et seq* (1976).
5. Health and Safety Commission, *Proposals for new legislation for pressurised systems*, Consultative Document, HMSO (1977).
6. National Safety Council, *Accident prevention manual for industrial operations*, 7th Ed., N.S.C., Chicago, 435 *et seq* (1974).
7. Garrett, A., 'Environmental Colour', Chap. 4 in McCullough, W., *Physical Working Conditions*, Gower Press (1969).
8. BS 1710: 1975, *Identification of Pipelines,* British Standards Institution.

3 THE WORKING ENVIRONMENT

3

The working environment

3.1 INTRODUCTION

When speaking of the working environment, we mean mainly the man-made environment. This is often imperfect and contains hazards to life and limb, but without it the human population of today could not survive nature's hazards for long. Before the inventions of fire, shelter and clothing, man could gain no more than a bare toehold in the latitudes of northern Europe.

The natural environment on this planet varies widely in its friendliness or hostility towards man. Ideas on heaven and hell and the Garden of Eden probably arose from comparison of different natural environments. Monte Cerboli in Tuscany, a volcano at the centre of the soffioni or blow holes from which steam laden with boric acid and sulphur issues into the earth's atmosphere was regarded by the local inhabitants as the entrance[1] to Hell, and was probably the inspiration for Dante's 'Inferno'. Climate, and variations in the composition of the earth's crust are but two factors which make for a good or bad natural environment for humans.

Hippocrates, father of modern medicine, in his treatise[2] *Airs, Waters, Places*, advised his pupils to study the major features of the environment of their potential patients when considering setting up in practice at some new locality. Hippocratic medicine rested on the concept of a balance between the living organism and its environment.

However, if the environment was considered important to health, it was only the health of the wealthy which concerned Hippocrates and a doctor could hardly make a living by caring for the health and environment of slave workers. The Romans recognising the dangers of mercurial poisoning, used only slaves to work their mines at Almaden in Spain, about which Ramazzini[3] was to write in 1713 'It is from mercury mines that there issues the most cruel bane of all that deals death and destruction to miners.'

Ramazzini was one of the first medical men to take a practical interest in the working environment, which is typified in the following advice:

'When a doctor visits a working class home ... to the questions recommended by Hippocrates, he should add one more — What is your occupation?'

239

Table 3.1 SPECIFIC ENVIRONMENTAL HAZARDS OF INDUSTRIAL OCCUPATIONS (LESS COMMON TODAY)

Category of hazard or disease	Common name of hazard or disease	Occupation or industry	Proper name or cause
Physical	Boiler makers' deafness	Boiler makers, rivetters	Exposure to excessive noise
	Cauliflower ears	Boxers	Fist blows of opponent
	Covent Garden tummy	Porters, market garden	
	Billingsgate hump	Porters, fish market	
	Tailors ankle	Tailors	Chronic bursitis caused by pressure, friction or repeated blows over a bursa.
	Hod carriers shoulder	Bricklayers and their mates	
	Weavers bottom	Weavers	
	Housemaids knee	Housewives	
	Pneumatic hammer disease	Operators of pneumatic hammers and portable powered tools	Vascular lesion and cyanosis of extremities caused by vibration.
	Raynaud phenomenon		
	Bends, staggers	Divers	Gas bubbles in blood during depressurising.
Metal poisoning	Painters colic	Decorators	Inhalation of dust containing lead
	Hatters shakes, Danbury shakes }	Felt hat makers	Inhalation of fibres containing mercury.
	Brass founders fever	Brass foundries	Inhalation of fumes containing zinc and other metals.
	Metal fume fever	Welding, oxy-fuel gas cutting }	
Other elemental poisoning	Phossy jaw	Rat poison and fireworks manufacture	Handling yellow phosphorus
Pneumoconioses	Card room asthma	Cotton workers	Byssinosis
	Printers asthma	Printers	Inhalation of gum acacia spray
	Cuttlers asthma	Cuttlers }	
	Potters consumption	Potters }	Silicosis and/or metal poisoning through inhaling hazardous dusts.
	Stonemakers disease	Stone masons }	
	Miners phthisis	Miners	Silicosis
	Grinders consumption	Grinder operators }	

Category	Disease	Occupation/Exposure	Cause
Biological	Glanders	Grooms, shoesmiths	Contact with infected horses
	Wool sorters disease	Sorters of wool and other fleeces and bristles }	Anthrax
	Maladie de Bradford		
	Miners anaemia	Exposure to warm, wet earth	Hook worm disease
Cancers	Mule spinners cancer	Contact with carcinogenic mineral oils including oil mists used in lubrication of machine tools.	Cancer of skin and scrotum.
	Cancer of bladder	Dye workers and dyestuffs manufacture	Contact with particular organic amines.
	Cancer of lung	Nickel refining / Chromate production / Asbestos industry / Sheepdip manufacture }	Inhalation of specific carcinogens.
	Cancer of breast	Nuns	
	Cancer of cervix	Prostitutes	
Miscellaneous	Dishpan hands	Housewives	Dermatitis, skin lesions caused by contact with cleaning solutions.
	Hop pickers gout	Hop pickers	Traumatic tenosynovitis
	Stokers cramp	Workers in high temperatures	Loss of sodium chloride through perspiration.
	Writers cramp	Lawyers clerks and others }	Psychoneurosis and/or Lactic acid build up.
	Money counters cramp	Bank clerks and others	
	Arc eye	Welders mates and others exposed to welding radiation	Photo-opthalmia
	Malt workers lung	Production of malt	Allergy leading to fibrosis.

Excellent though this advice was, it was too much to expect doctors of the day, whose livelihood depended largely on curing the ills, real or otherwise, of wealthy patients, to effect the social revolution needed for them to be able to follow it. The industrial revolution in Britain which largely took place between 1760 and 1830 was accompanied by the growth of a host of environmental hazards each characteristic of a particular occupation, and a fatalistic attitude towards them by their victims and onlookers. Typical is the comment of an Irish woman on work in a litharge factory as quoted by Dickens[4] in *The Uncommercial Traveller*:

'Some of them gets lead-pisoned soon, and some of them gets lead pisoned later, and some, but not many, niver: and 'tis all according to the constitooshun, sur; and some constitooshuns is strong and some is weak.'

So common were many industrial hazards or occupational diseases that they became household words[5]; a selection is given in *Table 3.1*. Although most of these have been largely eliminated or are under control today, the ingenuity of the research chemist has ensured that others take their place, and the rate at which new compounds are synthesised greatly exceeds that

Figure 3.1.1   Showing effects of phossy jaw (London Hospital Medical College Museum)

at which all their possible effects on humn beings who may be exposed to them can be tested.

These specific environmental hazards are more hazards to health than causes of accidents. The environment in which we work is, however, largely man made, and by careful control of environmental conditions, much can be done to remove both hazards to health and hazards liable to cause accidental injuries.

REFERENCES

1. Mellor, J. W. *A comprehensive treatise on inorganic and theoretical chemistry*, Longmans, London 5, 49 (1952).
2. Jones, W. H. S., (English Translation 4 vols), *Hippocrates*, 'Airs Waters Places', Heineman (1931).
3. Ramazzini, B., *De morbis artificium diatriba*, Geneva (1713). Translated by Wright, W. C., University of Chicago Press (1940).
4. Dickens, C., *The Uncommercial traveller*, Hazel, Watson & Viney, London.
5. Hunter, D., *The Diseases of occupations*, E.U.P. (1957).

3.2 AIR AND BREATHING

Contents

3.2.1 The air about us

The air surrounding the earth amounts to approximately 5.1×10^{15} tonnes (dry basis). Its composition at sea level is approximately 78% vol nitrogen, 21% vol oxygen and 1% vol argon plus other inert gases, carbon dioxide (0.03%) and some dust, as well as trace impurities such as sulphur dioxide and fluorinated hydrocarbon gases which man has introduced. Air also contains water vapour, the quantity of which varies over a wide range (up to about 4% vol in the tropics). The composition in terms of the main molecular species nitrogen, oxygen and argon is remarkably constant at all points of human habitation on the earth's surface, although reversible changes take place as a result of ultra-violet radiation in the rarefied upper atmosphere, with the formation of ozone and other species.

The carbon dioxide content shows more local variations and the sulphur dioxide content even more. The air's density, which is inversely proportional to its pressure decreases from about 1.2 kgm^{-3} at the surface to 0.7 kgm^{-3} at 5000 m which is close to the extreme limit of human habitation. Our respiratory processes suffer from the reduced partial pressure of oxygen at higher altitudes, and the general effect of this is shown in *Figure 3.2.1*.[2]

Most industrial processes operate on the earth's surface at heights between 0 and 1000 m above sea level. Altitude at these levels seldom presents a hazard, although above 2000 m, significant problems may arise.

The earth's atmosphere has probably had much the same composition for the last 100 million years. Its high oxygen content makes it unique among the atmospheres of planets. Two billion years ago there was a great deal more carbon dioxide in the atmosphere but very little oxygen. The oxygen has nearly all been formed by photosynthesis by plant life from carbon dioxide. Most of the original carbon dioxide has either been converted through plant life into organic matter which has been buried (the fossil fuels) or converted into insoluble mineral carbonates such as chalk.

The oxygen and carbon dioxide in the atmosphere play essential roles in plant, animal and human life, in combustion processes and in many industries.

The atmosphere contains approximately 2.4×10^{12} tonnes of carbon

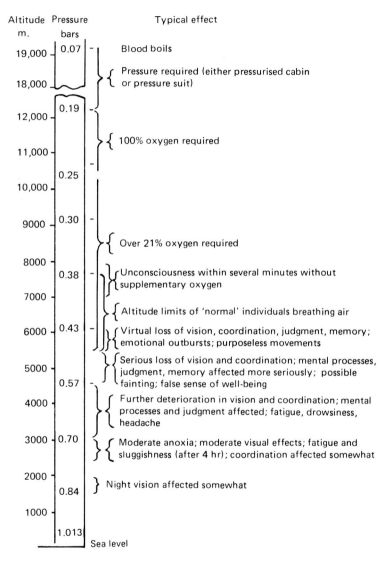

Figure 3.2.1   Effects of altitude on human performance (Based on data from 'Human Factors Engineering' by E.J. McCormick[2])

dioxide. In the natural course of events about 8×10^{10} tonnes of carbon dioxide (i.e. about 3% of the total) are removed annually by photosynthesis and converted into plants and trees, and an equal quantity is returned to the atmosphere as a result of animal and human respiration, and the burning or decay of vegetation (*Figure 3.2.2*). By extracting and burning fossil fuels man has seriously disturbed this natural Hippocratean balance. As a result of Man's activities, about 5×10^{10} tonnes of carbon dioxide (i.e. over half the amount formed naturally by respiration and decay of plant life) are being formed every year by combustion of fossil fuels. This might be expected to cause an increase in the carbon dioxide content of the atmosphere, and indeed this is what is happening. Between 1870 and 1970 the total quantity of atmospheric carbon dioxide is estimated to have increased by 11% (from 294 to 321 ppm), and a further increase to 370 ppm may be expected by AD 2000! The increase would be considerably greater were it not for the fact that the oceans form a large additional reservoir for carbon dioxide. The interference by Man with the balance of his natural environment appears to represent a major hazard which future generations will have to endure.

Carbon dioxide in the atmosphere absorbs part of the sun's rays which would otherwise be reflected, whereas it does not contribute to the same extent to heat loss from the earth to the surrounding space. One of the first effects of an increase in the carbon dioxide content of the atmosphere is warming of the atmosphere by a greenhouse effect. This may in time cause melting of large parts of the ice caps of Greenland and Antarctica and a general rise in the level of the oceans. The first ice sheet to melt would be that known as the Western Antarctic Ice Sheet. This rests on ground which is well below sea level and which would dissolve if the water became a little warmer.

Indeed, there has been an increase in ocean level of about 15 cm between 1930 and 1948, accompanied by a general rise in air temperature and the retreat of many of the glaciers in the northern hemisphere.[1,3]

The process seems to have been halted and even went into reverse between 1950 and 1970, but current indications suggest that it is now again in progress, and warnings from responsible scientists have been given that a significant raising of the ocean level (by as much as 5 m) could occur within the next generation.[3]

It may seem somewhat far-fetched for directors and safety specialists to worry about such matters. The carbon dioxide content of the atmosphere is not, after all, the only factor affecting its mean temperature. Particulate dust and sun spot activity are other factors, to mention only two. None-the-less, it does seem prudent that they should take such warnings into account.

At least it is worth considering the added possibilities of flooding when choosing the location of any plant or factory, or whether steps could be taken to protect existing plants and factories from flooding if the anticipated rise in the sea level occurs. Most present ports would be affected as well as low lying areas such as the Netherlands, Florida, and, in the UK the Cambridgeshire fens and Canvey Island (*Figure 3.2.3*).

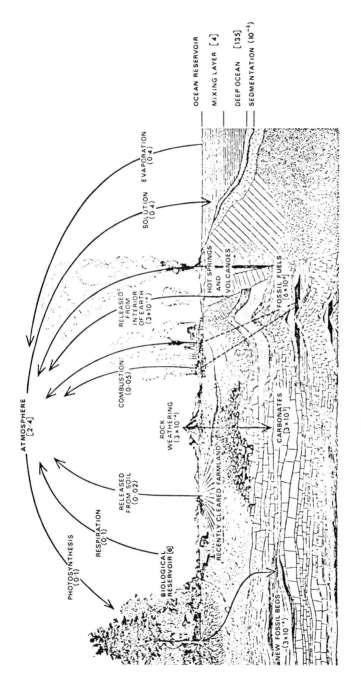

Figure 3.2.2 Carbon dioxide imbalance in the atmosphere (From 'Atmosphere, Weather and Climate' by Barry & Chorley published by Methuen)

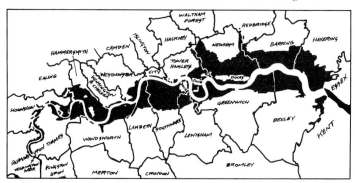

Figure 3.2.3   Showing how parts of London would be affected by flooding

Other ways in which Man has interfered on a grand scale with the atmospheric environment are:

1. Clearing forests.
2. Intensive soil cultivation with higher yielding crops.
3. Irrigation.
4. Increase in the amount of particulate matter released into the atmosphere.

The first three of these must clearly affect the carbon dioxide balance though it is difficult to know their overall effects. It seems that the cutting down of forests tends to make the regions concerned more arid and therefore increases the need for irrigation. Replacing trees which took many years to grow by crops which take a year or less between sowing and harvesting reduces the inventory of organic matter on the earth's surface. By removing this 'buffer capacity' the overall stability of the system is reduced. Increase in industry has increased the amount of atmospheric pollution by particulate dust, and reduction in forest cover and exposure of soil to winds has done the same, probably to an even greater extent. Evidence from the Soviet Union shows a sharp rise in dust-fall on mountain snowfields since 1930, and atmospheric turbidity has increased by 57% over Washington D.C. between 1905 and 1964 and by 85% over Davos, Switzerland (1920-1958).[1] These increases in particulate pollution of the atmosphere seem to be largely man-made.

By 1970, man's activities were said[1] to account for about 30% of the particulate dust in the atmosphere. Although the effect of these activities can sometimes be eclipsed temporarily by some natural phenomenon such as the Krakatoa eruption of August, 1883, which had world-wide effects on atmospheric turbidity, they are none the less more serious because they are more persistent.

Particulate matter in the atmosphere can only be injurious to human health, especially when it has a high silica content. The increase in concentration of atmospheric particulate matter which has occurred simultaneously with the increase in carbon dioxide content in fact has the opposite effect to carbon dioxide, since it increases the back scatter of short wave radiation. Therefore efforts to reduce particulate pollution will (if successful) probably bring forward the day when the waters rise and much of London's Underground system is flooded, along with other and worse calamities; that is unless at the same time we drastically cut down on our burning of fossil fuels. Increasing the rate of carbon dioxide extraction by photosynthesis will not help us since nearly all the additional plant growth will be either eaten, burnt or will decay. To restore the Hippocratean balance we have to bury it again deep underground.[1]

By the year 2000 it is said that the human population on this planet will be about 6 000 000 000.[4] This corresponds to 16 tonnes of carbon dioxide per year per man, woman and child converted into vegetable matter by photosynthesis. This is 11 tonnes per year of dry vegetable matter (assumed to be carbohydrate) which is surely none too much, considering:

1. Much of it is inaccessible to man and rots or is eaten by other species.
2. The animals as well as men need some.
3. Man has to keep his body warm, cook, and burn his bricks, not to mention making steel implements and bicycles and other means of transport.

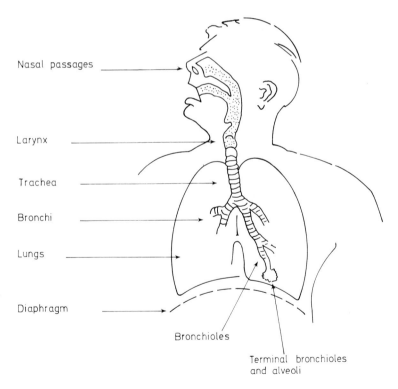

Nasal passages

Larynx

Trachea

Bronchi

Lungs

Diaphragm

Bronchioles

Terminal bronchioles and alveoli

Figure 3.2.4   The human respiratory system (The Chemical Engineer)

3.2.2 Respiration

Man's respiratory system is shown in simplified form in *Figure 3.2.4.*

As it is so reliable and because we are unconscious of it except when it is involved in trouble, we take it entirely for granted, usually without stopping to consider its limitations or whether contaminants in the air could harm us through its delicate organs. It carries oxygen to an enormous number (about a billion) of air cells or alveoli through the walls of which oxygen passes to the blood, and through which carbon dioxide is removed into the air which is expelled on the return trip.

3.2.2.1 Air and oxygen requirements

The human respiratory system has about a tenfold range in working capacity, and for much of the time is working at the lower end of its capacity. Oxygen consumption, lung ventilation, percentage of oxygen utilised and heart beat for various work loads are given in Table 3.2.1.

Table 3.2.1 LUNG AND HEART PERFORMANCE AT DIFFERENT WORK LOADS[6]

Work load	Oxygen consumption l/min	Lung ventilation l/min	% oxygen absorbed	Heat beat beats/min
Very low (resting)	0.25-0.3	6-7	20	60-70
Low	0.5-1.0	11-20	22	70-100
Medium	1.0-1.5	20-31	23	100-125
High	1.5-2.0	31-43	23	125-150
Very high	2.0-2.5	43-56	22	150-175
Extremely high (Sport)	2.5-4.0	60-100	20	over 175

The air expired has 16 to 17% by volume of oxygen and 4 to 5% by volume of carbon dioxide. If the oxygen content of the air inspired falls below 20% by volume, respiration becomes noticeably harder. Respiration becomes virtually impossible if the oxygen content falls below 14% volume.

The respiratory system is under the control of a centre in the brain which is normally automatic and works whether the subject is awake or unconscious, although it is subject as required to overall conscious control (e.g. as diving or swimming). When working under automatic control it responds to impulses from other regions of the body and to chemical stimulae such as the concentrations of carbonic or lactic acid in the blood reaching it. If the nerves which carry impulses from the brain to the respiratory muscles are cut, respiration ceases. (The heart whose rate is also controlled by a centre in the brain has its own built-in pacemaker which is sensitive to many agents in the blood such as carbon dioxide. It will continue to function even if the nerve connecting it to the brain is cut.)

From the above it follows that a man working under a medium work load needs to be supplied with a minimum of about 5 m³ per hour of fresh air for respiration only. The Factories Act 1961, section 2, requires that a minimum air space per employee in a factory of 400 ft³ or 11.5 m³, so that a complete air change every two hours would in theory suffice for respiratory purposes. In practice this would, in fact, be intolerably 'stuffy' due to the accumulation of body odours and build up of water vapour in the atmosphere.

In normal ventilation practice, assuming smoking were permitted an air flow rate of about 40 ft³ per person per minute (or slightly more than 1 m³/minute) would be expected. This would correspond to 6 air changes per hour with 400 ft³ per person, or twelve times that required for respiration only. For non-smoking adults the figure could be considerably less — about 12 ft³ per person per minute (i.e. 0.34 m³/minute or 20.4 m³/hour) although they would probably object to the body odours at that level.

Man's normal requirements of fresh air in factory conditions are several times higher than his requirements for respiration only, and are governed by the need to remove his own odours, as well as those of any toxic gases, vapours, fumes or dusts generated by the work.

3.2.3 Air contaminants and respiratory diseases

Man's respiratory system is singularly susceptible to the effects of various contaminants in the air he breathes — gases, vapours, fine droplets and suspended particles, often at concentrations well below those at which he is even aware of their presence. It is even more unfortunate that he is all too frequently unaware of the dangers he runs in breathing such contaminated atmospheres. Some of these contaminants and their effects were mentioned in the opening section of this chapter.

Lung diseases have always been prevalent among miners, although the type and severity has depended on the composition as well as the concentration and size of the dust particles.

The first occupational lung diseases were recognised as 'pneumoconioses' in about 1870, the term meaning 'dusty lungs'. These involve the inhalation of insoluble particles of minerals such as silica, asbestos, talc, mica, fuller's earth and coal dust. Other mineral dusts such as limestone, lime, cement, gypsum, and iron oxide are less harmful. Inorganic dusts which have caused pneumoconioses include aluminium powder, beryllium and silicon carbide, whereas tin oxide and barium compounds appear to have little effect. The main feature of pneumoconioses is that the particles accumulate inside the lung of the victim.

The size of the particles is important in determining whether they reach the alveoli and get trapped there. Particles of less than 5 μ can reach the alveoli and particles below 1 μ have the highest probability of being deposited there. Their fate then depends on their solubility and reactivity. One reason for the greater hazard of quarrying sandstone compared with limestone appears to be that the silica particles are not dissolved and simply accumulate in the lungs, whereas limestone particles are dissolved by the lung fluids and removed by them.

Some dusts of vegetable and animal origin also produce pneumoconioses, although these usually operate by producing an allergy rather than causing a large and harmful accumulation. Dusts such as pollen and fungus spores are large enough to be trapped in the upper respiratory passages, producing asthma and hay fever. Dusts of smaller particle size which reach the alveoli can produce chronic bronchitis, broncho-pneumonia and fibroid lung. These include the dusts of bagasse, cotton, derris, flax, flour, grain, gum arabic, hay, hemp, jute, linseed, malt, nuts, paprika, seeds, sisal, straw, tea, tobacco and wood.[7] They seem to have a fairly similar action.

The best known of these pneumoconioses is byssinosis or card room asthma which is found in cotton, hemp and flax workers. The initial symptoms are a tightness of the chest on Monday morning after returning to work after a weekend break. The bracts of the cotton ball contain an agent which causes the lungs to release histamine which causes the constriction.[9] The exposure on Monday morning releases histamine which has been formed and stored over the weekend, but for the rest of the week there is insufficient histamine available in the lungs to produce severe constriction. The constrictions are accompanied by a very aggravating cough and breathlessness. In the second stage the symptoms extend over the rest of the week and become permanent. In the third stage the worker becomes permanently disabled with chronic bronchitis and shortness of breath.[7]

In addition to the pneumoconioses, toxic poisoning commonly occurs through the inhalation of toxic gases (chlorine, carbon monoxide, hydrogen sulphide, arsine), toxic vapours (benzene, carbon disulphide, nickel carbonyl, metal compounds of lead, mercury, tellurium and many others), and toxic dusts, both inorganic and organic. With most of these, the lungs are merely the port of entry through which the toxic compound enters the body, passing first into the blood stream and then to its ultimate target. With gases, such as chlorine, the lining of the lungs themselves is the target.

Many of the toxic contaminants which enter the lung are carcinogenic. It has been estimated that more than 50% of workers certified as having asbestosis will die of cancer of the lung, and recent work has shown that this is not merely confined to crocidolite.[5]

Early detection of pneumoconiosis is very important, since when a certain stage is reached some occupational lung diseases (e.g. asbestosis, byssinosis) progress inexorably without further exposure. If they are diagnosed sufficiently early, before the onset of the progressive phase, much of the morbidity and mortality associated with them can be reduced.

3.2.4 Air contaminant control

The nature and toxicity of the contaminant should be known when planning measures of air contaminant control. These can best be built into the factory or plant at the design stage rather than added later. The first step is one of fact finding. A list of all materials expected or liable to become airborne in the factory or process should be made. This will include all gases, liquids and solids which may be present in a finely divided state. Toxic solids which may be present in bulk form should also be included to ensure that toxic materials which are normally present in bulk but become ground up or disintegrated to form a dust are included and steps taken to ensure their safe handling.

Next their toxicities should be studied with reference to published data. Toxicities are expressed in several ways:

LD 50 (Lethal Dose 50). This is the dose generally expressed as mg/kg bodyweight whereby 50% of the experimental animals die within a given time of administering the materials. The route of dosing and the experimental animal must also be stated.

LC 50 (Lethal Concentration 50). Lethal concentration 50 is that concentration whereby half the number of exposed animals die after a given time; the experimental animal must be stated. Concentration is in mg/kg^3 or ppm.

TLV (Threshold Limit Value). This is the maximum permissible concentration to which workers may be exposed for 8 hours a day and 5 days a week.

EPEL (Emergency Population Exposure Limit). This is the concentration which can be borne for a given time by workers in emergencies without effecting lasting damage to health but possibly causing nuisance or irritation or intoxication.

In addition to the above the Odour Threshold is important, being the lowest concentration in air at which the material may be detected by its odour by workers exposed to it. (Note that odour susceptibility becomes dulled through prolonged exposure.) TLV, EPEL and Odour Threshold are all expressed as mg/kg^3 of air or ppm.

Dangerous and toxic chemicals are also discussed in sections 3.9 and 6.2.

3.2.4.1 Sources of Information

The principal sources of information on toxic materials are:

1. Technical Data Note 2/75[10] — *Threshold Limit Values for 1975*, published by the Health and Safety Executive. This gives TLV values for about seven hundred chemicals and commonly used industrial materials which were drawn up by the American Conference of Government Industrial Hygenists. It also shows which materials are absorbed through the skin, and gives data on a number of mineral dusts, decomposition products of polytetrafluoroethylene, welding fumes, industrial carcinogens, nuisance particulates, simple asphyxiants and methods of calculating TLV's for mixtures.

2. *Documentation of threshold limit values for substances in workroom air*, obtainable from the American Conference of Governmental Industrial Hygienists, Cincinnatti, P.O. Box 1939, Ohio, 45201.
3. *Dangerous properties of industrial materials*, by N. Irving Sax.[11] Besides giving data on a large number of chemicals and industrial materials this has chapters on Toxicology, Industrial air contaminant control, Respiratory protection and personal hygiene, Environmental pollution, Radiation hazards, Storage and handling of hazardous materials and others. This book is of particular value for those concerned with safety and industrial hygiene in the chemical industry and industries which employ dangerous chemicals.

3.2.4.2 Control methods other than ventilation

While ventilation is the most commonly used method, the possibility of using a simpler method which may be more effective should always be explored first. This includes the following concepts:

1. Use of automatic operations which require few, if any, workers exposed to the contaminated atmosphere.
2. Changing the condition of a toxic material, i.e. using it in briquetted form rather than as a powder, or using the powder wet.
3. Substitution of a non-toxic or less toxic material for a more toxic one, e.g. replacement of white lead in paint pigments by zinc, barium or titanium oxides.
4. *Isolation in space or time.* In foundries castings may be removed from sand only on a special afternoon shift when day workers have left, by men wearing suitable respirators. In factories involving electro-plating or paint dipping, the process giving rise to contamination may be segregated in a separate room and provided with general ventilation only. Any exposed workers should then be supplied with respirators and trained how, when and why they should use them.
5. *Segregation of personnel.* Sometimes the operator in a contaminated environment such as a foundry can be placed in a completely enclosed cabin ventilated under positive pressure from outside the building. This applies also to operators of overhead travelling cranes.
6. *Local suppression of contaminants.* This includes the use of wet processing (drilling, crushing, grinding or milling), the use of baffles and evaporation suppressants on the surface of toxic and volatile liquids.
7. *Housekeeping.* This is very important as a positive method of reducing contaminant control. Many toxic materials in bulk if left on the floor are soon trodden into dust which easily becomes air borne. Cleaning by air blowing should be forbidden when toxic materials are involved. Only vacuum cleaning and wet washing should be permitted. Good housekeeping is greatly facilitated by design. Smooth impervious walls and floors make good housekeeping simpler and more effective.

3.2.4.3 Ventilation

Ventilation is the most widely used method of preventing dangerous concentrations of atmospheric contaminants in workrooms, factories and plants. The problems are best solved by an industrial hygienist and a ventilation engineer working together. The ventilation needs will be determined partly by the toxicity of the materials, as given in Table 3.2.2.

Table 3.2.2 MATERIAL TOXICITY AND TLV RANGES[10]

Toxicity	TLV range	
	ppm	mg/m^3
Slight	over 500	over 0.5
Moderate	101-500	0.1-0.5
High	0-100	0-0.1

In general, operations releasing highly toxic contaminants nearly always need control by local ventilation; operations releasing moderately toxic contaminants usually need similar control, while those releasing slightly toxic contaminants occasionally need control.

The TLV's of some common compounds and materials selected from reference 10 are given in *Table 3.2.3*. The reader should, however, be warned that changes in TLV's of materials are periodically made as more research is carried out, and is advised to apply, where necessary, to the Health and Safety Executive for the latest information.

The TLV's of selected compounds which are hazardous through absorption through the skin as well as the lungs, are given in *Table 3.10.3* under 'Protective clothing'.

Table 3.2.3 THRESHOLD LIMIT VALUES OF COMMON COMPOUNDS AND MATERIALS

Substance	ppm	mg/m^3	Substance	ppm	mg/m^3
Acetic acid	10	25	Chloroform	25	120
Acetone	1000	2400	Copper fume dusts		
Ammonia	25	18	and mists	—	1
Benzene — skin	10	30	Cyclohexane	300	1050
Beryllium	—	0.002	DDT	—	1
Boric oxide	—	10	Dibutyl phosphate	1	5
Bromine	0.1	0.7	Dibutyl phthalate	—	5
n-butyl acetate	150	710	Diquat	—	0.5
Cadmium (metal dust			Ethyl alcohol	1000	1900
and soluble salts	—	0.05	Ethyl ether	400	1200
Caprolactam (vapour)	5	20	Ethylene glycol		
Carbon dioxide	5000	9000	(vapour)	100	250
Carbon monoxide	50	55	Formaldehyde	2	3
Chlorine	1	3	Hydrogen sulphide	10	15

(cont.)

Substance	ppm	mg/m³	Substance	ppm	mg/m³
Iodine	0.1	1	Oil mist, particulate	—	5
Lead (inorganic fumes			Ozone	0.1	0.2
and dusts, as Pb)	—	0.15	Paraquat (skin)	—	0.5
LPG	1000	1800	Parathion (skin)	—	0.1
Mercury (excluding			Phenol (skin)	5	19
alkyl compounds)	—	0.05	Phosdrin (skin)	0.01	0.1
Methyl alcohol	200	260	Pyrethrum	—	5
Naphthalene	10	50	Pyridine	5	15
Nickel	—	1	Sodium hydroxide	—	2.0
Nicotine (skin)	—	0.5	Sulphur dioxide	5	13
Nitric acid	2	5	Tetraethyl lead (skin)	—	0.10
Nitric oxide	25	30	Toluene (skin)	100	375
Nitroglycerine (skin)	0.2	2	Trichloroethylene	100	535

Mineral dusts. For dusts containing quartz, the TLV is given by the formula (^{30}/% quartz$+3$) mg/m³.
For asbestos the TLV is 5 fibres/cc greater than 5 μ in length.
For bituminous coal containing <5% quartz, the TLV is 2 mg/m³.
For nuisance dusts containing <1% quartz, the TLV is 10 mg/m³.

Chemical plants producing highly toxic compounds are usually built in the open and rely entirely on natural dispersion by the wind. This economises both on building and ventilation costs, but it leads to very uneven dispersion and it has been the cause of several cases of serious contamination of the surrounding area, with resulting casualties and crop damage.

Building ventilation is of two kinds, local exhaust ventilation and general ventilation (dilution).

An important and often overlooked point with both methods but particularly with local exhaust ventilation, is the need to supply make up air, heated as required, to replace the air removed by the exhaust system. Natural infiltration of air may sometimes be sufficient, but very often it is not. In a room or building with tight walls and doors designed for fire protection and/or noise insulation, the exhaust system will not only be inadequate but will usually create cold draughts in unwanted places, with the result that the exhaust system may be switched off entirely. The make-up air system requires almost as much attention in design as the exhaust system itself.

1. Local exhaust ventilation. This system is designed to capture the airborne contaminant as close to its source as possible and remove it from the environment. It relies primarily on the use of specially designed hoods or canopies, ducting and fans and a means of disposing of the contaminated air stream (*Figure 3.2.5*). The contaminated air may sometimes be discharged into the atmosphere, but it is becoming increasingly necessary first to remove the bulk of the contaminants.

The design of exhaust systems is quite technical and is best left to qualified ventilation engineers with experience in this field. The method chosen for removing contaminants from the exhausted air depends entirely on the nature of the contaminants.

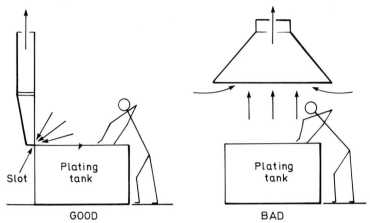

Figure 3.2.5 Direction of air flow around an industrial plating tank. The hood should be located so the contaminant is removed away from the breathing zone of the worker. (National Safety Council, U.S.A.)

For gases and vapours either adsorption on an activated adsorbent such as charcoal or silica gel may be suitable, or absorption in a tower with countercurrent flow of some absorptive liquid in which the contaminant is readily soluble, may be used. In both cases it is generally necessary to provide means of removing the contaminant from the adsorbent (or absorbent) to avoid excessive consumption of these materials. This results in recovering the contaminant in a pure and concentrated form; occasionally the value of the recovered contaminant will bear the cost of recovery.

Another method sometimes used with gases or vapours is reaction with some chemical, either as a solid or in solution. Yet another and perhaps more commonly used method, is incineration or catalytic combustion, where the contaminated air is passed through a flame or catalyst bed maintained at such a temperature that the gas or vapour is burnt harmlessly to carbon dioxide and water vapour.

Entirely different methods are called for when the contaminant is a particulate solid, but again the actual choice of method much depends on the nature, quantity and particle size of the contaminant. Coarse particles are generally separated in cyclones or chambers with baffle plates. Finer particles require more sophisticated methods of removal which include high efficiency cyclones, filters, impingement separators, specially designed liquid washers and electro-static precipitators. Mists and fine liquid droplets require yet other methods which are mainly based on causing the fine liquid droplets to coalesce, e.g. by passing through a pad made of a number of layers of knitted woven wire mesh.

Often it will pay to recirculate cleaned air back to the room, workshop or factory, especially if it has not been cooled in cleaning and the temperature outside is low. In this case it is necessary to monitor the cleaned air for concentration of contaminant to ensure that it is well below the allowable threshold limit value.

The whole subject of local exhaust ventilation and air cleaning, while of the utmost importance to the industrial hygienist and safety specialist, is too specialised to be discussed in any detail here. Each case must be studied by an expert on its own merits.

2. General dilution ventilation. This should be so designed that the contaminants released into the atmosphere are continuously diluted to

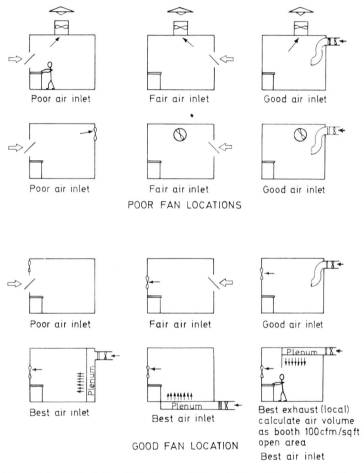

Figure 3.2.6   Principles of dilution ventilation. During winter months, inlet air requires tempering (National Safety Council, U.S.A.)

levels below the TLV at points where operatives are stationed. As with local exhaust ventilation, it must be designed to meet a particular situation, and this requires specialist assistance. One important point is the need to supply sufficient and properly directed make up air to replace the air exhausted. Another is to arrange that the distance which the contaminated air needs to travel from the source of contamination to the fan or other extraction device from the workroom should be as short as possible, and that it should not be drawn past workers or into their faces but away from them. Typical distribution conditions which illustrate this point are shown in *Figure 3.2.6*.

Ventilation is needed not only to reduce risks to workers but also to prevent the accumulation of dangerous concentrations of flammable vapours in the atmosphere. This is discussed later in chapter 4 under 'Fire and explosion hazards'.

3.2.4.4 Air contaminant monitoring

In plants and factories where air contamination by toxic substances may arise, it becomes necessary and often obligatory to sample and analyse the atmosphere for the hazardous contaminant. This requires the use of techniques which are specific to the contaminants concerned.

Sampling of air for contaminant gases and vapours is usually fairly straightforward, but sampling for dusts and mists is more difficult. Laboratory analyses are usually required for absolute determinations, although various instruments are available which give a continuous intermittent record of the concentrations of a number of impurities.

A number of booklets has been prepared by the Health and Safety Executive which describe simple and rapid means of measuring low concentrations of a number of contaminants in the atmosphere. These are:

Booklet number	Contaminant
1	Hydrogen sulphide
2	Hydrogen cyanide vapour
3	Sulphur dioxide
4	Benzene, toluene and xylene, styrene
5	Nitrous fumes
6	Carbon disulphide vapour
7	Carbon monoxide
8	Phosgene
9	Arsine
10	Chlorine
11	Aniline vapour
12	Organic halogen compounds
13	Mercury and compounds of mercury
14	Lead and compounds of lead
15	Trichloroethylene
16	Acrylonitrile

17   Chromic acid mist
18   Ozone in the presence of nitrous fumes
19   Hydrogen fluoride and other inorganic fluorides
20   Aromatic isocyanates
21   Iron oxide fume
22   Copper fume and dust
23   Acetone
24   Isophorone
25   Zinc oxide fume
26   Cyclohexanone and methyl cyclohexanone

Specialist analytical expertise should be consulted in setting up a programme of atmospheric testing. The value of traditional methods should also not be entirely ignored. In many steelworks for instance where carbon monoxide is a common atmospheric contaminant, caged canaries are still considered by many to be the most rapid and effective form of monitoring. The canary is more sensitive to carbon monoxide in the atmosphere than humans, and falls off its perch at concentrations lower than those at which humans begin to be affected.

3.2.4.5 Breathing apparatus

Often the only effective means of protection is through the use of breathing apparatus. A number of different types is available and these cover a wide range of situations. These devices are discussed in section 3.10.5.

However the factory, plant, etc should be designed so that operatives can work in health and safety without the use of special breathing apparatus, which many are reluctant to use.

3.2.4.6 Smoking

Smoking is both a health and fire hazard. The former is a relatively recent discovery so that this generation and the next have the problem of eradicating a habit, often thought pleasurable, which became inbuilt into western society over a period of some four hundred years.

The statistical correlation between smoking, especially of cigarettes and lung cancer is now irrefutably established. Doctors themselves were the first to recognise this. Between 1951 and 1965 about half the British doctors who used to smoke gave up. The mortality rate from lung cancer in doctors fell by 38%, whereas it increased by 6% in the rest of the male population. Each year about 32 500 men and women over 35 die of lung cancer. Of these 15 000 are under the age of 65.

A principal reason why people still continue to smoke is that nicotine is extremely addictive. This makes it difficult for the smoker to give up. After a few hours without a cigarette the nicotine content in the blood stream drops below a certain level.

Once the addiction is broken, the craving to smoke completely disappears. But although smokers are slaves to their chemistry, breaking the addiction is not all that easy. Smoking cures, such as Lobeline, a compound which fools the nervous system which it mistakes for nicotine, are of great help to smokers who sincerely want to break the habit. Besides the doctor, there are organisations to help the addict, notably ASH (Action on Smoking and Health).

To those who are still addicts, the writer's personal message is that life as a non-smoker is far more pleasurable.

The main problem facing anyone who gives up smoking after years of addiction is that of over-eating and putting on weight. This, however is usually far less serious and more easily dealt with than nicotine addiction itself. There are few non-smoking overweight persons who would take up smoking in the hope of losing weight!

Safety specialists who are convinced non-smokers should need little encouragement to appreciate the rights of other non-smokers in their organisation to a clean and healthy atmosphere, uncontaminated by smoke and unpleasant fumes and odours. They are entirely within their rights to make and enforce regulations to protect non-smokers from exposure, without their consent, to tobacco smoke in the working environment. This is also sound safety practice.

The fire hazards of smoking are dealt with in Chapter 4.

REFERENCES

1. Barry, R. G., and Chorley, R. J., *Atmosphere weather and climate*, 3rd ed., Methuen, 25 (1976).
2. McCormick, E. O., *Human factors engineering*, McGraw Hill, 537 (1967).
3. Revelle, R. R., and Geophysics Study Committee, *Energy and climate (Studies in Geophysics)*, US National Academy of Sciences, Washington D.C. (1977).
4. *The 'Readers Digest' Great World Atlas.*
5. Khogali, M., 'Occupational lung diseases', *The Chemical Engineer*, The Institution of Chemical Engineers, Rugby, 571, 323, Aug. (1977).
6. Grandjean, E., *Fitting the task to the man*, 2nd ed., Taylor and Francis, London, 54 (1975).
7. Hunter, D., *The diseases of occupations*, 2nd ed., English Universities Press, 124 (1959).
8. Marx, K., *Capital* (Facsimile reprint), Allen and Unwin, 230 (1946).
9. Page, J. A., and O'Brien, M. W., *Bitter wages*, Grossman, N.Y. (1973).
10. Health and Safety Executive, *Guidance Note EH 15/79, Threshold Limit Values*, 1979, HSE.
11. Sax, N. I., *Dangerous properties of industrial materials*, 4th ed., Van Nostrand, 24 (1975).
12. Royal College of Physicians, *Smoking and health now*, The Health Education Council, London (1971).
13. Mercer, J. N. 'West Antartic ice sheet and CO_2 greenhouse effect; a threat of disaster', *Nature*, 271 (No 5643) 331 (29 Jan, 1978).

3.3 LIGHTING AND VISION

Contents

Sight is the most important of our senses. The fulfillment of nearly every task depends on proper visual perception, and inadequate visual perception is one of the most serious causes (both direct and contributory) of accidents. Visual perception depends on a variety of factors, the most important of which are:

1. The eyesight of the person;
2. The degree and quality of the illumination of the objects seen;
3. The size and shape of the object in relation to its distance from the viewer;
4. The degree of colour contrast between the object and its background;
5. The speed of the object in relation to its distance from the viewer;
6. Obstructions in the path of the light rays between the object seen and the eye of the viewer. These include solid objects, airborn mist and dust, transparent screens which reflect or absorb some of the rays and dirt or condensation on the surface of the screens.

3.3.1 Eyesight of industrial workers

Studies made on the relation between eyesight and accident rates of industrial workers[1] have shown quite conclusively that most accident-prone workers do not have the efficient visual skills held by accident free workers. In one American study it was found that workers in a large factory who had extreme deviations from normal visual standards were found to have six times as many accidents in a given time as those with negligible variations from visual standards. Workers with marked but less extreme variations from normal visual standards were found to have between two and three times the accident rate as those with whose eyesight met normal standards.

Defects in vision are extremely common and in the majority of cases these may be brought up to normal standards by the prescription and wearing of spectacles. For work in which eye protection (i.e. from flying particles) is required, the worker who normally requires spectacles should be at no disadvantage compared with his fellow worker who has full eyesight without spectacles. Nevertheless, even after correction with spectacles, there is still a fair proportion of those whose eyesight is below normal standards and this proportion increases with age. Some guidance on the provision of safety spectacles is given in a booklet by the Association of Optical Practitioners.[4]

Special devices are available for routine testing of several aspects of vision including visual acuity, convergence (phoria), depth perception, colour discrimination and dark adaptation.[5]

1. *Visual acuity* is the ability to perceive black and white detail at various distances, which is largely controlled by the accommodation of the eyes. Inadequate accommodation leads to 'near sightedness' and 'far sightedness' conditions which can usually be corrected by glasses and lenses.

2. *Convergence* means that the images of an object on the retinas of both eyes are in corresponding positions so that we get the impression of a single object. It is controlled by muscles that surround the eyeball. Some individuals tend to converge too much and others not enough. Some individuals have one eye that tends to point up relative to the other. These conditions which result in double images are known as 'phorias', which are so visually uncomfortable that those suffering from them try to compensate for them by using the muscles around their eyes. This creates muscular

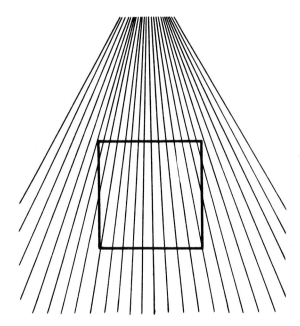

Figure 3.3.1   Optical illusion. Apparent distortion of a square to a trapezoid when a series of radiating lines were superimposed on it (From 'The Biology of Work' by O.G. Edholm, published by Weidenfield & Nicholson)

stresses and strains which bring about early fatigue in workers required to use their eyes continuously for relatively long periods.

3. *Depth perception* or *stereopsis* is the impression of depth gained from the fact that both eyes see an object from a slightly different angle. The depth perception in people varies considerably. Those without it often find other cues for judging depth or distance such as the relative size, clarity or position of objects, but even so they are at a visual disadvantage to those with normal depth perception.

4. *Dark adaptation* or the adaptation of the eye to different levels of brightness and darkness is brought about by two processes:

  (a) Enlargement (in the dark) or contraction (in bright light) of the pupil of the eye to increase or decrease the amount of light entering the eye. This takes a little time.
  (b) A physiological process in the retina in which 'visual purple' is built up as the level of illumination drops, so that the colour sensitive cones lose most of their sensitivity and the rods take over the visual functions.

The time required for dark adaptation is quite long — half an hour or more, although the time taken to adapt from darkness to light is shorter, usually less than a minute. However, during the process of adaptation, especially in its early stages, people with normal sight have partial temporary blindness.

5. *Colour vision* is due to the ability of the retina of the eye to distinguish between light of different wavelengths. It is not however equally sensitive to all wavelengths. The brightest spectral colour is at about 5500 Å which gives the impression of yellowish green. The brightness progressively diminishes as the wavelength approaches 7000 Å (red) or 4000 Å (purple). Complete colour blindness is rare, but some defect in colour vision is common. The most usual is an inability to distinguish between red and green. Colour blindness is largely confined to men. About one young person in ten has some colour vision fault, and the power to discriminate colours decreases with age, although those concerned are usually unaware that they suffer from any defect.

In addition to these visual qualities, tracking and spatial perception must be considered. Tracking involves hand (or body) and eye coordination in following a fast moving object such as a cricket ball. The eyes must be able to follow and focus on the fast moving object and the information must be received and processed rapidly by the brain. Ability in ball games is closely related to tracking ability.

Spatial perception and speed of perception were referred to in section 1.7. They are essentially combined functions of eye and brain. The total amount of information which can be obtained in a given time is limited, although this is influenced by lighting as well as by the characteristics of the object. Perception involves many processes, notably the ability to concentrate attention on certain parts of the visual field.

The most important quality required in difficult visual tasks, e.g. inspection, is to spot quickly an unusual detail in a familiar object. This is a power which can be cultivated, although it runs counter to man's natural instinct of seeing 'what he wants to see'. It is this instinct which lies at the basis of many familiar optical illusions, and in the difficulties, particularly of authors, in spotting mistakes when reading proofs of their own work. Often an outsider who has never seen the work before will pick up a number of mistakes to which the author who is thoroughly familiar with the text has become blind. However, this is properly the field of 'mental blind spots' which are outside the scope of this section.

It is clear from the foregoing that there can be large differences in the various visual skills of individuals. However, the insistance on high visual standards for all workers irrespective of their job can not only cause considerable hardship but also lead to the loss or wastage of many skills.

The recommended solution to this problem is two-fold:

1. To classify the visual standards required for the various jobs in the works or factory. This requires some preliminary fact finding, during the course of which the various other factors affecting visual perception of the worker may come to light and be capable of correction, i.e. inadequate illumination, excessive glare, poor colour contrast which could be improved by a different background colour, or the presence of persistent fog in some work areas resulting from steam leaks.
2. To submit all new employees to a sight test by a qualified optician as a precondition of starting work and to classify them into a small number of groups with differing standards of vision which should allow for faulty colour perception. Each main group should be split into two sub-groups corresponding to unaided vision and vision assisted by the use of glasses.

The visual standards of each worker and the date of examination should be recorded both centrally and on the pass or clockcard. The visual standards required for each task should also be recorded centrally and be known by the managers and supervisors in the area or department. It is then a simple matter to ensure that the jobs requiring high visual standards for safe as well as efficient execution are filled only by workers whose visual standards match or exceed those required for the jobs. Actually, the visual standards of the worker should exceed those required for the job by a safety margin which is explained later.

In addition to this initial examination and classification both of workers and jobs into various visual standards, regular periodic re-examination both of workers and jobs should be carried out. This should not be done too frequently or haphazardly, nor left to the whims or discretion of untrained persons, but should be carried out officially by properly trained persons with full management approval.

The benefits resulting from the adoption of such a scheme need to be carefully explained to workers and supervisors alike to secure their cooperation. A worker whose sight has altered or deteriorated to below the standard required for his job may be reluctant to be re-examined, but if the examination includes the exact prescription of lenses which he now requires, he is likely to accept re-examination more readily.

So far the sight of workers has been discussed under test conditions when the worker can be expected to be fit, fresh and free from strain. Sight is however significantly impaired by ill health, under-nourishment, fatigue and strain. Eyestrain can be caused partly by the task itself, but it is also caused and greatly intensified by poor lighting and glare, and persons suffering from defective vision are more subject to eyestrain than those with normal vision. Sight quality is not a static thing but is subject to temporary deterioration as the result of one or more adverse factors. To ensure an adequate safety margin, the visual standards of a worker under test conditions should not merely match those of the job he is required to do, but should exceed them by a margin sufficient to allow for temporary deterioration caused by one or more of the factors outlined.

3.3.2 Lighting quality

Good illumination, like adequate eyesight, is a most important factor in industrial accident prevention. Minimum standards are given in the Factories Act (1961)[2]. These are based on regulations made in 1941 before fluorescent tubes were readily available and are below standards thought necessary today for the safe and efficient operation of many factory processes.

The subject is now so complex that it is difficult to discuss in simple language and requires specialised knowledge to achieve the best results. However, there are several examples of poor or inadequate illumination to be found in most works or factories which may be recognised by workers, supervisors and managers. These can be much improved by simple and inexpensive means. Typical of these are the following:

1. *Flickering tubes.* These may be caused by faulty electrical supply or connections or more usually by the fact that the tube has reached the end of its working life. The condition produces eyestrain and should be promptly reported and corrected.
2. *Reflectors misadjusted.* This can result in glare which interferes with normal vision.
3. *Poor maintenance.* This includes not only the replacement of defective tubes and bulbs, but also the cleaning of reflectors and windows for daylight illumination. Poor maintenance may reduce illumination by as much as 50%, leaving insufficient for safe and efficient work.
4. *Operation at voltage outside rated range of tube or bulb.* This is easily checked by a qualified electrician. Variations in the voltage supply to a works or plant have an immediate and serious effect on illumination as also on the operation of motors and machinery, thereby increasing the hazard. Every effort should be made in the design of the electrical

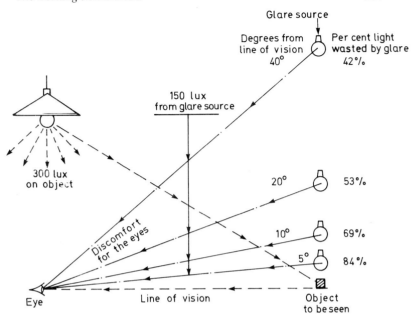

Glare source

Degrees from
line of vision
40°

Per cent light
wasted by glare
42%

150 lux
from glare source

300 lux
on object

Discomfort
for the eyes

20°   53%

10°   69%

5°   84%

Eye   Line of vision   Object
to be seen

Figure 3.3.2   Decrease in visibility caused by glare (From 'Industrial Accident Prevention' by H. W. Heinrich. Reproduced by permission of McGraw-Hill Book Co.)

supply system to ensure that voltage reductions are kept to a minimum and that warning of them is given in advance.

5. *Installation of lights within line of vision.* Wherever possible, lights should not be installed within the normal line of workers vision. Where this is unavoidable however, the lights should have proper shades or louvres to protect the workers eyes from the direct glare of the light. The light should not be so bright as to cause glare or eyestrain from brightly illuminated objects near the light.

6. *Lack of uniformity.* Wide variations in the intensity of lighting in adjacent areas, for example a corridor and a machine shop in which fine work is carried out, lead to many accidents through the inability of workers' eyes to adjust quickly enough from one lighting condition to another. This danger is intensified when operators of trucks, overhead travelling cranes and other vehicles moving at speed are exposed to this hazard. The danger should not arise in the first place if the lighting design has been carried out by an experienced and qualified illuminating engineer. Some improvement on a bad situation can be achieved on a trial and error basis by anyone of normal intelligence.

7. *Shadows on important objects and spotty illumination.* This condition tends to reduce safety and working efficiency. It is generally quite easy to cure.

8. *Inadequate illumination.* The level of illumination required for different jobs varies a great deal, depending on the visual attention required for the task. Visually exacting tasks which require sustained attention to fine detail require much higher illumination levels than tasks which are visually simple. Typical levels required for various jobs are given in *Table 3.3.1*. Further details are given in Reference 3.

3.3.2.1 Light sources

Light sources most used in industry may be classified under three main headings:

Incandescent
Fluorescent
Metal vapour lamps

3.3.2.2 Incandescent bulbs

Incandescent bulbs contain a tungsten wire which is heated to incandescence by passage of electric current. Their light has a wide band spectrum with a higher concentration of red light and a lower concentration of blue light than natural daylight. They are commonly designed for about 1000 hours life although longer life bulbs are available (with lower efficiencies). They are now mainly used as a back-up to fluorescent lighting to give local higher light intensities for particular machines and operations and for coloured illuminations and signals. They are also used for beams and spotlights (protective lighting), for portable lamps and for lamps in hazardous areas (with special fittings).

Clear and white-coated bulbs are available of approximately the same efficiency — the latter giving a more diffused light. Coloured bulbs are available including the so called 'daylight' incandescent lamp which has an efficiency of about 65% of a plain incandescent lamp.

The following special types of incandescent bulbs should be considered for particular applications:

A. Reflector bulbs which have a self contained reflector built into them. They are more expensive than plain bulbs but are generally designed for a life of 2000 hours. Their advantage over plain bulbs fitted into a reflective housing is that the reflector needs no cleaning.
B. Rough service and vibration service bulbs are made with extra filament supports to withstand mechanical shocks. They are principally used with extension cords.
C. Thermal shock resistant and special service bulbs are available for applications where drops of water may fall on the hot bulb.
D. Quartz iodine bulbs resist thermal shock and give continuous and high efficiencies over a working life of 2000 hours or more. They are

useful for spotlights and searchlights where their small size is conducive to good beam control.

E. Bulbs for hazardous areas (flameproof, explosion proof and dust-proof). Illumination in hazardous areas generally requires a fitting with a totally enclosed bulb inside a glass cover. It is essential to check that the bulb is suited for the application since normal bulbs tend to become overheated due to lack of air circulation round them.

3.3.2.3 Fluorescent lamps

Fluorescent lamps are the most widely used type of lamp for industrial lighting, and produce light from the fluorescence of 'phosphors' contained in the glass of the tube which are activated by ultraviolet radiation from a low pressure mercury vapour discharge inside the tube. They have about three times the efficiency of incandescent bulbs and about ten times their life.

Lamps are available with a continuous band spectrum which closely resembles daylight, and other modified spectra are available — cool white being most popular for industrial lighting (see also section 2.11.5). Their performance is, however, markedly dependent on the ambient temperature and air movements around the tube, falling off both at low and high temperatures.

Special lamps are available for operation in unheated buildings or in refrigerated rooms.

3.3.2.4 Metal vapour discharge lamps (mercury or sodium)

Today, these are mostly mercury vapour discharge lamps which operate at a much higher pressure than fluorescent lamps and produce much of their energy as visible light. Clear mercury vapour lamps produce light which is rich in yellow and green tones and largely lacking in red. Their efficiencies lie in the same range as fluorescent lamps, but they are less affected by ambient temperature. Their colour spectrum can be improved in two ways:

1. By the use of 'halide lamps' which contain in addition to mercury, the iodides of thallium, indium and sodium; these dissociate under the action of the electric discharge and emit light of complementary colours.
2. By the incorporation of solid phosphors in the envelope of the tube similar to those used in fluorescent lamps.

Metal vapour discharge lamps tend to be cheaper, smaller and brighter than fluorescent lamps of the same output but they require special shielding both to protect them and because of their brightness. Both fluorescent and mercury lamps when operated on alternating current circuits give most of their light on one half of the cycle and much less on the other (depending on the phosphors used). This creates a stroboscopic effect on revolving and reciprocating objects which may cause them to appear to be running slowly, stationary or even backwards. This hazard can be largely overcome by operating adjacent fixtures on different phases of a multi-phase system.

3.3.2.5 Codes of practice, units, and recommended levels of illumination

Recommended codes for the illumination of various building interiors are published by the Illuminating Engineering Society of New York, and the Illuminating Engineering Society of London[3] (two entirely separate bodies).

Their recommendations are in practice very similar, but the unit of illumination used by the American Society is 'the foot candle' on the object or 30 in above the floor, while the units used by the British Society are known as SI units. The unit of illumination is the lux which is produced by a luminous flux of one lumen on one square metre of surface. One footcandle is equal to 10.76 lux.

Table 3.3.1 RECOMMENDED MINIMUM ILLUMINATION LEVELS AND LIMITING GLARE INDICES FOR VARIOUS ROOMS AND BUILDINGS

	Minimum illumination level, lux	*Limiting glare index*
General building areas		
Entrance halls, waiting rooms and gatehouses	200	
Entrance gates, locker rooms, lavatories	100	
Corridors, passageways	100	22
Stairs and escalators	100	13
First aid rooms, treatment and consulting rooms	400	
Rest rooms	50	
Industrial buildings		
Turbine houses, switch rooms, heavy industrial buildings	200	25
Medium bench and machine work, rough grinding, buffing and polishing	400	25
Laboratories and test rooms	600	19
Gauge and tool rooms	900	19
Outdoor		
Car parks	10	
Stores, stockyards	20	
Boiler platforms	50	
Offices		
General	400	19
Drawing and tracing boards	600	
Store rooms, vaults	200	25

The British code specifies a limiting glare index for each interior for which a minimum illumination level is given. The glare index is a measure of the visual discomfort experienced by workers from bright light sources, reflecting surfaces and low light fittings. The glare index can be readily calculated by a qualified engineer with access to all relevant data and its value can be reduced where necessary by repositioning light sources, using different light fittings and changing the surface texture of walls, floors and ceilings. The index is useful in expressing quantitatively a source of accidents and visual discomfort which is generally felt only in subjective terms and provides means of calculating the degree of discomfort and reducing it to tolerable levels.

Some of the recommended minimum illumination levels and limiting glare indices recommended by the British IES are given in *Table 3.3.1*.

Another factor which has to be considered when determining illumination levels is whether workers in the building are required to wear tinted goggles or spectacles to protect their eyes from radiation such as that produced by electric arc welding. Where tinted spectacles are worn, the illumination levels should be increased so that the apparent illumination level viewed through tinted spectacles is no less than that recommended for normal viewing.

3.3.2.6 Adequacy of lighting

In spite of a general increase over the past twenty-five years in the level of illumination in industrial buildings, this is still usually far less than that given by daylight. Performance and safety can thus generally be improved by better lighting.

Windows in industrial buildings not only provide admission of daylight but also provide a distant focus which relaxes the eye muscles. They also relieve a claustrophobic feeling felt by many people in a windowless room. All window areas should however be provided with devices to control brightness and particularly the admission of sunlight.

Electric lighting is generally considered as made up of three parts:

1. General lighting which produces uniform illumination throughout the area involved. This should not deviate anywhere in the area by more than 17% above or below the average level.
2. Localised general lighting to reinforce the general lighting in specific areas.
3. Supplementary lighting to provide higher illumination levels for small or restricted areas, or specific brightness or colour or aiming of light sources.

Protective lighting is a special form of lighting used for policing outside areas at night and to reduce fire risk. It produces adequate light on border

Spectral energy distribution

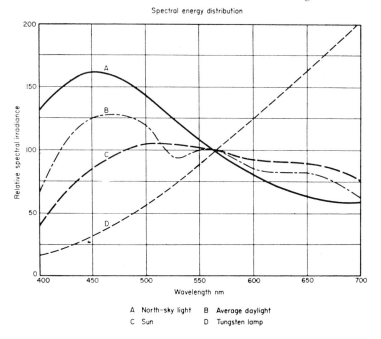

Figure 3.3.3   Spectrum of sunlight and various light sources (From 'Physical Working Conditions' edited by McCullough and published by The Industrial Society)

areas and dazzles intruders while keeping the guards in darkness. Protective lighting may also be combined with emergency lighting for speedy evacuation of a plant or building when it is coupled up to batteries and/or a generator. The adequacy of these for the likely period of use in case of a power failure should be checked and such systems must be frequently inspected and properly maintained ready for use. Protective and emergency lighting is not sufficient for the operation of plant and machinery.

Strong contrasts in the illumination levels in buildings tire the eyes and should be avoided[4]. The area on which most light is needed should not have more than three times the illumination level of the area just surrounding it. This, in turn, should not be more than three times as bright as that in the rest of the room or building, giving an overall contrast of not more than 10 to 1.

3.3.3 Colour contrast and decoration (see also section 2.11.5)

Safety and efficiency are improved by the use of suitably coloured walls, floors and ceilings, by highlighting vital machine parts with special colours and the use of colour standards for identifying danger spots, exits, gas cylinders, electrical circuits and pipelines. Light surfaces everywhere are recommended, taking care to avoid shining surfaces with high reflectance. Some recommendations on the use of colour are given in Appendix A.

When colours are used in this way, it is most important to ensure that the type of lighting employed allows the colours to be readily identified. Some of the cheapest and most efficient forms of lighting measured from the stand-point of electrical conversion (such as metal vapour lamps, mercury or sodium) are unfortunately made up of a very limited range of colours and allow very little differentiation by colour of objects illuminated by them.

A high pressure fluorescent mercury vapour lamp, in which part of the light is obtained by exitation of a layer of fluorescent material on the bulb of the lamp, gives better colour rendering than a plain mercury vapour lamp. Likewise halide lamps in which light is produced by an electric discharge through a mixture of metal vapour (usually mercury) and the products of dissociation of iodides of thallium, indium and sodium give better colour rendering than pure metal vapour lamps.

3.3.4 Maintenance

Maintenance is as important for the windows and interior surfaces of buildings as for the lighting fittings themselves.

Side windows of heavy industrial buildings should be cleaned at intervals of one to two months and roof lights at intervals of not more than three months. Room surfaces should be cleaned and redecorated as required. Spectacles, goggles and transparent shields and the windows of protective cubicles used by workers should be cleaned as frequently as necessary to keep them free of dirt, grease and condensation. Glass surfaces liable to mist may be treated with a special surface active agent which causes the mist to spread as a thin film and prevents the formation of droplets that obscure vision.

Fog, smoke and dust clouds may, among other evils, obscure important objects and sometimes prevent their being seen altogether. The problem is also often one of proper maintenance, though sometimes it may call for engineering revision.

The question often arises whether to replace faulty bulbs and other light sources as they burn out or all together after they have been in operation for a specified number of hours, when some failures will have occurred. In cases where a number of light sources with the same life expectancy have been installed and used together, it is usually found most economical in a factory to replace them altogether, whether they have failed or not.

Disposal of failed light sources, particularly metal vapour discharge tubes and fluorescent tubes, can present some safety hazards, e.g. cutting hands and fingers with broken glass containing toxic compounds and contaminated by mercury. If these have to be broken to fit in the refuse container, they should first be wrapped in several layers of paper or cloth, then broken mechanically. This should be done out of doors and the remains placed in a refuse container in the open air.

REFERENCES

1. Stump, N.F. quoted by Heinrich, H.W., *Industrial accident prevention*, 4th ed., McGraw Hill, New York, 351 (1959).
2. Fife, I., and Machin, E.A., *Redgrave's health and safety in factories*, Factories Act 1961 Part 1, section 5, Butterworths, London, 38 (1976).
3. Illuminating Engineering Society, (London), *Interior lighting*, IES (1977).
4. Association of Optical Practitioners, *Eyes at work*, A.O.P. London (1970).
5. Edholm, O.G., *The biology of work*, Weidenfeld and Nicolson, London, 102 (1967).

3.4 NOISE AND HEARING

Contents

Most machines and many manual operations produce noise, which is an almost inevitable accompaniment to production. Noise is unwanted sound, which can be described as rapid pulsations in air pressure produced by a vibrating source.

Noise is generally acceptable up to certain levels and some sounds emitted by machinery may even be of value to the worker in warning of malfunction or in judging speed or performance. Excessive noise is worse than a nuisance and it is hazardous in at least three ways:

1. It interferes with spoken communication and warning signals and thus causes accidents.
2. It reduces the morale, efficiency and general awareness of workers and this too leads to accidents.
3. High noise levels injure workers directly by impairing their hearing and causing deafness in severe cases.

Occupational deafness can result from excessive noise and this is one of the most neglected areas of occupational health. The only statutory limits on noise exposure in industry are those[1] contained in the Woodworking Machines Regulations 1974 made under the Factories Act 1961. Section 44 of these regulations reads:

'*Noise*. Where any factory or any part thereof is mainly used for work carried out on woodworking machines, the following provisions shall apply to that factory or part, as the case may be:
(a) Where on any day any person employed is likely to be exposed continuously for 8 hours to a sound level of 90bA or is likely to be subject to an equivalent or greater exposure to sound, (i) such measures as are reasonably practicable shall be taken to reduce noise to the greatest extent which is reasonably practicable, and (ii) suitable ear protectors shall be provided and made readily available for the use of every such person:
(b) All ear protectors provided in pursuance of the foregoing paragraph shall be maintained, and shall be used by the person for whom they are provided in any of the circumstances specified in paragraph (a) of this Regulation:

(c) For the purposes of paragraph (a) of this Regulation the level of exposure which is equivalent to or greater than continuous exposure for 8 hours to a sound level of 90 dB(A) shall be determined by an approved method.'

A *Code of practice for reducing the exposure of exposed persons to noise*[2], prepared on behalf of the Industrial Health Advisory Committee was published by the Department of Employment in 1972. But, according to the foreword of a 1975 sub-committee report[3] *Framing noise legislation*, 'Compliance with the noise code, however, is voluntary, and although many firms have made big efforts to put it into practice, there is still a disturbingly high number of workers exposed to noise levels high enough to damage their hearing.'

Regulations for the protection of people's hearing from noise can be made under section 15 of HASAWA (1974), and the sub-committee report quoted above recommends that these regulations should contain six basic provisions:

On employers
'1. To carry out noise surveys under the 1972 Code of Practice.
2. To take all reasonably practicable steps to reduce the sound level and/or duration of exposure in cases where it is shown from the survey that the codes provisions are being exceeded (or reduce them to the greatest practicable extent).
3. To identify and mark areas where the provisions are, or are likely to be, exceeded.
4. To prohibit people from entering areas so marked without wearing approved ear protection but allowing certain exceptions.

On employees
5. To wear the ear protection provided when in the marked areas and to cooperate with their employer on the application of the protective measures.

On machinery designers, manufacturers, etc.
6. To provide a warning to purchasers of their machines when used for the purpose for which they are sold if they are likely to produce hazardous noise (details yet to be formulated).'

It appears also from the sub-committee's report and subsequent utterances from the Chairman of the Health and Safety Commission that the 1972 code is likely to be converted to an 'approved code' (see section 2.2) under section 16 of HASAWA.

Noise levels of building sites can be a menace to the local community as well as workers on the sites. The subject comes under sections 60 and 61 of the Control of Pollution Act, 1974.[4] Three regulations[5] have been made under this Act. These refer to a British Standard Code of Practice.[6]

3.4.1 Effect of noise on hearing

Contrary to general belief, the ear drum is not usually damaged by high noise levels. The damage occurs in the inner ear, which is located deep inside the skull. A sound wave causes the eardrum to vibrate; the vibration passes to the middle ear, and from this to the inner ear; here the vibration strikes against a delicate structure of hair-like cells which form part of the so-called organ of Corti.[7] These auditory hairs or otiliths are connected to delicate nerve filaments which pass through the spiral ganglion to the auditory nerve and thence to the brain. Vibration of the auditory hairs generates nerve impulses which pass to the brain and result in hearing (*Figure 3.4.1*).

Hearing mechanisms are sensitive and easily damaged in a number of ways. Temporary deafness may be caused by accumulation of wax in the ears, or a boil in the auditory canal. Adenoids cause deafness in children while osteosclerosis may cause deafness in middle age. The middle ear is liable to a number of infections resulting in some degree of deafness, these being associated with catarrh, measles, scarlet fever and tonsilitis.

Deafness is also caused by exposure to high noise levels (from 90 dB upwards) which overstimulate the auditory nerves and result eventually in deafness. One needs here to distinguish between prolonged exposure to high noise levels which result in gradual impairment of hearing and exposure to single events such as explosions, both of which produce deafness.

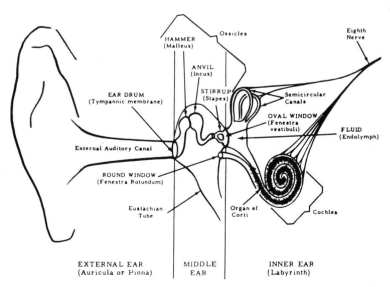

Figure 3.4.1   Structure of the Ear (Courtesy American Foundrymen's Society and the National Safety Council, U.S.A.)

Deafness produced by prolonged exposure to high noise levels has long been recognised as an occupational disease, first among boilermakers, and subsequently among blacksmiths, cotton weavers, air crew and submarine personnel.[7] Heavy forge operators were once so widely affected that a man without partial deafness could hardly hope to qualify as experienced.

Individuals with a hereditary tendency to deafness as well as those suffering from a disease or infection of one or both ears are more susceptible to noise damage than others. Workers exposed to high noise levels are seldom aware of their deafness until it is severe. Their first complaint is commonly that they are unable to carry on a conversation in which several people take part — so-called *society deafness*.

3.4.1.1 First symptoms of occupational deafness

Tests with an audiometer have shown[7] that the earliest defects are confined to a comparatively small area of high pitched tones around C_5 (4.096 Hz), which is above the range of normal speech (300 to 3000 Hz).

Figure 3.4.2 shows typical audiograms of hearing loss due to age, noise and middle ear disease. The C_5 dip, which on prolonged exposure to noise gradually extends through the lower frequency range, is specially characteristic of noise induced deafness and traumatic deafness such as that following an explosion and cerebral concussion. It can also be caused by toxic agents, such as alcohol, quinine, cocaine and carbon disulphide. Whilst the use of an audiometer is the surest way of detecting hearing damage, some simple tests can provide useful clues to the safety specialist.

Perhaps the most useful of these it to check the distance at which the individual can hear a whispered sentence (which is normally in the C_5 area).

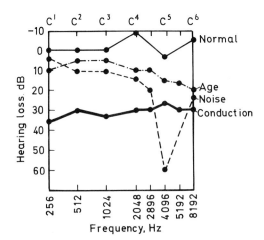

Figure 3.4.2   Audiograms in hearing loss due to age. Noise and middle ear disease (From 'The Diseases of Occupations' by Donald Hunter, published by E.U.P.)

Someone with good hearing can hear a whisper at 8 m in a quiet room, whilst someone in the early stages of occupational (noise induced) deafness cannot hear the whisper until much closer — even as little as 1 m. Other symptoms which a recent H & SE booklet[8] warns readers to be on the lookout for are:

1. Inability to hear and understand speech in a noisy (working) environment.
2. Sensations of head noises (such as buzzing, uproar, droning or growling) and ringing or whizzing in the ears after working for hours in a noisy environment. (The medical term for this is tinnitus).
3. Temporary deafness after exposure to high noise levels for short periods (which may cause individuals to seek doctor's advice).
4. Individuals being told by their families that they are going deaf.
5. High labour turnover in workshops and sections with high noise levels.
6. Feeling by managers that noise is affecting production.

There should be little doubt that a company where these symptoms are manifest has a noise problem. However, it must also be realised that the workplace is not the only location where noise can damage hearing. The noise levels in many public places, especially certain places of entertainment such as discotheques, can only be described as 'Indescribable'!

3.4.1.2 Use of audiograms before and during employment

Companies who are seriously concerned with the health of their staff check the hearing of employees on engagement by audiograms. Staff who are exposed to high noise levels at work should have their audiograms repeated after one month's employment and again after 6 month's employment. If impairment of perception at any frequency, particularly at frequencies below 3000 Hz has increased at either time, the individual should be removed to a quieter location, and steps should be taken to reduce the noise level at that location.

The widespread use of audiometric testing in industry has its protagonists and opponents.[9] Opposition is based mainly on grounds of expense and the variability of results of tests on the same person at close time intervals. Its use can, however, be expected to increase as industry and the nation at large becomes more aware of the noise problem and its effect on hearing.

3.4.2 Noise level and sound frequency

The most significant properties of sound are its noise level and its frequency. Noise levels are measured with various types of meter which indicate the level in decibel units.

The number of decibels of any sound is the ratio of the pressure caused by that sound to the pressure caused by the lowest sound that the average person can hear. Thus the faintest sound that the average person can hear has a sound level of one decibel.

Some typical noise levels close to the source of noise are:

average conversation	60 to 65 dB
circular saw	100 dB
power press	120 dB
jet engine	130 dB

An increase of 3 dB doubles the noise level. In open air, the noise level produced by any source attenuates as the distance from the source increases. By doubling the distance, the noise level is reduced by about 6 dB.

The frequency or number of sound vibrations per second is a measure of the note or pitch of the noise, a high frequency corresponding to a high note.

The human ear is sensitive to sound frequencies from 20-16 000 Hz. Normal conversation has a frequency range from 250 to 3000 Hz.

Most factory noise has a frequency range from 600-10 000 Hz. Noise in the range of 800-5000 Hz appears to be more damaging to the human ear than noise of lower frequency.

It is possible to make a single measurement of the noise level over the entire range of audible frequencies. This measurement, however, is of limited use since the ear is more sensitive to some frequencies than others. It is more useful to know how the noise is distributed throughout different frequency ranges or octave bands. An octave has a range of frequencies of 2:1, so that 90-180 Hz is one octave and so also is a range of 1400-2800 Hz. When noise surveys are made for assessing hearing hazard and devising protective measures, the noise level in each of 8 octave bands covering the range from 45-12 500 Hz is measured.

For rapid assessment of hearing hazard and nuisance levels, a weighted sound level can be made by a meter containing selective filters such as is shown in Figure 3.4.3. This level corresponds as closely as possible to the noise intensity sensed by the human ear over the entire frequency range. Various weighting scales are used by sound specialists, but the scale most widely used in industry is the 'A' scale. The decibel unit on the weighted 'A' scale is abbreviated dBA, and all noise levels quoted in this section refer to the dBA scale. Some difficulties are found in measuring impulse noise, that is to say the noise caused by single percussions at widely spaced intervals. This is best done by a sound specialist.

3.4.3 Tolerable noise levels

Noise levels from 85-100 dBA over prolonged periods of exposure can harm the hearing of sensitive people. Noise levels from 100-120 dBA will damage the hearing of most people. Noise levels over 120 dBA create immediate discomfort and may do permanent damage to the hearing of normal people even after short exposure. Noise levels over 130 dBA produce pain.

The Health and Safety Executive[8] stipulates various maximum sound levels to which workers may be exposed for various durations of exposure per day without ear protection.

Table 3.4.1 MAXIMUM DAILY EXPOSURES AT VARIOUS NOISE LEVELS

Exposure duration (hours per day)	Maximum sounds (level dBA)
8	90
4	93
2	96
1	99
$\frac{1}{2}$	102
$\frac{1}{4}$	105

Frequently workers are exposed to noise which fluctuates in level throughout the day. A simple method for estimating whether exposure to fluctuating noise is excessive is given below.

1. Estimate the daily exposure at each sound level.
2. Express each duration as the fraction of the duration permitted by Table 3.4.1.
3. Add together all the fractions; if the total is more than 1, the exposure is excessive.

This is clarified by the following example.
Exposure at 90 dBA for 4 hours followed by 99 dBA for ¾ hour.
Exposure at 90 dBA = ½ that permitted by Table 4.2.1.
Exposure at 99 dBA = ¾ that permitted by Table 4.2.1.
Total ½ + ¾ = 1¼. Hence exposure is excessive.

Impulse noise consists of widely spaced impulses such as the noise from shears and chop hammers. A preliminary assessment may be made using an ordinary sound level meter set to the fast response. If the meter needle passes 90 dBA the noise should be regarded as potentially hazardous, and further assessed by a specialist. If a sound meter is not available, the following simple checks may be made:

Where workers must raise their voices to carry on a conversation, the noise is sufficient to affect their morale and efficiency.
Where the noise level is so high that two people at arms' length cannot hear each other when talking loudly, supervisors can anticipate that the hearing of some people in the noisy area will be damaged. If either of these checks indicates a problem, supervisors should take corrective action and advise the safety manager.

The impression should not however be formed that most hearing loss results from exposure to high noise levels. Diseases in childhood, ear injuries, abscesses and advancing age are other factors which may impair hearing.

3.4.4 Remedial measures

Remedial measures in an established works or factory usually take one of three forms.

1. Reduction of noise at source.
2. Reduction of noise transmitted through air or building structures.
3. Use of personal hearing protectors.

When a new factory is being planned or an existing one modified, noise should be considered. The sites of new buildings should be examined to see whether noise from external sources such as road and rail traffic and aircraft will be acceptable. If it is not, steps should be taken to find a better site or to screen the noise source.

Figure 3.4.3   Precision octave band sound level meter (Dawe Instruments Ltd.)

When replanning work in existing buildings, work requiring quiet should be sited as far away from noise sources (including busy roads and railways) as possible. In works and factories, the noisy processes should be placed well away from other processes. Consideration should be given to the use of sound-absorbing materials and the siting of structures to act as noise screens.

3.4.4.1 Reduction of noise at source

Reduction of noise at source is a problem to which management, designers and works supervisors or foremen should all contribute.

In the design stage, it is most important to consider the noise levels that will result from all sources and to assess probable noise levels at various work places. When machinery or equipment known to be inherently noisy has to be installed, reliable figures on noise levels from machinery available from different suppliers should be obtained. Before a final choice of supplier is made, satisfactory guarantees on noise levels should be obtained from the maker, backed-up if necessary by noise surveys on identical machines in operation elsewhere.

Transmission of noise through structures may be reduced by supporting machines on resilient mountings. The noise radiated from metal plates and other parts is intensified by resonance in the plate. This can be reduced by stiffening the plate (e.g. by adding ribs) or by coating the surface with a sound damping compound. Panels likely to vibrate can be made from a highly damped material such as 'sound deadened steel' which is made from two sheets of metal bonded with a special adhesive. Frames used to support vibrating equipment radiate less noise than structures and machine guards made of perforated plate radiate less noise than unperforated plate.

Noise can usually be reduced by substitution of plant or equipment. Examples of this are:

Use of electric trucks in place of internal combustion engine trucks;
Compression riveting in place of pneumatic riveting;
Grinding instead of chipping;
Use of electric tools in place of pneumatic tools where the pneumatic tools are too noisy;
Pressing instead of forging;
Welding instead of riveting;
Mechanical ejection in place of air-blast ejection;
Rubber tyres, preferably pneumatic, on trucks in place of solid and all metal wheels;
Plastic gears and fibre bearings in place of metal gears and bearings;
Rubber or rubber lined chutes, buckets and tote boxes in place of metal ones;
Covering hard surfaces on which materials will be dropped with a resilient material or the use of 'sound deadened steel'.

Many machines are noisy because of worn parts, poor maintenance, inadequate lubrication, or because they are out of balance. Planned maintenance, replacement of worn parts and regular lubrication of all machines will reduce noise and increase efficiency.

Equipment operated by compressed air tends to be noisy. Leaking air valves should be replaced. Reduction in air pressure and enlarging air openings discharging to atmosphere reduce noise. The use of automatic air shut-off valves, which ensure that air is available to ejectors and similar equipment only when required, instead of passing through continuously and the use of mufflers or silencers on steam and air exhausts from hoists and other equipment should be considered.

Noise from the fans of electric motors and ventilating systems may sometimes be reduced by reducing the air flow or fitting silencers.

3.4.4.2 Reduction in noise transmission

If it is impracticable to reduce a serious noise at source to an acceptable level, it may be possible to isolate the noise source or to move workers away from the noise.

The effect of distance has already been mentioned. A machine producing a noise level of 112 dBA at a distance of 1 m will give a noise level of approximately 106 dBA at 2 m, 100 dBA at 4 m, 94 dBA at 8 m and 88 dBA at 16 m, all assuming 'open air' conditions. It is seldom possible to rely on distance alone for reducing noise level, but it is sometimes possible to locate automatic machines which require little attention in a storage area or out-building where no one is working continuously.

When an enclosure is to be built round a noisy machine, it should be designed by someone with training and experience in acoustics. The following basic principles of soundproof enclosure design are given to enable supervisors to make best use of such enclosures.

Lightweight porous materials such as mineral wool, cloth, foam rubber, cork and fibreboard reflect very little sound, absorb a fairly high proportion of it, and allow most of the rest of it to pass through, depending on the frequency. Sound of high frequency is absorbed to a greater extent than sound of low frequency, most of which is transmitted or passes straight through.

Hard massive materials such as brick or stone reflect a high proportion of the sound waves reaching them and thereby prevent their transmission to the other side. But little of the sound is absorbed unless the barrier is very thick or heavy. The fraction of sound transmitted is inversely proportional to the mass of the wall material per unit area.

The ideal structure for enclosing or isolating a noise source would have walls of a heavy, hard material such as steel, brick or concrete, to reflect most of the sound and thereby prevent transmission, and a soft inner lining such as cork or foam rubber to reduce internal reflection and absorb sound reaching it directly as well as sound reflected from the wall.

Even small openings (e.g. keyholes) in an enclosure reduce its effectiveness considerably. Doors, covers and hatches in the enclosure

Figure 3.4.4   Helmet and ear protection used whilst rock drilling (American Optical Safety International)

should be kept tightly closed. An enclosure may be built round a set of machines, a single machine or even a part of a machine.

Where the machine cannot easily be enclosed, it may be possible to enclose the operator. For example, one man in a steel rolling mill controls the operation of several noisy machines. The operator in one such mill was completely enclosed in a sound-proof box ('pulpit') with its own air supply. This reduced the noise reaching him by approximately 25 dBA.

In other cases where complete enclosure is impossible, barriers may be used. Brick walls are particularly effective and a suitably plastered brick wall can reduce the noise by 50 dBA. Doors and windows in enclosures and barriers need special attention. Either double doors or windows or single doors and windows of extra heavy construction should be used, and it is most important that they fit tightly without gaps or cracks.

Where fixed walls are not possible, mobile screens help to reduce noise. They should preferably extend to the roof or to the structural floor above. The use of a layer of sound absorbent material (rock wool, foamed rubber or foamed plastic) on the inside of walls and ceilings of buildings containing noisy machinery also reduces the overall noise level by reducing noise reflected from the walls.

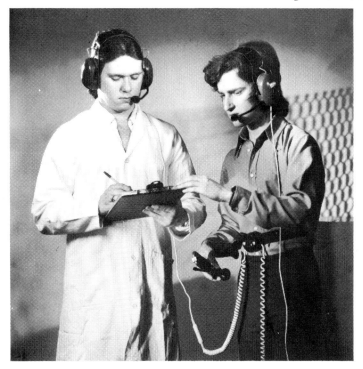

Figure 3.4.5   Example of ear muffs in use (Chubb Fire Security Ltd.)

One point should not be overlooked. All measures discussed in this section greatly improve the situation of workers who are separated from noisy machinery by the enclosure or barrier but they do little for a machine operator who has to work alongside the machine inside the enclosure or barrier.

3.4.4.3 Personal ear protectors

It is seldom possible to protect all employees in a works which has many noisy machines by the measures outlined in the previous sections. For employees whose hearing cannot be so protected, personal protection is the only answer.

One frequent and ill-founded objection often made about ear protectors is that the wearer is unable to hear spoken messages or audible warnings. This is incorrect. Personal ear protectors do not cut out all sound, they merely attenuate it. The human ear can hear oral messages quite distinctly over a wide range of sound levels, i.e. from 20 dBA or less to 90 dBA or more, provided the ear is not 'deafened' or desensitized by very loud background noise, i.e. 100 dBA or more.

The personal ear protector gives the same decibel reduction for the human voice as for the background noise. If the background noise amounts to 110 dBA, a person without ear protectors will be unable to hear loud conversation at 75 dBA because his ear is 'deafened' by the loud background noise. If he is now provided with ear protectors giving a 25 dBA reduction, the background noise will appear as 85 dBA and the human voice as 50 dBA. His ear is more sensitive and no longer deafened and he is able to hear the human voice through the protectors in spite of the background noise.

Two types of ear protection are available, ear plugs and ear muffs *(Figure 3.4.5)*. The protection afforded by different types of ear protector is given in *Table 3.10.4*. When it is necessary for employees to wear ear plugs or muffs to protect their hearing, the foreman or supervisor should do all in his power to see that this is done.

The first difficulty will be to convince employees that protection is necessary. This is particularly true of very noisy industries, where many of the workers have already suffered impairment of their hearing, and are not prepared to face the truth. Also since loss of hearing is a slow, gradual process, the average employee will have little awareness of the effectiveness of his ear plugs. Unless the noise level is so high, i.e. over 120 dBA, as to cause actual discomfort, many employees will wear ear plugs for a short time only, and then discard them. The types of ear protection available are discussed in section 3.10.8.

Before selecting ear protectors for any situation, a full noise survey under typical working conditions should be carried out, in which the noise levels in the various octave bands are measured and not just the overall noise level. The reason for this is that ear protectors give different degrees of protection for noise of different frequencies.

REFERENCES

1. Fife, I., and Machin, E.A., *Redgrave's Health and Safety in Factories*, 'The Woodworking Machines Regulations, 1974, S.I. 1974, No.903, Part X, Regulation 44, Noise', 1537 (1976).
2. Department of Employment, *Code of practice for reducing the exposure of employed persons to noise*, HMSO, London (1972).
3. Health and Safety Executive, *Framing noise legislation*, H & SE, London (1975).
4. *The control of pollution act,* HMSO, London (1974).
5. *The Control of Noise (Code of Practice for Construction Sites) Order,* 1975 (S.I. 1975, No. 2115).
   The Control of Noise (Appeals) Regulations, 1975 (S.I. 1975, No.2116).
   The Control of Noise (Measurement and Registers) Regulations, 1976 (S.I. 1976, No.37).
6. British Standards Institute, BS 5228: 1975, *Noise control on construction and demolition sites*, B.S.I. (1975).
7. Hunter, D., *The diseases of occupations*, 2nd ed., English Universities Press, London, 790-796 (1957).
8. Health and Safety Executive, Booklet 25, *Noise and the worker*, H & SE, London (1976).
9. Atherley, G., 'The value of audiometry in industry', *Journal of the Society of Occupational Medicine*, 23, 19-21 (1973).

3.5 VIBRATION

Contents

Vibration occurs to a greater or lesser extent in all moving machinery and in most buildings and structures. Excessive vibration is a hazard both to health and safety. Vibration in metals, particularly aircraft frames results in fatigue, which caused the two well-known Comet disasters at Rome airport in 1954. Less spectacularly, vibration can loosen nuts and screws on engines and machines and has been responsible for many an accident. Vibration in audio frequencies, from 16 to 20 000 Hz causes sound and noise, the hazards of which were discussed in section 3.4.

Whilst we generally try to minimise vibration in machinery, a number of machines make deliberate use of controlled vibration — screens, 'grizzlies', and vibratory conveyors and tools. Reciprocating engines and machines are chronic sources of vibration. The analysis of vibrations and sound from machines is used for detecting incipient faults while the machine is running (section 2.10, Inspection and Condition Monitoring). Very high frequency vibrations or sound (ultrasonics) are used for determining metal thicknesses, particularly for inspecting pressure vessels, pipes and tubes in the process industries.

Many of the hazards to plant, buildings and machinery caused by vibration are dealt with by mechanical engineers rather than safety specialists. Their activities do, however, need checking and monitoring and they are not always aware of the health hazards of vibration. Safety specialists need to be on the alert for unusual or excessive vibration in machinery, or increases in vibration levels of familiar machines. They should also be on the look out for loose fixtures, methods of fixture (the need for spring washers on nuts and bolts), and for objects which may be dislodged by vibration from surfaces on which they are placed, thereby falling accidentally, causing possible injury or damage.

An elevated motor may be mounted on a shelf with some heavy object resting close to it. The motor normally runs quietly and without vibration. One day something happens to affect the balance of the shaft and armature, the motor vibrates slightly, but not enough to attract immediate attention, and the heavy object moves until it falls off the shelf — perhaps on someone's head, hand or foot.

The safety manager should examine all shelving or flat surfaces on which objects are placed to see whether they are subject to possible vibration from any source. If vibration is a likely hazard, a lip or raised edge to the shelf or surface should be provided to prevent objects being shaken off and falling. Even unlikely sources can cause vibration — e.g. electric cables and small coils through which an alternating current passes, or a vibrating hand-tool may be used for maintenance purposes which causes a particular shelf or table top to vibrate in resonance with it.

The safety specialist also needs to study, and be familiar with, various types of anti-vibration mounting, their installation and maintenance. Many anti-vibration mountings make use of rubber cushions or washers which in time perish or harden so that they cease to be effective.

Another hazard of vibration is its effect on dial instruments and visual acuity. A vibrating pointer on an instrument cannot be read accurately, and such vibrations (caused, for example, by reciprocating pumps and compressors) must be suitably damped. Sometimes the whole panel may vibrate, or the operator may be sitting on a vibrating seat with the panel vibrating out of phase with, or with a different amplitude to that of his head. The eye attempts to track the moving dial, but this is fatiguing. Tracking is possible at a frequency of 1 Hz, but becomes very difficult at 3 Hz, which is the most dangerous frequency for instrument reading.[1] At frequencies of 5 Hz and higher, tracking is not attempted and the dial or meter is read from visual images which are formed at either end of the vibrational swing, (like watching a cinematograph film). Under conditions of vertical vibration horizontal scales are preferred, and *vice versa*. The figures on the scale may need to be larger than normal (see section 3.7).

We are all subject to a greater or lesser extent to spontaneous involuntary vibration, without any assistance from machinery. We can tremble with fear, shiver with cold, or get the shakes from various causes. Parkinson's disease or shaking palsy is a condition which until recently has been virtually incurable. Hatters shakes or Danbury shakes was an occupational disease of hat makers caused by mercurial poisoning. Various parts of the body have their own natural frequencies of vibration — the whole head between 1.2 and 2 Hz, the lumbar region of the spine about 4 Hz; the internal organs of the head and body and various suspended organs have higher natural frequencies. All forms of involuntary vibration are potential causes of accidents, the seriousness of which is related to the job. Medical examination of workers for jobs where involuntary vibration is hazardous (e.g. electricians, dentists) should include tests for this propensity.

3.5.1 Vibration as a health hazard

The vibrational exposure of workers is generally classified as 'whole body vibration' and 'hand/arm vibration'. A guide[2] to the subject published by the British Society for Social Responsibility in Science has been found valuable in preparing this section, and contains many useful references.

Whole body vibration occurs when the subject sits, stands or lies on a vibrating seat, floor, structure, bunk, etc. In severe cases the result may be giddiness and sea-sickness, spinal disorders, varicose veins, piles, headaches and constipation. Most organs of the body can be affected. Visual and mental capacity are also affected. The effects depend largely on the frequency and amplitude of the vibrations.

Hand and arm vibration is suffered by the users of reciprocating and some rotary hand tools — rotary saws, pneumatic hammers, rivetting, chipping, caulking and fettling tools. This can lead to several types[3] of occupational disorders:

1. Small areas of decalcification of the bones as shown by X-ray photographs. (These do not appear to lead to serious consequences).
2. Injury to, and hardening of, the soft tissues of the hands — particularly the palms of the hands.
3. Osteoarthritis of the arm joints, especially the elbows. This has been compensatable in Germany since 1929, but it does not yet appear to be compensatable in the UK.
4. Vascular disturbance, known as Raynauds phenomena, 'white fingers', 'dead fingers', 'banana fingers', and pneumatic hammer disease. This is the commonest and most serious effect on health of hand/arm vibration. It results in cyanosis of the fingers, loss of circulation and chilling which start usually in cold weather — when getting up in the morning and going to work. In time the disease is found in warm weather as well as cold, and leads to loss of feeling and partial loss of the use of fingers. In extreme cases gangrene, amputation and death have followed.

The only known remedies for hazardous hand/arm vibration are complete mechanisation of the operations or radical redesign of the hand-tools to reduce vibration.

3.5.2 Types and frequency of vibration

One of the oldest forms of human vibration exposure is sea-sickness — caused by low frequencies of less than 1 Hz. This vibration is often termed 'motion' or 'movement'. It affects the human balancing organ which consists of three semi-circular canals set at right angles to each other, close to the inner ears (*Figure 3.5.1*). The canals are filled with fluid of high density and supplied with nerve endings at each end which are sensitive to changes in pressure caused by movement of the head and whole body. Movement of a ship at sea and sometimes of aeroplanes and cars cause strong movement of the fluid in the canals which stimulate the nerve endings violently; giddiness and sickness result.

Reciprocating engines, especially diesel, give rise to vibrations with a range of frequencies to which people are subjected in ships, submarines, aircraft, lorries, buses, cars and tractors. Divers and aircrew are some of those most affected. Vibration of the natural frequency of the spine (about 4 Hz) which is transmitted through the driver's seat is a major cause of back trouble in tractor and lorry drivers.[4]

At this frequency the movement of the head is amplified by the resonance

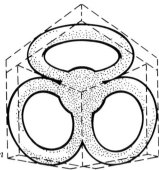

Figure 3.5.1   The organ of balance. Three circular canals attached to the inner ear (From 'The Biology of Work' by E.G. Edholm published by Weidenfield & Nicholson)

of the spine, and a very rough ride is experienced. At about 10 Hz the amplitude of the head and seat vibrations are about equal. A person standing on a platform vibrating at 4 Hz is less affected than a seated person because his legs absorb most of the vibration. Excessive vibration at frequencies of 10 to 20 Hz can cause damage to the lungs, rectal bleeding, blood in the urine, constipation and heart failure in extreme cases. Vibration at 20 Hz can cause resonance of internal head organs with consequent headache and psycho neurotic symptoms.[4]

Methods of preventing vibration in transport in order of importance are:

1. Elimination or reduction of vibration at the design stage.
2. Isolation of the vibrations from the engine or surface over which the vehicle moves by good suspension of the cab or cabin.
3. Isolation of some of the vertical vibration by the use of a suspension seat.
4. Reduction in exposure time by good breaks and shorter working hours.

Farm tractors pose one of the most difficult problems. A suspended cab has been developed by the National Institute of Agricultural Engineering in England. This is said to cut vibration by half, although it is not expected to be available commercially for several years.[2] Suspended cabs and seats are gradually becoming available on large trucks and lorries, although not all have come up to their makers' expectations. A suspended seat only reduces vertical vibrations whereas a properly suspended cab should reduce vibrations in all directions.

Requirements of a good suspension seat are said to be:

1. Adjustability to the driver's weight.
2. Adjustability of vertical height.
3. Adjustability of driver's position forwards and backwards.
4. Adjustability of back angle (0 to 20°, except for sleeping, when 40° is required).
5. Suitably placed in relation to controls.

6. Vibration data obtained by an independent testing organisation should be available.

Much hand/arm vibration is at frequencies considerably higher than 20 Hz. Many sources of excessive vibration are also sources of excessive noise, but this can by no means be certain.

Up to 1945, disabilities due to excessive vibration caused through the use of hand-tools had been confined to workers using tools which delivered hammer-like blows.[3] Since then, following the introduction of portable hand-held tools powered by small electric motors, disabilities have been experienced by men using portable grinders, chain saws, and rock drills. Other operatives whose job involves holding a workpiece pressed hard against a moving tool, or *vice versa* (stationary grinding wheels, swaging and spinning operations) have also suffered disabilities. The most dangerous frequencies appear to be between 40 and 120 Hz, although higher frequencies or rotational speeds have also caused disabilities perhaps through the formation of harmonics at lower frequencies. A vibration analyser is shown in *Figure 3.5.2*.

3.5.3 Protection against vibration and noise

Since the machines which cause hand/arm vibration disabilities are also usually intolerably noisy, steps taken to reduce vibration disabilities usually reduce noise levels as well.

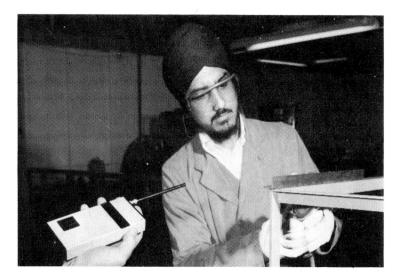

Figure 3.5.2   Metrologger db 306 vibration analyser (Dawe Instruments Ltd. under licence from Metrosonics Incs., USA)

The best form of reduction is at sources in the design stage, but good maintenance is also most important.

Vibrations from established machines can be reduced by insulating them from the surrounding surfaces. The principal methods are:

1. Mounting the machine on a heavy base.
2. Using insulating mountings.
3. A combination of (1) and (2).
4. Using (1) and (2) and sinking the machine in a pit.

3.5.3.1 Chain saws

Chain saws have been tested by the British Forestry Commission and by research institutes in Sweden and Finland.[2] The only saws coming near the safe limits set by the Forestry Commission in tests reported in 1975 were the Swedish Husquana 180S and 160S and the Jonsored 55AU.[2]

Exposure times should, however, be strictly limited (3 to 5 hours per day depending on the saw), and operators should not start work on these saws with cold hands. Disabilities from hand/arm vibration are significantly increased when the operators' hands are cold.

The following health and safety check list has been given for chain saw operators.[2]

Health	*Safety*
Anti-vibration handles	Dead hand trigger
Large muffler	Chain brake, automatic activation
CD ignition	Safety chain
Weight less than 9 kg	Sprocket nosed bar
Heated handles with adjustable setting	Power mate sprocket
	Chain catcher
	Chain brake guard (rear handle)
	Positive on-off switch
	Smooth underside

3.5.3.2 Pneumatic tools

The first question which should always be asked is 'Can the use of a pneumatic tool be eliminated or cut down?' In the case of castings, the work can sometimes be done as satisfactorily by manual broaching or by using modified lathes. Again the casting may be better designed or preformed to reduce the amount of chipping or fettling required.

A Swedish firm, Atlas Copco is developing manipulators and robot pneumatic tool systems which hold and apply force to the tool, so that the operative has merely to guide it[2]. The vibrations are taken up by the pneumatic cylinder and suspension frame of the manipulator. The same firm has also developed a new and improved chipping hammer design which they claim reduces vibration levels by 90%, halves the noise level with a new silencer and reduces the weight by 25%.

3.5.3.3 Grinding

Dr Pelmear of GKN (Guest, Keen and Nettlefolds) who has studied the problems of hand/arm vibration arising from grinding has recommended[2]:

1. The abrasive wheel should be as soft as possible.
2. Wheels should be dressed using an anti-vibration dressing tool on a fixed location bar to give an eccentricity of less than five thousandths of an inch.
3. Turning or milling should, where possible, be used instead of grinding.
4. Anti-vibration devices should be used on vibrating tools and holding tongs.

3.5.3.4 Swaging, spinning

The tube or workpiece should be held by a mechanical clamp and not by hand.

3.5.4 Vibration measurement and standards

The most important factors in vibration are the frequency (Hertz), the acceleration (metres/second^2) of the moving surface and the duration of exposure. At one time the amplitude was quoted instead of acceleration, but the permissible amplitude is so dependent on frequency that acceleration is now preferred. Acceleration is a function of frequency and amplitude.

Measurement of vibration requires an accelerometer, an expensive instrument which is available for use in laboratories and as a field instrument. Its use requires skill and training.

Attempts have been made by the BSI to develop standards; and two drafts are available:

1. DD 32:1974 *Guide to the evaluation of human exposure to whole body vibration*[5] and
2. DD 43:1975 *Guide to the evaluation of exposure of the human hand-arm system to vibration.*[6]

The draft limits for whole body vibration are set out graphically with scales representing acceleration and frequency for both vertical and horizontal vibration. These are given in *Figure 3.5.3*. Each graph shows three curves, a reduced comfort boundary line, a lowered proficiency boundary and an exposure limit — all based on eight hours exposure. The minimum tolerable acceleration for vertical acceleration is found between frequencies of 4 and 8 Hz (corresponding to spinal vibration). That for horizontal vibration is found between 1 and 2 Hz, corresponding to vibration of the whole head.

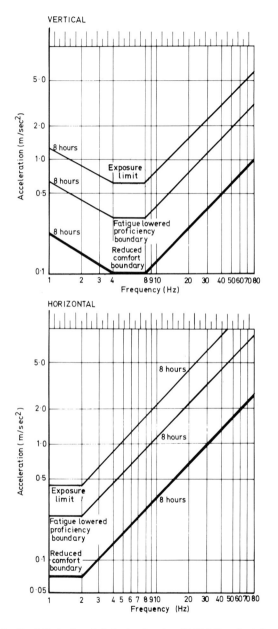

Figure 3.5.3   Draft limits for whole body vibration (British Standards Institution[5])

Criticism of the proposed acceleration limits of the standard on whole body vibration have been made by an American expert[7] on vibration on the grounds that they are too high at the most sensitive frequencies.

REFERENCES

1. Edholm, E.G., *The biology of work,* Weidenfeld and Nicolson, 121 (1967).
2. Dalton, A.J.P., *et al, A workers guide to the health hazards of vibration and their prevention*, British Society for Social Responsibility in Science (BSSRS) London (1977).
3. Hunter, D., *The diseases of occupations*, 2nd ed., The English Universities Press, 805 (1957).
4. Murrell, K.F.H., *Ergonomics*, Chapman & Hall, London, 348-9 (1965).
5. British Standards Institution, *Draft for development, guide to the evaluation of human exposure to whole-body vibration*, DD 32 (1974).
6. British Standards Institution, *Draft for development, guide to the evaluation of exposure of the human hand-arm system to vibration*, DD 43, (1975).
7. Janeway, R.N., 'Human vibration tolerance criteria and application to ride evaluation', paper presented at SAE, Automotive Engineering Congress, February (1975).

3.6 HEAT AND COMFORT

Contents

Man is an adaptable creature; he can endure both tropical and arctic climates, yet he has always sought protection from extremes of terrestrial temperature.

Maintenance of a comfortable working temperature is important for efficiency and productivity and for safety and freedom from errors.

The importance of temperature was recognised in the Factories Act, 1961, Part 1,[1] section 3, which reads as follows:

'3. Temperature. (1) Effective provision shall be made for securing and maintaining a reasonable temperature in each workroom, but no method shall be employed which results in the escape into the air of any workroom of such a character and to such extent as to be likely to be injurious or offensive to persons employed therein.

(2) In every workroom in which a substantial proportion of the work is done sitting and does not involve serious physical effort a temperature of less than 60°F shall not be deemed, after the first hour, to be a reasonable temperature while work is going on, and at least one thermometer shall be provided and maintained in a suitable position in every such workroom.'

Temperature alone, whether measured in the shade or in direct sunlight, is not the only factor affecting Man's sensation of warmth and cold. This sensation is dependent both on environmental factors and on factors concerned with the Man himself, his clothing and the nature of his activities at the time.

The environmental factors are air temperature, radiation, humidity and air movement. A humid place such as Calcutta may provide more severe conditions than Cairo, although the air temperature in Cairo may be several degrees higher.

Climatologists, ergonomists and heating and ventilation engineers have sought to obtain a single parameter by which to define these environmental factors.[2]

In addition to dry bulb shade temperature, which is what we normally mean by air temperature, there are two other important temperatures, the wet bulb temperature and the black globe temperature. The wet bulb temperature can never be higher than the dry bulb temperature and is equal to it when the relative humidity of the air is 100%. The lower the relative humidity the larger is the difference between the wet and the dry bulb temperature.

The black globe temperature is used to measure thermal radiation. The bulb of the thermometer is located at the centre of a copper globe which surrounds it, the outside of the globe being painted matt black. Under standard measuring conditions it is suspended in the open approximately 1.5 m above the ground. Various indices, derived by combining these three temperatures, attempt to describe environmental factors of man's sensation of heat or cold by means of a single figure. Perhaps the best known is the wet-bulb globe temperature index (WBGT) which is derived from the formula:

WBGT = 0.7 globe temperature + 0.2 wet bulb temperature + 0.1 dry bulb temperature.

This makes some allowance for air movement as well as radiation, as a black bulb exposed to thermal radiation is also cooled by air movement. The WBGT is discussed further in section 3.9.1.

The value of the wet bulb globe temperature was well demonstrated during the training of members of the United States Marine Corps. The 'safe' upper body temperature limit for an acclimatised man is about 39.5°C. Body temperature above 41°C are very dangerous, and death is usual at temperatures of 42°C and over. Before this temperature is reached, work performance will have been impaired and at or above it many people will collapse. There must of course be sufficient difference in temperature between the body and its surroundings for heat produced by the body's metabolism to be removed, and heat is lost more easily when the humidity is low because of faster evaporation of perspiration.

At one time the Marine Corps stopped or limited training whenever an air temperature of 32.5°C was reached, but in spite of this heat casualties were not uncommon. The wet bulb globe temperature is now used, and training ceases when a WBGT of 31°C is reached. There are now virtually no heat casualties, and there is less loss of training time. This is due to the fact that a dry bulb temperature of 32.5°C can be reached and exceeded when the WBGT is considerably below 31°C, if the humidity and/or radiation are low.

A scale used to assess the severity of conditions in hot mines, factories and foundries where there may be serious heat stress is the four-hour sweat rate (P4SR). This was obtained by studying the sweat rates of subjects exposed to particular conditions in a climatic chamber and performing a standard task.

For work in severe cold, the Wind Chill Scale has been derived. This is the time taken for a given amount of water hung up in a standard tin to freeze under different conditions of cold. It measures a combination of temperature, wind speed and radiation loss. A man suitably dressed can work in still air at −40°C with his face exposed, but would suffer severe frostbite in even a slight breeze at that temperature.

3.6.1 The comfort zone

First, we must differentiate between different rates of bodily activity and the amount of clothing worn. An average man dressed in a lightweight suit will be comfortable working sitting at a desk in an office at a temperature of 21°C (70°F). But he would also be comfortable in the same clothes when walking outside at a speed of 3 m.p.h. at a temperature of 10°C or when running outside at 6 m.p.h. at a temperature of 0°C. This is because the rate at which we generate body heat (which has to be lost to the environment) is roughly proportional to the rate at which our muscles do work and expend energy. The efficiency of the human body, considered as an engine, is about 20%, which means that for every unit of mechanical work done, four similar units are converted to heat and have to be removed.

Next we have to realise that different human beings vary considerably in their ideas of a comfortable temperature, even when they are similarly clad and exerting the same amount of bodily energy. This leads to endless arguments. Morin in 1863 described[4] a squabble in the Theatre Lyrique in Paris in a warm day in May when the temperature on stage was about 22°C, the manager and prima donna complaining of the high temperature while the director of the opera grumbled that he was cold.

The recognised comfort zone for homes and offices in the UK[2] is generally considered to be:

Temperature range	15-20°C
Air movement	25 cm/second
Relative humidity	50-70%

A higher temperature range is preferred in the USA, i.e. 18 to 23°C, where lighter clothing is customary.

As lighter clothing, especially among our womenfolk, has become more popular in England, so it seems our ideas on comfortable indoor temperatures have moved upwards. Women and elderly people seem to prefer higher indoor temperatures than men and younger people.

3.6.2 Body temperature regulation

A constant body core temperature of around 37°C, with a swing of about 0.5°C between night and day, is maintained in the brain, the heart and the abdomen by a temperature regulating centre in the brain. The temperatures of the skin and bodily extremities are lower and can vary considerably, i.e. from 30° to 35°C. The mechanism which controls the core temperature is shown schematically in *Figure 3.6.1*.

The three methods of temperature control, in order of importance are:

1. Control of blood supply to the skin and extremities; by increasing the supply the skin temperature rises, and heat is lost faster from the skin to the surroundings.
2. By secretion of sweat which assists body cooling by evaporation.
3. By shivering, which creates heat.

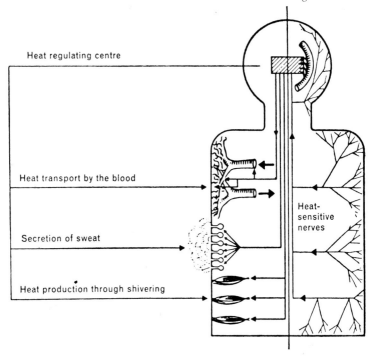

Figure 3.6.1   Physiological mechanisms regulating body temperature (This illustration and Figures 3.7.1, and 3.7.3 are taken from 'Fitting the Task to the Man' by E. Grandjean published by Taylor & Francis and reproduced by permission of the Publishers)

Heat is transferred from the body to its surroundings in four different ways: conduction; convection; evaporation; and radiation. Conduction can take place locally through objects of high thermal conductivity (especially metals) in contact with the skin. Local loss of heat by conduction to cold objects is not only unpleasant, but can reduce the efficiency of the organ in question and lead to disabling ailments (such as arthritis) in severe cases. Thus, wherever possible, workrooms should have floors with good insulating qualities and table tops, handles on machines, levers and tools should be covered with an insulating material such as wood, rubber, plastic or leather.

Convection under normal conditions amounts to about 25 to 30% of the heat loss from the body to the environment.

About 25% of the heat loss of the body is lost normally by insensible perspiration, corresponding to the evaporation of about one litre of water per day. But at temperatures above about 25°C the clothed human body can lose little heat by convection and radiation, and hence more has to be lost by evaporation through stimulation of the sweat glands.

Heat exchange by radiation depends on the difference between the temperature of the skin and surrounding surfaces. In temperate climates, 40 to 60% of our heat loss is due to radiation, but in hot conditions where the surrounding surfaces are hot (e.g. in deep gold mines) little radiation can occur and almost the entire loss of body heat must occur by perspiration and evaporation of sweat. We become aware of radiation loss, for example by the feeling of chill when standing in a heated room near a large window.

3.6.3 Limits of comfort and effects of discomfort[3]

It is only when conditions deviate from comfort limits that one becomes aware of them, the sensation varying from annoyance to agony.

Overheating causes sensations of tiredness and sleepiness which make performance more tedious and *increases the frequency of errors*. This slowing down of activity reduces the rate of heat production in the body. Overcooling results in restlessness and reduced attention which affects mental work in particular. The rate of heat production in the body has to be increased by increased muscular activity. Thus a comfortable environmental climate is essential for full efficiency and safe performance.

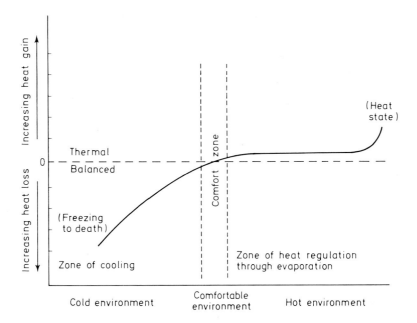

Figure 3.6.2   Heat balance of the body under different climatic conditions

The environmental temperature zone (e.g. as measured by the WBGT) in which a person can perform the same task and wear the same clothing in comfort is quite narrow — i.e. only 2-3°C wide. Although adaptation to a wider range of temperature is possible through acclimatisation, this creates stress in the regulatory mechanism, and it is doubtful whether people can feel as comfortable or work as efficiently and safely outside this range.

Figure 3.6.2 shows the typical heat balance of the body under different environmental conditions.

3.6.4 Air speed and air temperature

The same 'perceived temperature' (as approximately measured by the WBGT) is experienced at various combinations of air speed and dry or wet bulb temperature. Thus a dry bulb temperature of 20°C in still air will give the same impression of temperature to a person as one of 23.5°C with an air speed of 2.0 m/sec.

Air speeds more than 0.2 m/sec however, feel uncomfortable to a seated person, and should be avoided in offices. For very precise work where the body must be held quite still, air movements should be less than 0.1 m/sec. However, higher air velocities, up to about 0.5 m/sec can be allowed for standing work which demands considerable physical effort.

3.6.5 Relative humidity in heated buildings

A great deal of work has been done on the effect of relative humidity on health and comfort in heated rooms. The general conclusion is that a relative humidity of 40 to 50% is desirable for comfort and hygiene. Relative humidities below 30% are undesirable and cause dehydration of the mucous membranes of the nose and the respiratory tracts. Dry air also encourages the formation of static electricity and may cause damage to wooden furniture and other objects.

The relative humidity in many heated industrial buildings and offices in winter is too low, and many ear, nose and throat specialists attribute the frequency of colds in winter months to this cause, as the dried mucous membrane is more easily infected by the organisms which cause colds and influenzas.

3.6.6 Environmental climate for physical work

Heavy physical work results in higher rates of heat production and to achieve a comfortable heat balance with the surroundings a lower environmental temperature is required. *Table 3.6.1* gives the preferred room, building or environmental temperature for various types of physical work.

Table 3.6.1 PREFERRED ENVIRONMENTAL TEMPERATURES FOR VARIOUS TASKS

Type of work	Heat production, (calories per 24 hours) Men	Women	Preferred environmental temperature (°C)
Sedentary, mental	2200	1800	20-23
Sedentary, light	2700	2200	19
Standing, light	3000	2500	18
Standing, medium	3600	3000	17
Standing, heavy	4000 and over	—	10-15

3.6.7 Room temperatures in summer, hot climates and air conditioning

Although the human mind and body can, to some extent, adapt and acclimatise to higher temperatures, there can be little doubt that they are under greater strain and operate less efficiently and with more errors.

Table 3.6.2, reproduced from Professor Grandjean's book already cited, spells out clearly the effects on Man's performance indoors as the ambient temperature is increased. The use of air conditioned cars, offices and living quarters in hot climates is ample testimony to the benefits of a comfortable working temperature.

Table 3.6.2 EFFECTS OF DEVIATIONS FROM A COMFORTABLE ROOM CLIMATE
(Reproduced from 'Fitting the task to the man' by Professor E. Grandjean).

20°C	1. Comfortable temperature	Fully effective
	2. Discomfort, Increased irritability Lack of concentration Decline in performance of mental work.	Psychological disturbance
At 50% relative humidity	3. Increase in errors Decline in efficiency of skilled work Increase in accidents	Psycho-physiological disturbance
	4. Decrease in efficiency in heavy work Disturbance in water and salt circulation Heavy stress on heart and circulation Heavy fatigue and threatening exhaustion.	
35-40°C	5. Highest tolerable temperature	

Professor Grandjean notes a difference in the American and European approaches to air conditioning. Americans, he says, prefer to keep the temperature of the air conditioned room near to the optimum and tolerate a rapid change in temperature on entering it, whilst Europeans prefer a smaller temperature drop and consider that the difference between the exterior and interior temperature should never exceed 4°C.

3.6.8 Modern office buildings

Modern office buildings tend to have lower ceilings and a larger percentage of window area on the outer walls than older ones. This tendency accentuates the climatic problems and increases the need for ventilation and air conditioning in modern buildings.

3.6.9 Heat stroke and hot industrial work

Workers are exposed to high temperatures even in temperate climates in many industries, e.g. foundries, mining, steel and glass making. Their comfort can be improved either by general air conditioning, or where this is not possible by wearing special jackets and hoods which are ventilated by cold air supplied by flexible hoses.

If workers are required to work at wet bulb or wet bulb globe temperatures above 30°C, the time of exposure should be strictly limited, depending on the wet bulb temperature. Grandjean gives recommended maximum exposure times at various temperatures.

Without protection, the exposed worker is liable to suffer from heat stroke. The first symptoms are general listlessness, red skin, rapid heart rate and thin pulse. Further heat leads to severe headache, vertigo, shortness of breath, vomiting, and muscular cramps arising from salt depletion. Finally the person loses consciousness and often dies within 24 hours despite all medical efforts. Fat people are more at risk than thin ones, and age, food intake, amount of physical work and degree of acclimatisation all play a part. Heat stroke among workers has been found even at 30°C at high humidities.

Grandjean makes a number of recommendations regarding drinks, rest periods, protective clothing and conditions for workers exposed to hot conditions. Rather surprisingly he regards the provison of extra salt as debatable.

REFERENCES

1. Fife, I., and Machin, E.A., *Redgraves health and safety in factories*, Butterworths (1976).
2. Edholm, O.G., *The biology of work*, Chapters 3 & 4, Weidenfeld and Nicolson, (1967).
3. Grandjean, E., *Fitting the task to the man (English Edition)*, pp 136-151 Taylor and Francis, London, (1975).
4. Bedford, T., *Basic principles of ventilation and heating*, 2nd ed., H.K. Lewis, London (1964).
5. McCormick, E.J., *Human factors engineering*, Chapter 15, McGraw Hill, London. (1964).
6. Murrell, K.F.H., *Ergonomics*, Chapter 12, Chapman and Hall, London (1975).

3.7 ERGONOMICS

Contents

Ergonomics is the scientific study of the relation between man and his working environment. The word, coined from the Greek '*ergos*' (work) and '*nomos*' (natural law) was first used in 1949 when a society was formed to bring together workers in various fields who were concerned with different aspects of the subject. These included, anatomists, physiologists, psychologists, industrial medical officers, industrial hygienists, design engineers, work study engineers, architects, illuminating engineers and others. This society, the Ergonomics Research Society, founded in Britain, led to the formation of the International Ergonomics Association which held its first meeting in Stockholm in 1961.[1]

Most of the activities included in the science of ergonomics had been studied for many years before the word was born. In the USA the terms 'Human Engineering' and 'Human Factors Engineering' are used to cover basically the same field. Biomechanics and Engineering Psychology are other expressions used. In its broadest sense, ergonomics covers virtually all of the subjects discussed in this chapter, as well as much of sections 5.2 'The slipped disc syndrome' and 5.3 'Manual handling'.

Ergonomics arose primarily from the military need during World War II to ensure that the weapons used were compatible with and properly matched to those who had to use them. This need brought together people of the various disciplines mentioned above and the realisation by them that an inter-disciplinary approach and grouping was needed. During the post-war years ergonomics has been increasingly applied in industry to improve performance, reduce human stresses, and reduce or eliminate hazards at the man/machine interface.

While ergonomics thus covers a variety of subjects dealt with by older disciplines, its hard core remains the physical matching of artefacts and workers for safe and efficient performance. It is this aspect which is discussed here.

Professor Grandjean's book *Fitting the task to the man*[2] has been largely followed, mainly because the treatment is condensed and definitive and highlights the essential points. Readers requiring to explore the subject in greater depth are recommended to study in addition the excellent books by Edholm[3], Murrell[1], and McCormick.[4]

3.7.1 All shapes and sizes

It is odd that while we take pains to ensure that our shoes and clothing fit our individual bodily dimensions, we customarily expect a standard size of chair, table, desk, lathe and a variety of machines to match men and women whose bodily dimensions show great variations. If the standard artefact matches a person of average size, then those much taller, shorter, fatter or stiffer than average are likely to experience considerable discomfort in using it, and it is only man's exceptional adaptability that allows them to do so at all. Chairs, tables and machines cannot really be designed to match the whole range of human shapes and sizes.

The first step to be taken before designing an artefact is to make an anthropomorphic survey of the population for which it is being designed. Men are taller on average than women, and north Americans are taller than Japanese. This survey will give not only the average main bodily dimensions, but also the frequency distribution of particular dimensions throughout the population. One must then consider what proportion of the population can be satisfactorily matched to the artefact designed for a man or woman of average size in that population.

Sometimes attempts are made to match the middle 90% of the population in all bodily dimensions surveyed, excluding the 5% with smaller and 5% with larger dimensions. At other times attempts are made to cater for people with dimensions from, say 20% smaller to 20% larger than the average. In most cases the artefact needs to be adjustable to cater for people whose dimensions are near the ends of the range covered.

The chief compensation of those whose dimensions are outside this range is that they will often qualify for special tasks for which their exceptional dimensions are advantageous. Very tall people are at an advantage for certain work in crowds — television cameramen or bookies' 'tic-tac' men — whereas very short men may be employed as jockeys or for work in confined spaces.

A survey of 1008 Swiss factory workers gave the range of body sizes shown in *Figure 3.7.1*. The averages of a number of body dimensions are given in *Table 3.7.1*.

Table 3.7.1 AVERAGE BODILY DIMENSIONS (CM) (SWISS WORKERS)

Part of body	Men	Women
Body height	169.0	158.8
Shoulder height	140.8	131.9
Hip height	102.6	100.1
Arm span	173.7	158.8
Arm length	70.4	63.6
Forearm plus hand length	47.5	43.8
Upper arm length	36.3	33.7
Knee height (sitting)	52.2	47.1
Back of knee to sole (sitting)	45.4	37.4
Back of knee to back (sitting)	46.8	46.6

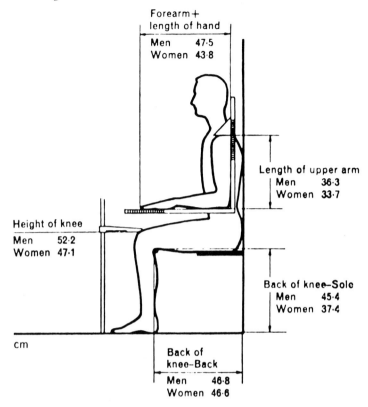

Figure 3.7.1 Body measurements when seated

3.7.2 Body size and the workplace

Grandjean[2] who has made an extensive ergonomic study in Swiss industry makes the recommendations given in the following paragraphs.

3.7.2.1 Standing work at benches

The optimum height of the working area (bench top) is:

5 to 10 cm below the height of the elbow for light work.
15 to 20 cm below the height of the elbow for heavy work,
5 to 10 cm above the height of the elbow for precision work, (where the forearms rest on the bench).

On average, the height of the elbow above the floor is 104.5 cm for men and 98 cm for women. Corresponding bench heights are given in *Table 3.7.2.*

Table 3.7.2 HEIGHTS OF WORK BENCHES (CM) FOR STANDING WORKERS

Type of work	Men	Women
With supported elbows	100-110	95-105
Skilled work on small objects	90-95	85-90
Heavy work, filing, pressing, etc.	75-90	70-83

3.7.2.2 Seated work at desks and tables

A working area several centimetres below elbow height is usually preferred, but two limiting factors arise:

1. *Visual distance and head position.* For very delicate work involving small parts (e.g. watch making and repairing) which require close visual distances, a special working bench with a raised central portion and sloping arm rests is recommended *(Figure 3.7.2)*. When sitting, the normal inclination of the head is such that the eyes are looking downward at an angle between 35° and 42° to the horizontal.
2. *Knee room.* The lower side of the desk or table top must leave room for the knees. The average height of the knee from the floor is 52.2 cm for men and 47.1 cm for women. After allowing for tall people, high heel shoes, and the thickness of the table top, the minimum heights of the table top are 65 cm for women and 68 cm for men.

Table heights for various types of sedentary work are given in *Table 3.7.3.*

Table 3.7.3 TABLE HEIGHTS (CM) FOR SEDENTARY WORK

Type of work	Men	Women
Precision work with short visual distances	90-110	80-100
Writing and reading	70-78	70-74
Typing and heavy manual work	68	65

Desks and tables should be chosen for the taller rather than the shorter worker, and foot rests supplied where needed for the latter. Seat heights should generally be between 27 and 30 cm below the height of the desk surface or typewriter keyboard.

Physical complaints during sedentary work were investigated by Grandjean and results are given in *Table 3.7.4.*

Figure 3.7.2 Special workbench used by watch repairer. The height of the work must allow the body to incline slightly forward while providing the optimal visual distance. The workbench shown in this illustration allows the arm to take up a natural position so that static stress is avoided.

Table 3.7.4 PHYSICAL COMPLAINTS DURING SEDENTARY WORK

Discomfort of:	%
Head	14
Back	57
Neck and shoulders	24
Buttocks	16
Arms and hands	15
Knees and feet	29
Thighs	19
No complaints	15
Total	100

Back trouble is by far the commonest complaint of sedentary workers and most people will endure cramped and unsuitable leg positions in order to allow a comfortable trunk posture. While doctors and ergonomists generally recommend that workers be seated, there can be disadvantages — slackened abdominal muscles, rounded back and pressure on the digestive tract and lungs.

Grandjean's recommendations for office chairs are given below:

1. Seats should be adjustable between 40 cm and 53 cm in height.
2. Back rests should be adjustable vertically between 14 cm and 24 cm above seat height.
3. Back rests should be adjustable in depth between 34 cm and 44 cm from the front edge of the seat.
4. The depth of the seat should be at least 35 cm.
5. The chair should be stable, neither tilting nor sliding.
6. The chair should allow sufficient freedom of movement for the specific activities of the user.

Four-legged chairs are preferable for stability; the back rest should not be wider than 32 cm.

3.7.2.3 Conditions for safe skilled work

The following recommendations are made:

Information transmission	When information is transmitted by speech or sound signals, the noise level in the room should be at least 10 dB below that of the information.
Visual control	All movements should be made under visual control. Levers, switches, hand wheels and instruments should be placed where they can be readily seen and operated without changing body position. The visibility of all work elements and the speed of perception should be enhanced by the use of suitable lighting, shapes and colour.
Unwanted stimuli	Unexpected and discontinuous noises should be eliminated as far as possible ('now look at what you have made me do'). Visual distractions by third persons, moving machine parts, brightly coloured objects, lights or reflections should be avoided.
Unnecessary manual work	Manual work of secondary importance should be kept to a minimum. Clamps, hangers and jigs should be used to hold work, and shutes used for the supply and removal of work pieces. Pedal-operated switches and controls can reduce less essential hand movements.

Control arrangements	Controls and instruments should be arranged to match the sequence of operations so that movements follow one another logically.
Effort and stress	Skilled operations should not require great physical effort and simultaneous static stress should be avoided.
Hand and arm movements	Horizontal movements are more precise than vertical ones. Circular movements are preferred to zig-zag ones. Movements towards the body are easier to control than movements away from it.
	Serial movements should flow rhythmically.
	Sudden and jerky movements are difficult to control and increase fatigue.
	After completion of a series or cycle, the operator should be close to his starting position.
Arrangement of materials, components and tools	Materials, components and tools should be arranged in a half circle in front of the operator so placed as to allow a rhythmic flowing movement.
	The working area should lie inside this half circle as shown in *Figure 3.7.3*.
Two handed operations	The effort should be divided as equally as possible between both hands, the movements for both hands starting and ending together.
Rhythm	A free unpaced rhythm is better than any kind of imposed rhythm.

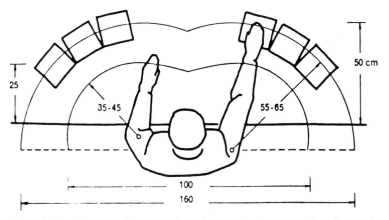

Figure 3.7.3 Grasping distance and work area. Optimum grasping distance is 350 — 450 mm from lowered elbow. Maximum distance is 550 — 650 mm from the shoulder.

3.7.3 Control of machines

In many operations machines are employed as tools, using levers, knobs, switches, wheels, buttons and pedals to control the machine and receiving feed-back information from the machine about the operation from display instruments with dials.

3.7.3.1 Dial instruments

There are three common types of dial instrument:

1. Open window where a numerical value is displayed (counter).
2. Round dial with a movable pointer.
3. Fixed pointer with a movable dial.

If the sole object of the instrument is to give the figure, the open window is preferred, providing that only the required number is visible.

If a change in value is to be observed, or if a control has to be set at a definite figure (e.g. voltage, pressure), then a fixed dial with movable pointer is generally best. If absolute values are not required, but the process has merely to be kept within certain limits (e.g. of temperature), numbers may be omitted and the range merely shown in an appropriate colour.

The shape of the dial may influence reading accuracy. The percentage of errors on reading different types of dial with a display time of 0.12 sec at the same distance from the eye is shown in *Figure 3.7.5*.

The following principles should be followed when selecting instrument dials:

1. The degree of accuracy shown should match the accuracy required. Dials giving a greater accuracy than necessary make reading more difficult and increase errors.
2. The information should be given in the simplest way to the operator and superfluous information avoided.

Figure 3.7.4   Simple display instrument

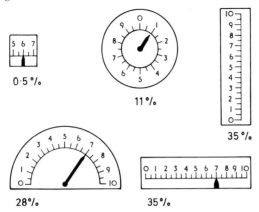

Figure 3.7.5   Effects of dial types on reading precision

3. The information should be easily understood and converted. Only simple conversion factors such as 10 to 100 should be used.
4. Sub-divisions should correspond only to values of 1, 2 or 5.
5. Figures should only be attached to the large scale markings.
6. Figures on a fixed circular scale should be upright. On a movable circular scale, they should be tangential (i.e. upright when opposite the fixed pointer).
7. The sizes of scale markings must be appropriate to the maximum expected reading distance. Recommended minimum dimensions are given in *Table 3.7.5.*
8. The tip of the pointer should cover neither the figures nor the scale markings, and it should have the same thickness as the marks. Correct and incorrect arrangements are shown in *Figure 3.7.6.*
9. The plane of pointer should lie as close as possible to the plane of the scale markings to reduce reading error.

Table 3.7.5 MINIMUM SIZES OF SCALE MARKINGS FOR MAXIMUM PROBABLE READING DISTANCE 'a'

Type of marking or letter	Size of marking or letter
Height — large scale	a/90
Height — medium scale	a/125
Height — small scale	a/200
Thickness of scale markings	a/5000
Distance between markings — small scale	a/600
Distance between markings — large scale	a/50
Height of small letters or figures	a/200
Height of large letters or figures	a/133

Divisions of scale

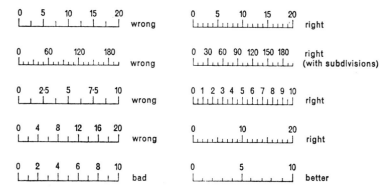

Figure 3.7.6   Correct and incorrect scale divisions. Subdivisions should correspond to values of 5, 2 or 1

3.7.3.2 Types and choice of controls

Knobs, levers, handwheels and pedals are the main types of control which may be used. The correct choice is important for safety, and the following principles should be followed:

1. Fast and precise operations should be performed by fingers or hands; **operations requiring force, by arms or feet.**
2. Hand controls should be within easy grasping distance between elbow and shoulder height, and be clearly visible.
3. The minimum distance between finger controls (knobs) should be 15 mm and between hand controls 50 mm.
4. For operations requiring minimum force, and for stepwise or continuous adjustment of small amplitude and high precision, manually operated push buttons, toggle switches or knobs are recommended.
5. For operations requiring considerable force, of great amplitude and of little precision, long arm lever, cranks, hand-wheels and pedals are preferred.

3.7.3.3 Controls for precision work requiring little force

1. *Push buttons.* These take up least room and may be distinguished by colour and other markings. The surface should be slightly concave and the diameter sufficient for the finger tip to touch it lightly without slipping. Recommended features are:

Diameter — simple buttons   12-15 mm

Diameter — emergency stop (for entire machine)	30-40 mm
Movement	3-10 mm
Resistance	250-500 g

2. *Toggle switches.* These should be clearly marked and should preferably have only two positions. When used for three positions, the angle of movement between two adjacent positions must be at least 40°.

3. *Knobs.* These exist in many shapes — circular, bar shaped and pointed — while combinations of two or more knobs on a common axis are sometimes used. They must be easy to feel and provide a reliable grip. Any movement must be clearly visible. The pointer and scale should be on circles of sufficient diameter to avoid being obscured by the fingers.

Pointed bar knobs are preferred for stepwise control, and circular knobs for continuous control. Knobs of different shapes and/or surface texture which can be distinguished by feel as well as sight are recommended for complex operations. Recommended dimensions for circular knobs are given in *Table 3.7.6.* A knob can be turned through an angle of 120° without changing grasp.

Table 3.7.6 RECOMMENDED DIMENSIONS FOR CIRCULAR KNOBS

Dimension	Two or three finger knobs	Whole hand knobs
Diameter, mm	10-30	35-75
Depth, mm	12-25	20-50
Maximum torque, gm/cm	450	2500
Preferred maximum angle turned through		120°

3.7.3.4 Controls requiring force over wide range with little precision

1. *Cranks.* These are suitable for continuous movements over a wide range. The following lever arm measurements are recommended:

Low torque — up to 200 rpm	60-120 mm
High torque — up to 160 rpm	150-220 mm
Quick movement — torque between 1 and 2.5 kg	up to 120 mm
Exact positioning — torque between 1 and 3.5 kg	120-200 mm

2. *Handwheels.* These are recommended where large turning forces are required and necessitate using both hands. The inner rim of the handwheel should be ridged to allow a firm grip.

3. *Levers.* These are mainly used for gear changing. For large manual gearshift levers, the maximum forces and movements are given in *Table 3.7.7.*

4. *Pedals*. Nowadays pedals are seldom used as continuous power sources, but for positioning an object or for speed control or braking. For standing workers a wide pedal at floor level which can be used by either foot is preferable.

Table 3.7.7 LARGE MANUAL GEAR CHANGE LEVERS

	Maximum force (kg)	Maximum movement (mm)
Forwards and backwards	13	350
Sideways	9	150

3.7.3.5 Inter-relation between controls and display instruments

Display instruments and controls should be designed so as not to violate certain natural human reactions. The most important of these being:

1. The display pointer should move in the same direction as the control.
2. A clockwise movement should lead to increase in the process controlled, except liquid flow which is often reduced by clockwise movement.
3. Scales and knobs concerned with a given function should be grouped together, preferably scale above and knob below. Where the display panel is separate from the switchboard, the arrangement of the controls should match that of the dials.

For complex and critical operations a link analysis should be carried out to determine the most satisfactory arrangement of the control panel. This involves preparing a drawing of the proposed panel then going through the motions of operation, drawing links between the various dials and controls which need to be used simultaneously or consecutively.

The frequency of use and importance of each link are then evaluated, and a value assigned to each link by multiplying frequency of use by its importance. The dials and controls are then re-arranged to minimise the lengths of the links with the highest use-importance value.

An example of the use of this method to modify a radar control panel is given in *Figure 3.7.7*. Panels so arranged reduce training time, lead to improved safety and operability, and reduce the time taken to find an element in an emergency when normal sequences are disrupted.

Instrument panels for process plants and power stations are frequently combined with mimic diagrams which show the flow of materials through the plant. These may be combined with warning lights to show potentially dangerous or undesirable conditions in various parts of the plant.

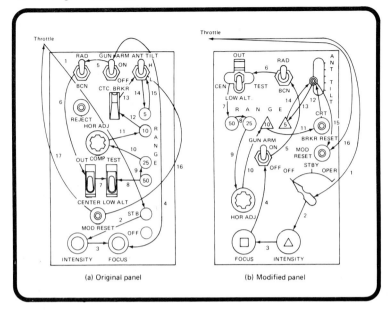

*Figure 3.7.7 Example of modification of hypothetical airborne radar-control panel.
The numbered lines show normal sequence of movements (C. E. Cornell 'Minimizing
Human Errors' Space/Aeronautics, March 1968, p.79)*

Acknowledgement

Figures 3.7.1 to 3.7.6 originally appeared in the book *'Fitting the task to the
Man'*[2] and have been reproduced by permission of the publishers.

REFERENCES

1. Murrell, K.F.H., *Ergonomics*. Chapman and Hall, London (1965).
2. Grandjean, E., translated by Davis, P.R., *Fitting the task to the man,* Taylor
and Francis, London (1969).
3. Edholm, O.G., *The biology of work*, Weidenfeld and Nicolson, London
(1967).
4. McCormick, E.J., *Human factors engineering*, 2nd ed., McGraw Hill, New
York, (1964).
5. Cornell, C.E., 'Minimising human errors', *Space/Aeronautics*, 79, (March,
1968).

3.8 FATIGUE AND REST BREAKS

Contents

The subject of fatigue has already been introduced in sub-section 1.8.1 under 'Rest Pauses, Fatigue and Boredom'. Here we consider it in more detail.

Fatigue occurs both in manual and in mental work, yet its sensations are appreciated in the mind only — the mind of the fatigued worker or the mind of the onlooker of fatigued workers or animals. Different onlookers can receive very different impressions; the subject is highly emotive and evocative of exploitation and slavery as the accompanying illustration from Robert Cruikshank's book[1], published in 1833 shows. This is partly because a fatigued human or animal can be prodded into further activity by fear of pain — hence the expression 'flogging a dead horse'.

The discovery that a fatigued worker is generally both inefficient and accident prone is a fairly recent one and not always appreciated.

Here we look first at the subject of muscular stress and fatigue, then at mental fatigue, then at recuperation and the needs for rest breaks.

3.8.1 Muscular stress and fatigue

All human movement is brought about by muscles each of which consists of a very large number of parallel collagen fibres. These fibres have the ability to contract to about half their original length when stimulated by nervous impulses which are initiated by motor neurons in the brain. The greatest force is exerted at the beginning of the contraction.

In prolonged contraction of a muscle, groups of fibres contract alternately to produce continuous contractions of the whole muscle, so that each fibre is relaxed for part of the time.

The mechanical work done by muscular contraction is derived from chemical reactions in which high energy organic phosphates are converted to low energy products. The low energy phosphates are then reconverted to the high energy compounds by degradation of glucose to lactic acid. Most of the lactic acid is next regenerated to glucose, the remainder forming water and carbon dioxide. The regeneration of glucose from lactic acid requires oxygen which is supplied by the bloodstream. Muscular performance can therefore be limited by the supply of blood to the working muscle. This supply is controlled by the heart rate, blood pressure and the dilation of blood vessels leading to the muscle.

Figure 3.8.1 English Factory Slaves — Engraving from a book entitled 'The condition of the West Indian Slave contrasted with that of the infant slave in our English Factories' by Robert Cruikshank. Published about 1833 by W. Kidd, London

3.8.1.1 Dynamic and static use of muscles

Muscles can be used dynamically to do work by alternate contraction and relaxation of the whole muscle, or in static contraction to hold a limb or object in position. The static use of contracted muscles is akin to the action of an electromagnet, where a current must flow to keep the magnet energised. When standing, a series of muscles in our legs, hips, back and neck are in continuous contraction. These are relaxed when we lie down.

During static use of muscles, the blood vessels are compressed by a rise of pressure in the muscles which reduces the flow of blood to them, whereas when a muscle is used to do work dynamically it acts as a booster pump on the blood and accelerates flow of blood through it. A muscle is strong and prolonged static contraction receives neither sugar not oxygen from the blood and has to consume its own reserves. Waste products also accumulate in the muscle causing the pain of muscular fatigue which eventually forces one to relax the muscle. Static use of a muscle which requires 50% or more of the maximum force which the muscle can exert can only be kept up for a minute, whereas static use requiring 20% or less of the maximum force can be maintained for much longer, since blood flow through the muscle is then scarcely impeded. But dynamic use of muscles where they may be used

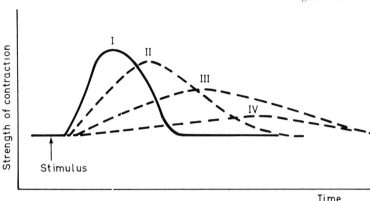

Figure 3.8.2   *Fatigue in an isolated frog's muscle.*
1. Contraction and relaxation of a fresh muscle
2. The same with the muscle slightly fatigued by moderate activity
3. The same after severe activity
4. The same after excessive activity

repeatedly to develop a high proportion of their maximum force may be kept up for a long time without fatigue if a suitable rhythm is maintained.

The main lesson from this is that bodily positions which cause static muscular stress, such as leaning or bending over objects or holding an arm outstretched should be avoided if possible or otherwise reduced to a minimum.

Fatigue in a muscle of a frog or other animal is frequently demonstrated by exciting the muscle electrically and observing the height to which the muscle lifts a given weight, the time of contraction and relaxation, and the time lag between stimulation and the start of contraction (see *Figure 3.8.2*). The performance in all three aspects deteriorates as a function of the amount of previous activity or stress which the muscle has recently undergone. This decrease in performance is the result of chemical fatigue. Humans usually experience muscular fatigue or exhaustion some considerable time before this point of ultimate chemical fatigue is reached. This is proved by the finding that after a muscle has been driven to apparent exhaustion by voluntary contractions, electric stimulation of the motor nerve causes it to contract again normally. This indicates that the fatigue felt by the person originates from his central nervous system or brain and is felt some time before ultimate chemical exhaustion of the muscle takes place. Chemical changes in the muscle which is becoming fatigued stimulate sensory fibres which conduct the impulses to the brain as pain, where reflex inhibition of the motor centres controlling movement occurs. According to this theory of muscular fatigue, the incipient chemical changes in the muscles trigger off nervous activity which results in all the symptoms of fatigue.

3.8.2 Mental and nervous fatigue

The nerves and brain are also subject to fatigue. It is a common experience that a person fatigues sooner when he has to devote conscious attention to a task than when he has been so trained than he does it unconsciously. In this partly lies the value of learning and training. These cause new circuits to be formed in the brain which function mostly as conditioned reflexes outside the conscious sphere. The higher the degree of automatic physiological control, the less stressful and fatiguing is the task.

Training especially of the young is largely acquired through imitation. It is therefore of great importance that new and young employees should work alongside safe and skilled workers who may confidently be imitated. It is seldom realised how dependent we are for safe survival on unconscious habits and conditioned reflexes which we have acquired through imitation.

3.8.2.1 Forms and mechanism of mental and nervous fatigue

The following different types of mental and nervous fatigue have been recognised:

1. Visual fatigue — caused through stress on the visual apparatus.
2. General fatigue — caused by stress on the whole organism, (including muscular stress).
3. Mental fatigue.
4. Nervous fatigue — caused by one sided stress of psycho-motor functions.
5. Boredom fatigue — caused by monotonous work or dull environment.
6. Chronic fatigue — caused by several persistent fatiguing factors.

It is now believed that all sensations of fatigue have in common certain regulating mechanisms in the brain.

Basically the brain appears to function in a similar way to a control room in a modern chemical plant or power station. The highest point of consciousness which is located in the cerebral cortex is connected to the senses and muscles by two sets of nerves which correspond to impulse lines — ascending nerves which bring messages to the consciousness and descending nerves from consciousness which control various movements and reactions.

These signalling systems are self-activated and can quickly generate a high level of activity by forms of closed loop feedback control. Many of the nerves are linked with the automatic system, also located in the brain, which controls the involuntary organs.

In addition, there appears to be a centre in the mid brain which is responsible for our various moods and functions responsible for the preservation of life and also for sleep. Part of this centre is concerned with inhibition of the incoming and outgoing signals between the cerebral cortex and the senses, muscles and the autonomic system. This inhibitory system which may well be a comprehensive and entirely independent system from the others, is thought to be responsible for our sensations of fatigue and for the effects of fatigue on our behaviour.

There is a continuous balance in the level of our activity caused by creation of fresh activity by the activating system and cancelling out of activity by the inhibitory system. This may help to explain why the feeling of fatigue can disappear suddenly when we are alarmed by an unexpected external stimulus or an alarming thought. Rest, which allows the electrical and chemical balance of our nerves to be restored, reduces the activity of the inhibitory system so that fatigue disappears.

3.8.2.2 Boredom and chronic fatigue[2]

Boredom arises when the environment is lacking in external stimuli or when those that exist are weak. The result is that the initial activation of the cerebral cortex is inadequate to overcome even a low level of the inhibitary system. It leads to dullness, tiredness and sometimes 'falling asleep on the job'. Vigilance is decreased and errors and accidents increase.

Monotony and boredom are promoted by prescribed or paced work on conveyor belts, on machines, and are less common in activities which allow a free and individual rhythm. Even continuous repetitive work which requires skill and attention does not necessarily lead to boredom provided motivation exists (e.g. profit), or constant impulses are present which stimulate the activating system. But work which only needs occasional unskilled activity or attention with long time intervals between is most likely to cause boredom.

Chronic fatigue results from severe and continuous daily fatigue. Symptoms are:

Tiredness which persists after work well into the evening.
Increased irritability and intolerance.
Lack of drive and dislike of work.
Depression.

There is also a tendency to psychosomatic illness, such as headache, loss of appetite, indigestion and insomnia, increased absenteeism and a need for longer rest periods.

Psychological conflicts and difficulties tend to promote a state of chronic fatigue. It is often difficult to distinguish between cause and effect and get to the root of the problem.

3.8.2.3 Causes and measurement of fatigue

A number of different causes thus contribute to fatigue, not all connected with the work done. The most important are:

Environment — climate, lighting and noise.

Psychological — responsibility, worries and conflicts.
Intensity and duration of physical and mental work.
Monotony.
Illness, pain, malnutrition.

Fatigue may be measured by a number of tests which include reading, numerical tests, lower efficiency in tests of skill, increase in errors, increase in reaction time, poorer balancing, and many others. But most methods of testing suffer from one inherent drawback. Measurements which demand some performance from the subject stimulate the activating system and thereby tend to overcome existing fatigue.

Another suggested method is to ask subjects to indicate their sensations at the time, e.g. interested or bored; relaxed or tensed; irritated or calm; fresh or tired; sleepy or awake; attentive or inattentive.

A questionnaire on these lines is easily prepared. It is probably best to ask the subject to give his sensations at several times, both during working and leisure hours. This helps to build up a full picture of his fatigue-recovery profile in a particular situation.

3.8.3 Rest pauses[2]

Working hours have been discussed in section 1.8 where it was concluded that a working week of forty-eight hours cannot be exceeded without detriment to safety or productivity, and that a shorter working week is preferable. Studies of workers show that rest pauses are of four types:

1. Spontaneous
2. Disguised
3. Work conditioned
4. Prescribed

Rest pauses are indispensible as a physiological requirement for preserving working efficiency, and are necessary for occupations imposing mental and nervous stress as well as for physical work. Repetitive operations carried out at great speed particularly require rest pauses as they impose special stress on the nervous system.

Before considering what prescribed rest pauses are required for any job, it is necessary to consider the extent of the other pauses. Spontaneous rest pauses seldom last long although they frequently appear during strenuous physical work.

Disguised pauses are periods of secondary activity not required for the task in hand. Examples are lighting a pipe, leaving the workplace with the pretext of consulting a colleague or superior, cleaning a machine part or altering the position of ones chair. These disguised pauses are generally necessary from a physiological viewpoint.

Work-conditioned pauses are waiting periods which result from the organisation of the work or movements of machinery. They occur in a machine shop when waiting for a component or for another operator to repair or adjust a machine. Office employees and shop assistants often have long waiting periods. Operatives on conveyor belts have pauses which depend on the speed of their own performance.

Some investigators have found that the interpolation of short prescribed rest pauses reduces the length of disguised and spontaneous pauses, whilst others have found that the introduction of prescribed rest pauses reduces absenteeism in heavy industries.

For heavy work, obligatory rest pauses should be prescribed, evenly distributed over the working day. The lengths of the pauses depends on the nature of the work. Workers left to choose for themselves tend to work continuously and save up their rest pauses till the end of the day to allow them to leave earlier. This leads to overstress.

For lighter work of medium intensity with waiting times created by the working process, one pause in the morning and one in the afternoon, each of ten to fifteen minutes duration and including the provision of light refreshments, are recommended.

For light work of high intensity without waiting time, one or two short pauses of five minutes each during each half of the day are recommended as well as the mid morning and afternoon pauses recommended for less intense work. The times and size of meals taken should be considered in relation to the working day. Fatigue increases after a heavy meal while digestion is in progress, and also three to four hours after a meal, such as breakfast.

Workers who take no breakfast or who have breakfasted before 7.30 a.m. are likely to need a snack with their mid morning break if their attention is not to wander by mid-day.

A lunch break of forty-five to sixty minutes is normally sufficient provided that lunch (taken in the works canteen or in a nearby pub or cafe) is only a light meal and not the main one of the day. This gives adequate time both to eat and digest the meal before restarting work in the afternoon.

3.8.4 Insomnia

Researchers Luce and Segal[3] report that at least 14% of the US population have frequent sleep problems. One result is that many workers start the day tired. This is not always the direct result of loss of sleep.

There are three types of insomnia:

A. Difficulty in getting to sleep
B. Disturbed sleep
C. Early morning arousal.

Many poor sleepers wake early, at say 4 a.m. Thinking they have not had enough sleep they reach out for a couple of sleeping pills and are still suffering from the effects when they start work by 8 or 9 a.m. later that morning.

Sleep is a complex phenomenom and has various phases, including 'D sleep' in which dreams occur, when the different parts of the brain communicate with each other; rather like a computer clearing itself before starting another operation. Most sufferers from insomnia after trying various simple cures such as counting sheep, alcohol or even transcendental meditation, finish up by getting prescriptions for sleeping pills. Hoskisson[4] estimates that 10% of the population take hypnotics and sedatives — this figure rising later in life to 50% in elderly females.

There are many different types of sleeping pills and the subject is too difficult to tackle here. Seldom, however, are drugs the simple solution hoped for by the sufferer and most have unfortunate side effects. Some, particularly the barbiturates, are addictive and produce unpleasant withdrawal symptoms. In many cases the system builds up a resistance to the drug, so that larger and larger doses are required. Most sleeping drugs whilst having the desired hypnotic effect of rendering the sufferer unconsious, at the same time inhibit 'D sleep', prevent dreaming and leave some of the important natural functions of sleep unfulfilled. Those taking such pills regularly usually find when they stop for a few days that they then suffer on several nights from disturbing dreams — making up for many nights of dream inhibition.

Since most doctors do not have time to study the sleeping problems of their patients in the depth required, safety specialists should interest themselves in the problem, even when they themselves are sound sleepers. The best way of doing this is to read a few good books on the subject, and in turn recommend them to staff and employees who have sleep problems. These are nearly always potential accident hazards.

REFERENCES

1. Cruikshank, R., *The condition of the West Indian slave contrasted with that of the infant slave in our English factories*, W. Kidd, London (1833).
2. Grandjean, E., *Fitting the task to the man*, 2nd ed., Translated, P.R. Davis, Taylor and Francis, London (1975).
3. Luce, G.G., and Segal, J., *Insomnia*, Doubleday, New York (1969).
4. Hoskisson, J.B., *What is this thing called sleep?* Davis-Poynter, London (1976). This contains an extensive bibliography.

3.9 INDUSTRIAL HYGIENE AND TOXICOLOGY

Contents

"All things are poisons, for there is nothing without poisonous qualities. It is only the dose which makes a thing a poison." *Paracelsus*.

Many facets of industrial hygiene and toxicology are discussed elsewhere in this book under other headings. Every section of this chapter and chapter 6 is concerned in some way with industrial hygiene. There are seldom sharp dividing lines between the work of the industrial hygienist (where there is one), the safety specialist, the ergonomist, the works medical officer, the ventilating engineer and the works chemist (where there is one).

Large companies, particularly those whose operations are most liable to present toxic hazards, often have occupational health units with qualified industrial hygienists specialising in biological, chemical, physical and engineering aspects.[1] In smaller companies, the safety manager who may have little training in the monitoring of toxic hazards is often responsible for dealing with a range of industrial hygiene problems as one of several activities.

It is thus not surprising that industry has a rather patchy record in dealing with hygiene problems; many preventable deaths, caused for example by exposure of workers to atmospheres contaminated by lead compounds and asbestos fibres, have occurred. Most of these could have been prevented had knowledge of these hazards which existed at the time of exposure been more widely known and the knowledge acted upon. This is not so much a criticism of professional safety specialists as a recognition of the limitations of their training and authority.

In defining the roles of industrial hygienists and toxicologists to distinguish their fields from those of other professionals, it is helpful to consider the situation in the USA. Here, particularly since the passing of the Occupational Health and Safety Act of 1970, a cadré of professional industrial hygienists has been trained and organised to monitor the main industrial health hazards and initiate preventative and protective measures. Industrial Hygiene has been defined by the American Industrial Hygiene Association[2] as

'that science and art devoted to the recognition, evaluation and control of those environmental factors or stresses arising in or from the work place which may cause sickness, impaired health and well-being, or significant discomfort and inefficiency among workers or among the citizens of the community.'

'An Industrial Hygienist is a person having a college or university degree or degrees in engineering, chemistry, physics or medicine or related biological sciences who, by virtue of special studies and training, has acquired competence in industrial hygiene. Such special studies and training must have been sufficient in all of the above cognate sciences to provide the abilities: (a) to recognise the environmental factors and stresses associated with work and work operations and to understand their effect on man and his well-being; (b) to evaluate, on the basis of experience and with the aid of quantitative measurement techniques, the magnitude of these stresses in terms of ability to impair man's health and well-being; and (c) to prescribe methods to eliminate, control or reduce such stresses where necessary to alleviate their effects.'

The main stresses which the industrial hygienist is required to recognise and evaluate are chemical and physical. 'Chemical stresses' are caused by the presence of various chemical substances — gases, vapours, fumes and dusts present in the work room air, by a deficiency of oxygen and by chemical substances which might be ingested or enter the body by skin contact.

'Physical stresses' are caused by heat and cold, various electro-magnetic radiation hazards — ionising radiation, lasers, microwaves and ultra violet radiation, as well as noise and vibration.

Specialised industrial hygienists fall mainly in the following categories.[3]

1. *Industrial or occupational hygiene chemists*
Tasks include sampling work room air, assaying biological samples from exposed workers, identifying materials thought to be harmful. When newly introduced materials are thought to be hazardous, assay methods must be developed. Industrial hygiene chemists are usually graduates of chemistry.

2. *Industrial or occupational hygiene physicists, radiation protection officer or health physicists*
This field has become quite specialised. The work includes radiation monitoring, heat and noise monitoring, and the development of instruments to monitor noise, electro-magnetic radiation of different types, and concentrations of various chemical contaminants in the atmosphere. Industrial hygiene physicists may be graduates in physics, electrical engineering or chemistry (with training in physics).

3. *Industrial or occupational hygiene engineers*
These design equipment to protect employees from physical and chemical stresses and cooperate with other engineers to eliminate the stresses at source. Examples are the design of local exhaust ventilation systems, acoustic barriers and screens to absorb ionising radiation. Their work often overlaps with that of industrial hygiene chemists and physicists in that they may be required to identify air-borne dusts and study their properties in order to design appropriate protection for employees. Industrial hygiene

engineers are also responsible for the selection and testing of respirators, breathing apparatus, ear protectors and the whole range of personnel protective equipment for particular applications. They are frequently graduates in chemical engineering.

It is clear from these job descriptions that in the absence of a recognised industrial hygienist, much of the work described might be undertaken by members of other professions, e.g. analytical chemists, instrument engineers or ventilation engineers. The work may perhaps be instigated by the safety manager unless he has the means and capability to do it himself.

The appointment of industrial hygienists together with a statutory requirement to meet clearly specified standards of industrial hygiene should result in placing the work on an organised and systematic basis, rather than waiting for adverse publicity from a coroner's inquest before an improvised programme is mounted under panic conditions.

Toxicology has been defined[4] as 'the science of poisons, their effects, antidotes and detection'. Few industrial firms unless they are in the business of manufacturing food, drugs or food additives will employ their own full-time toxicologists. The toxicological problems present in most factories should be dealt with by the industrial hygiene chemist when he exists. If there is no industrial chemist then these problems become the responsibility of the director for health and safety who may delegate the work to a committee which includes the safety specialist, the works medical officer, the works analytical chemist, the production manager of the department concerned, and the trade union safety representative. The leading role will probably be taken by the safety specialist providing he has a reasonable grounding in chemistry. He will be well advised to discuss the problem with the suppliers of the materials giving rise to the toxic hazards, and with the local branch of the Health and Safety Executive, who can call in the aid of specialists where needed.

Most safety specialists need to have a good practical grounding in the essentials of industrial hygiene and to know where further information on particular subjects may be found. It is, for instance, quite useless for the safety specialist merely to dispense respirators of a particular type to protect against say carbon monoxide poisoning, without

(a) checking that every respirator given out fits the person to whom it is supplied closely and with less than the maximum allowable leak;
(b) that the respirator can be used in comfort, without causing excessive interference with breathing or vision;
(c) that effective and easily understood means are available for checking the activity of the cartridge or cannister and that they are changed before they become exhausted;
(d) that each respirator is reserved for one particular user, marked with his name, and disinfected, maintained and stored under safe and hygienic conditions.

The fact that respirators have to be issued at all is often an admission of failure, since the work and working conditions should be so engineered that the atmosphere is non-hazardous and safe to breathe.

3.9.1 Industrial hygiene physics

Among the physical conditions which industrial hygiene physicists may be required to monitor, noise has been discussed in section 3.4, vibration in section 3.5, temperature in section 3.6, and electro-magnetic and other radiation hazards in section 6.9 and 6.10.

Useful guidelines are published by the American Conference of Governmental Industrial Hygienists as Threshold Limit Values for Physical Agents.[6] These are reviewed annually.

These physical agents are:

Heat stress
Ionising radiation
Lasers
Microwaves
Ultra-violet radiation
Noise.

While special instruments are required to monitor most of these agents, the environmental factors which contribute to heat stress are more easily measured as the Wet Bulb Globe Temperature Index (WBGT)[8]. Outdoors, when the sun is shining, this is given by the equation:

$$WBGT = 0.7\,WB + 0.2\,GT + 0.1\,DB$$

Indoors, or outdoors when there is no sun, it is given by:

$$WBGT = 0.7\,WB + 0.3\,GT$$

where

WBGT = Wet bulb globe temperature index
WB = Natural wet bulb temperature
DB = Dry bulb temperature
GT = Globe thermometer temperature

All three temperatures which are used in the calculation of WBGT may be measured simply by three reliable mercury in glass thermometers mounted on a retort stand *(Figure 3.9.1)*. The bulb of the thermometer measuring the wet bulb temperature is covered by a wick, the lower extremity of which is immersed in a 125 ml flask containing distilled water. The bulb of the thermometer measuring the globe temperature is placed at the centre of a 6 inch diameter copper sphere, such as a toilet float, the outside of which is painted matt black. Permissible limits of the WBGT for exposed workers depend on the work load, the work-rest regimen, water and salt supplementation, clothing and acclimatisation. Information on these points is given in the sources already quoted.

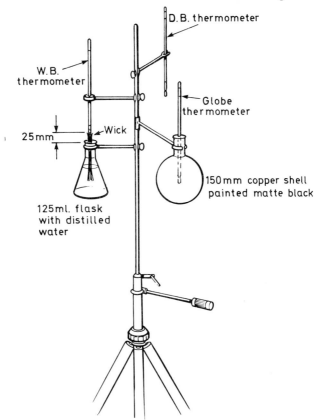

W.B.
thermometer

D.B. thermometer

Globe
thermometer

Wick

25mm

150mm copper shell
painted matte black

125ml. flask
with distilled
water

Figure 3.9.1   Wet bulb globe temperature index (National Safety Council, USA)

3.9.2 Toxic hazards and industrial disease[4, 9]

History has shown that materials may be in common use in industry over many years before their adverse effects on human health are fully recognised. This applies particularly to carcinogenic substances[10, 11] whose effects on the human organism are slow to develop. It also applies to other slow acting poisons such as lead and fibrogenic dusts such as silica. Many factors combine to mask their ill effects — trade secrecy, fear of litigation and often a pathetic desire on the part of the employee whose health is being undermined to hide the true state of his health from himself, his employer and from other potential employers.

Economic pressures often cause workers to continue to do a job although suffering from an active occupational disease. Home debts encourage fast production on piece-work and overtime working and discourage the use of protective measures which might slow down production and earnings.

The positive diagnosis of the cause of any disease, whether natural or occupational in origin is a slow process requiring unusual qualities of patience, perseverance and humility. Scientists and physicians with these qualities are usualy most reluctant to be treated as hostile witnesses under the adversary system when a legal showdown over compensation hinges on their findings. Nature however abhors a vacuum, and as a result a less scrupulous though plausible brand of scientific advocate has stepped in as 'professional expert witness' to fill the gap. How often is an expert witness found giving evidence which might jeopardise the interests of the party by whose solicitors he has been retained?

The result of all these factors is that many materials accepted as safe have, in the past, been used in industry over long periods without special protection for employees until mortality and sickness statistics indicating that something was wrong stuck out like a sore thumb. Examples are asbestos, benzene (which produces severe anaemia[9] and, until recently, was in common use as an industrial solvent) and beryllium used in the manufacture of phosphors for fluorescent lamps until 1950 and which causes acute pulmonary disorders at very low concentrations in the air.[9]

3.9.2.1 States and classes of toxic substances present in the atmosphere

It is important to distinguish between the different states in which toxic substances may be present in the atmosphere; these states are revealed by the common class names. It should be pointed out that the same substance may sometimes be present in one state or class, sometimes in another. The classes are: dusts, fumes, smokes, aerosols, mists, vapours and gases.[4, 5]

Dusts are solid particles generated by mechanical disintegration of solid materials such as rock, ore, coal, wood and grain. They generally range from 0.1 to 25 microns in size: those above 5 microns tend to settle rapidly and usually cause less serious respiration problems than the smaller particles. Dusts do not as a rule flocculate except under electrostatic forces.

Fumes are solid particles generated by condensation from the gaseous state (e.g. metals vapourised during welding) and often oxidised in the process. They are very fine, generally less than one micron in size; they often flocculate and coalesce in the absence of an applied electrostatic field.

Smokes are formed during the combustion of organic materials and may range from nearly pure carbon (black smoke) resulting from incomplete combustion to inorganic non-combustible residues which are usually gray or white. Smokes often contain tars and liquid droplets. The particle size is generally very small — from 0.1 to 0.25 microns.

Aerosols are made deliberately by spraying solutions of the material in a liquefied gas or low boiling solvent through a fine nozzle into the atmosphere. They may consist of liquid or solid particles.

Mists are suspended liquid droplets produced in almost any way, whether by condensation of a vapour or dispersion of a liquid. Liquid aerosols are a type of mist.

Gases are defined in elementary science courses as 'formless fluids'. They differ from vapours in that the boiling points of the pure materials under atmospheric pressure are below the ambient temperature.

Vapours have the same state as gases but are derived from materials whose boiling or sublimation points under atmospheric pressure are above the ambient temperature. All gases and vapours mix completely with one another and once mixed will not separate under gravity. They also diffuse.

Dusts, fumes, smokes and mists may be removed from the air by filtering, impingement or flocculation, but they are not removed by adsorption on solids of high surface area such as charcoal or silica. They may sometimes be removed by absorption in liquids in which they are soluble, but less easily than is possible with vapours.

Vapours and gases cannot be removed by filtration, impingement or flocculation but they can often be removed by selective adsorption on the surface of solids for which they have an affinity or by absorption into liquids in which they are soluble or with which they react.

Vapours often condense to mists or fumes, so that a particular contaminant may be present partly as a vapour and partly as a mist.

3.9.2.2 Threshold limit values[12]

Threshold limit values of toxic materials in the atmosphere have been set by the American Conference of Governmental Hygienists in the USA and are reviewed annually. They have been adopted by many other countries, sometimes with variations. They are the maximum concentrations, expressed either as parts per million by volume or as milligrams per cubic metre to which occupants may be continuously exposed for a 7 or 8 hour working day and a 40 hour working week. Since they are time weighted concentrations, some temporary excursions above them may be allowed if these are balanced by long periods where much lower concentrations prevail.

Decisions on allowable excursions are usually made by rule of thumb. Threshold limits for fibrogenic dusts are given on different bases which differ from one dust to another. For asbestos they are 5 fibres per cubic centimetre greater than 5 microns in length. The determination of concentrations of toxic materials in the atmosphere often presents difficult sampling and analytical problems. This applies particularly to dusts, fumes, mists and smokes, the particles of which may separate from the atmosphere in the sampling process itself.

The sampling and analysis of the atmosphere for toxic gases and vapours (*Figure 3.9.2*) presents fewer problems, although accurate and reliable results cannot be obtained without proper training and apparatus. Some

Figure 3.9.2   Personal air sampling equipment in use for toxic dusts (F. F. Casella & Co.)

selected TLV values are given elsewhere in this book and more complete lists are given in references quoted.

In addition to the Threshold Limit Values originated by the ACGHI, the American National Standards Institute (ANSI) publishes the Z.37 series of Acceptable Concentrations (AC)[12] for a number of substances and the American Industrial Hygiene Association publishes a series of Emergency Exposure Limits (EEL) which define single brief accidental exposures to air-borne contaminants that can be tolerated without permanent toxic effects.[13] Further information about specific contaminants is given in the *Hygiene Guides* published by the Hygienic Guides Committee of the American Industrial Hygiene Association.

3.9.2.3 Toxicity ratings

TLV values which furnish a good indication of the toxicity ratings for materials which may become air-borne have only been assessed for a relatively small number of common toxic materials. There are many more whose toxicities are quite unknown or have only been determined qualitatively. The following system of ratings is used by the American Environmental Research Centre[4] and has been adopted widely elsewhere.

U = Unknown. This includes materials on which no tests have been made and also materials on which the results of the tests are of questionable validity.
O = No toxicity. This covers materials which could produce toxic effects on humans only under very exceptional conditions.
1 = Slight toxicity.
2 = Moderate toxicity.
3 = Severe toxicity.

In assessing the toxicity rating of any material, four types of exposure are recognised:

(a) acute local;
(b) acute systematic;
(c) chronic local;
(d) chronic systematic.

Some materials are more likely to have local effects, e.g. strong acids and caustic alkalis which are splashed on the skin; whilst others which are inhaled or ingested and pass into the blood stream will have systematic effects.

Acute exposure usually means 'of short duration' and usually refers to a single exposure which may last for seconds, minutes or perhaps a few hours. Chronic exposure refers to prolonged or repeated exposure measured in months or years. Chronic exposure is often confused with chronic symptoms. Some materials produce chronic or lasting symptoms after only a single exposure. Others to which workers may be exposed chronically over a long period or repeatedly may produce symptoms only while the exposure lasts. The symptoms quickly clear up when exposure to the materials ceases.

Sax[4] lists a large number of chemical compounds with their toxicity ratings, while Hamilton and Hardy[8] deal systematically with the toxicity of metals and metalloids, chemical compounds, high polymers, pesticides, mineral dusts, and biological hazards.

3.9.2.4 Modes of entry

There are four modes of entry for toxic materials in to the human body.

1. By contact with or through the respiratory tract.
2. By contact with or through the skin.
3. By contact with or absorption through the digestive tract.
4. By subcutaneous injection (innoculation).

1. *Absorption through the respiratory tract*
Some 90% of all cases of industrial diseases are caused by the entry of toxic materials, usually finely divided dusts, through the respiratory tract[9]. Some experts believe that some industrial diseases previously attributed to ingestion were really caused by inhalation; thus leukemia found in luminous paint workers was more likely to have been caused by inhaling radio-active materials in the atmosphere than by putting contaminated brushes in their mouths to tip them.[9]

Whilst it is obviously wrong to bring sandwiches into a room where there is a significant amount of dust containing lead in the atmosphere, workers in the room are more likely to suffer from inhaling the dust than through eating the sandwiches. The influence of the particle size of the contaminant on its likely point of attack in the respiratory system is discussed in sections 3.2 and 6.3.

The absorption of harmful materials by the lungs is very complex and depends on many factors including the solubility in body fluids, the permeability of the lungs and the particle size. The harm done in inhaling contaminated air depends not so much on the amount of contaminant inhaled but rather on the amount of contaminant which enters the body. Determination of the concentration of the contaminant in the air may thus give misleading information, and it is often necessary to sample and analyse the blood and/or urine of affected workers.

2. *Absorption through the skin*
The skin acts as an almost impermeable barrier to most inorganic compounds and their aqueous solutions, but some organics, particularly solvents, hydrogen cyanide and nitriles, can penetrate the skin to such an extent that respiratory protection alone is insufficient (*see Table 3.10.3*). Washing hands in organic solvents and wearing clothing containing them should be avoided.

In addition, a great many materials cause dermatitis when in contact with the skin. This may be caused by irritation or by sensitisation — this is the result of an allergic reaction to a substance and becomes established over a relatively long induction period.

In industry, chemicals are the main cause of dermatitis. Cutting oils and similar substances are another important cause. Other materials affect the skin by removing natural oils, by removing water, by coagulating or hardening the outer layers of the skin, by combining with water or by oxidising the protein of the skin. Solvents such as hydrocarbons, chlorinated hydrocarbons and ketones not only enter the body through the skin, but also remove natural oils and fats, thereby increasing the risks of dermatitis and making it easier for even more toxic compounds to enter the body in this way.

3. *Ingestion*
Whilst it is always possible for toxic materials to be mistakenly placed in the mouth or to contaminate sweets or foodstuffs brought into the workplace, this is a less common cause of industrial disease than inhalation.

Food poisoning caused by low hygiene standards in canteens and other places is another matter, but it is not included here as an industrial disease. Where workers are handling materials, care should be taken to prevent the materials contaminating sweets and cigarettes. A check should be made to examine all cases where these materials might enter the mouth, and steps taken to prevent this occurrence.

Sometimes, highly toxic materials are handled which create pleasurable sensations when inhaled or imbibed. Iso-propyl alcohol, di-ethyl ether and chloroform are examples; some of these can be addictive. Care must be taken to warn workers of these temptations and strict disciplinary action taken against anyone found indulging in such dangerous habits.

4. *Sub-cutaneous injection*
Care must, of course, be taken to avoid the risks of sharp objects contaminated by toxic substances penetrating the skin. Such dangers are generally quite specific, e.g. splinters from wood treated with toxic preservatives or metal turnings wet with cutting oils.

3.9.2.5 Site of action and detoxification[3, 4, 9]

Different poisons act on different parts of the body. Benzene affects the bone marrow; carbon monoxide the haemoglobin in the blood; arsenic affects the blood, the nervous system, the liver, kidneys and skin, and organic phosphates inhibit the enzyme cholinesterase which is present in red blood cells.

Apart from external skin irritants, toxic substances must generally be absorbed into the body and distributed by the blood for poisoning to occur. The body possesses a system of protective mechanisms which eliminate many toxins. The kidneys and liver play key roles in detoxification. Some foreign materials are excreted unchanged through the urine and faeces. Toxic gases which have been absorbed may be re-excreted through the lungs. Some compounds are changed chemically in the body to less toxic ones, whilst others may be converted to more toxic ones.

Different people also respond differently under similar conditions of exposure, and this can be so marked that the term 'individual susceptibility' has been coined to express these differences. Sex, age, nutritional factors and speed of working all affect susceptibility. The young and the fat are generally more susceptible to toxins than the middle aged and the slim.

Sometimes supervisors may be faced with the unexpected. A young industrial chemist appointed to supervise the operation of a new nitroglycerine plant in World War II was confronted by a delegation of newly recruited workers. One of the effects of exposure to the nitroglycerine contaminated atmosphere was they claimed that their sexual potency had declined, causing complaints from their wives who suspected they were 'going out' with other girls. 'What we need', said their spokesman 'are certificates which we can show our wives which explain that our impotence is caused by the materials present at our workplace.' Whether the workers' complaint was justified or not, this story illustrates the embarrassing nature of many of the unrecognised industrial diseases which workers have endured. A century ago child chimney sweeps suffering from painful and usually fatal scrotal cancer through exposure to carcinogenic soot were branded as suffering from venereal disease and berated for their loose morals.

3.9.2.6 Toxic hazard survey

A survey[3] of the toxic hazards for any industrial process has many sides to it. The main headings for such a survey are:

1. *Medical history of employees, present and past*
Records of illness, absenteeism, sickness benefits and insurance claims should be examined for any patterns which differ from the 'norms' of the general population in the area.

2. *The process and process materials*
The process should be examined and the flow diagram and composition of materials entering, leaving and at various stages in the process should be studied. Special attention needs to be given to those parts in the process where airborne dusts may be formed or where vapours and gases may enter the atmosphere — particularly where workers may be exposed to them.

Manufacturers' literature should be examined for toxicities and carcinogenities of all materials used in the process. The compositions of materials employed under trade names should be established as well as their toxicities.

3. *Physical examination of work area and working conditions*
The work area should be examined for 'good housekeeping' and spills of solvent and the presence of solvent saturated rags, dust, ventilation and air patterns noted. Proximity of operations which may produce a hazardous

interaction such as welding and the use of chlorinated hydrocarbons should be noted. Severe working conditions which necessitate limited exposure of workers or short working hours should be specially noted.

4. *Detailed sampling and analytical survey*
Once it has been established that a potential hazard may exist, a programme of sampling and analysis, both of the factory atmosphere in the breathing zones of workers and of workers blood and urine is needed to evaluate it.

3.9.2.7 Prevention and cure[3, 4, 5]

In trying to eliminate toxic hazards the slogan 'prevention is better than cure' should be literally followed as far as possible.

1. Non-toxic materials should be substituted for toxic ones wherever possible.
2. Where the use of toxic materials is unavoidable, they should be contained so that they cannot escape into the workroom atmosphere, and local exhaust ventilation (with suitable exhaust gas scrubbing where necessary) employed at danger points to avoid this.
3. Personal respiratory protection should not be required by workers during normal working. Conditions should be so improved by measures (1) and (2) that this is unnecessary. However, for certain emergency operations and maintenance, such protection should be provided, making sure that it is suitable for the conditions and the tasks.
4. A cadré of workers trained to give first aid to anyone suffering from inhalation, ingestion or skin contact with toxic materials should always be present when they are handled. First aid is discussed in Appendix B. Poisons which have been ingested should, where possible, be diluted by administering water, milk, soapy water or saline water. Vomiting is generally but not always desirable. Some chlorinated hydrocarbon solvents act as systematic poisons by ingestion whereas inhalation produces anaesthesia. If the patient is made to vomit, solvent vapour will enter the lungs so that he suffers from both effects simultaneously. A 'universal antidote' consisting of powdered burnt toast, strong tea and milk of magnesia will absorb and neutralise many poisons.

REFERENCES

1. Carter, J.T., 'The biological basis for chemical environmental hygiene standards', *Symposium Series No.47*. The Institution of Chemical Engineers, Rugby, (1976).
2. *The American Industrial Hygiene Association Journal*, 20, 428-430 (1959).
3. Herrick, R., 'The historical perspective', Section 1 from Save, N.I., *Dangerous Properties of Industrial Materials*, 4th ed., 6-7 (1975).

4. Durham, W.F., 'Toxicology' Section 9, ref 3, 289.

5. National Safety Council, *Accident prevention manual for industrial operations*, 7th ed., N.S.C. Chicago, 1095-1107 (1974).

6. American Conference of Governmental Industrial Hygienists, *Threshold Limit Values for Physical Agents*, PO Box 1937, Cincinnati, Ohio 45201, USA (1973 or later edition).

7. Ref. 6, page 1053.

8. Hamilton, A., and Hardy, H.L., *Industrial toxicology*, 3rd ed., Publishing Sciences Group Inc. (1974).

9. Weisburger, E.K., Industrial Cancer Risks, Section 8, ref 3.

10. Arcos, J.C., Argus, M.F., and Wolf, G., *Chemical induction of cancer*, Academic Press, New York, **1** (1968).

11. American Conference of Governmental Industrial Hygienists, *Threshold limit values for chemical substances in workroom air*, PO Box 1937, Cincinnati, Ohio 45201, USA (1973 or later edition).

12. American National Standards Institute, *Acceptable Concentrations* Z.37 Series, ANSI, 1430 Broadway, New York, N.Y. 10018.

13. American Industrial Hygiene Association, *Hygiene guides*, AIHA, 66 S Miller Road, Akron, Ohio 44313, USA.

3.10 PERSONAL PROTECTIVE CLOTHING AND DEVICES

Contents

This section deals with safety clothing and special protective devices worn by personnel and covers:

1. The philosophy behind the use of protective clothing and devices and the problems encountered in their adoption.
2. The British legal requirements to provide and ensure the use of such clothing and devices in specific industries.
3. Protective clothing and devices available and guidelines to their selection.

3.10.1 Philosophy and problems of protective clothing and devices

Most protective clothing and devices cause some inconvenience to the wearer and restrict his movements or sensory perception. The fact that they are needed is an admission that hazards exist which have not been eliminated by better methods. Every effort should be made to plan, engineer and arrange operations so that protective clothing and devices are unnecessary. Unfortunately, there are frequent limits to how far this can be achieved. Personal protective clothing and devices are also required sometimes as a last line of defence in case other means of protection fail or have to be removed, e.g. during maintenance.

Protective clothing and devices need to be carefully selected to meet the following requirements:

1. They must give adequate protection against the specific hazard or hazards to which the worker will be exposed.
2. Clothing should be of minimum weight and cause minimum discomfort compatible with protective efficiency.
3. Attachment to the body should be flexible but effective.
4. The weight should be carried by a part of the body well able to support it.
5. The wearer should not be restricted in movement or sensory perception essential to the job.
6. Clothing should be attractive.
7. They should be durable.
8. Parts should be accessible so that they can be adequately inspected and maintained on the premises by the staff of the organisation using them.
9. Protective devices should not introduce additional hazards through their form, materials of construction or probable misuse.
10. Construction should be in accordance with accepted standards for the duty required.

In some situations the need for personal protective devices is obvious, e.g. the use of eye protection by electric welders (*Figure 3.10.1*), hard hats by construction workers or gloves for workers handling hot or cold metal objects. In these cases the worker does not need much convincing that he will get hurt if he does not wear appropriate protection. It is far more difficult to ensure that protection is worn where the hazard exists but the

Figure 3.10.1 'Optoshield'. A one-piece moulded plastic eye protection unit (American Optical Safety International)

probability of injury is less evident. Thus, the first problem is to identify situations where the hazard justifies the use of personal protective devices, the second is to select a satisfactory device or range of devices and the third is to bring home to workers the chances they are taking if they do not wear a suitable device.

Three main questions should be considered:

1. *Is there a hazard which warrants the wearing of some protective device? What is the probability of the hazard being realised?* Even when the hazard has been or is about to be eliminated by an engineering modification, there might be circumstances such as valve failure or fire, where the modification failed and the hazard materialised. These factors should be considered and some quantitative or semi-quantitative hazard analysis carried out before the question can be answered in the negative. (See section 2.9 for hazard analysis).

2. *What degree of protection is required and what clothing or equipment would be most suitable?* To answer this question satisfactorily may require detailed examination of several examples of how the hazard might materialise, and the carrying out of appropriate tests.

3. *How can one ensure that the devices are not only worn, but worn correctly?* This involves the comfort, appearance and, above all, the confidence of the worker in the selected equipment or range of equipment. Where possible the worker should be allowed to make his own choice from a selected range of equipment which fulfills the expected requirements.

Once a particular form of personal protection has been prescribed for a given task because of the existence of a persistent hazard, it is recommended that the wearing of the device for that task should be made a condition of employment. Penalties may be imposed for not wearing it unless there is a powerful and convincing reason for not doing so.

It should also be mandatory for managers and other visitors who may be exposed, even for a short time, to the particular hazard, to wear similar protection. Besides protecting the visitor, this sets an example to the workers and shows them that management takes the problem seriously. While personal protective devices should not be prescribed lightly, once the decision has been taken, then compulsory use has generally been found to be the only satisfactory way of ensuring that they are used and that the workers are properly protected. Again precise instructions are needed as to when and in what circumstances the devices are to be worn. Have they to be worn continuously on the job, or just when a particular task which is part of the job is being performed? If only the latter is necessary, there should be a written job instruction making this crystal clear, and the message should be reinforced by one or more posters displayed at the spot where the task is performed.

In most cases the protective device should be allocated for the personal use of one worker only. It should be marked with his name or works number. He should be made responsible for its safe care and be required to contribute to the cost of replacing it if he loses it. At the same time, the management is responsible for providing a locker or other accommodation where the employee can keep the device securely on the premises. Management must also arrange for its regular cleaning, disinfection, testing, maintenance and replacement.

Visitors' needs for protective clothing and/or devices must be provided for. Someone, generally the security guard at the gate, must be made responsible for fitting and issuing any devices considered necessary (usually hard hats) on a loan basis as the visitors enters the works, collecting them again from the visitor as he or she leaves. The same person should also be responsible for cleaning and disinfecting the hat and/or clothing before they are used again.

Sometimes a worker may request a particular device for use on a task where it had not hitherto been prescribed. Provided the device is readily available, it is generally advisable to issue to the worker the device requested as a temporary measure until the need for it has been objectively assessed. The worker is unlikely to have requested the device without sound reason, and such a request would often lead to the prescription of the device for that particular task. However, if close examination shows that the request was unjustified and based on a non-existent or utterly remote hazard, it is probably best to transfer the worker to some other job.

3.10.2 Legal requirements for protective clothing and devices[1]

Various statutes and regulations made over the years prescribe protective clothing and devices for tasks in particular industries. Usually the employer or occupant of the premises is obliged to provide the clothing or protective devices, although some earlier regulations made before 1910 simply state that certain workers should wear specified clothing which they might be presumed to purchase themselves or get their women folk to make for them.

The employer or occupier of the premises is also usually required to maintain, wash and disinfect the clothing or devices regularly and to provide a place where they can be kept safely and in good condition.

Some regulations spell out the clothing or device in some detail; others merely stipulate that suitable clothing or devices shall be provided. Most (but not all) regulations order particular workers to wear the protective clothing or devices (when engaged in prescribed tasks), but most welfare orders merely obligate the employer or occupier of the premises to provide the clothing or device whilst leaving it to the employee whether he uses it or not.

A list of these legal requirements to provide protective clothing or personal protective devices is given in *Table 3.10.1*. The conditions attached to such requirements vary considerably, and employers and workers liable to such requirements should study the wording of the relevant regulations carefully. A number of these requirements serve mainly to protect the worker or his clothing from wet, dirt and the elements, but others serve to protect against more serious hazards, toxic vapours and dusts and ionising radiation.

Table 3.10.1 BRITISH REGULATIONS AND ORDERS PRESCRIBING PROTECTIVE
CLOTHING AND PERSONAL PROTECTIVE DEVICES

Title of Regulation or Order	Reference No.	No. of prescribing Reg.	Task or persons for whom protection is prescribed
The Aerated Water Regulations 1921	SR & O 1921 No. 1932	3 4	Filling bottles or syphons Corking, crowning, screwing, wiring, foiling, capsuling, sighting or labelling.
		5	All persons exposed to wet.
The Asbestos Regulations, 1969	SI 1969 No. 690	2	Work where asbestos dust may be present in atmosphere where it cannot be removed by exhaust ventilation.
The Blasting (Castings and other Articles) Special Regulations, 1949	SI 1949 No. 2225	12	All workers employed in blasting chambers.
The Bronzing Regulations 1912	SR & O 1912 No. 361	4	All persons employed in bronzing.
The Construction (Health and Welfare) Regulations 1966	SI 1966 No. 95	15	All persons required to work in open air during rain, snow, sleet or hail.
The Cement Works Welfare Order, 1930	SR & O 1930 No. 94	1	Workers standing in slurry mud or water. Persons exposed to coal or cement dust. Persons required to work in open air. Females employed in cleaning or repairing sacks
The Chemical Works Regulations 1922	SR & O 1922 No. 731 as amended by SI 1973, No. 36	25	Nitro or amido processes.

Protection prescribed	Party liable to provide & maintain protection	Is wearing obligatory?	Reg. No.	Remarks
Face guards to protect face, neck and throat and gauntlets for exposed hands & arms.	Employer	Yes	7	To protect against flying glass from exploding bottles. Exceptions allowed for fully automatic machines or other means of reducing exposure at discretion of HMF Inspector.
Waterproof aprons with bibs. Waterproof boots or clogs.	Employer	Yes	9	Beer bottle filling excluded.
Approved respiratory protective equipment and protective clothing.	Employers & occupiers of factories	Yes	3	Gives no protection to public from dust produced through use or demolition of asbestos artefacts. See report by Advisory Committee on Asbestos.
Protective helmet supplied with clean air for each exposed worker and suitable gauntlets and overalls.	Not apparent. Employer presumed.	Yes	12	Applies to all works or factories where blasting is carried out.
Suitable overalls for everyone and suitable head coverings for females.	Occupiers of factories and workshops.	Yes	5	Applies to use of dry metal powders in letterpress printing lithographic printing and coating metal sheets.
Adequate and suitable protective clothing.	Contractors	No		
Waterproof thigh boots.	The occupier	No		
Goggles.	The occupier	No		
Waterproof coats.	The occupier	No		
Overalls and head covering.	The occupier	No		
Overalls or working clothes & protective footwear.	The occupier	Yes	31	The law has clearly not kept pace with the chemical industry!

Table 3.10.1 (cont.)

Title of Regulation or Order	Reference No.	No. of prescribing Reg.	Task or persons for whom protection is prescribed
The Chemical Works Regulations 1922 (cont)			Grinding raw materials in a chrome process. Chlorate crystallisation & handling.
			Caustic grinding. Work in bleaching powder chambers.
The Chromium Plating Regulations 1931	SR & O 1931 No. 455 as amended by SI 1973 No. 9	3	All persons employed. Workers at baths.
The Clay Works (Welfare) Special Regulations 1948	SI 1948 No. 1547	5	1. Workers employed wholly or partly in open air or on work exposed to dust; and 2. Work at machines where oil is used on the brick material.
The Dyeing (use of Bichromate of Potassium or Sodium) Welfare Order 1918	SR & O 1918 No. 369	1	1. All persons coming into contact with chrome solutions. 2. Persons handling the crystals or immersing their hands in chrome solutions or handling textile material saturated with chrome solution.
The Electric Accumulator Regulations 1925	SR & O 1925 No. 28 as amended by SI 1973 No.36	14	1. Manipulation of raw oxide of lead. 2. Pasting. 3. Work in the formation room.
The Electricity Regulations 1908	SR & O 1908 No. 1312 as amended by SR & O 1944 No. 739	24	Where necessary.
The File Cutting by Hand Regulations 1903	SR & O 1903 No. 507	10	Every file cutter at work.

Protection prescribed	Party liable to provide & maintain protection	Is wearing obligatory?	Reg. No.	Remarks
Overall suits.				
Woollen clothing. Boots or overshoes whose soles contain no metal. Gloves & goggles. Flannel or other suitable respirators.				
Aprons with bibs. Loose fitting rubber gloves of suitable length and rubber boots or other water-proof footwear.	The occupier The occupier	Yes Yes	12 12	
Suitable protective clothing to include oil-proof aprons for (2).	The occupier	No		
Suitable protective clothing. Loose fitting rubber gloves of suitable length.	The occupier	No		
An overall, apron impervious to water and clogs or suitable footwear and for (1) and (2) head covering.	The occupier	Yes	21	
Insulating boots and gloves	The occupier	Yes	24	
A long apron reaching from shoulder and neck below the knees.	Not clear. Perhaps the worker.	Yes	10	The main hazard was inhalation of lead dust from pulverised blocks of lead on which file was supported. Hazard removed by mechanisation of process.

Table 3.10.1 (cont)

Title of Regulation or Order	Reference No.	No. of prescribing Reg.	Task or persons for whom protection is prescribed
The Flax and Tow Spinning & Weaving Regulations 1906	SR & O 1906 No. 117	9	Workers on wet spinning frames who are not protected by efficient splash guards.
		12	Persons employed in machine hacking, preparing and carding.
The Fruit Preserving Welfare Order 1919	SR & O 1919 No. 1136	1	Preparing & boiling fruit, filling, finishing & covering filled vessels, spinning on tops and any wet process.
The Glass Bevelling Welfare Order 1921	SR & O 1921 No. 288	1	All employed in process of bevelling glass, except where suitable splash guards are provided.
The Gut Scraping Tripe Dressing, etc. Welfare Order 1920	SR & O 1920 No. 1437	1	All employees. Persons employed in wet processes.
The Hollow-ware and Galvanising Welfare Order 1921	SR & O 1921 No. 2032	1	Employment in wet process & coming into contact with acid or acid solutions.
The Horsehair Regulations 1907	SR & O 1907 No. 984	9	All employed on material which has not undergone disinfection.
		10	Persons exposed to dust at time of exposure.
The India Rubber Regulations 1922	SR & O 1922 as amended by SI 1973 No. 36	9	All employed in a lead process.
The Ionising Radiations (Unsealed Radioactive Substances) Regulations 1968	SI 1968 No. 780 as amended by SI 1973 No. 36	40	All workers in active areas tracer areas, decontamina tion areas, total enclosures or at or in fume cupboards

Protection prescribed	Party liable to provide & maintain protection	Is wearing obligatory?	Reg. No.	Remarks
Waterproof skirts and bibs.	The occupier	Yes	13	
Suitable and efficient respirators.	The occupier	No		
Suitable protective clothing.	The occupier	No		
Suitable protective clothing.	The occupier	No		
Suitable overalls Waterproof aprons and waterproof boots or clogs.	The occupier	No		
Finger stalls or (where necessary) rubber or other suitable gloves, acid-proof aprons, clogs.	The occupier	No		
Suitable overalls and head covering.	Employer	Yes	16	
suitable respirator for every person.	Employer	Yes	17	
Suitable overalls	Occupiers	Yes	16	
Suitable personal protective equipment to include: (a) Clothing to protect body from contamination.	Occupier	Yes	40	The regulations prescribe monitoring of persons and clothing, and the disposal of contaminated clothing.
(b) Enough breathing apparatus to prevent risk of inhaling or ingesting any radioactive substances.		Yes	40	

Table 3.10.1 (cont)

Title of Regulation or Order	Reference No.	No. of prescribing Reg.	Task or persons for whom protection is prescribed
The Iron and Steel Foundries Regulations 1953	SI 1953 No. 1464 as amended by SI 1974 No. 1681	8	Workers handling hot, rough and sharp materials. Workers exposed to heavy dust concentrations.
The Foundries (Protective Footwear and Gaiters) Regulations 1971.	SI 1971 No. 476	5	Workers exposed to risk of contact with hot or molten metal.
The Jute (Safety, Health & Welfare) Regulations 1948	SI 1948 No. 1696	10	Workers engaged in cleaning and removing dust.
The Laundries Welfare Order 1920	SR & O 1920 No. 654	1	Persons employed in process involving exposure to wet. Persons engaged in handling soiled linen.
Factories Act 1961 Part IV		75	Women and young persons employed in processes involving use of lead compounds.
The Lead Compounds Manufacture Regulations 1921	SR & O 1921 No. 1443 as amended by SI 1973 No. 36	12	Complex lists of tasks.
The Lead Smelting & Manufacture Regulations 1911	SR & O 1911 No. 752 as amended by SI 1973 No. 36	8	Complex lists of tasks.
The Magnesium (Grinding of Castings & other Articles) Special Regulations 1946	SR & O 1946 No. 2107	12	Persons engaged in process involving substantial exposure to magnesium dust.

Protection prescribed	Party liable to provide & maintain protection	Is wearing obligatory?	Reg. No.	Remarks
Suitable gloves or other hand protection.	Occupier	Yes	8	
Suitable respirators.		Yes	8	
Suitable protective footwear and gaiters.	Employer	Yes	7	
Suitable respirators, overalls and head covering.	Occupier	Yes	12	
Suitable protective clothing including waterproof boots or clogs. Suitable overalls or aprons with bibs and armlets from waist to elbow.	Occupier	No		
Suitable protective clothing.	Occupier	Yes	75	
Suitable overalls, head coverings and clean respirators.	Occupier	Yes	24	
Suitable overalls & respirators	Occupier	Yes	14	
Suitable and sufficient overalls of fireproof material of smooth surface & close texture & leather aprons with leather bibs, which can be quickly removed and brushed down.	Occupier	No		

Table 3.10.1 (cont)

Title of Regulation or Order	Reference No.	No. of prescribing Reg.	Task or persons for whom protection is prescribed
Non Ferrous Metals (Melting & Bunding) Regulations 1962	SI 1962 No. 1667 as amended by SI 1974 No. 1681	13	Operations or work covered by the regulations. Areas where there is a heavy dust concentration.
The Oil Cake Welfare Order 1929	SR & O 1929 No. 534	1	All employees.
The Paints & Colours Manufacture Regulations 1907	SR & O 1907 No. 17 as amended by SI 1973 No. 36	7	All persons engaged in lead processes or at the roller mills.
The Vehicle Painting Regulations 1926	SR & O 1926 No. 299 as amended by SI 1973 No. 36	12	Every person engaged in painting (vehicles)
The Lead Paint Regulations 1927	SR & O 1927 No. 847 as amended by SI 1973 No. 36	9	Every person employed in or in connection with the painting of buildings.
The Pottery (Health & Welfare) Special Regulations 1950	SI 1950 No. 65 as amended by SI 1963 No. 879 & SI 1973 No. 36	9 and second schedule.	See Regulations. 27 tasks in all are scheduled, each requiring one or more of the protective items prescribed.
The Sacks (Cleaning & Repairing) Welfare Order 1927	SR & O 1927 No. 860	21 & 30

1 | All persons entering dust storage arks and others exposed to dust. All persons employed. |
| The Shipbuilding & Ship Repairing Regulations 1960 | SI 1960 No. 1932 as amended by SI 1969 No. 690 and SI 1974 No. 1681 | 50

73 | Workers in confined spaces where there may be fumes or shortage of oxygen. All persons employed when using gas cutting or welding apparatus or when engaged in machine caulking or machine riveting or in transporting, stacking or handling plates at machines |

Protection prescribed	Party liable to provide & maintain protection	Is wearing obligatory?	Reg. No.	Remarks
Suitable gloves or hand protection.	Employer (presumed)	Yes	13	
Respirators of an approved type.	Employer	Yes	13	
Suitable protective clothing.	Occupier	No		
Overalls	Occupier (presumed)	Yes	13	
Overalls	Employee	Yes	12	
Overalls	Employee	Yes	9	
Washable overalls. Washable aprons. Aprons with bibs & made of material impervious to water. Washable head coverings.	Occupier	Yes	10	
Approved respirators.	Occupier	Apparently Yes	30	
Suitable protective clothing.	Occupier	No		
Approved breathing apparatus	Every Employer	Yes	50	
Adequate hand protection, including suitable gauntlets to protect the hands and forearms from hot metal and rays likely to be injurious.	Every Employer	Apparently Not.		

354

Table 3.10.1 (cont)

Title of Regulation or Order	Reference No.	No. of prescribing Reg.	Task or persons for whom protection is prescribed
The Shipbuilding & Ship Repairing Regulations 1960 *(cont)*		76	Scaling, scurfing or cleaning boilers, combustion chambers or smoke boxes where injurious dust occurs.
The Tanning (Two-Bath Process) Welfare Order 1918	SR & O 1918 No. 368	1	All persons coming into contact with chrome solution.

Those who are continuously immersing their hands in the solution. |
The Tin or Terne Plates Manufacture Welfare Order 1917	SR & O 1917 No. 1035	1	All engaged in pickling or handling wet plates.
The East Indian Wool Regulations 1908	SR & O 1908 No. 1287	3	Persons engaged in collecting and removing the dust.
The Wool, Coat-Hair & Camel-Hair Regulations 1905	SR & O 1905 No. 1293	11	Persons engaged in collecting and removing the dust.
The Yarn (Dyed by Lead Compounds) Heading Regulations 1907	SR & O 1907 No. 616 as amended by SI 1973 No. 36	5	All persons employed if required by notice in writing from the Chief Inspector of Factories.
The Factories Act 1961		30	Work inside any chamber, tank, vat, pit, pipe, flue, or similar confined space in which dangerous fumes are present or in which the proportion of oxygen in the air is liable to have been substantially reduced.

Protection prescribed	Party liable to provide & maintain protection	Is wearing obligatory?	Reg. No.	Remarks
Breathing apparatus of a type approved for the purpose of this regulation.	Every Employer	Yes	76	
Rubber or leather aprons & bib and rubber boots or leather leggings which protect open tops and lace holes of clogs & shoes.	Occupier	No		
Loose-fitting rubber gloves of suitable length.		No		
Sufficient and suitable aprons of waterproof material and clogs.	Occupier	No		
Suitable overalls & respirators.	Occupier	Yes	5	
Suitable overalls & respirators.	Occupier	Yes No	22	Wearing of respirators is not obligatory on wearer.
Suitable overalls and head covering.	Employer	Yes	7	
Breathing apparatus of a type approved by the Chief Inspector, i.e. (a) Self-contained breathing apparatus certified as approved under the Chemical Works regs. 1922, or (b) A properly fitted helmet or facepiece with necessary connections by means of which a person breathes ordinary air.	Occupier	Yes	30	

Table 3.10.1 (cont)

Title of Regulation or Order	Reference No.	No. of prescribing Reg.	Task or persons for whom protection is prescribed
The Hemp Spinning & Weaving Regulations 1907	SR & O 1907 No. 660	7	Persons engaged in the opening of bales, machine hackling preparing & carding if dust is generated & inhaled to an extent likely to cause injury to health.
The Factories Act 1961. The Protection of Eyes Regulations 1974	Part IV SI 1974 No. 1681 as amended by SI 1975 No. 303	65	See text.
The Patent Fuel Manufacture (Health & Welfare) Special Regulations 1946	SR & O 1946 No. 258 as amended by SI 1973 No. 36	17	Persons employed in the factory (making coal briquettes or ovoids).
The Ionising Radiations (Sealed Sources) Regulations 1969	SI 1969 No. 808 as amended by SI 1973 No. 36	18	Every classified worker during any working period in which he is liable to be exposed to ionising radiations.

Protection prescribed	Party liable to provide & maintain protection	Is wearing obligatory?	Reg. No.	Remarks
Suitable and sufficient respirators.	Occupier	No		
Eye protectors, meaning goggles, visors, spectacles and face screens.	Employer	Yes	11	These regulations make the provision and wearing of eye protectors obligatory on employer and employee alike over a wide range of processes unless exempted in writing by the Chief Inspector.
Suitable barrier cream for skin protection & suitable goggles or other means of protecting the eyes.	Occupier	No		
Suitable photographic films in an appropriate holder or a suitable dosemeter of an approved type.	Occupier	Yes	18	

The person or persons liable to provide protection is sometimes referred to as 'the employer' sometimes 'the occupier' and sometimes the words are used in the plural. Sometimes 'a suitable overall' is specified, at other times 'suitable overalls', and at yet other times 'overall suits'. The reasons, if any, for these variations are not always clear. If the occupier of a factory requiring the provision, say, of a special type of breathing apparatus, turned out to be a group of squatters, a knotty legal conundrum might arise.

There seems to be more reluctance to make the wearing of breathing apparatus obligatory on the employee than, say, the wearing of overalls, even in cases where there is a known danger of toxic dust or vapour. The prescription of clogs is common in many regulations which are still on the statute book.

In most cases there is provision for exemption from the requirement to wear protective clothing or equipment, by the Chief Inspector (of the Health and Safety Executive) providing the employer or occupier can demonstrate that he has engineered the hazard which requires the protection out of his plant or process. Legal regulations invariably lag behind industrial progress, and the hazards against which protective clothing is prescribed in the chemical industry barely scratch the surface of those for which protection is required.

The tasks requiring protection in the smelting of lead, the manufacture of lead compounds and in potteries are spelt out in such detail in the regulations that it has been impossible to include them in *Table 3.10.1*. However, it appears that many tasks are now obsolete, having been superseded either by other processes, by mechanisation or by the use of compounds which do not contain lead.

Eye protection receives special attention in the Protection of Eyes Regulations, 1974, which were made under the provision of section 65 of the Factories Act 1961. Schedule 1 of these regulations contains a list of specified processes for which approved eye protectors are required and Schedule 2 contains a list of processes where protection is required for persons at risk but not actually employed in the processes.

The processes in these schedules are so widely used that it is felt justified to summarise them below. In nearly all cases the process listed is qualified in the regulations by words such as 'where there is a reasonably foreseeable risk of injury to the eyes of any person engaged in the work'.

Schedule 1. Specified Processes

Part 1. Processes for which approved eye protectors are required.

1. Shot blasting of concrete.
2. Shot cleaning of buildings or structures.
3. Cleaning by high pressure water jets.
4. Striking masonry nails (by hand or power tool).
5. All work with hand held cartridge tools.
6. All work on metal involving the use of a chisel, punch or similar tool by means of a hammer or power tool.

7. The chipping or scarifying of paint, scale, slag, rust or corrosion from metal and other hard surfaces by a hand or power tool.

8. The use of power driven high speed metal cutting saws, abrasive cutting-off wheels or discs.

9. The pouring or skimming of molten metal in foundries.

10. Work at open top molten salt baths.

11. Any work on plant which contains or has contained acids, alkalis, corrosive substances or substances harmful to the eyes, unless the plant has been treated, designed or constructed to prevent risk of eye injury.

12. The handling in open vessels or manipulation of substances referred to in (11).

13. Driving in or on of bolts, pins or collars to a structure or plant by a hammer, chisel, punch or portable hand tool.

14. Injection by pressure of liquids into buildings or structures which could result in eye injury.

15. Breaking up of metal by a hand or power driven hammer or tup.

16. Breaking, cutting, dressing, carving, or drilling by a hand or portable power tool of any of the following:

(a) glass, hard plastics, concrete, fired clay, plaster, slag or stone or similar materials or articles consisting wholly or partly of them.

(b) bricks, tiles or blocks of brickwork, stonework or block work (except wooden blocks).

17. Use of compressed air to remove swarf, dust, dirt or other particles.

18. Work at furnaces containing molten metal and pouring or skimming molten metal.

19. Foundry work where there is a danger of hot sand being thrown off.

20. Wire and wire rope manufacture where there is a risk of eye injury.

21. Coiling wire and similar operations where there is a risk of eye injury.

22. Cutting wire or metal straps under tension.

23. Manufacture and processing of glass and handling cullet where there is a risk of eye injury.

Part II. Processes in which approved shields or approved fixed shields are required

24. Processes using exposed electric arcs or exposed arc plasma streams.

Part III. Processes in which approved eye protectors or approved shields or approved fixed shields are required

25. Oxy-gas metal welding.

26. Hot fettling of steel castings by a flux injected burner or air carbon torch, and deseaming of metal.

27. Hot cutting, boring, cleaning, surface conditioning or spraying of metal by an air-gas or oxy-gas burner.

28. Instruments such as lasers which produce light radiation which can cause eye injury.

Part IV. Processes in which approved eye protectors or approved shields or fixed shields are required

29. Truing or dressing abrasive wheels.
30. Work with drop hammers, power hammers, horizontal forging machines and forging presses other than hydraulic.
31. Dry grinding of materials by applying them by hand to a wheel, disc or band or by applying a power driven portable grinding tool to them.
32. The fettling of metal castings.
33. Operation of pressure die casting machines.
34. Machining of metals including any dry grinding process not elsewhere specified.
35. Electric resistance and submerged electric arc welding of metals.

Schedule 2. Cases in which protection is required for persons at risk but not employed in the specified processes

1. Item 6 of schedule 1.
2. Item 24 of schedule 1.
3. Item 30 of schedule 1.
4. Item 32 of schedule 1.
5. Item 28 of schedule 1.

3.10.3 Protective clothing and personal protective devices available and guidelines to their selection

The BSI has issued a number of standards covering a wide range of protective clothing and personal devices used in British industry. These standards have for the most part been drawn up by the Personal Safety Equipment Standards Committee which represents Government departments, scientific and industrial organisations.

British equipment manufactured, tested and used in accordance with these standards supplies the main UK requirements for protective clothing and personal devices. The equipment described in this section is mainly confined to that covered by British Standards.

In the USA a wider range of approved equipment covered by national standards is available. Details of American equipment are given in the Accident Prevention Manual published by the National Safety Council of Chicago.

Care must be taken when using personal protective equipment from more than one country of origin to ensure that the different codes are understood, and that confusion does not arise from the use of different colour markings for equipment serving the same purpose when supplied from different countries of origin. Canisters for gas masks for protection against different toxic gases and vapours are an example where such care is needed, since the USA and UK colour codes for canisters for different toxic gases are different.

The parts of the person protected, taken in order, are:

Eyes
Respiratory system
Face (generally included in the above)
Head
Ears
Hands
Feet
Body, including neck, legs and arms.

A list of the main British Standards for such protection is given in *Table 3.10.2.*

Table 3.10.2 BRITISH STANDARDS FOR PROTECTIVE CLOTHING AND EQUIPMENT

Part protected	Short title or description	BS No.
Eyes	Industrial — general (non radiation)	2092
	Filters for welding, etc	679
	Green, for steel workers	1729
	Filters for intense sun	2724
	X-ray protection	4031
Respiratory system	Guide to selection	4275
	Breathing apparatus (air line & self contained — in 4 parts)	4667
	Respirators, general dust & chemical	2091
	Dust respirators, high efficiency	4555
	Dust respirators, powered	4558
Face and eyes, etc	Radiation protection, welding	1542
	Dust hoods and blouses, powered	4771
Head	Industial safety helmets	5240
	Scalp protection, light duty	4033
	Welders helmets	1542
Ears		
Hands	Industrial gloves	1651
	Rubber gloves — for electricians	697
Feet	Safety footwear (3 parts)	1870
	Womens safety footwear	4972
	Hearing protectors	S108
	Antistatic footwear/Conducting footwear	5451
	Footwear & gaiters, for foundries	4676
Body (clothing)	Air & liquid impermeable	4724
	For construction workers	4679
	For intense heat	3791
	Waterproof	4170
	Welders	2653
	X-ray protecting aprons	3783

3.10.4 Eye protection

The industrial hazards for which eye protection is required are:

1. Impact (from flying particles).
2. Molten metal.
3. Dust.
4. Gas.
5. Chemicals.
6. Radiation.
7. Any combination of these.

The first step needed when considering eye protection is to identify the tasks with eye hazards. The legal minimum here is to check every task carried out to see whether it comes under the requirements of the Protection of Eyes Regulations 1974. Having done this, a careful check is needed to see if there are any other tasks not covered by these regulations which need eye protection. The second step is to identify the hazard of each task, as listed above, so that correct protection can be provided. If the exposure to any of these hazards is moderate to severe, protectors meeting the requirements of BS 2092 should be specified.

The main types of protection available are:

Spectacles (with or without side shields).
Goggles (cup type or box type).
Face screens.

Spectacles are the most comfortable and generally acceptable form of protection from small flying particles and radiation. They are available with toughened or laminated glass lenses, to prescription if required, with various tints in strong frames to meet the rigorous requirements of BS 2092, 1967.

Where injurious dusts, gases or chemicals are present which could endanger the eyes, dust goggles, gas tight goggles or chemical goggles may be required. But since these hazards generally require respiratory and often face protection as well, a full face respirator or breathing apparatus or a special hood supplied with clean air is more likely to be required.

BS 2092 covers the testing and marking of eye protectors for all hazards listed above, except radiation.

Due to discomfort and their tendency to mist up ordinary goggles are less satisfactory to wear than spectacles. They are not, as a rule, supplied with prescription lenses and are usually held in place with an elastic headband. Since goggles, when required, generally have to enclose the eyes completely without ventilation, the misting tendency is increased. Lenses are made of various transparent plastics, but the most satisfactory are polycarbonates (particularly CR 39) which give highest impact resistance. Goggles are available with lenses which are treated with a hydrophilic coating which greatly reduces misting.

Lightweight goggles, sometimes known as eye shields, are designed to fit over prescription spectacles and are used by visitors to hazardous areas and for 'light duties'. Plastic frames of good heat resistance are preferred to metal frames for spectacles exposed to radiant heat since they cause less discomfort to the wearer.

Clip-on and hinged flip-up tinted eye shades are used for attachment to spectacles with clear lenses to protect against welding and other radiation.

Spectacles with lead glass are required for exposure to X-rays. Glass lenses for protection against most lasers can be obtained on special order from spectacle suppliers. It is, however, most important that the wavelength of the laser be specified and the glass be guaranteed to provide a specified attenuation at that wavelength.

Face shields consist of curved, generally transparent, polycarbonate screens supported from above by a plastic harness fitted over and round the top of the head, and protect the face entirely. Heavier types used for welding are made of fibre glass or metal with a rectangular window of dark green toughened glass in front of the eyes. When the device covers the ears and top of the head and neck, it is known as a welding helmet.

Fixed glass shields should be utilised as far as possible for eye protection, as they cause no discomfort and minimum inconvenience, and there is less difficulty in ensuring that they are used.

3.10.5 Protection of respiratory system

The hazards of airborne dusts and vapours to the respiratory system were discussed in section 3.2 'Air and breathing'. Whilst every effort must be made to prevent their presence in the atmosphere where people work it is seldom possible to protect all workers from exposure to atmospheres containing harmful impurities. Paint sprayer and grinder operatives are nearly always exposed. One must also be prepared for emergencies when men are obliged to enter and work in a dangerous atmosphere.

Protection of the respiratory system is unfortunately one of the most difficult and intractable problems of personal protection. This is caused by the wide variety of hazards needing different types of equipment, the handicaps which workers suffer from when wearing such equipment, and the difficulties found in ensuring that contaminated air does not leak through gaps between the equipment and the wearer and so reach his respiratory system.

The selection of different types of equipment is dealt with in detail in BS 4275: 1974. This should be consulted as only an outline of the problem can be given here.

Atmospheres may be hazardous for any one or more of the following reasons:

(a) Oxygen deficiency.
(b) Contamination by hazardous dust, mist or fume.
(c) Contamination by toxic gas or vapour.

Respiratory protective apparatus falls under two main types:

1. Respirators, which remove contaminants from the surrounding air,
2. Breathing apparatus, which provides the wearer with a supply of clean uncontaminated air or oxygen.

Respirators cannot be used in atmospheres deficient in oxygen (less than 16% volume oxygen) and are specifically designed for selected ranges of contaminants of types (b) or (c). They also have limited service life before the canister or cartridge, where the contaminant is removed, becomes exhausted.

The different types of respirator and breathing apparatus are covered by the following British Standards.

BS 2091. Respirators for protection against harmful dusts, gases and scheduled agricultural chemicals.
BS 4555. High-efficiency dust respirators.
BS 4558. Positive pressure, powered dust respirators.
BS 4667. Breathing apparatus. Part 1. Closed-circuit breathing apparatus; Part 2. Open-circuit breathing apparatus; Part 3. Fresh air hose and compressed air line breathing apparatus; Part 4. Escape breathing apparatus.
BS 4771. Positive pressure powered dust hoods and blouses.

Respirators are used mainly for the removal of hazardous dusts, mists and fumes, and for some toxic gases and vapours. Where there is any doubt about the suitability of a respirator, breathing apparatus should be used.

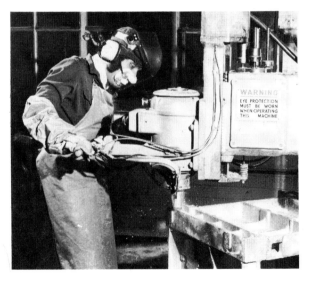

Figure 3.10.2   Eyes, ears and head protection with dust mask (American Optical Safety International)

Figure 3.10.3   Dust respirator in use (American Optical Safety International)

Both respirators and breathing apparatus require a means of connection to the nose and mouth of the wearer which prevents ingress of the contaminated atmosphere. Two types of facepiece are available for both respirators and breathing apparatus:

(a) The half mask facepiece covering nose and mouth only.
(b) The full facepiece, covering eyes, nose and throat.

The facepieces must make a good seal against the face. This is generally difficult or impossible to achieve for people wearing spectacles, beards, whiskers or 'sideboards' or even half a day's growth of 'stubble'. All facepieces have non-return exhalation valves, in which moisture may condense and even freeze in very cold weather, causing the valve to fail. The transparent windows of full facepieces are prone to fogging through condensation of moisture in the breath. Speech is impossible when most types are worn, although some incorporate a speech diaphragm which transmits speech. The act of speaking is, however, liable to disturb the face seal and cause inward leakage. Special facepieces for self-contained breathing apparatus incorporate a microphone which is connected to a portable transistorised transmitter.

Breathing apparatus and positive pressure-powered respirators can also be used with hoods. When used alone they cover the head, neck and parts of the shoulders. When used with blouses they cover the entire upper half of the body. They may be used with complete air-tight suits which cover the entire body and which are supplied with clean or decontaminated air under a positive pressure. Facepieces may also be supplied with air under a slight positive pressure, but when used with respirators other than powered respirators, they depend on the suction produced by the wearer to draw in clean air through a non-return valve and the canister or cartridge that removes the contaminant.

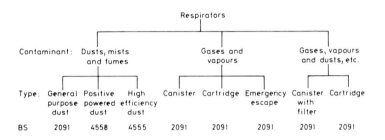

The different types of respirator are shown in the chart above.

The cartridge type contains the purifying agent (a filter for dusts or an absorbent or reactive chemical for gases and vapours) in one or two relatively small cartridges which are attached to and supported by the face-piece. The canister type contains the purifying chemical in a larger canister which is attached to the body and connected to the face-piece by a large diameter flexible rubber tube.

Most respirators are covered by BS 2091. The positive powered dust respirator has been developed to reduce the effort of the wearer in inhaling air through the filter of the respirator, as well as reducing leakage of contaminated air between the face piece and the face. It can be used either with a general purpose type dust respirator or with a high efficiency dust respirator. The high efficiency dust respirator is used for very fine and toxic dusts. The different types of breathing apparatus are shown below.

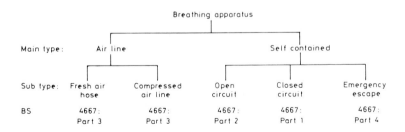

3.10.5.1 Respirators

All respirators require careful fitting to the wearer by a skilled person. This requires the following steps:

1. Remove the seal from the canister or cartridge and put on the facepiece.
2. Adjust the head straps until the facepiece fits closely and comfortably.
3. For canister types, check there are no kinks in the connecting tube.
4. Breathe naturally and observe the resistance to inhalation.
5. Close off the air supply by appropriate means (i.e. squeezing the air tube of a canister respirator, or sealing the openings on a cartridge type). Inhale deeply until the face piece collapses. It should remain collapsed until the intake is re-opened.

Fitting should be checked by a test in which the wearer enters a room containing a low concentration of an irritating contaminant which his apparatus is designed to remove.

Canisters and cartridges should normally be kept sealed and out of contact with air. Some method of testing cartridges and canisters that have become partly exhausted is necessary. The maker or supplier should advise on this, and also on re-sealing cartridges and canisters that have had limited use. Cartridges and canisters have limited shelf life, and the last date by which they should be used should be stamped on the outside.

Respirators fall into two general types, those designed for specific gases and vapours, and those designed for dusts, although respirators designed for gases and vapours can where necessary also incorporate a filter which provides some protection against dusts.

3.10.5.2 BS 2091. Respirators for gases and vapours and specific agricultural chemicals

BS 2091 covers three main types of respirator: dust respirators; canister and cartridge types of gas respirator. All have a face-piece and harness and two non-return valves, one for inhalation through the purifier and one for exhalation.

3.10.5.3 Dust respirators

Two types of dust respirator are covered:

Type A dust respirator has a low resistance to inhalation (20 mm water gauge) when tested under standard conditions. It is intended for use against dust of low toxicity under conditions where low breathing resistance is important.

Type B dust respirator has a higher inhalation resistance (32 mm water gauge) and is more efficient in stopping particles of fine dust than is type A.

The maximum allowable inward leakage rate is 15% for type A and 10% for type B, about half of this being due to leakage between the face and face piece.

Neither is suitable for moderately toxic dusts or for toxic gases and vapours of any kind. The filter elements are replaced when the amount of solids collected is sufficient to cause a significant increase in inhalation resistance.

Many different filter media are available — paper, resin impregnated wool and felt pads being the most common. When the type of dust can be identified, the manufacturer should be consulted on the most suitable type. Usually much of the dust to be removed is quite coarse. This can be trapped by a prefilter pad placed in front of the filter proper. The prefilters should be replaceable.

3.10.5.4 Gas respirators, canister type

These should not be used in atmospheres immediately dangerous to life. The canister-type gas respirators allow seven different types of canister to be used for protection against different toxic gases and vapours. These are distinguished by colour markings. The type of canister with the widest range of application is type CC and has a black marking with a grey stripe. This is suitable for most organic and acid gases and vapours and incorporates a filter which will handle most fine smoke particles. It will also handle the vapour of mercury and lead tetra ethyl.

A special type A canister with blue markings is used against ammonia vapour, and type O (black with orange stripes) is used for halogenated hydrocarbons such as mono and dichloro methane. These canisters achieve virtually complete removal of the contaminating gas or vapour and are designed to last for about thirty minutes with a concentration of 1% by volume of the contaminant in the incoming air. The facepieces are specially designed to give a first class seal with the face, and a maximum inward leakage of contaminant is 0.25% by volume.

3.10.5.5 Gas respirators, cartridge type

These are designed to provide protection against low concentrations of certain toxic gases from the inhaled air. Only one type of cartridge is available for the same substances that type CC canister respirators protect against. They should not be used for atmospheres containing contaminants whose threshold limit values exceed 100 parts per million, nor should they be used when the contaminant concentration exceeds 0.1% by volume.

The cartridges are marked black. The maximum inward leakage rate with this type of respirator is 5% by volume. The cartridges have a limited life and need frequent replacement.

3.10.5.6 BS 4558. Positive powered dust respirators

These respirators comprise a mask supplied with filtered air from a battery-operated power pack carried by the wearer, which reduces respiratory load and virtually eliminates inward leakage between the face and face piece during inhalation. They can be used with fairly ordinary types of dust filter, such as those covered in BS 2091, or with a high efficiency type of dust filter for hazardous dusts. They supply filtered air at a constant rate to the face piece in excess of the maximum demand rate during inhalation.

Exhaled and surplus air pass out to the atmosphere through a non-return valve which maintains a slight positive pressure inside the face-piece at all times. While this type of respirator has obvious advantages in reducing the stress of inhalation against a negative pressure on the wearer, and allowing more efficient filters to be used, the amount of air which has to be filtered is several times that actually required for breathing. Thus larger filters are required, and there is a significant weight and convenience penalty as a result of the blower and batteries and the larger filter.

3.10.5.7 BS 4555. High efficiency dust respirators

These respirators are designed for protection against highly toxic particulate dusts, including radioactive materials. They do not give the same degree of protection as a positive powered dust respirator with a high efficiency filter, but they give considerably better protection than the general purpose dust respirators meeting BS 2091. They differ from the general purpose dust respirators in having a larger and more efficient filter (often two filters, one on either side of the facepiece), and a more efficient seal between the facepiece and the face.

This type of respirator has a full facepiece with a wide curved plastic window, and usually a second inner facepiece which makes an efficient seal which covers the mouth and nose. These respirators are designed so that the leakage of contaminant through the face seal or seals on inhalation does not exceed 0.1 % by volume. The pressure drop through the filters on inhalation should not exceed 50 mm water gauge on a standard test.

These respirators are normally provided with a speech diaphragm and special steps are taken to prevent misting of the window. They also allow spectacles of special design to be worn and need to be very carefully fitted to the wearer. When used to protect against radioactive dusts, the filter elements should be tested regularly to check the quantity of radioactive material which has accumulated since radioactive material in such close proximity to the face is dangerous.

3.10.5.8 Other respirators

Several simple respirators and dust masks are available to remove non-toxic nuisance dusts. These should not, however, be used on toxic dusts since they are not usually efficient for fine particles (about 1 micron size) which are most damaging to the lungs.

Powered dust hoods and blouses covered by BS 4771 have the same type of battery operated blower and filter as positive powered dust respirators, but the exhaust and surplus air escapes from the bottom of the hood or blouse where it fits round the body.

These are required for toxic dusts which can cause damage through contact with the skin.

3.10.5.9 Breathing apparatus

The types of breathing apparatus discussed are all covered by one of the four parts of BS 4667. Air line apparatus which employs an air hose or compressed air line is covered by Part 3 and is used mainly for regular planned tasks such as paint spraying or work inside a tank or pit with a hazardous atmosphere. Self-contained breathing apparatus tends to be used more for emergency operations such as fire fighting and rescue.

3.10.5.10 Air-line breathing apparatus

This is covered by Part 3 of BS 4667, and includes:

1. Fresh air hose apparatus
   (a) without blower
   (b) with hand blower
   (c) with motor operated blower
2. Compressed air line apparatus
   (a) constant flow
   (b) demand valve type.

1. Fresh air hose apparatus

This consists of a facepiece (full or half mask) and valve system with a reinforced rubber hose of adequate diameter (about 1 in) which extends to a point where uncontaminated air may be drawn in.

For distances up to 9 m, the air may be drawn in by the inhalation of the wearer.

For distances up to 36 m the air supply should be provided by a blower at the air inlet which may be either hand or motor operated. When used in atmospheres immediately hazardous to life, the blower type should be used, and the wearer should also wear a safety belt attached to a safety rope. This should be held by a second person located at the other end of the air hose in a safe atmosphere who is able to haul the wearer out in an emergency.

2. Compressed air line apparatus

This type of apparatus differs from that previously described in that the air is supplied under pressure through a small diameter line from an air compressor, a compressed air cylinder or a works compressed air supply. The compressed air line is connected to a coupling and pressure reducer attached to the wearer. The air supply may be continuous or operate only on demand when the wearer inhales, causing a drop in pressure in the face piece.

Compressed air line respirators are used for paint spraying, welding and cutting and for exposure to fumes, dusts, gases and vapours which pose no *immediate* danger to life. They are not recommended for tank entry or in circumstances where the worker's life would be in jeopardy should the air supply fail. They are used with hoods for shot blasting or heavy paint spraying to protect workers from rebounding particles. Since they offer little or no resistance to breathing, they can be used for long periods without fatigue.

Air line respirator helmets may be combined with complete air supplied suits or jackets to provide body comfort in hot atmospheres (e.g. steel works) or in corrosive chemical atmospheres.

When used with a compressor, care must be taken that the air is not contaminated with oil droplets, or that the compressor does not become overheated thereby decomposing the lubricating oil and causing dangerous fumes. Non-lubricated compressors of the diaphragm or carbon ring type are preferable. When used with a works compressed air supply the quality of the compressed air should be checked, and traps, filters and drains provided to ensure that the air is clean and dry. Where a new installation is planned which will call for the use of compressed air line apparatus, consideration should be given to installing a special compressor and compressed air main to be used solely for this purpose.

3.10.5.11 Self-contained breathing apparatus

Apparatus of this type should only be used by men in good physical condition who have been thoroughly trained in its use. Within a works its use will generally be restricted to selected members of the works fire brigade. The training must be specific to the type of apparatus used.

1. Open circuit breathing apparatus

This is covered by BS 4667, Part 2. It comprises a portable cylinder of compressed air carried by the wearer by a body harness, a full face-piece and harness with window and means for speech transmission, an exhalation valve operated automatically by the pressure in the breathing circuit, compressed air hoses and tubes and suitable automatic valves which either supply a constant air flow to the mouth and nose of the user or supply air on demand.

Figure 3.10.4   Self-contained breathing apparatus and gas and chemical splash-proof suit (Draeger)

The weight of the compressed air cylinder makes the whole apparatus rather heavy. The maximum time for which the apparatus can be used without changing the cylinder usually varies between 20 and 40 minutes.

The specification lays down stringent tests of the apparatus when worn by subjects undergoing commando type exercises, such as walking on a treadmill at 6.5 km/h while carrying out various head movements and reciting the alphabet!

2. *Closed circuit breathing apparatus*

This is covered by BS 4667, Part 1. The apparatus differs from the open circuit apparatus in the following respects:

    1. In place of compressed air, compressed oxygen or liquid oxygen (or oxygen enriched air) is used.

2. The exhaled air from the face-piece passes into a purifier which contains chemicals which absorb the exhaled carbon dioxide. The oxygen mixes with the purified gases which pass to a breathing bag and from this through a cooler to the face-piece.

The apparatus is considerably more complicated than open circuit apparatus, but the size of the gas cylinder is very much less, so that the overall weight is reduced. This type of apparatus generally can be worn and used continuously for considerably longer than the open circuit type before the gas cyliner is empty.

At least one American type of closed circuit breathing apparatus carries no oxygen as such, but has in its place a chemical which reacts with the carbon dioxide exhaled and liberates oxygen in its place.

The specifications for closed circuit breathing apparatus include satisfactory performance when worn by fit persons carrying out commando like exercises similar to those employed in testing open circuit apparatus.

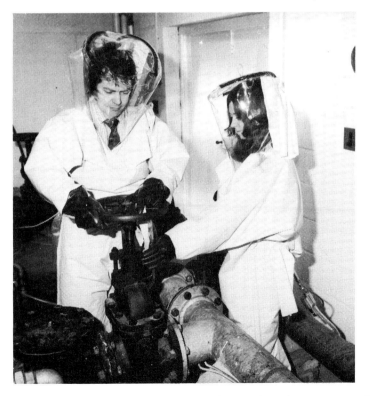

Figure 3.10.5   Industrial standard full suit with internal ventilation system and exhaust filters (Plysu)

Figure 3.10.6   Air-line breathing apparatus (Draeger)

3. *Escape breathing apparatus*

This is covered by BS 4667, Part 4. The apparatus may be either of open circuit or closed circuit type, and differs mainly from normal open and closed circuit types in that the air or oxygen cylinders are smaller. The apparatus is intended simply to allow persons to escape if trapped by an emergency in an area such as a coal mine, where the atmosphere is hazardous or would cause an immediate danger to life.

3.10.6 Face protection

Face protection has already been largely covered under eye and respiratory system protection. It is provided by various types of shields and hoods. In atmospheres containing toxic gases and vapours which enter the body through the skin or attack the skin, face protection alone is insufficient, and completely impervious air tight suits are required, usually supplied with a stream of air inside (in combination with a compressed air line breathing apparatus).

A short list of common substances whose vapours are absorbed through the skin as well as by the respiratory system is given in Table 3.10.3. While people exposed to these vapours should wear impervious clothing as well as respirators, in practice it is usually possible to keep their concentrations below danger levels by proper ventilation.

Table 3.10.3 COMMON SUBSTANCES WHOSE VAPOURS ARE HAZARDOUS BOTH THROUGH RESPIRATORY SYSTEM AND SKIN

Substance	*T.L.V.*	
	p.p.m.	mg/m³
Aniline	5	19
Carbon disulphide	20	60
Cresol	5	22
Dichloroethylene 1.1		
Decalin		
Di-ethylene oxide (dioxane 1.4)	50	180
Di-ethyl phthalate		5
Dimethyl aniline	5	25
Di-nitrobenzene	0.15	1
Di-nitrochlorobenzene		
Hydrogen cyanide	10	11
Iodoform		
Lead tetraethyl		0.100
Mercury and its compounds	0.001 & over	0.01 to 0.05
Nicotine		0.5
Nitrobenzene	1	5
Nitroglycerine	0.2	2
Organic phosphate insecticides	about 0.01	about 0.1
Phenyl hydrazine	5	22
Tetralin		

Face and skin protection is most necessary for a number of less common but super-toxic materials which are hazardous at concentrations of a few parts per billion only.

3.10.7 Head protection

Protective headware serves two main functions:

1. to protect the head from impact from falling objects and bumping stationary ones,

2. to contain the hair and prevent it from becoming entangled with moving machinery.

Headgear designed for the first function usually fulfils the second also, but the converse does not apply.

Protective headgear may also serve the functions of normal headgear, i.e. to protect the wearer from rain and extremes of temperature, to shade the eyes, and to provide a symbol and means of easy identification of the class or group of persons to which the wearer belongs.

Protective headgear is also sometimes required and designed to provide a means of support for other things, i.e. shields to protect the eyes and face of the wearer from radiation and flying objects, and ear muffs to protect the ears from excessive noise.

Hoods which cover the head, face and neck and protective suits designed for fire fighting or protection against substances which cause injury to the skin, incorporate suitable protective headgear in their design.

Sometimes protective headgear must be insulating to protect the wearer against contact with high voltage sources.

3.10.7.1 Protection from impact

The range of safety helmets available in Britain whilst narrower than in the USA, appears to serve most needs.

Safety helmets in the UK generally conform to BS 5240, *General purpose industrial safety helmets*. This does not restrict the choice of materials so long as they pass the mechanical, electrical and weight specifications laid down. They are generally brimless caps with a peak, moulded from ABS resin. Glass reinforced plastic helmets (fibreglass) are also available. The shell of the hat is attached to a cradle and adjustable headband which fits the wearer. This holds the shell so that there is a minimum clearance everywhere between the shell and the skull.

These safety helmets are available in a wide range of colours, which are useful for identifying personnel of different grades and departments. Extra markings can be added as required. Chin-straps and felt or other warm linings are usually available as optional extras, but this should be checked before ordering. Safety helmets can be cold and draughty, especially if the wearer is working aloft in a strong north east wind. Special safety helmets can be supplied with detachable ear muffs or face shields, although neither of these attachments can be fitted to most standard helmets.

Light-weight 'bump caps' or scalp protectors are covered by BS 4033. Their prime object is to afford some protection against bumping into projecting pipes and steelwork, and not against falling objects. They consist of a smooth shell fitted with protective padding, and they may have a rigid or pliable peak. They are generally smaller, lighter and more comfortable than safety helmets, and there are many instances where the risks of falling objects are very low where their use is preferable.

Most safety equipment suppliers offer a range of hair nets to prevent long hair from becoming entangled in machinery or around shafts. There are many machine operations where their use is essential on safety grounds, and careful persuasion is needed to ensure that employees wear them.

3.10.8 Ear protection[4] (see section 3.4)

The adverse effects on human hearing of exposure to excessive noise were discussed in section 3.4. Two types of ear protectors are available for use when it is impossible to reduce the noise by engineering methods, ear plugs and ear muffs.

3.10.8.1 Ear plugs

These produce maximum attenuation at the high frequency end of the spectrum (about 30 dB) but far less attenuation at the low frequency end. They are available either as permanent plastic or rubber plugs, designed to fit into the ear canal, or as disposable plugs made of very fine glass wool, special wax or wax impregnated cotton wool. Ordinary dry cotton wool is not effective.

When a decision is made to use plugs, it is preferable to employ permanent plugs, available in a range of sizes, and have them fitted by a trained person. They are often supplied in pairs with a connecting neck cord to prevent their getting lost. Unfortunately, plugs are generally uncomfortable to wear, and it is often difficult for supervisors to know when they are being worn.

3.10.8.2 Ear muffs

Ear muffs have a hard shell which completely covers the ear and are sealed to the head with a soft cushion. They are held in position by a sprung headband. Although heavier and bulkier than ear plugs, they give much

Table 3.10.4 PROTECTION GIVEN BY VARIOUS EAR PROTECTORS

	Frequency (Hertz)				
	250	500	1000	2000	4000
Plastic ear plugs	12	14	18	27	31
Mean attenuation, dBA	12	14	18	27	31
Standard deviation, dBA	7	7	8	8	8
Ear muffs, lightweight					
Mean attenuation, dBA	13	20	27	35	42
Ear muffs, heavyweight					
Mean attenuation, dBA	25	33	39	40	45
Ear muffs, all types					
Standard deviation, dBA	5	5	5	6	6

better protection and are more easily fitted. Ear muffs are available with frames which will fit over a safety helmet. There is generally less difficulty in persuading employees to wear ear muffs than ear plugs, and it is, of course, easy to observe if they are being used. *Table 3.10.4* shows the protection given by various ear protectors at different frequencies.

In selecting ear protectors for different locations, the principle should be followed of reducing the sound to less than the danger level, (i.e. 85 to 90 dBA). While it is not always easy to persuade employees to wear ear protectors when needed, there is usually less resistance to the wearing of muffs than ear plugs.

3.10.9 Hand protection

A great variety of gloves, gauntlets and mitts is available of different materials and styles, with the result that quite often types are chosen which are not the most suitable for the task or hazard.

BS 1651, *Specification for industrial gloves* does a great deal to assist safety specialists and others concerned with selecting gloves by classifying and recommending suitable types of gloves for these hazards. The following types of gloves are recommended.

Type 1.   Chrome leather wrist gloves.
Type 2.   Chrome leather inseam mitts and one finger mitts, wrist and gauntlet style.
Type 3.   Chrome leather gauntlet gloves.
Type 4.   Chrome leather inseam gauntlet gloves, with canvas or leather cuffs, with or without reinforcement between thumb and forefinger.
Type 5.   Discarded.
Type 6.   Chrome leather stapled double palm wrist gloves.
Type 7.   Horse hide or cattle hide inseam gauntlet gloves, with vein patches and aprons covering palm to first joints of fingers.
Type 8.   Horse hide or cattle hide palm, chrome leather back, inseam wrist gloves.
Type 9.   Felt mitts, palms faced with canvas, mole or chrome leather, wrist and gauntlet style.
Type 10. Polyvinyl chloride wrist gloves and gauntlet gloves, lined.
Type 11. Polyvinyl chloride wrist gloves and gauntlet gloves, unlined.
Type 12. Lightweight natural and synthetic rubber wrist gloves and gauntlet gloves, unlined.
Type 13. Medium weight, as type 12.
Type 14. Heavy weight, as type 12.
Type 15. Handguards, rubber, leather or fabricated with rubber.
Type 16. Cotton drill gloves.
Type 17. Cotton drill gloves with chrome leather palms.
Type 18. Lightweight natural and synthetic rubber wrist gloves and gauntlet gloves, lined.
Type 19. Medium weights, as type 18.
Type 20. Heavy weight, as type 18.

The types of rubber used in rubber gloves are designated by the letters A, B or C.

A = Natural rubber latex or solution.

B = Chloroprene latex.

C = Nitrile latex.

The letter R refers to reinforcement. The classification of hazards is given in Table 3.10.5. Where flame resistance is required BS 3120 should be referred to.

At least one glove supplier provides a chart showing the resistance of different types of rubber and PVC gloves to various chemicals, and most suppliers have a technical department which will advise on particular problems. For further details and methods of ordering gloves BS 1651 should be referred to.

BS 697 'Rubber Gloves for Electrical Purposes' gives details of 4 classes of rubber gloves rated according to voltage to earth — 650, 1100, 3300 and 4000.

Table 3.10.5

CLASSIFICATION OF HAND HAZARDS AND RECOMMENDED GLOVE TYPES
(BS 1651)

Hazard Group	Protection against	Typical operations	Recommended types of gloves
A	Heat, but wear not serious and no irritant substances present	Furnace work, drop-stamping, casting and forging, handling hot tyres and similar operations	1, 2, 3, 4, 9.
B	Heat, when wear more serious or irritant substances present	Stoking gas retorts, riveting, holding-up, hot chipping	2, 3, 9.
C	Heat, when fair degree of sensitivity is required, and splashes or spatters of molten metal may occur	Welding, case hardening in cyanide bath	2, 3, 4.
D	Sharp materials or objects	Swarf, metal after guillotining, blanking or machining	2, 4, 7, 9, 10, 14A, 14A(R), 15, 20A.
E	Sharp materials or objects in an alkaline de-greasing bath		10, 11, 14A, 14A(R), 19BC, 20BC.

Hazard Group	Protection against	Typical operations	Recommended types of gloves
F	Glass or timber with splintered edges		15, 17.
G	Abrasion	Handling cold castings or forgings, precast concrete, bags of cement and bricks	2, 6, 8, 10, 12A(R), 13, 13AC, 14AC, 15, 17, 18AC, 19AC, 20AC.
H	Gross abrasion	Shot blasting	2, 9, 12A(R), 13A(R), 14A.
J	Light abrasion	Light handling operations	1, 2, 9, 10, 12ABC, 13ABC, 15, 16, 17.
K	Chemicals	Acids, alkalis, dyes and general chemical hazards not involving contact with solvents or oils	10, 11, 12ABC, 13ABC, 14ABC, 19ABC, 20ABC, 12A(R), 13A(R), 14A(R).
L	Solvents, oils and grease	General chemical hazards involving contact with solvents or oils	10, 11, 12BC, 13BC, 14BC, 18BC, 19BC, 20BC.
M	Electrolytic deposition	Plating and subsequent operations	10, 11, 12A(R), 12ABC, 13A(R), 13ABC, 14A(R), 14ABC, 19ABC, 20ABC.
N	Hot alkaline cleaning baths		10, 11, 13A(R), 13ABC, 14ABC, 19ABC, 20ABC.
O	Spraying paints or cellulose lacquers		1, 9, 11, 12BC, 13BC, 18BC, 19BC.
P	Special hazards: lead tetraethyl, mercury lead salts		10, 13ABC, 14ABC.
Q	Classification discarded		
R	Electric shock	See BS 697 *Rubber gloves for electrical purposes*	

3.10.10 Foot protection

'Safety Footwear' is designed primarily to protect the wearer's toes and forefeet from damage by falling objects. Such footwear also has strong soles with non-slip treads, and a wide choice of shoes and boots is available which provide ankle support and leg protection against impact, hot objects and various liquids as required. It is covered by BS 1870.

Part 1. Safety boots and shoes other than rubber.

Part 2. Lined rubber safety boots.

Footwear meeting these specifications is available from most safety footwear suppliers.

For women, who may be reluctant for reasons of appearance and comfort to wear the rather heavy footwear described in BS 1870, a lighter range of protective footwear is given in BS 4972 'Women's Protective Footwear'.

Anti-static rubber footwear designed to reduce hazards of static electricity whilst yet sufficiently insulating to prevent serious danger of electric shock through contact with mains voltage, is covered by B.S. 2506. Conducting footwear, giving extra protection against static for workers handling detonators, explosives and unstable compounds is covered by BS 3825.

When safety footwear is made available to employees (by sale or otherwise) through company channels, arrangements should be made to inspect, service and repair worn or damaged shoes, and to ensure that safety shoes that have been worn or damaged beyond repair are withdrawn from service and destroyed.

When selecting a supplier of safety shoes, two points should be checked:

1. Are the shoes supplied in an adequate range of widths as well as foot lengths?
2. Is adequate protection to the instep provided?

For workers whose feet are exposed to hot and molten metal, a range of suitable protective footwear is specified in BS 4676 'Footwear and gaiters for foundries'.

One difficulty sometimes faced by safety specialists is the danger of some 'normal' footwear worn by workers for jobs which hardly justify the issue of protective footwear. If tactful persuasion is not enough, it may be necessary to ban particularly dangerous footwear.

3.10.11 Protective clothing

Nearly all clothing is protective in one sense or another, and in particular provides insulation against too rapid loss of body heat in most environments. Special work clothing is, however, usual in most industrial occupations, much of it being quite specialised, and job related. The principal reasons for the use of job related protective clothing are:

1. To prevent wear, tear and contamination of the employee's usual clothing.
2. To obviate job related risks inherent in some of the usual clothing — such as entanglement of sashes and bows with tools or machines.
3. To provide a rapid means of identifying different categories of employees, and to assist identification of the employee with his job. The effect is partly psychological.
4. To protect against particular work related exposure — heat, cold, wet, radiation, mechanical injury and contact with substances that could injure or mark the employee.

British standards are available for many different types of clothing, e.g.

BS 4724 Air and liquid impermeable clothing.
BS 4679 Construction workers' clothing.
BS 3791 Clothing for protection against intense heat.
BS 4170 Waterproof clothing.
BS 2653 Welders' clothing.
BS 3783 Lead rubber aprons for protection against X-rays.

When selecting protective clothing, insufficient thought is sometimes given as to how it will be cleaned and maintained (which sometimes includes fire proofing of fabrics).

3.10.11.1 Protection against heat and hot metal

Leather is frequently the best choice. The garments and fastenings should be made so that gaps do not appear during body movement, and so that the garment can be easily and quickly removed. The clothing should be designed to prevent hot particles lodging in pockets, turn-ups, etc.

The use of asbestos clothing has declined greatly in recent years, due partly to greater knowledge of the dangers of inhaling asbestos fibres and partly to advances in the technology of other materials. Clothing with a glass fibre or polyester base and aluminised coating is largely used by workers exposed to intense thermal radiation. This may consist of separate trousers, coats, gloves and hoods. One piece units from head to foot are also available. Some suits are supplied with air for cooling and comfort.

Most ordinary fabrics can be treated with flame proofing or flame retarding compounds. A solution of borax and boric acid may be applied to clothing after laundering to render it flameproof. This however must be re-applied each time after laundering. Clothing can also be treated with flame retarding compounds which withstand laundering.

Many synthetic fibres, nylon, polyolefins and, to some extent, polyesters, melt and form a hot sticky mass when exposed to high temperatures. This can cause or aggravate skin burns and injuries. Nylon shirts, socks and underwear should not be worn where there is a significant risk of fire or burns.

3.10.11.2 Protection against impact and cuts

Padded clothing and aprons are available for jobs where there is risk of impact. Metal and hard fibre guards and shields are also sometimes used. The hazards of the job or task must be carefully identified when specifying particular guards or padding.

3.10.11.3 Protection against toxic and corrosive materials

Clothing of many impervious materials is available. These include natural and synthetic rubbers, PVC, polyethylene and polypropylene film, and many treated fabrics. The clothing cannot be specified until the hazardous agent has been identified. Suppliers of such clothing have technical experts who will assist in selecting the most suitable for protection against particular compounds. In some cases it will be found more economic to employ disposable clothing, e.g. of polythene fibre, reinforced paper or film rather than more durable clothing which has to be laundered.

Other special clothing includes:

high visibility clothing, for workers exposed to traffic hazards, leaded clothing, for workers exposed to X-ray and gamma radiation, conductive clothing, for use by electric linesmen working on high voltage conductors.

REFERENCES

1. Fife, I., and Machin, E.A., *Redgrave's health and safety in factories*, Butterworths, London (1976).
2. British Standards — See Tables 3.10.1, 3.10.2 and 3.10.5.
3. Health and Safety Executive, Technical Data Note. *Threshold limit values,* Health and Safety Executive (1976). Published Annually.
4. Health and Safety Executive, Booklet 25, *Noise and the worker*, Health and Safety Executive, London (1976).

3.11 LOCKER, DRESSING AND WASHING ROOMS

Contents

Proper facilities are needed for:

1. Safe-keeping and drying garments which employees shed before starting work,
2. Safe-keeping, drying, cleaning and maintaining employees working clothing and special protective equipment while they are away from work,
3. Changing before and after work, washing and in many cases taking a shower after work, and washing hands before meal breaks.

The extent of these needs are largely related to the nature of the work and the contaminants to which employees and their clothing are exposed.

The subject is dealt with in Health and Safety at Work Booklet 5, *'Cloakroom Accommodation and Washing Facilities'*.

There are various legal requirements to provide such facilities, and these are discussed here first.

3.11.1 Legal requirements[1]

3.11.1.1 Clothing accommodation

The Factories Act, 1961, section 59(1) states that in any factory 'there shall be provided and maintained for the use of the employed persons adequate and suitable accommodation for clothing not worn during working hours; and such arrangements as are reasonably practicable . . . shall be made for drying such clothing.'

The Docks and Harbours Act, 1966, section 35, makes similar provision not only for clothing not worn during working hours, but also for special working clothes.

The Offices, Shops and Railway Premises Act, 1963, makes similar provisions to the Docks and Harbours Act, 1966.

In addition to the main provisions of the Factories Act, 1961 a number of special regulations and welfare orders made under it require the employer to provide accommodation for, clean, disinfect and generally maintain special working or protective clothing called for (see *Table 3.10.1*). Most of these require that the accommodation provided for the protective clothing be separate from that provided for ordinary clothing not worn during working hours, to avoid contamination of the employees' ordinary clothing. Such regulations and welfare orders are listed in *Table 3.11.1*.

The most common contaminants implied here are powdered lead derivatives, although others with strong odours are also implied.

3.11.1.2 Washing accommodation

The Factories Act, 1961, section 58(1) requires all factories to provide adequate washing facilities for employees. These must include a supply of clean running hot and cold water, soap and clean towels or other means of cleaning and drying. In the absence of other mandatory standards, Factory Inspectors recommend at least one wash basin for 20 people doing clean work, 10 people doing dirty work, or 5 people handling poisonous substances.

The Docks and Harbours Act, 1966 includes similar washing facilities in approved welfare amenities to be operated in certain specified ports.

Special washing facilities are required in several special regulations and welfare orders made under the Factories Act, 1961, and these are also shown in *Table 3.11.1*. In a number of cases it is necessary for the employer to provide nail brushes, and in some it is obligatory on the employee to use the special washing facilities.

3.11.1.3 Bathing facilities

Whilst the Factories Act, 1961 has no general requirements for bathing facilities, such facilities are legally required under several special regulations and welfare orders made under the Act. These also are shown in *Table 3.11.1*.

3.11.2 Practical requirements

The legal requirements are minimum ones which have arisen in response to past needs that have come to government attention. Industry however develops much faster than governments legislate and in planning locker, dressing and washing rooms, real needs must be considered. It is not simply morally wrong to fail to provide facilities where there is a real need; it weakens employee morale and motivation and jeopardises the entire health, hygiene and safety programme. It is also usually more economical to provide proper facilities when the factory or plant is being planned and built than to add them later as an afterthought, when the proper sites for them are no longer available.

The best way to determine practical requirements is to examine each job carefully (if this has not been done already) and answer a number of questions A sample questionnaire is given at the top of page 388.

Table 3.11.1 SPECIAL REGULATIONS AND WELFARE ORDERS MADE UNDER THE FACTORIES ACT, 1961

These require the provision of separate accommodation for working clothing, special washing facilities and bathing facilities.

Short title of Regulation or Order	SR & O or SI Number	Separate Accommodation for working clothes required	Special washing facilities			Bathing facilities required
			Required	Employees must use	Nail brushes to be provided	
Bakehouses	1927:191		X			
Biscuit factories	1927:872		X			
Blasting (castings & other articles)	1949:2225	X				
Bronzing	1912:361		X	X	X	
Chemical works	1922:731	X	X	X	X	X (in certain processes)
Chroming plating	1931:455		X		X	
Clay works	1948:1547		X		X	
Construction	1966:95		X		X	
Electric accumulators	1925:28	X	X	X	X	X (in certain processes)
Enamelling, vitreous of metal or glass	1908:1258	X	X	X	X	
File cutting	1903:507		X			

Process/Industry	Reference	1	2	3	4
Glass bevelling	1921:288		X		X
Gut scraping	1920:1437	X			
Hides and skins	1921:2076		X	X	X
Horse hair	1907:984	X	X	X	X
India rubber	1922:329		X	X	X
Iron and steel foundries	1953:1464				X
Jute	1948:1696				
Lead compounds	1921:1443		X	X	X
Lead smelting	1911:752	X	X		X
Non-ferrous metals (melting and founding)	1962:1667		X		X
Oil cake (if application made by 50% of employees)	1929:534				X
Painting of buildings	1927:847		X	X	X
Painting of vehicles	1926:299		X	X	X
Paints and colours	1907:17	X	X	X	
Patent fuel	1946:258	X	X	X	X
Pottery	1950:65	X			
Sugar	1931:684		X		X
Tanning	1930:312	X			
Tinning of holloware	1909:720		X	X	
Wool, goat hair, etc.	1905:1293		X	X	X
Yarn heading	1907:616	X	X	X	X

---

1. What personal working or protective clothing or protective devices are required by the worker for this job?
2. Is the clothing or the worker liable to contamination by:
   Dust?
   Tacky or tarry substances?
   Liquids?
   Toxic substances?
   Radio-active substances?
   Odorous materials?
   Body perspiration?
3. How frequently will the clothing or equipment need cleaning (laundering, dry cleaning)?
4. How frequently will the clothing or material need other treatment such as dressing, fire proofing or polishing (apart from maintenance)?
5. What spare clothing or equipment needs to be provided to allow for the needs of (3) and (4) above?

The use of such a questionnaire will allow decisions to be made on whether separate lockers are required for normal and working clothing, what size they should be, how many wash places or wash basins are needed per 100 workers, what special washing facilities are required, and whether showers or baths are needed. It should also enable a cleaning and maintenance programme for protective clothing and equipment to be planned and budgeted for.

3.11.3 Accommodation and facilities recommended[1,2]

For a small plant or factory, the changing and washing accommodation is best sited near the entrance. For large plants it may be preferable to locate it centrally, or in several buildings adjacent to the work areas served.

If showers are not required, a single combined room for lockers, dressing and washing for each sex is often adequate. If showers are needed, it is better to have two separate locker/dressing rooms with the shower/wash room between them, so that the lockers for normal and working clothing are segregated. The floor space should be sufficient for dressing, undressing and washing in comfort and without congestion.

Whilst pegs and clothes hangers may be appropriate in small factories with only a handful of workers, it is better to provide a separate locker (or two lockers) for each worker to reduce the risk of pilfering. The name or works number of the worker should be displayed on the locker.

Lockers should have sloped tops to prevent material being placed on top of them, should be large enough for clothing to be hung up to dry inside them, and should have room for footwear, headwear and such personal protective equipment as may be issued to the worker, with separate compartments for them as necessary. Lockers should be perforated for ventilation. Forced circulation of warm air through the base of each locker

Figure 3.11.1   Locker room with 'boxed in' warm-air circulation through lockers (Link Steel Furniture Ltd.)

and out through the top is the best method of drying and will be needed if the clothing is at all wet or heavy. The placing of heating pipes under the perforated bases of the lockers may in some cases be sufficient.

Lockers are best placed on metal frames to keep them clear of the ground, with a minimum of floor supports which impede cleaning. They should be anchored to prevent overturning.

The floors of locker/dressing wash rooms should be durable, well drained, impervious to water and non-slip. Glazed brick, non-slip tiles and concrete with an abrasive grain surface are preferable to terrazo, marble, polished concrete and smooth tiles. The flooring material should continue up to the wall as a cove to a height of six inches before there is a joint. The walls should be impervious to water for a height of at least five feet.

Locker/dressing/washing rooms should be kept at a temperature of at least 20°C and a relative humidity of 50%.

If shoes, trousers or underwear have to be changed, seats or benches (with a minimum of supports) should be provided close to the lockers. The seat may be an extension of the metal frame on which the locker is supported.

Floors must be regularly cleaned and mopped dry.

Washing facilities must comply with the Washing Facilities Regulations 1964 (SI 1964 No. 965). Individual wash basins or troughs capable of being

Figure 3.11.2   Locker room showing acceptable lockers with sloping tops (H. & S.W. Booklet[5])

used by several people at the same time are acceptable. Wash basins should have a plug on the outlet pipe. Where troughs are installed, jets or sprays providing water at a suitable temperature should be sited at intervals of approximately two feet. Circular troughs with a central water supply at which several workers may wash their hands simultaneously are acceptable. The water supply is better actuated by a foot pedal than by hand taps. A worker's hands will be recontaminated if, after washing, he closes a tap which he opened with a dirty hand.

Mirrors should be on a separate wall, and not above wash basins where they make for congestion. Where tablet soap is preferred, it is better to give each worker his own tablet; otherwise liquid or jelly soap supplied from dispensers is preferable. Solvents and alkalis should not be used for removing dirt and stains from the hands as they are liable to cause dermatitis.

Where such staining is apparently unavoidable, reputable suppliers of barrier creams should be consulted with a view to providing a suitable barrier cream which can be applied to the hands and affected parts before starting work.

Common towels which may be used by several workers are best avoided. Absorbent paper towels or warm air blowers are preferable, or individual towels may be supplied to each worker. The employer is, however, responsible for the laundering and changing of towels, which should be done at least once a week.

Waste bins should be provided, and adequately sized door mats at each entrance are useful. Proper arrangements must be made for cleaning, maintenance and replacement of soap, towels, etc.

Shower baths provided with hot and cold water are the most practical form of baths and should be provided for workers engaged in hot and dusty work even where there is no legal requirement. One shower should be provided per 5 to 15 workers depending upon the nature of the work. Each shower cubicle should have a door or waterproof curtain to ensure privacy and prevent water splashing out into the rest of the room. The floor space within a shower bath should not be less than three feet by four feet, and the partitions between adjacent baths should not be less than six feet high. The floors should have an impervious non-slip surface and slope slightly to a drain at the back of the shower bath. The hot water is best thermostatically controlled at a maximum temperature of 60°C, but employees should be able to turn off the hot water and finish with a cold shower. Individual bath towels are essential where shower baths are employed. Proper arrangements must be made for their issue, return and laundering and drying.

If the canteen is close to the locker/dressing/wash rooms, it may be possible for employees to use the washing facilities there before meal breaks, but if not it is essential to have proper wash rooms located between the workplace and the canteen. For workers handling dirty or toxic materials, wash basins should be provided in or close to the workrooms. These points are covered in several regulations which affect particular industries.

In areas where chemicals are stored, handled or used, emergency flood showers and eye wash fountains should be available close to the points where accidental spillages are most likely. The showers should have a specially designed deluge head, a quick-opening valve, and their location should be clearly visible and indicated by a green light above the shower. The water pipe must be protected against freezing, and the showers must be regularly inspected and tested to ensure that they function properly. Access to these emergency showers must be as direct as possible and free from tripping hazards and other obstructions. A person who needs to use an emergency shower is usually in a state of shock and often partially blinded by the acid or chemical with which he has been in contact. In several recorded cases where men were injured by acid splashed and sprays, the emergency shower failed to operate when required, thus aggravating the injury.[3]

REFERENCES

1. Health and Safety Executive, *Booklet 5, Cloakroom accommodation and washing facilities*, 2nd ed., HMSO, London, (1968).
2. National Safety Council, *Accident prevention manual for industrial operations*, 7th ed., National Safety Council, Chicago, (1976).
3. King, R., and Taylor, M., 'Post accident investigation', Symposium paper 18, Eurochem Conference, *Chemical Engineering in a Hostile World*, (June 1977).

4 FIRE, EXPLOSION AND IMPLOSION HAZARDS

4

Fire, Explosion and Implosion Hazards

INTRODUCTION

The hazards of fires and explosions are often grouped together, perhaps because they are often initiated in the same way, i.e. an officer in ordering his men to shoot shouts 'fire!' However, fires and explosions are very different things, fire being related to heat and combustion while explosion is related to blast and shock waves. Fire can be fought by individuals often successfully, but fighting explosions is best left to automatic suppression systems.

Two other words which are often confused and used interchangeably are 'prevention' and 'protection'. In theory at least, preventative measures are those taken to eliminate the hazard (whether it be fire or explosion) at source, and protective measures are those taken to mitigate its consequences once it has materialised. Of course there is no hard and fast borderline, since protective measures will include those intended to prevent the fire or explosion from spreading. But it is worth making the effort to use them correctly. Fire has been used by man since early times and was probably the principal factor which enabled man to spread after the Old Stone Age to the colder regions of the earth.[1] Explosives are a more recent invention. Hazards arise through the unintentional or malicious initiation of fires and explosions or from those started for some useful and peaceful purpose which get out of control.

It seems unncessary to define fire other than as a combustion process involving the transport of atmospheric oxygen to the combustion zone, its reaction with combustible materials there, and the removal of gaseous products of combustion. There is perhaps a somewhat ill-defined borderline between lower temperature oxidative degradations, cold flames and slow combustion processes and true fires. It is also true that fires can happen in atmospheres free of oxygen, i.e. a chlorine flame in hydrogen. These refinements, however, will be ignored for the present.

Explosion is a fairly loosely defined word, and when considering it we should not forget the word coined to express its opposite, *'implosion'*. The latter was coined by Sir Wyville Thomas in about 1880 following the results of an experiment in which he lowered a sealed glass tube containing only air and surrounded by a copper cage to a depth of 2000 fathoms in the middle of the ocean on the end of a long piece of string. On raising this to the surface he found in place of the glass tube a small heap of powdered glass and fragments.[2] From this he deduced that the glass tube had burst inwards as the result of external pressure and coined the term 'implosion'.

There were no witnesses to the event and the evidence was purely circumstantial; nevertheless his theory was quietly accepted as fact.

Since the hazards of implosions are easier to describe and deal with than those of explosions or fires, they will be dealt with first (section 4.1) followed by those of explosions (section 4.2) and finally fires (section 4.3).

REFERENCES

1. Bernal, J.D., *Science in History,* Penguin Books, London, 183 (1964).
2. *The Oxford English Dictionary.*

4.1 IMPLOSION AND SIMILAR HAZARDS

Contents

Quite a few implosions occur in closed tanks, and even pressure vessels, on *terra firma* as the result of partial vacuum caused by the condensation of steam or other vapours. The worst injuries suffered are usually to the pride of the person who allowed the accident to happen.

Such accidents to pipes or vessels will occur whenever the external pressure *exceeds* the internal pressure by a margin greater than that which the vessel is capable of withstanding. Many people seeing a sturdy vessel capable of withstanding a pressure of several atmospheres automatically assume that it is equally capable of withstanding a vacuum. This is far from true as any pressure vessel designer will tell you.

A metal pipe or vessel may be immensely strong under internal pressure, where the metal is under tension, yet fail easily under external pressure due to buckling. This is accentuated if the pipe or vessel is at all out of round, and the buckling force increases as distortion progresses. Two simple rules will generally prevent this happening:

1. A check should be made on all closed vessels and large pipes in the works or factory as to whether they are certified to withstand full vacuum at maximum and minimum operating temperature and at ambient temperature. Certified equipment should be recertified by an authorised inspector whenever the vessel is inspected under various legislation or to meet insurance requirements.

2. All vessels and pipes which are not certified as capable of withstanding full vacuum should be fitted either with a permanent vent, without isolation valve, or with vacuum valves designed to admit air when the internal pressure falls below the external pressure by more than the safe working vacuum which the vessel is certified to withstand. The vents and/or vacuum valves should be designed to cope with the maximum rate of vacuum formation in the vessel or pipe, and be checked and tested in accordance with the same stringent procedures that apply to pressure relief valves. They must be connected to the top of the vessel and not to some point below the liquid level.

If the contents of the vessel or tank are sensitive to air and deteriorate in contact with it, then some inert gas blanketing system should be used. This however is not infallible since the capacity of the system to supply inert gas may not be equal to the rate at which pressure falls in the vessel.

The following incidents are typical of vessel or pipe collapse under vacuum.

1. A low pressure storage vessel was used in an oil refinery in the Middle East to hold isopentane, a hydrocarbon with a boiling point of 28°C used as a gasoline blending component. One cold winter night when the contents were being pumped out, the vessel completely imploded with a loud report.
2. A large diameter sea water main made of reinforced plastic and running above ground with an internal pressure of 3 atmospheres, was fed by five large centrifugal pumps in a tidal basin. A power failure caused all pumps to fail. The momentum of water in the pipeline created a vacuum which caused the pipe to collapse, in spite of air admittance valves which had been fitted to cater for such an emergency. They were either too small or had become blocked by marine growth.
3. A large distillation column operating at atmospheric pressure had been steamed out to remove flammable vapours prior to inspection and maintenance *via* a valved vent and drain line. Other connections which led through a condenser to the atmosphere had been spaded off. Through faulty instruction, the operator closed the vent and drain lines when he shut off the steam, and the column collapsed.

Pipes which collapse under water, like toothpaste tubes, are one of the many difficulties faced by North Sea operators, as a recent article has shown.[1] A critical factor causing failure is the eccentricity or out-of-roundness of the pipe.

4.1.1 Cavitation

Cavitation is a phenomenon which is very similar to implosion. It occurs inside the body of a liquid where there is a fast-moving propeller or impeller or an ultrasonic generator which creates localised regions where the pressure in the liquid is lower than its vapour pressure. Small bubbles or voids filled only with the vapour of the liquid are formed. These collapse as they are swept into higher pressure regions of the liquid.

High pressures are created at the point where the opposite sides of the bubble meet suddenly due to abrupt changes of momentum of the liquid at this point. Shock waves and acoustic waves are created, the former being capable of causing severe local damage, especially to the fast-moving propeller or other turbo mechanism. Such parts fail rapidly by 'erosion', although there is no abrasive solid present. *Figure 4.1.1* illustrates a pump rotor weakened by cavitation.

Although the principle of cavitation is employed deliberately in ultrasonic cleaning equipment, the results of unwanted cavitation are nearly always harmful and often dangerous because of failure of the parts. Cavitation is avoided as far as possible by limiting the speed of the rotating part and placing it at a low point in the system or at sufficient depth to provide a net positive suction head (NPSH) in the fluid as it approaches the rotating part.

Figure 4.1.1 Pump rotor weakened by cavitation (Reproduced by courtesy of the Director, National Engineering Laboratory)

In situations where cavitation appears inevitable, only materials possessing good resistance to cavitation damage should be employed. The requirements are:

High hardness.
Good fatigue and corrosion resistance.

Small grain size.
Ability to work harden under repeated stressing.

Cathodic protection is of help in reducing cavitation damage through generating hydrogen which cushions the collapse of the bubbles. The injection of air or other gases into a cavitated region has a similar effect and is used to reduce cavitation damage in some machinery.

Cavitation can also occur in pipes and valves through which hot liquids are flowing, where its effects are very similar to those of water hammer. It also occurs whenever live steam or vapour is injected below the surface of water or some other liquid in a vat, tank or vessel. Its effects are minimised by distributing the steam through a large number of small holes, by carefully securing the vessel and injection tube and by cathodic protection or air injection as mentioned above.

REFERENCES

1. Haagsma, S.J., 'Research and test study. Collapse of sub-sea pipeline', *Oil and Gas Journal*, 54-62 (November 1 1976).
2. *The Encyclopaedic Dictionary of Physics,* (see entries on 'Cavitation in turbo machinery', 'Cavitation, mechanism of' and 'Cavitation pressure and damage'), Pergamon Press (1961).

4.2 EXPLOSION HAZARDS

Contents

Explosion is quite a loose term. According to the Oxford English Dictionary 'to explode' means (among other things) 'to go off with a loud bang, of gas, gunpowder, etc; to expand violently with a loud report under the influence of suddenly developed internal energy, of a charged jar, mine, boiler, gun; to fly in pieces or burst'. Explosions may be variously classed as:

Chemical explosions.
Physical explosions.
Mechanical explosions.
Electrical explosions.
Nuclear explosions.

Chemical explosions — sometimes referred to as thermal explosions — may involve the decomposition of an unstable compound or explosive (which may be a mixture or a single compound), or the explosion of a mixture of air and some fuel, as in the cylinder of an internal combustion engine. The fuel may be a gas or vapour, a cloud of small droplets, or a dust cloud.

A further important distinction in the type of chemical explosion lies in whether it is a deflagration or detonation. A deflagration is propagated at a low velocity (up to about 100 m/sec in a solid) and the burnt products stream in the opposite direction to that of the combustion wave.[1] The term 'low explosives' is used for those that deflagrate. They are mainly used as propellants and have low shattering power unless strongly confined. Examples are, cordite, gunpowder and the gas mixture in the cylinder of an internal combustion engine.

A detonation is propagated at a high velocity in the range 1 to 10 km/ sec and always faster than the speed of sound in the medium concerned. A detonation is a strong shock wave accompanied by a chemical reaction which supplies energy to maintain the wave. The products of combustion stream in the same direction as the wave, and high pressures are developed. Detonating explosives are termed high explosives. Most high explosives have shattering power even when unconfined. They also have characteristic detonating velocities.[1] In a gas the shock wave compresses and heats the unburnt reactant in a way which causes it to react very quickly.

Many substances — gases, liquids and solids — which deflagrate when ignited weakly, will detonate under sufficiently strong ignition. A deflagration wave may also accelerate into a detonation wave.

Flammable gas-air mixtures have a greater tendency to detonate and develop higher pressures when the explosion is confined, but escapes of large quantities of flammable gases and vapours when mixed with air can undergo surprisingly violent and forceful explosions when unconfined. This is the 'unconfined vapour cloud explosion' and is typified by the explosion at Flixborough 1974, Pernis (Holland) 1968, Feyzin (France) 1966 and a number of others in the USA and elsewhere.[2] These have mostly occurred in oil refineries and petrochemical works and in the storage and transport of liquefied flammable gases.

Some unstable chemicals such as acetylene can undergo a double explosion; first one occurs in the material when it is inside a container, giving rise to hydrogen and carbon particles; this bursts the container. The escaping products mix with the surrounding air and can undergo a further explosion if the quantity is large enough or if the mixture is confined in a building.

Physical explosions result from the sudden formation of large quantities of vapour caused by water or some other liquid coming into contact with some hotter material, e.g. molten metal, hot metal or hot oil.

A particularly disastrous physical explosion occurred at Scunthorpe on 4th November 1975[3] when molten steel from a furnace was being run into a transport container which water had also entered through a leak in the cooling system. Molten steel was ejected over a considerable area. Nineteen men were injured, of whom four died immediately and seven later. Considerable damage was caused. The largest physical explosion to be recorded occurred during the Krakatoa eruption in 1883 when the sea poured into a submarine crater containing millions of tons of molten lava.

Physical explosions may take place in oil refineries and chemical plants,[4] especially during the start up of vacuum distillation columns when pockets of water lying in low parts of the columns come into contact with hot oil circulating above them. Damage is usually confined to a few distillation trays inside the column.

It is even possible for physical explosions to begin at the interface between water and a light hydrocarbon lying above it since the mixture may begin to boil at the interface at a lower temperature than the temperature at which either or both of the bulk liquid phases are present.[5] The most familiar example of a physical explosion occurs when water is tipped into a

Figure 4.2.1   The Feyzin explosion (United Press International)

large pan of hot frying oil.

Mechanical explosions simply result from the rupture of a container under pressure such as a boiler, pressure vessel or tyre. Boiler explosions were quite common in the mid-nineteenth century, and resulted in several legislative acts on the design, construction, inspection and use of steam boilers. As a rule, there needs to be a cause such as a transient pressure rise inside a pressure vessel to cause a mechanical explosion. If the pressure vessel contains a hot liquid hydrocarbon at a temperature above its atmospheric boiling point, it is possible to have three different types of explosion in succession:

1. A physical explosion inside the vessel (due for instance to the presence of water).
2. The increase in pressure due to (1) causes the vessel to fracture — a mechanical explosion.
3. The escaping vapour and liquid droplets form an explosive vapour cloud in the air which reaches a source of ignition producing an unconfined vapour cloud explosion.

Another type of mechanical explosion to which liquefied petroleum gas storage and transport vessels are prone is the BLEVE (boiling liquid expanding vapour explosion). This may happen when a partially full vessel is involved in and exposed to a fire. While the lower part of the vessel is kept relatively cool by the boiling liquid inside it, the upper uncooled part is heated to a temperature at which the steel is much weaker. Even when such vessels are fitted with safety valves which open and prevent excessive pressure build-up, the upper part of the vessel may split because of the weakened state of the steel. The pressure of the hot vapourising liquid then propels one (or both) parts of the vessel which may take off like a rocket. The vapour escaping to the atmosphere usually ignites either as a large flash fire or an explosion. One of the worst BLEVEs was that at the Feyzin refinery in France in 1966 when a 2000 m³ storage sphere containing liquid propane under pressure split as a result of a fire round it.[6] Forty-five people, many of them firemen, were killed, at distances of several hundred metres from the sphere which exploded *(Figure 4.2.1)*.

Electrical explosions result from the sudden conversion of electrical energy into heat, generally as a result of the breakdown of an insulator. Perhaps the most common are those reported from time to time in electrical switchrooms where work is being done while some of the circuits are live. The air then generally breaks down as an insulator becomes ionised and conducting. An electrical explosion is nothing more than a very large spark or series of sparks, typified in nature by lightning.

Nuclear explosions are of course the most devastating and feared of all. It is enough to say that they are of two types, depending on whether the energy is derived from the fission or fusion of atomic nuclei.

A fission explosion results when a quantity of fissile materials, Uranium 235 or Plutonium 239, is brought together suddenly into a critical condition. The energy yield may be equivalent to the explosion of up to 100 000 tons of TNT or even more.

A fusion explosion occurs when a light element such as tritium is heated to such a high temperature that the rate of heat evolution through nuclear fusion exceeds the rate of heat dissipation. A fission bomb may give the high temperature required for initiation. The yield may be equivalent to many millions of tons of TNT.

The control of the manufacture and handling of such materials is subject to strict government licensing and security arrangements which place them outside the scope of this book.

4.2.1 Chemical explosions

Chemical explosions can be classified into two groups:

1. Those which involve substances which are legally classed as explosives.
2. Other chemical explosions.

An explosive in the chemical sense is surprisingly hard to define, and some definitions given even by experts are sometimes wide of the mark. Thus Henri Muraeur[7] defines an explosive as 'a definite compound or mixture of compounds capable of giving off in an extremely short time a great volume of gas raised to high temperatures'.

This definition would deny that silver acetylide which decomposes violently when struck to give silver and carbon (both solids) is an explosive.

Watts[8] quotes an unnamed authority in stating the following qualifying conditions for an explosion:

'The reaction liberates heat.
The transformation is one of complete decomposition and hence the explosive system is extremely sensitive.
The decomposition is of such a character as to render an explosion possible by means of any suitable initial impulse or shock'.

Watts[8] goes on to remark that

'There will be borderline cases and it is difficult to give any hard and fast rule whether a substance is or is not an explosive'.

The legal definition of an explosive is contained in Section 3 of the Explosives Act 1875[8], namely:

(1) Gunpowder, nitroglycerine, dynamite, guncotton, blasting powders, fulminate of mercury or the fulminates of other metals, coloured fires and every other substance whether similar to those above-mentioned or not, used or manufactured with a view to produce a practical effect by explosion or a pyrotechnic effect; and
(2) Fog signals, fireworks, fuses, rockets, percussion caps, detonators, cartridges, ammunition of all descriptions and every adaptation or preparation of an explosive as above defined.

4.2.1.1 Manufacture and use of legally-recognised explosives

The complexity and legal ramifications of these subjects are such as to render it impossible to deal with their hazards and safety aspects within the scope of this book. The manufacture of explosives and fireworks is carried out only under special licence and is controlled by the Chief Inspector of Explosives. This Inspectorate is a separate department under the Health and Safety Executive, and firms and personnel affected may be expected to have a sound knowledge of the legal and technical safety aspects. We thus confine ourselves here to those cases where explosives (in the legal sense) are employed in other industries.

The principal regulations[9] which affect such storage and use are:

1. *The Factories Act 1961, Section 146.*
This empowers an inspector 'to enter, inspect and examine at all reasonable times by day and night a factory and every part thereof when he has reasonable cause to believe . . . that explosive materials are stored or used.'

2. *The Construction (General Provisions) Regulations 1961.*
Section 19. Explosives. Explosives shall not be handled or used except by or under the immediate control of a competent person with adequate knowledge of the dangers connected with their use and steps shall be taken to see that, when a charge is fired, persons shall be employed in positions in which, so far as can reasonably be anticipated, they are not exposed to risk of injury from the explosion or from flying material.

3. *The Health and Safety at Work Act etc, 1974.*
Section 33, sub-section 4(c). This makes it an offence to acquire, attempt to acquire, possess or use an explosive article or substance (within the meaning of any of the relevant statutory provisions) in contravention of any of the statutory provisions. The penalty may be a fine of up to £400 or two years imprisonment or both.

Section 44, Subsection 4 of the same Act provides for the forfeiture or destruction of the article referred to above.

4.2.1.2 Cartridge operated fixing tools

Fixing tools operated by explosive cartridges are commonly used in building and civil engineering (*Figure 4.2.2*). Accidents caused through their use are generally serious; this has prompted the Health and Safety Executive to publish a Guidance Note.[10]

Cartridge-operated fixing tools are used to drive metal pins into brick and concrete walls and mild and structural steel plates and sections. Their energy is about the same as that of a bullet from a point 22 rifle. To prevent the pin being projected like a bullet, they should be designed so that the cartridge can only be fired when the nozzle is pressed hard against a surface into which the pin will be fired. Their design and construction are covered in BS 4078[11].

Figure 4.2.2   Cartridge operated fixing tool (Ramset Fasteners Ltd.)

There are two types of tool:

1. *Direct-acting tools.* These are similar to a pistol. The pin is propelled by the compressed gases from the cartridge, which is fired by the action of a spring released by a trigger.
2. *Indirect-acting tools.* The gases released when the cartridge is fired act on a piston which transfers the force to a pin. The cartridge may be fired by a spring and trigger or by a heavy hammer. Hammer-operated tools are quite distinct in their design, construction and use from spring operated tools.

Training of operators who may have to use either type of tool is vital and mandatory. Some manufacturers of these tools not only train operators of their customers but examine their competence and certify them.

The main hazards of cartridge operated tools are dealt with in the following paragraphs.

1. Tool held at angle to working surface, causing pin to ricochet. Most tools are designed so that they can only be fired when the axis of the tool is nearly perpendicular to the surface.

2. Pin hits a hard object within the surface and ricochets internally, causing splinters or the pin itself to fly out. The usual remedy for this hazard is by fitting a splinter guard in the form of a wide ring round the end of the muzzle of the tool.

3. *Wrong cartridge used.* BS 4078:1966 specifies a colour code to indicate the strength of the charge in a cartridge. Under this, brown signifies 'extra light charge'. Colour codes of tools and cartridges made in other countries differ. Imported cartridges for one foreign tool are marked brown to signify 'extra high charge'.

Safety specialists must be alert to this hazard and try to ensure that only one make of tools and cartridges is used within their organisation. Operators should not rely on the colour marking but check the charge from the packing label.

Apart from this, the surface may differ from that which the maker had in mind when recommending certain cartridges. Whenever a new surface-pin combination arises, the maker's recommendations should be checked with trial fixes using the weakest cartridge first.

4. *Use of unsuitable pins.* A wide range of pins of different shape and hardness is available for use on different surfaces. An unsuitable pin may shatter instead of penetrating the working surface and produce dangerous fragments. The maker's recommendations should be studied, and the maker's choice of pin put to the test before being regularly adopted for a particular job.

5. *Pin passes through wall, panel, etc hitting someone on the far side.* This is generally the result of an unsuitable choice of cartridge on a thin material. To guard against such mishaps, the area on the far side of the wall, etc where a cartridge tool is being used should be fenced off and a warning notice posted.

6. *Hard or brittle materials.* Cartridge tools are unsuitable for use on hard or brittle materials with little ductility — i.e. tile, marble, or cast iron — as well as on many heat-treated steels.

7. *Maintenance, cleaning and misfires.* Maintenance and cleaning of tools should only be carried out by someone who has undergone special instruction on the tools in question. To ensure that his skills are recognised he may be given the special status of armourer. Tools which have misfired should be dealt with in the same way.

8. *Storage.* Each cartridge-operated tool should be stored in its own rigid locked case or box with compartments for the splinter guard, operating and maintenance instructions, goggles, cleaning tools, cartridge boxes and pins. Bulk supplies must be stored in accordance with the requirements of the Explosives Act and local and/or Police Authorities.

9. *General.* If a tool misfires, the operator should hold the tool in position

for 30 sec and then try to fire it a second time. If it still misfires he should hold it in position for another 30 sec before following the maker's instructions to remove the cartridge. Pins should not be driven into drilled holes unless a special adaptor is used. Anyone using a cartridge tool must do so from a firm and stable position, since the tool may have considerable recoil.

Cartridge tools should not be used in areas where there is flammable vapour or risk of dust explosion. Operators should be provided with a safety helmet, eye and ear protectors.

4.2.1.3 Spontaneous explosions of chemicals not recognised as explosives

This section deals mainly with explosions of chemicals during transport, storage and use. An even wider range of explosive possibilities (e.g. runaway reactions) occurs in chemical manufacture. These are discussed briefly in section 6.2.2.3.

The incidents referred to here are those in which some unstable compound, which may have been produced either accidentally or on purpose, explodes spontaneously without the involvement of atmospheric oxygen or some other chemical. It is also assumed that the incident takes place under normal or nearly normal conditions of storage, transport or use, without the intervention of fire or other comparable hazards, and that the substance is not legally classified as an explosive.

The whole subject is fairly specialised and to appreciate it properly a sound knowledge of chemistry is required. This category includes a large number of compounds most of which are fortunately encountered only in laboratories. A number are also encountered as intermediates or by-products in various chemical industries and a few are found in other industries.

Some compounds are stable at low pressures but liable to explode at high pressures. Others may decompose spontaneously under normal conditions at a very low and almost negligible rate. When present in great bulk however, the heat generated by this slow decomposition is unable to escape and raises the temperature of the material in the centre to a point where the speed of decomposition accelerates to an explosion. The decomposition of some compounds is promoted by small traces of catalysts, e.g. acids, and retarded by inhibitors. Some normally stable compounds can be transformed by the influence of air, sunlight, etc into explosively unstable materials. Sometimes the products of decomposition catalyse the decomposition reaction, thereby accelerating the reaction until it becomes an explosion. Some, again, may be exploded by an extremely powerful impact.

Some of these materials which were once common household articles have been replaced by safer products. There is the cryptic story of the flamenco dancer in the twenties, wearing artificial silk stockings of nitrocellulose which were dyed with picric acid, whose act terminated in an unexpected burst of flame which was initiated by a particularly vigorous heel tap! Some of the more common materials which can present this hazard are given below.

Acetylene. This is a gas used widely for welding and cutting. Alone, it can be detonated at pressures of about 2 atm and higher. Acetylene also forms copper acetylide in contact with moist copper and copper alloys. This compound is a sensitive explosive which is readily set off by friction when it is dry. Thus, if acetylene were kept in contact with moist copper at moderate pressures, and the moisture then removed, an explosion would be probable.

Acetylene is actually a prohibited explosive under Sections 104 and 43 of the Explosives Act, Order in Council No.30 amended by Statutory Rules and Orders 1947, No.805 (Appendix 6), with certain exceptions.[8] These exceptions include the limited range of conditions under which acetylene is customarily stored and used in industry for cutting and welding. Those responsible for the safe use of acetylene in industry should be familiar with the limits to the conditions for its safe use.

Besides common acetylene, there are several other acetylenes which are explosive. Perhaps the most hazardous one encountered in industry is vinyl acetylene which is made by dimerisation of acetylene and as a by-product of ethylene manufacture, and used in the manufacture of some synthetic rubbers.

Ammonium Nitrate. Ammonium nitrate is the major component of many blasting explosives and a principal component of many fertilisers. Although not classified as an explosive in the pure state, it can decompose explosively. There have been several major explosions of ammonium nitrate in bulk storage and in transport in ship's holds.

Ammonium nitrate apparently decomposes very slowly under ordinary conditions, but the heat build up when it is stored in bulk can initiate an explosion. There is a critical mass above which this is liable to happen. Special precautions for storage must be taken. These include limitation of stack size, adequate air spaces and ventilation. No hydrocarbons or other oils may be allowed near ammonium nitrate.[12]

Benzoyl peroxide. See Organic Peroxides.

Butadiene and other dienes (isoprene and cyclopentadiene). These low boiling hydrocarbons, used in the manufacture of synthetic rubbers and resins, readily form explosive peroxides when in contact with air. The hydrocarbons boil off rapidly leaving the concentrated peroxide.

Dinitro-compounds (Dinitro benzene, dinitrotoluene, dinitro cresol and dinitro phenol). These compounds which are made as intermediates in the manufacture of fine chemicals decompose explosively under a strong stimulus. Dinitro phenol which is classified as an explosive is exempted by Order in Council No. 27 if it is handled as a mixture with water containing not less than 15% weight of water.

Ethers. Many ethers, particularly di-ethyl ether, form explosive peroxides when left in contact with air and light.

Ethylene oxide is handled as a liquefied gas under pressure and undergoes explosively violent reactions with many compounds. It is capable of exploding by itself when heated, and also of spontaneous exothermic polymerisation which can create the temperature required for an explosion.

Hydrogen peroxide is generally handled as a 20% — sometimes 50% — solution in water which has been stabilised by the addition of a tin compound. It reacts violently, sometimes explosively with many organic compounds, the violence depending on its concentration. It is also decomposed vigorously by finely divided platinum and manganese oxide with formation of oxygen gas. At concentrations above 80% in water, hydrogen peroxide can explode.

Nitroparaffins (Nitro methane, nitro ethane and nitro propane). These liquids used as paint and lacquer solvents and as strippers can deflagrate explosively, nitro methane being the most explosive. They are also mild lung irritants.

Organic peroxides.[13] Organic peroxides may be classified as (a) those manufactured, stored and used deliberately — often as a polymerisation catalyst — and (b) those formed accidentally by the reaction of some organic compound with air.

Most organic peroxides are hazardous and some are explosive. They are classified according to their fire and explosion hazard as follows:[14]

Class I. These decompose explosively with little provocation. This group includes peroxides which can be handled safely at certain controlled temperatures or in solution but which become highly sensitive if they crystallise or become concentrated.
Class II. These peroxides present an intermediate explosion hazard which is less violent than Class I.
Class III. Have moderate explosion but severe fire hazards, burning rapidly with evolution of intense heat. Their decomposition products may form explosive mixtures with air.
Class IV. Do not explode and have moderate fire hazards which can be dealt with by normal sprinkler systems.

The manufacture, storage and use of peroxides in Classes I to III require stringent safeguards which are in every way as stringent as those required with explosives. Lists of peroxides showing their grouping and safeguards are given in references 14 and 15.

Strict precautions must also be taken in handling materials such as ethers and dienes which form peroxides by contact with the air. These may include the addition of inhibitors to suppress peroxide formation and storage under an inert gas blanketting system. There must also be procedures for dealing with concentrated peroxides which have been allowed to form inadvertently. Most dangerous peroxides are decomposed by alkalis. The most common way of dealing with them is to dissolve them in a suitable solvent to which an alkali has been added.

Sodium chlorate. Sodium chlorate on its own, although a strong oxidising agent which liberates oxygen on heating, is not generally considered as an explosive. However, it readily forms explosive mixtures with most organic materials such as sawdust, coal dust, flour or sugar. These mixtures may be detonated by friction or shock. If packed in paper, this may become explosive if wetted and then dried, so all packing materials should be promptly burnt. Contact with mineral acids liberates chloric acid which is spontaneously explosive.

4.2.1.4 Chemicals which react violently or explosively with water[16, 17]

A number of chemicals react violently with water. Some liberate flammable gases which instantly burn or explode, others liberate flammable gases which may burn or explode later, and others react with great evolution of heat which may cause the mixture to boil.

A short list of the most common chemicals which react violently with water, and the effects produced are given in *Table 4.2.1*. Strict precautions must be taken to keep all such chemicals sealed in strong air and watertight containers and stored in a dry place which is not subject to flooding. They should only be handled by trained operators, wearing appropriate protection for the eyes and body and following carefully worked out safe procedures. A common danger with these materials is to either leave a half-full container improperly sealed and exposed to wet, or leave the chemicals in places where people who are not aware of the hazards may find them.

All such containers should of course be properly marked with warning and identification signs; but there is often a danger that these can become detached or obscured through exposure to wet or corrosive atmospheres.

4.2.1.5 Explosions in mixtures of chemicals not recognised as explosives [16, 17, 18, 19]

A great many reactive chemicals which are not themselves explosives are liable to explode when mixed with some other chemical with which they can react. Most rocket fuels are in this category. Explosions between hydrogen and chlorine are familiar from school chemistry courses.

Unfamiliar chemicals should, therefore, never be mixed. Qualified chemists who are familiar with their properties should always be consulted first, and in case of doubt, asked to carry out the proposed operation personally first on a small scale in a laboratory.

The handling of liquid oxygen is hazardous in many ways, not least through the ease with which combustible materials — oils, greases and flammable gases and liquids can become inadvertently mixed with it.

Another common hazard lies in the reaction between finely-divided aluminium powder used for example as a paint, and iron oxide which may be present as rust. This can sometimes be initiated by a sharp blow causing the material to react and become red hot and igniting any combustible

Table 4.2.1 MATERIALS WHICH REACT VIOLENTLY OR EXPLOSIVELY WITH WATER

A. MATERIALS WHICH LIBERATE HYDROGEN

Material	Formula	State	Action
Calcium	Ca	Solid	Moderate
Lithium	Li	Solid	Moderate
Sodium	Na	Solid	Vigorous, may explode
Potassium	K	Solid	Usually explodes
Calcium hydride	CaH_2	Solid	Moderate to vigorous
Lithium hydride	LiH	Solid	Moderate to vigorous

B. MATERIALS WHICH LIBERATE HYDROCARBON

Aluminium tri-ethyl	$Al(C_2H_5)_3$	Liquid	Vigorous, ethane formed, generally explodes.
Aluminium di-ethyl mono chloride	$Al(C_2H_5)_2Cl$	Liquid	
Calcium carbide	CaC_2	Solid	Moderate to vigorous, acetylene formed which may detonate if confined.

C. MATERIALS WHICH LIBERATE OXYGEN

Fluorine	F_2	Gas	Violent, ozone and acid fumes formed.
Sodium peroxide	Na_2O_2	Solid	Moderate on heating

D. MATERIALS WHICH GENERATE STEAM AND ACID FUMES

Acetyl chloride	Liquid	Silicon tetrachloride	Liquid
Aluminium chloride	Solid	Sulphuric acid	Liquid
Phosphorus pentachloride	Solid	Sulphuryl chloride	Liquid
Phosphorus pentoxide	Solid	Thionyl chloride	Liquid
Phosphorus trichloride	Liquid	Titanium tetrachloride	Liquid

E. MATERIALS WHICH GENERATE STEAM ONLY

Calcium oxide — quicklime	Solid	Activated silica	Solid
Activated alumina	Solid	Activated molecular sieves	Solid
Sodium hydroxide	Solid	Potassium hydroxide	Solid

material in contact with it. Some clue that a chemical may react violently (with air, water or other chemicals) is often given by the label. This is shown by the code letters P, S, W or Z on the Hazchem label (see section 4.3.2.6) and by a high number (3 or 4) in the appropriate box of the NFPA label.

4.2.2 Explosions of flammable gas or vapour mixtures with air[20]

Explosions of this type are frequently accompanied by fires and occur under similar circumstances. They may be sub-divided into:

1. Confined explosions in buildings, boilers, process plants, etc.
2. Unconfined vapour cloud explosions.

Dust explosions will be discussed in a subsequent section.

Both confined and unconfined vapour cloud explosions usually originate in a leak of gas or vapour or vapourising liquid into the atmosphere. However whilst quite small leaks can cause explosions inside buildings, an extremely large leak is needed to result in an unconfined vapour cloud explosion out of doors.

4.2.2.1 Confined gas or vapour explosions in buildings[21]

The usual culprit here is town gas, which is mainly methane although any flammable gas or vapour which is present inside pipes, vessels, etc and which escapes may explode.

Table 4.2.2 lists a number of common flammable gases and volatile liquids which may become involved in gas and vapour explosions, together with some of their properties which affect their behaviour in an explosion, i.e. their densities, relative to air, their lower and upper flammability limits in air, their ignition temperatures and their gross calorific values which give some indication of the relative force of an explosion. The maximum pressure generated by the explosion of gas-air mixture in a confined space lies between 6 and 9 atm when the mixture is at atmospheric pressure before the explosion.

The gases and vapours are listed in order of increasing gas density. The first three are lighter than air and tend to rise, the second three have nearly the same density as air and show little tendency to rise or fall, and the remainder all have higher densities and tend to form a layer near the ground and accumulate in pits, trenches, basements and other depressions. The most dangerous gases and vapours are those with low ignition temperatures (readily ignited), wide flammability limits, high calorific values and those which cannot readily be detected by smell.

The flammability limits define the concentration limits within which the gas-air mixture will deflagrate. If the gas concentration is just within these limits, the explosion will be a deflagration which will cause sufficient rise in pressure to break windows and doors and usually walls as well. However, if the gas concentration is well within the flammability limits, a detonation becomes possible. This has far greater shattering power than a deflagration.

Flammable gas-air mixtures have limits of detonability which have in many cases been determined and which lie within the flammability limits. For example the flammability limits for hydrogen lie at 4.0 and 94% hydrogen by volume and the detonability limits lie at 15 and 90% hydrogen by volume in air.

The gas most commonly involved in explosions as an air mixture in confined spaces is natural gas, which is predominately methane. This is followed by the liquefied petroleum gases propane and butane which are widely used as piped gases, usually premixed with some air, in areas where natural gas is not available. All these fuel gases are normally 'stenched' and

Table 4.2.2 COMMON FLAMMABLE GASES AND VAPOURS. EXPLOSIVE LIMITS AND OTHER PROPERTIES IN ORDER OF INCREASING VAPOUR DENSITY

Compound	B.P. °C at 760 mg Hg	Sp. Gr. air = 1	Flammability Limits in air (% volume)		Ignition temp. (°C)	% of LEL for full scale reading	Gross calorific value (k cal/kg)
			Lower	Upper			
Hydrogen	−253	0.07	4.0	74	560		33 940
Ammonia	−33	0.5	15.0	27	630		5 370
Methane	−161	0.55	5.0	15	538	88	13 260
Acetylene	−84	0.9	1.5	100	305		11 930
Carbon monoxide	−205	1.0	12.5	74	605		2 430
Methanol	65	1.1	6.0	36.5	455		5 420
Propane		1.5	2.1	9.5	466	62	12 030
Ethylene oxide		1.5	3.0	1000	429		
Ethanol	78	1.6	3.3	19	425		7 100
Butane	−1	2.0	1.5	8.5	365	60	11 830
Acetone	56	2.0	2.1	13.0	535		7 370
Vinyl chloride	−14	2.1	3.6	33	472		
Di-ethyl ether	35	2.5	1.85	48	170		8 800
Carbon disulphide	46	2.6	1.0	60	100		3 240
Benzene	80	2.8	1.2	8	560		9 990
Hexane	69	3.0	1.1	7.5	233		11 540
Toluene	111	3.2	1.2	7.0	508		11 150
Styrene	145	3.6	1.1	8.0	490		9 685
Xylenes: mixed	142	3.6	1.0	7.6	480		10 280
Octane	127	3.9	0.84	4.7	210		11 140
Petrol (gasoline)		4.0	Variable				

Figure 4.2.3   Explosion damage at Brentford, London in December 1976 after a gas explosion which injured twenty-eight people (Popperfoto)

contain sufficient of a foul smelling additive to make a person with a normal sense of smell aware of the gas at a concentration between 10 and 20% of the lower explosive limit.

A number of serious explosions of natural gas-air mixtures have occurred in buildings in recent years — particularly after long holidays (Christmas and New Year) when the building was locked and closed and unattended. *Figure 4.2.3* shows explosion damage at Brentford in 1976. An explosive atmosphere has gradually built up as a result of a small escape which may have been the result of a gas leak or the extinction of a flame which had no flame failure device. Such explosions are far less common when the building is occupied since the escape is normally detected by smell long before an explosive concentration has built up. However, if the leak is small enough, it is theoretically possible for the gas to build up so slowly that the sense of smell becomes accustomed to the stenching agent before it reaches the threshold. In this case an explosive concentration can build up un-noticed.

Good standards of plumbing and housekeeping and the implementation of a strict procedure to close valves at the incoming main whenever a building is left unoccupied for longer than a specified period should prevent most risks of such gas explosions.

4.2.2.2 Flammable gas detectors

In some plants where buildings are lightly manned or where the type of an equipment makes an escape more likely (e.g. gas turbines), flammable gas

*Figure 4.2.4 Flammable gas detector showing control module and detector head
(Dectection Instruments Ltd.)*

detectors should be installed (see *Figure 4.2.4*). Paint spraying, drying
painted or lacquered articles, the mixing and use of flammable liquids,
battery charging and solvent extraction are among the other processes
where they should be considered. Guidance is given in the Health and
Safety Executive's Guidance Note CS 1[23]. Most flammable gas detectors
are of the catalytic combustion type. There are a number of points regarding their use which require careful attention[24]:

1. The detector heads, wiring and monitoring and alarm instruments
   should comply with the electrical safety requirements of the areas
   where they are to be installed.
2. Most detector heads only give a linear response with increasing
   concentration of flammable vapour up to about the lower explosive
   limit. Above the concentration required for complete combustion, the
   instrument reading actually falls, thus giving a completely false sense
   of security. (Fortunately such high concentrations are seldom
   encountered).
3. The positioning of the heads requires careful study, which takes into
   account the possible sources of escape and the movement of gas after
   escape. Often too few heads are installed to pick up a strong but
   directional escape at an early stage.
4. Instruments should be calibrated to give 100% reading at the lower
   explosive limit for the particular gas they are intended to detect.

5. A check should be made to ensure that no vapours which might poison the pellistor are liable to be present — examples are organo compounds of lead, silicon and phosphorus.
6. When specifying detector heads, the response time required should be carefully considered and included in the specification. With some detectors the response time is too slow to deal with the emergency envisaged.
7. When planning an installation which employs flammable gas detectors, provision should be made for their regular testing and servicing by competent people.

Flammable gas detectors may be used not only to give warning when a certain percentage (usually 30%) of the lower explosive limit has been reached, but also to initiate positive measures such as increasing ventilation, closing valves, shutting down machinery and blanketting with inert gas or halons when a second step on the hazard ladder — say 60% of the LEL — has been reached. Such precautions are justified on expensive, highly automated and generally compact installations such as oil and gas production platforms. A regular emergency procedure must in any case be worked out for action when a dangerous gas concentration is detected. It is no use waiting until a dangerous gas concentration has been reached before thinking what action to take.

4.2.2.3 Explosions inside gas or oil-fired plant

Serious explosions all too often occur in gas or oil-fired industrial plant, caused either by the ignition of accumulations of unburnt fuel or by the ignition of flammable solvent vapours released from the articles heated. The danger is greatest when the working temperature in the plant is not high enough to ignite the fuel. Most of these explosions occur when the plant is being started up. Safety depends on sound design and construction, careful training of operators and the adherence to clear lighting and shutting down procedures.

Detailed guidance which covers a number of situations is given in the Health and Safety at Work Booklet No. 46[22]. Vapour explosions in this type of equipment may occur with fuel oils whose flash points are well above ambient temperature, when the temperature inside the plant is considerably higher and above the flash point of the oil.

General design features which should be checked whenever a low temperature gas or oil fired plant is procured or designed are:

1. The operator should be able to see the burners clearly without performing acrobatic feats. In awkward cases, mirrors may have to be installed.
2. The plant must be designed so that it can be lit safely by a definite means of ignition which is preferably incorporated into the design.
3. Appropriate flame-guards (sometimes known as flame protection devices) should be incorporated to ensure that no fuel can be

delivered to a burner unless a flame or other source of ignition is present and sufficiently close to the burner to ignite the fuel.
4. Explosion relief devices should be provided wherever practicable, particularly when the enclosed volume, where an explosive mixture may be present, exceeds 2 m³.
5. The number of burners should be kept to the minimum; where several burners have to be employed a specially-designed single pilot burner which will ignite all of them should be used wherever possible.
6. The plant should be designed with adequate combustion ventilation.

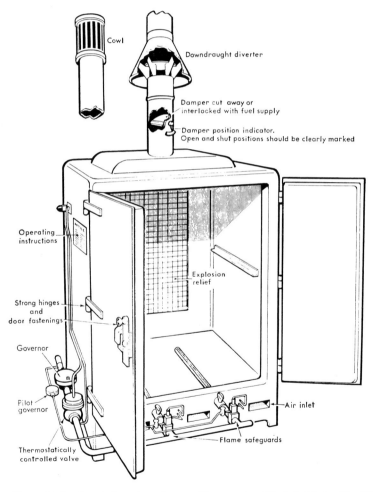

Figure 4.2.5   A typical gas heated oven with protection (H. & S. W. Booklet 46)

This must allow the fuel to be burned completely and the products of combustion removed both during normal operation and during start-up while the plant is still relatively cold. It must be possible to purge the plant with five times its volume of air before lighting the burners.

7. For ovens where solvents are removed by evaporation, mechanical ventilation should be provided, so designed that the concentration of vapour in the oven atmosphere is always well below the lower flammable limit (see *Figure 4.2.5*).
8. Fuel gas and oil lines should as far as possible be welded steel, and have the least number of joints. These should as far as possible be flanged. Screwed joints should only be permitted on small lines (1 in and less) with low operating pressures and sufficient unions should be supplied for proper installation.
9. Positive leak tight isolation valves either of the plug cock or ball cock type must be provided on the fuel supply. They should be securely mounted, using brackets where necessary, to ensure that a joint in the fuel supply line is not loosened when the lever handle is moved to open

4.2.2.4 Unconfined vapour cloud explosions

An unconfined vapour cloud explosion is the aerial explosion of a large cloud of flammable vapour which has been suddenly released from an appropriate source.

Such sources occur in oil refineries, petrochemical works and in the transport of liquefied flammable gases. They have involved many tons of flammable vapour with explosive forces equivalent to between 2 and 50 tons of TNT. An unconfined vapour cloud explosion is usually a major emergency. About twenty of the largest explosions to have so far occurred caused damage estimated on 1977 prices at over 10 000 000 dollars per explosion. The Flixborough disaster of 1974 was one of these (see *Figure 4.2.6*).

Most installations, from which an unconfined vapour cloud explosion might originate, are classed as 'notifiable installations' and would be subject to the scrutiny of the Risk Appraisal Group of the Health and Safety Executive. The following types of installation have each featured in more than one vapour cloud explosion:

Crude oil tanks which were fitted with steam coils in the base.
Large pressurised LPG storage vessels — spheres, etc.
Large road and rail LPG tank cars.
Ethylene plants.
High pressure polyethylene plants.
Cyclohexane oxidation plants.

The types of material which have most commonly given rise to unconfined vapour cloud explosions are:

Figure 4.2.6   Some effects of Flixborough explosion

1. Liquefied petroleum gases.
2. Light hydrocarbon gases present in processes at high pressures (over 35 atm) and in large quantity.
3. Hydrocarbons containing from 5 to 9 carbon atoms per molecule which are contained as liquids in vessels under moderate pressures and at temperatures substantially above their boiling points at atmospheric pressure.

Steps which may be taken to minimise the probability and consequences of an unconfined vapour cloud explosion include:

1. Redesigning the plant or equipment to minimise the quantities of LPG and other flammable liquids contained under pressure at temperatures above their atmospheric boiling points.
2. Avoiding the use of steam coils or other means of heating in crude oil tanks.
3. High standards of design, construction, inspection, monitoring and operation for hazardous plants.
4. Operability and hazard analyses for hazardous plants.
5. Generous spacing between hazardous plants, and blast-proof construction for control buildings which have to be located near them.

The above are mainly general measures. Elimination of sources of ignition is usually impractical in preventing vapour cloud explosions, mainly because of the distance an explosive vapour cloud may travel before meeting a source of ignition and exploding. This has amounted to about half a mile in some cases, and probably much further in one or two. The most it is likely to achieve is to ensure that 'someone else catches it'. Elimination of possible causes of escape — most of which involve the initial rupture of a vessel — is a more fruitful type of attack. Some of the causes of vessel rupture are given in the following paragraphs:

1. *Exposure of vessel to fire*. The fitting of relief valves calculated to relieve all the vapour formed inside a vessel at a certain pressure when the vessel is surrounded by fire is not enough. One effect of the fire will be to heat and thereby weaken the upper walls of the vessel which are not in contact with liquid inside. So the vessel can fail at a pressure below that at which the relief valve is set to open.

2. *External mechanical damage*. This applies particularly to LPG in transit. At least three vapour cloud explosions occurred as a result of accidents to LPG rail cars in shunting (or 'humping') operations in US goods yards. The moral of this is that LPG containers in transit should be handled with every bit as much care and attention as that given to high explosives.

3. *Sudden rapid pressure rise in container with which the pressure relief system cannot cope*. Such sudden rises are known technically as 'transients'. They may last for only a few milleseconds — which had lulled many people into a false sense of security. (It only takes a millesecond for a bullet to pass through a man's heart.) There are three principal causes of such transient pressure rises:

(a) *Internal chemical explosions or very fast chemical reactions caused by the release of accumulated chemical energy*. These could be due to the sudden decomposition of an accumulation of some unstable compound such as a peroxide or acetylide which has formed perhaps as an unwanted by-product of the process. Another cause is the sudden but delayed start of a normal and wanted reaction which, for some reason, had failed to take place when the reactants were first brought into contact, e.g. loss of stirring or catalyst.

(b) *Physical explosions or eruption caused by the sudden boiling at the interface between two liquid phases in a situation of 'latent superheat'*. This hazard may be introduced by fitting a steam coil in a crude oil storage tank (Pernis 1968) or removing a stirrer from a reactor (Flixborough 1974).

(c) *Water or liquid hammer*. This is discussed briefly in a subsequent section.

4. *Earthquakes and other Acts of God*. The damage caused by a severe earthquake generally overshadows even that of an unconfined vapour cloud explosion. If building in earthquake zones is unavoidable, then special engineering standards should be used. The building of hazardous plants in such areas should be avoided It is not unknown for large LPG

containers to have been overturned and ruptured by a hurricane or typhoon, although this is one of the few instances where an unconfined vapour cloud explosion is unlikely, because of the speed with which the escape is diluted and dispersed.

4.2.3 Dust explosions[25,26]

Most combustible solids in the form of a fine dust are capable of forming an explosive mixture when dispersed in air. As in the case of flammable gases and vapours, there are lower and upper dust concentrations in air within which an explosion is likely to occur. The extreme limits for a wide range of dusts are between 20 and 500 grams per litre, although the average lower explosive limit for most combustible dusts is about 40 grams per litre. In appearance such a dust cloud resembles a very thick fog.

Dust explosions generate pressures up to about 8 atm gauge; (there is no meaningful lower limit). The size of the dust particles is important in determining the force of the explosion and whether one will occur. The most severe explosions occur with particle diameters between 10 and 50 microns. Coarser dusts with particle diameters above 200 microns present little or no explosion risk (see *Figure 4.2.7*).

Almost all combustible materials can form explosive dust clouds. These include light metals such as aluminium and magnesium, plastics such

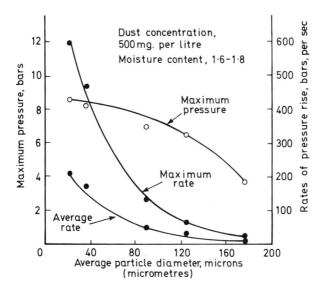

Figure 4.2.7   Effects of particle size on explosion pressures generated by clouds of starch dust (Based on US Bureau of Mines Report of Investigation 4725)

as polystyrene, cellulose acetate, urea formaldehyde resins and most others, agricultural products such as flour, sugar, cocoa, coffee and dust from grain, and many miscellaneous materials and chemicals such as coal dust, wood flour, sulphur, aluminium stearate, rubber and cork dust.

Explosive dusts are normally encountered inside plant and processes and not in work rooms themselves, although a fire or explosion in a work room may readily 'stir up' thick dust deposits which have settled on beams, ledges and other surfaces into a dust cloud which can explode. These are usually secondary explosions; their elimination lies largely in good housekeeping and cleaning and on good design which makes such deposits less likely to form and easier to remove. The use of unsuitable cleaning methods such as compressed air does on the other hand increase the probability of a dust explosion.

The causes and elimination of dust explosions are well described in Booklet 22 in the Health and Safety at Work Series, which includes a useful bibliography.[25]

Since the first dust explosion generally occurs inside a well-defined piece of plant or equipment where its possibility can be expected, the elimination of ignition sources within the equipment is important. A great deal can also be done to design equipment and instal protection systems which minimise both the likelihood of dust explosions and the damage caused when they occur.

Dust clouds are less easily ignited than most flammable vapours, and although they can be ignited by a powerful spark, a flame or hot surface is generally needed.[27] Ignition temperatures for dust clouds range from little over 300°C for products such as cellulose acetate to over 700°C for coal tar pitch. Sulphur has an exceptionally low ignition temperature of 190°C. If, however, the plant in which a dust cloud is usually present has a hot surface (e.g. a steam pipe) on which dust can settle, slow combustion can begin in the settled layer at a much lower temperature than the ignition temperature of the dust cloud. A local fire then results which can initiate a dust cloud explosion.

The types of plant and equipment in which dust cloud explosions have occurred or are feasible include:

Milling, grinding, pulverising, disintegrating and stamping machines.
Kilns, pneumatic dryers, rotary drum dryers, spray dryers and fluidised bed dryers.
Screens, classifiers, bag filters and dust collectors.
Conveyors and elevators of various types.
Cyclones and settling chambers.
Storage bins.

Sometimes it is economically feasible to prevent the possibility of a dust explosion entirely, e.g. by avoidance of dust, by use of an inert gas in place of air, thereby avoiding oxygen, or by the use of wet instead of dry methods of processing. Where this cannot be done, suitable protective measures should be taken such as making the equipment strong enough to withstand a dust explosion — placing it out of doors where it will cause no injury or further damage, by the use of explosion relief panels and by the use of explosion suppression systems.

4.2.3.1 Prevention of dust explosions

Wherever practical, plant which does not produce dust clouds should be used.

For conveying dusts and fine powders, drag-link type conveyors which transfer the material in a solid mass are preferable to bucket elevators, screw conveyors, pneumatic conveyors and vibro-conveyors. The return leg of a drag link conveyor can be reduced to a narrow section in which the plates or links fold up. In grinding, pulverising and disintegrating operations, wet processes in which the material is handled as a paste or slurry are usually dust free.

Wet dust collectors and washers of the Venturi type can often be used where the dust or fine material is not required in a dry state, and are in most cases just as effective and economic as dry ones. Wet type collectors are obligatory for grinding magnesium and its alloys (see section 6.5).

Sometimes an inert gas system can be used in place of air, e.g. for drying. The inert gas may be nitrogen, less frequently carbon dioxide, and sometimes the products of combustion of a fuel in air from an inert gas generator. Where inerting is used, it is not necessary to use an entirely oxygen free atmosphere. There is a minimum oxygen concentration for most powders below which explosion is impossible. For some metals this is as low as 2% volume, but for most organic materials it is about 10%. The use of inert gas in dryers tends, however, to be expensive and it is generally necessary to condense water from the exhaust gas and recycle the gas through the dryer. Sometimes the use of an inert gas in place of air improves the quality of the dried product appreciably, so that its use can be justified on the dual grounds of improved quality and safety.

Magnetic and gravity separators remove tramp iron and stones etc. from powders in conveyors and also help to reduce ignition risks. Tramp iron not only increases the ignition hazard, but usually also reduces the value of the product.

4.2.3.2 Protection against dust explosions

Some types of equipment where dust is unavoidably present, such as air classified rod mills, are normally built sufficiently strongly to withstand the effects of a dust explosion (although this might still damage the inlet and outlet ducting). Large equipment which cannot be built to withstand the pressure of a dust explosion such as cyclones and bag filters should be located separately from the rest of the plant, where possible on the roof of a building to which access is restricted.

The spread of an explosion in a dust handling system can in many instances be restricted by the use of chokes, e.g. a rotary star valve between a hopper and a bin below, or a horizontal screw conveyor with a flight

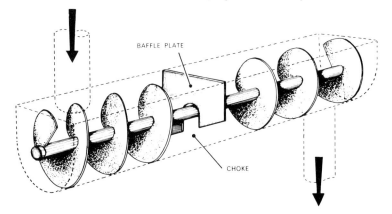

Figure 4.2.8   Restriction of dust explosions in screw conveyors by chokes. Screw conveyors may be used to prevent an explosion from propagating from one part of the plant to another. They are effective only if the dust itself forms a choke within the casing. One flight of the norm should be removed on inclined conveyors. On horizontal conveyors a baffle plate is also necessary (H. & S. W Booklet 24)

removed and a stationary segment baffle plate in its place to form a choke (see *Figure 4.2.8*).

Plants should be designed to minimise the volume within which a dust explosion can occur, and also to minimise the quantity of combustible dust present.

With many inorganic dusts the ease of ignition and the force and speed of a dust explosion are critically dependent on the moisture content of the dust. The moisture content of the material should in such cases be controlled at the maximum economic level. In dryers for materials such as flour, it is possible to install relatively cheap instruments working on the wet bulb principle which will control the moisture content to any level desired. Their use also results in fuel economy as well as reducing the dangers of sparking from static electricity.

Dust-tight electrical systems, especially for lighting the interior of bins and silos, are necessary where combustible dusts are handled. But even where all these precautions are taken, further protection is often necessary. There are two principal methods:

1. Bursting panels or explosion doors.
2. Explosion suppression systems.

1. Bursting panels or explosion doors
As a rough guide, a relief area of 0.2 m² per m³ of plant volume is required for organic dusts, and 0.4 m² per m³ for metal powders. The design of bursting panels and explosion doors have evolved both from theory and practice, and Reference 25 should be consulted. Manufacturers of the

equipment which requires such protection will generally incorporate it in the design. Bursting panels and explosion doors should vent to the open air in a restricted area where personnel are not present.

2. Explosion suppression systems (*Figures 4.2.9 and 4.2.10*)
These are based on the principle that a dust explosion is not instantaneous but is preceded by a slower initial rise in pressure which lasts for perhaps 10 or 15 ms. A fast acting pressure sensitive element forming part of the explosion suppression system detects this initial pressure rise and transmits an electrical signal to one or more suppressors. The suppressor contains a

Figure 4.2.9 Suppression system on cyclonics of spray drier plant (Fenwall)

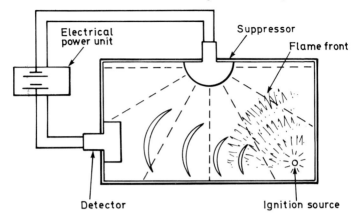

Figure 4.2.10   Basic explosion suppression system. The detector senses the the incipient explosion and the suppressing agent is dispersed explosively. The agent extinguishes the explosion flame by chemical action and cooling and also inerts the unburnt explosive mixture.

cartridge filled with compressed and liquefied carbon dioxide or Halon or a dry powder. This is released by a detonator which is incorporated into the suppressor. Sufficient inert gas must be released from the suppressor to dilute the oxygen in the dust cloud below the minimum explosive concentration.

The release of the suppressant even in the absence of a dust explosion will in many cases over-pressurise the equipment, so an explosion relief panel, door or bursting disc may still be required. The explosion relief device may be opened or (in the case of a bursting disc) broken by a detonator operated by the same detector that initiates the suppressor.

All joints, inspection doors, slide valves, etc. on plant handling flammable dusts should be dust-tight, with suitable flanged joints and packing.

4.2.3.3   Dust explosions in the free volume of buildings

Plant handling combustible dusts should as far as possible be located in the open air; where this is not possible, it should be located on the top floor of the building, with a light roof which incorporates adequate explosion relief. Explosion relief of at least 0.1 m² per 2 m³ of plant volume should be provided wherever combustible dusts are stored. This must vent to the open air. Many materials when emptying from a bin or silo tend to form bridges inside the bin or silo. These have to be cleared. Suitable means must be provided for doing this from outside the bin or silo. People should not have to enter the bin or silo to clear such blockages.

In order to minimise the escape of dust from the plant into the building, the plants should operate under a slight negative pressure.

The building should be carefully designed to facilitate cleaning and minimise surfaces on which dust can settle. Window sills, door frames and projecting fittings should be flush with the wall, and other horizontal surfaces such as girders and beams should be bevelled to prevent settlement. Walls should be smooth and corners rounded. Vacuum cleaning is recommended, taking care that the electrical equipment is safe for areas with flammable dust. The use of compressed air for cleaning should be avoided.

The handling of sacks containing fine powders is often a main cause of dust in buildings. Local exhaust ventilation should be provided at charging and bagging-off points. The minimum number of sacks should be kept in rooms and buildings where plant and machinery are installed and a separate sack store or warehouse should be provided.

4.2.3.4 Silos and bins

All silos and bins should be of fire resisting construction and enclosed to prevent escape of dust.

Explosion relief of at least 0.1 m^2 per 2 m^3 of plant volume should be provided wherever combustible dusts are stored. This must vent to the open air. Means must be provided for clearing bridges in the material being emptied from outside the bin, without the need for anyone to enter the bin.

4.2.4 Physical explosions

Physical explosions as noted earlier can occur whenever vapour is formed very rapidly by the physical interaction between two separate liquids which are present initially at different temperatures. They can also occur when a liquid comes into contact with a large mass of hot solid *(Figure 4.2.11)*. At first sight it might appear that the phenomenom applies only to situations where there is a large difference between the boiling points of the two phases and where the phase of lower boiling point (generally water) comes into contact with a much hotter one (liquid or solid) so that it boils rapidly. The danger of water in moulds and ladles in foundry operations is discussed in section 7. The Scunthorpe steel plant disaster[3] caused by water coming into contact with molten iron is referred to in section 4.2, as are some of the explosions caused by water coming into contact with a hot oil of much higher boiling point. Most of these led to an eruption of steam and hot oil droplets sometimes partly vapourised from a tank or process vessel, and in some cases the oil escaping caught fire or exploded. Thus we are dealing with a very serious hazard.

The accident nearly always arises when a plant or process is being started up, or, in the case of a batch operation, during a 'heating up' stage. The lessons to be learnt partly affect design and partly operation. It is impossible to summarise all the design lessons of so serious a problem here, for although the principles are simple enough and seemingly obvious, there are all too many instances where they should be applied yet get overlooked.

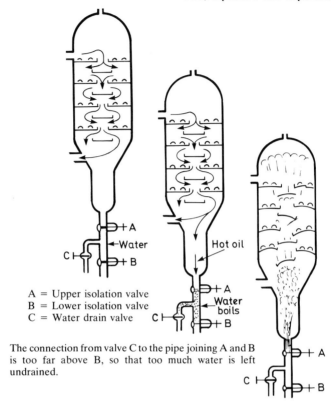

A = Upper isolation valve
B = Lower isolation valve
C = Water drain valve

The connection from valve C to the pipe joining A and B is too far above B, so that too much water is left undrained.

Figure 4.2.11   Physical explosion caused by mixing water and hot oil inside oil distillation column

Often it is simply a question of the correct location of valves. Anyone involved in designing a system involving hot oil is seriously recommended to read References 4 and 31. Three examples are illustrated here.

One concerns the position of a drain valve on a bottom offtake line from a crude oil distillation column, which allowed water to collect in a pocket below it. As the column was being started up, hot oil in the column came into contact with the water causing it to mix with the hot oil and boil explosively, fortunately wrecking merely the internal trays of the column but without damaging the shell. In the second, a valve wrongly located on a large vapour line from the top of a reactor to a fractionator allowed water to condense in the line and collect above the valve. When the latter was opened the water entered the fractionator which contained hot oil. Trays, support trusses and internal piping were badly damaged.

Two major disasters have occurred during attempts to heat the contents

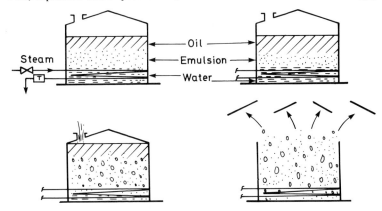

Figure 4.2.12   Cause of Pernis explosion 1968

of a tank or vessel which contained separate water and oil layers when the heat was applied to only one of the layers, and the contents were unstirred.

The basis for the Pernis disaster of 1968[28] is illustrated in *Figure 4.2.12*. Here a crude oil tank containing crude oil and slops with emulsified water was being heated by a steam coil in the base. This caused the emulsion to break or invert so that the heating coil was surrounded by water or a thin water in oil emulsion. The temperature of the water rose, probably nearly to its boiling point, while the oil above it remained much cooler. As the oil got hotter, it started to boil at the oil water interface. This caused the contents of the tank to mix. There was sufficient heat in the hot water to vapourise the light hydrocarbons in the oil. Many tons of a mixture of hydrocarbon vapour and oil droplets were ejected from the tank as a cloud, which exploded.

In a disaster in New Zealand[29] almost the opposite occurred. A stirred vessel containing a viscous oil with water below it was being heated with steam coils which did not extend down into the water layer. Because the stirrer was ineffective, the oil was heated to a temperature considerably above 100°C while the water was still cool. But once the water became hot, it mixed rapidly with the oil, vapourised, and discharged most of the oil to the surroundings, causing a catastrophic fire in which much of the works was destroyed.

This phenomenom is not confined to cases of water and heavy oils, but it can occur also with water and a light hydrocarbon such as cyclohexane. Here the vapour formed is a mixture of water and light hydrocarbon. This mixture boils at a lower temperature than the boiling points of either water or cyclohexane. Despite the fact that this did not come out at all in the official inquiry, there is as the writer has shown, strong circumstantial evidence that it was the underlying cause of the Flixborough disaster; had a stirrer not been removed from one of the reactors some six months earlier, it is unlikely that the disaster would have occurred.[30]

Safety specialists are advised to be on their guard against allowing water and oil which are at different temperatures to come into contact.[31] Whether the water or the oil are the hotter of the two, the danger still arises. Mixtures of water and oil should never be heated unless they are effectively stirred. Once the mixture starts to boil, the rate of heating should be reduced to allow the water to vapourise without excessive foaming, and only increased again when all the water has evaporated. They must recognise that the vapourisation of a given amount of water produces a much greater volume of vapour than does the same amount of oil under the same conditions.

Operators of refinery and other plants involving hot oils must be made aware of the need to drain water regularly, especially during start up, from the bottoms of vessels, low points in lines, pumps and heat exchangers. Water and steam lines connected to vessels containing hot oil which are used only for steaming out or washing out at times of shut down are best 'blanked off' when not in use. The number of water and steam connections to hot oil or hydrocarbon plant should be reduced to the essential minimum; hazards which might arise in their use should be considered by safety and operating personnel, and procedures devised and formalised in writing to reduce them.

4.2.5 Mechanical explosions

The bursting of any pressurised container is a mechanical explosion. The simplest and most familiar is the pricking of a balloon. The subject is a vast one and includes boilers, pressure vessels, tyres, gas cylinders and containers and bombs. Mechanical explosions may be caused by gradual overpressurisation and over-stressing of the container, by weakening of the container as a result of heat, corrosion or mechanical damage, or by an internal explosion (chemical or physical) within the container causing at least transient overpressurisation. Phenomena sufficiently akin to mechanical explosions to be included here are:

Damage caused by water hammer.
Injuries and damage caused by the inadvertent opening of pressurised containers such as autoclaves and pressure cookers.

The damage caused by the bursting of any pressurised container depends very much on its contents. The bursting of a compressed air container will do far more damage than that of a similar one filled only with water, because of the far greater compressibility of air. The fragments of the air-filled container will be accelerated to high velocities and behave like missiles or bomb fragments. However, the bursting of a container full of a liquid such as water itself absorbs most of the energy stored in the compressed water; the energy stored in the compressed liquid is quickly released, so that the force of the compressed liquid has insufficient time to act on the fragments to impart a very high velocity to them. This applies if the liquid is cold and its vapour pressure is low compared to the total pressure. But if the contents comprise a gas liquefied by pressure or a hot liquid at a temperature above its boiling point at atmospheric pressure, then the boiling liquid will continue to exert a force on the fragments for some time after the burst, so that high velocity fragments are produced.

4.2.5.1 Tyres and air bags

Serious injuries have been caused when inflating heavy duty tyres in garages and workshops by blow-outs or by the locking rim blowing off. Tyres should only be inflated in steel cages which will restrain flying objects should a blow-out occur.[32] All employees who handle, inspect, repair or inflate tyres should be trained to recognise the hazards and unsafe working methods.

Air bags are commonly used as an aid to 'righting' overturned equipment, particularly vehicles which have been involved in road accidents.[33] They can cause serious injuries if they burst under load as a result of pressing against a sharp edge. There is a further danger from the collapse of the vehicle which the bag was supporting.

4.2.5.2 Bottles

Bottles containing aerated beverages occasionally break; since their contents are liquid, the fragments are seldom propelled far or fast enough to hit anyone in the eye, but if someone happens to be holding a bottle when it breaks, he or she is liable to receive a severe cut from a broken edge, which may sever an artery. The use of suitable gloves is recommended for any persons engaged in handling bottles and other glass articles.

4.2.5.3 Autoclaves and pressure cookers

Several explosions at steam autoclaves prompted the Department of Employment to issue a Technical Data Note 46 on their hazards.[34] These lie largely in the fact that autoclaves and pressure cookers have to be frequently opened to load and unload them, and are therefore fitted with quick opening doors. If the door is not closed and locked properly, it may be dislodged and blown off by the internal pressure. If it is opened before the pressure in the autoclave has been fully released, the person opening it is liable to be hit and killed. The main safety measures required are:

1. Provision of interlocking devices which make it impossible to heat or pressurise the autoclave unless the door is properly closed, or to open the door unless the autoclave is fully depressurised.
2. Regular examination and inspection.
3. Adequate training and supervision of operators, and conformity with established procedures.

4.2.5.4 Boilers

Nowadays boiler explosions are less frequent than 100 years ago, when operating temperatures and pressures were very much lower. Improvements in nearly every aspect of boiler operation and technology were needed to bring about today's better safety record, and it is really impossible to pinpoint any one aspect as predominating.

Conversely, it must be recognised that without the considerable skill and training that goes with the design and construction of boilers, their regular inspection and maintenance, boiler feed water treatment, boiler operation and firing and the testing of relief valves, the record would turn sour and we would be back in the days of frequent boiler explosions — but more disastrous because of their size, pressure and explosive potential. Today fire-box explosions cause more damage than bursting of boiler shells or tubes.[35]

All safety aspects of boiler construction, operation, and testing are covered in considerable detail in regulations 32 to 35 respectively of the Factories Act 1961.[9] Their main headings are:

Section 32. Steam boilers — attachments and construction.
Section 33. Steam boilers — maintenance, examination and use.
Section 34. Steam boilers — restrictions on entry.
Section 35. Steam receivers and steam containers.

The regulations are so detailed that any selection of 'essential points' which might be made here would almost certainly give an unbalanced impression. An expert summary of the main points by R.G. Warwick, Chief Engineer of National Vulcan Engineering Insurance Group is recommended to anyone who has difficulty in digesting the regulations.[36]

Boilers and water heaters come in all shapes and sizes, and one particular type that has in recent years given cause for concern is the unattended automatic water heater in self-service launderettes. Several serious explosions in these prompted the Health and Safety Executive to issue Technical Data Note 34.[37] Its main points which apply generally to larger industrial boilers as well, are summarised below:

Examination. Every installation must be thoroughly examined by a competent person when it is installed and the results of the examination recorded and certified. It should be re-examined again at intervals of no more than 14 months, the examination to include a cold inspection of the interior of the boiler and a hot inspection when the boiler is in operation to check safety valves, overheat and other control devices.

Overheat controls. An independent overheat control which shuts the boiler down automatically when a specified temperature is reached should be fitted to every hot water boiler in addition to its normal thermostatic control.

Feed water failure. If water can be drawn off the system when the feed water supply fails, a means of shutting down the boiler in the event of feed water failure must be provided.

Syphon breaker and vent. The cold feed line to the boiler should incorporate a high level syphon break which is open to atmosphere. This serves the dual purpose of preventing syphoning of hot water from the boiler and of relieving excess pressure in the system if all else fails.

Safety valve. A suitable safety valve provided with means to test it and protection against tampering by unauthorised persons should be fitted to every boiler.

Thermostatic valve. A thermostatic valve should be fitted in the flow pipe of every boiler close to and before the hot water tank. It should discharge to drain.

Thermometer. Every boiler should have at least one suitable thermometer.

Flame safeguards. Every gas or oil fired boiler should be fitted with a manually reset flame safeguard which cuts off the supply of fuel in case of flame failure.

4.2.5.5 Air receivers and unfired pressure vessels

While air receivers, i.e. compressed air reservoirs are well covered by section 36 of the Factories Act 1961, other unfired pressure vessels, e.g. those incorporated into an oil or chemical process, have in the UK received far scanter attention by legislation.

This situation is quite anomalous, since an air receiver which contains only compressed air is subject to far less exacting conditions than a process pressure vessel which may contain corrosive, erosive or unstable chemicals or be subjected to extremes of temperature and pressure. Thus a compressed air receiver which operates at a pressure of, say, 7 atm must be thoroughly cleaned and examined at least once in a period of 26 months, whereas if the receiver contained say a mixture of oxygen, chlorine and other nasties and was in continuous operation at a temperature of 500°C, it would be exempt from this mandatory inspection requirement. The absurdity of this situation is now officially recognised and steps which are currently being taken to correct it are described in section 2.11.2.1. There are, however, two main difficulties:

1. The conditions inside many process vessels in use today within the chemical industry are so many and varied as to make it almost impossible to legislate with any degree of certainty on the type and frequency of inspection required. Furthermore, any such legislation if it is to be effective is likely to necessitate divulging information about processes which their owners regard as confidential.
2. Following the report of the Robens Committee and the Health and

Safety at Work, etc. Act 1974, the onus on setting safe standards of design, inspection and operation has been largely shifted onto the shoulders of the companies and organisations who design and operate the plant.

Thus the duty to ensure the safety of an oil or chemical plant operating at high pressures is a general responsibility on the employer, under Section 2 of the Health and Safety at Work etc. Act 1974. There are no detailed regulations to state how this safety is to be achieved.

The current British design standard for such pressure vessels[38] is BS 5500:1976 *Unfired fusion welded pressure vessels*; this largely replaces earlier design standards BS 1500 *Fusion welded pressure vessels for general purposes* and BS 1515 *Fusion welded pressure vessels for use in the chemical, petroleum and allied industries*. These do not, however, have the status of 'Approved Codes' (see section 2.1) nor does it appear likely that any approved code will be issued in the near future.

Unfired pressure vessels designed in accordance with other national codes are also in use in the UK. The most notable of these is the American ASME Boiler and Pressure Vessel Code,[39] compliance with which is a legal requirement throughout most of the United States. This however has two alternate sets of design rules; Division 2 rules allow thinner wall sections than Division 1:

Division 1 (the old rules) covers vessels with pressure ratings up to 3000 psi. Designs are calculated according to the principal stress theory, and a safety factor of 4 is provided.
Division 2 entitled *Alternate rules for pressure vessels*, covers vessels for use at pressures above and below 3000 psi. The vessel design is based on a detailed stress analysis and a factor of safety of 3 is provided.

Explosions and failures of unfired pressure vessels are usually the result of one or more of the following causes:

1. Error in design, construction and installation.
2. Improper operation and improper education of operators.
3. Corrosion or erosion of the metal.
4. Failure or blocking of safety devices and/or automatic control devices.
5. Water hammer.
6. Failure to inspect thoroughly, properly and frequently.
7. Improper application of equipment.
8. Lack of planned preventive maintenance.
9. Internal explosion, either chemical or physical.

A point worth making here is that unfired pressure vessels employed in a process should receive every bit as careful and regular inspection as an air receiver and, in most cases, as a boiler. Many large oil and chemical companies have their own inspection departments which are entirely divorced from production and which have developed through inspection procedures. Two sidelights of this emerge from the Flixborough disaster:

1. Regular visits were being made to the works by an authorised firm of insurance inspectors to carry out statutory inspections of boilers and/or air receivers during the time that the ill-fated bridge pipe was installed. The firm however had no authority to inspect the infinitely more hazardous cyclohexane oxidation plant or the modifications made to it.

2. Whilst the Report[40] of the Court of Inquiry severely criticised the installation of the temporary bridge pipe between two thin walled bellows pieces which failed, it had curiously little to say on the appropriateness or otherwise of such bellows for this duty in the first place. The bellows had a diameter of 700 mm and a wall thickness of little more than 1 mm and were highly stressed in operation. This wall thickness is considerably less than the usual corrosion allowance which is added to the design wall thickness for most pressure vessels. Thin walled stainless steel bellows certainly have their uses and are appropriate and safe in some circumstances. But it is clearly inadvisable to employ large diameter and highly stressed thin walled bellows on such hazardous service, even when there is a regular metal inspection programme.

4.2.5.6 Pressure relief

Unfired pressure vessels should always be protected against failure by a properly-designed system of pressure relief valves. This is a mandatory requirement under the Factories Act 1961 for most boilers and air receivers, although the Factories Act has nothing to say about chemical reaction vessels. Process plants containing several pressure vessels also require relief lines with liquid knock out vessels and drains and a safe means of disposing of the gas relieved, e.g. a vent or flare. There has been a certain amount of perhaps justified criticism in the UK upon over-reliance on pressure relief systems, since the wall strength of a pressure vessel may be so weakened in the case of fire that the pressure relief system is ineffective. But equally the UK can be criticised in not having an adequate code for the pressure relief of plants containing a number of pressure vessels. In fact the UK is largely reliant on the use of American Codes;[41] in particular API. RP 520 *Recommended practice for the design and installation of pressure-relieving systems in refineries*, and API, RP 521 *Guide for pressure-relief and depressurising systems*.

These codes which are generally followed by designers and contractors for oil and petrochemical plants are excellent, but since they are not mandatory here, there is nothing to stop corner cutting by companies who would like to appear to be doing the right thing yet are dismayed at the cost of an effective pressure relief and disposal system. Compliance with these codes also requires a considerable enginering effort in sizing valves and lines, choosing correct materials for relief lines, and laying out and arranging the lines correctly.

This exercise tends to be the Cinderella of the process engineer, who sees little interest in it or profit to the manufacturer. It can also be quite a major exercise to check whether the pressure relief system on any but the simplest of installations has been designed correctly. Another hazard lies in plant additions and extensions whose safety valves are all connected into a relief system designed only for the original plant whose relief requirements were far less. It is all too easy to think just 'pipe it into the relief header and everything will be all right'.

At least two major releases of flammable vapours followed by explosions[2] were caused by the use of incorrect materials of construction for pressure relief systems. These are, in many cases, required to operate immediately at sub-zero temperatures (e.g. $-30°C$) caused by the cooling effect of expansion and vapourisation when the gas or vapour passes the relief valve. Most mild steel lines and fittings will suffer embrittlement at such low temperatures. This combined with the stresses caused by contraction can readily rupture a relief line. Materials of construction and stresses due to low temperature and high discharge rates need to be studied when the system is designed.

API RP 521 systematically considers the various causes of overpressure which may arise and which have to be guarded against, and it also notes those causes of transient pressure rise which the relief system cannot safely protect against. The most important of these are:

1. Water hammer.
2. Steam hammer.
3. Contact between water and hot oil (physical explosions).
4. Internal (chemical) explosions.
5. Rapid exothermic reactions (not quite fast enough to class as explosions).

Physical and chemical explosions have already been discussed, so we are left with the hazards of water and steam hammer.

4.2.5.7 Water and steam hammer

Water hammer is the result of hydraulic shock waves in a liquid filled system, and generally results from a valve in a long line through which liquid is flowing being slammed shut, or simply shut too fast.[42] Steam hammer can similarly occur in piping which contains a compressible fluid, or sometimes a mixture of liquid and vapour. It is due to a kind of oscillation of the steam or other fluid in the pipe and can cause vibration and violent pipe movement, and possible rupture of a pipe joint or equipment. Steam hammer is more difficult to predict, analyse and quantify than water hammer, but efficient trapping of condensate in steam lines and the avoidance of quick closing steam valves generally helps.

Water hammer can be avoided by giving strict instructions to close valves gradually, by fitting valves which can only be closed gradually, or when both of these are impossible by fitting air chambers, surge tanks or other forms of pulsation dampeners.[41,42] The pressure that can be developed in an air chamber installed to relieve water hammer can however be surprisingly high. The head of one such chamber was ruptured by the water hammer which the chamber was intended to prevent and was hurled upwards making a hole in the ceiling above it.

4.2.6 Electrical explosions

Electrical explosions caused by man made electricity are discussed in section 5.6.3. Details on precautions against lightning are given in Reference 43.

REFERENCES

1. Holmes, H.H., 'Explosives', *The Encyclopedia Americana*, International Edition, Americana Corporation, N.Y. (10) 788-791 (1972).
2. Davenport, J.A., 'A study of vapour cloud incidents', Paper 24.a 83rd National Meeting, American Institute of Chemical Engineers, Houston, Texas, (March 1977).
3. *The explosion at Appleby-Frodingham Steel Works, Scunthorpe on 4th November 1975*, HMSO (1976).
4. Jennings, A.J.D., 'The physical chemistry of safety', *The Chemical Engineer*, London 637 (October 1974).
5. King, R.W., 'The dangers of sudden boiling of superheated liquids', *FPA, Fire Prevention Science and Technology*, (15), 17-21 (August 1976).
6. Kletz, T., 'Protect vessels from fire', *Hydrocarbon Processing*, International Edition (56), 98 (August 1977).
7. Miraeur, H., *Powders and explosives*, University Press of France, Pt. 8 (1947).
8. Watts, H.E., (HM Chief Inspector of Explosives) *The Law relating to explosives*, Charles Griffin & Co, London, 25 (1954).
9. Fife, I., and Machin, E.A., *Redgrave's Health and Safety in Factories*, Butterworth, London (1976).
10. Health and Safety Executive, *Safety in the use of cartridge operated fixing tools*, Guidance Note PM 14, HMSO, London (1975).
11. BS 4078:1968, *Cartridge-operated fixing tools*, BSI, London.
12. Information Sheet H33, *Ammonium Nitrate*, Fire Protection Association London.
13. National Safety Council, *Accident prevention manual for industrial operations*, 7th ed., NSC, Chicago, 1131 (1974).
14. Factory Mutual System, *Loss prevention data sheet 7-30 peroxides*, 1151 Boston-Providence Turnpike, Norwood, Mass. 06062, USA.
15. Swern, D., *Organic Peroxides*, (3 Vols) Wiley Interscience, N.Y. (1970).
16. National Fire Protection Association, *Standard 49, Hazardous Chemicals Data*, NFPA, Boston, Mass, USA.
17. Sax, N.I., Dangerous Properties of Industrial Materials, 3rd ed., Van Nostrand (1968).
18. Factory Mutual System, *Loss prevention data sheet 7-43, loss prevention in chemical plant*, 1151 Boston-Providence Turnpike, Norwood, Mass, 06062, USA.
19. Fire Protection Association, *Fire and related properties of industrial chemicals*, FPA, London.
20. Fire Protection Association, *Flammable liquids and gases, explosion hazards*, FPA, London.

21. See reference 13, Pages 1338-1340.
22. Health and Safety Executive, *Booklet 46, Low temperature gas and oil fired plant and solvent evaporating ovens*, HMSO, London.
23. Health and Safety Executive, *Guidance Note CS 1, industrial use of flammable gas detectors*, HMSO, London.
24. King, R., 'Flammable gas detectors', *Control and Instrumentation* **8** 34-35 (March 1976).
25. Health and Safety Executive, *Booklet 22, Dust Explosions in Factories*, HMSO.
26. Fire Protection Association, Booklets 'Explosible dusts, flammable liquids and gases: explosion suppression', 'Explosible dusts: the hazards', 'Explosive dusts: control of explosions', 'Explosible dusts: the elimination of ignition sources'.
27. Hay, D.M., and Napier, D.H., 'Minimum ignition energy of dust explosions' Symposium Paper, *Chemical Process Hazards with Special Reference to Plant Design – VI*, April 1977, Institution of Chemical Engineers, Rugby.
28. Ministry of Social Affairs and Public Health, *Report of the cause of the explosion at Shell Pernis 20th January 1968*, State Publishing House, The Hague, Holland.
29. *Report of Commission of Inquiry into the Explosion and Fire which Occurred at the Factory of Chemical Manufacturing Company Ltd. on 26th September 1974*, Govt. Printer, Wellington, New Zealand (1975).
30. King, R.W., 'A Mechanism for a transient pressure rise', Institute of Chemical Engineers Symposium, The Technical Lessons of Flixborough, Nottingham (December 1975).
31. The American Oil Company, Booklet No. 1, *Hazard of water*, 5th ed., AMOCO, Chicago (1964).
32. See Reference 13, page 1428.
33. Grice, R.J., *Vehicle recovery*, Newnes-Butterworths, 24 (1977).
34. Department of Employment, *Technical data note 46, safety at quick opening and other doors of autoclaves*, HMSO, London.
35. Oil Insurance Association, *Boiler safety*, Industrial Oil Insurers, Chicago (1971).
36. Warwick, R.G., 'Steam boilers and pressure vessels' in *Industrial Safety Handbook*, editor Handley, W., 2nd ed., McGraw Hill, 202 (1977).
37. Health and Safety Executive, *Technical data note 34, prevention of explosions of water heating systems in launderettes*, HMSO, London.
38. British Standards 5500:1976, 1500 and 1515 (see text).
39. American Society of Mechanical Engineers, *Boiler and pressure vessel code*.
40. Department of Employment, *The Flixborough disaster, Report of the Court of Inquiry*, HMSO (1975).
41. American Petroleum Institute, API. RP 520 and API. RP 521 (see text).
42. Baumeister, T., *Marks standard handbook for mechanical engineers*, 7th ed., McGraw Hill, 3-57, 3-72, 9-201 (1967).
43. Golde, R. H., *Lightning Protection,* Edward Arnold, London (1973).

4.3 FIRE HAZARDS

Contents

4.3.1 Statistics and causes

Fire hazards are of two kinds; hazards to life and health, and hazards to property. Data on the causes of death and injury resulting from fires in the UK are given in Table 4.3.1.[1] These show that gas and smoke claim more lives than burns and scalds, although burns and scalds cause the majority of non-fatal injuries.

Hazards to life frequently arise some distance from the fire itself and result from smoke, toxic gases, oxygen deficiency in the air and heat. Smoke, even when it is non-toxic, serves as a trap, blinding and confusing people, hindering escape from the more lethal hazards that follow.

Fires and explosions are often closely associated. Therefore, most loss statistics cannot discriminate between the two, but treat both under the broad heading of fire.

Casualties and damage from major fires (property loss in excess of £10 000 per fire) in buildings of different types of occupancy are given in *Table 4.3.2*. This shows that while casualties resulting from fires in industrial buildings represent well under 10% of total fire casualties in

Table 4.3.1   NATURE OF FATAL AND NON-FATAL* UK FIRE CASUALTIES

Year	1977	1978	1979
Non Fatal			
Burns and Scalds	2 596	3 105	3 292
Overcome by gas or smoke	1 353	1 671	2 093
Combination of above 2	80	110	134
Physical injuries	638	1 003	1 182
Shock only	531	1 615	2 080
Other	350	88	94
Not recorded	821	639	8
Total	*6 369*	*8 231*	*8 883*

Note: 1978 was the year of the fire service strike.

*Non-fatal casualties comprise all injuries which require further treatment than first aid.
**Published by the Home Office.

buildings, property loss from industrial fires (over £10 000 loss) represents about 50% of the total property loss from major building fires in the UK. In the years 1971 to 1973, there was about £2 million of property damage from major fires in industrial buildings for every life lost. There are thus powerful economic motives for fire prevention and protection in industry as well as the vital question of human safety.

The data presented in *Table 4.3.2* for various types of building occupancy is repeated in *Table 4.3.3* for various types of industry. It is difficult to draw valid comparisons between the different industries without comparative figures for manpower and capital employed. Chemical and allied industries and the manufacture of metal and metal goods would appear to show higher than average ratios of casualties to property damage, while textiles and paper show the reverse.

Sources of ignition of fires in industrial premises are given in *Table 4.3.4*. The two most common causes are electrical ignition and rubbish burning. Whilst 'children playing' has declined as a source, this is matched by an increase in 'malicious ignition'; arson is now a serious cause.

Whilst a great deal of study has been given to the behaviour of fires and fire protection systems, it appears that the way people behave in fires has received less attention until recently. Work which is still continuing in the Psychology Department of Surrey University[2] has brought a number of common human failings to light:

1. Failure to detect a fire sufficiently early.
2. Frequent inability to use even the simplest fire extinguisher.

Table 4.3.2 CASUALTIES AND MAJOR PROPERTY LOSSES FROM UK BUILDING FIRES, CLASSED BY TYPE OF OCCUPANCY

Type of occupancy	1977			1978			1979		
	Casualties Fatal	Non-fatal	Property loss from major fires (£ thousands)	Casualties Fatal	Non-fatal	Property loss from major fires (£ thousands)	Casualties Fatal	Non-fatal	Property loss from major fires (£ thousands)
Agriculture, forestry, fishing	1	43	2 863	5	69	3 629	7	109	3 100
Industrial premises	7	335	78 884	15	392	75 758	13	332	114 123
Public utilities	4	28	6 465	2	16	4 000	—	9	1 400
Distributive trades	10	187	30 007	6	160	35 192	14	213	52 394
Miscellaneous services*	60	578	26 762	42	500	33 168	40	882	40 530
Public administration and defence	—	32	1 088	3	71	625	2	56	5 114
Dwellings	652	4 261	5 094	733	5 446	4 736	865	7 964	8 964
Undefined & unoccupied sheds and garages	3	8	330	—	22	379	—	14	N.A.
Total	747	5 372	151 493	806	6 676	157 487	941	8 579	225 625

*Includes hospitals, schools, colleges, hotels, clubs, boarding houses, places of public entertainment and non-industrial offices.

Table 4.3.3 CASUALTIES AND MAJOR PROPERTY LOSSES FROM UK INDUSTRIAL BUILDING FIRES, CLASSED BY TYPE OF INDUSTRY

Type of industry	1977 Casualties Fatal	Non-fatal	Property loss from major fires (£ thousand)	1978 Casualties Fatal	Non-fatal	Property loss from major fires (£ thousand)	1979 Casualties Fatal	Non-fatal	Property loss from major fires (£ thousand)
Mining and quarrying	4	—	435	—	1	—	—	2	316
Food, drink and tobacco	—	13	14 299	1	17	8 928	2	36	13 024
Coal and petroleum products	—	1	113	—	2	335	—	4	—
Chemical and allied industries	1	48	12 513	2	71	6 389	1	36	4 894
Metal manufacture	1	54	3 577	4	52	3 237	1	59	4 402
Mechanical engineering	—	35	1 213	1	15	2 358	—	20	4 939
Instrument engineering	—	1	692	—	7	352	—	3	131
Electrical engineering	—	20	3 265	—	24	8 010	3	20	4 979
Shipbuilding and marine engineering	1	3	30	—	8	76	1	3	1 110
Vehicles	—	47	2 341	2	14	6 563	—	7	29 495
Metal goods not elsewhere specified	—	36	4 298	—	40	13 008	—	45	10 708
Textiles	—	13	11 870	2	17	8 377	—	29	16 146
Leather, leather goods and fur	—	5	463	—	1	1 061	—	1	664
Clothing and footwear	1	3	4 222	—	13	2 508	—	10	4 115
Bricks, pottery, glass, cement, etc.	—	8	1 937	2	11	818	—	9	2 938
Timber, furniture, etc.	3	17	4 052	1	46	2 629	—	15	4 069
Paper, printing & publishing	—	9	9 811	—	16	6 189	—	18	8 649
Other manufacturing industries	—	17	4 188	—	38	4 920	5	17	3 860
Construction	2	34	694	—	69	1 453	2	37	1 366
Total	13	926	65 279	15	924	99 211	15	331	150 545

Table 4.3.4 SOURCES OF IGNITION OF FIRES IN INDUSTRIAL PREMISES 1977-79

	1977*	%	1978	%	1979	%
Children playing with fire	803	(5.75)	923	(6.87)	896	(6.41)
Malicious or doubtful	1 364	(9.76)	1 341	(9.98)	1 467	(10.49)
Smoker's materials and matches	748	(5.35)	954	(7.10)	1 039	(7.43)
Cooking appliances	213	(1.52)	247	(1.84)	260	(1.86)
Space heating ⎫ Central heating installations ⎭	643	(4.60)	698 / 127	(5.19) / (0.94)	780 / 174	(5.58) / (1.24)
Water heating installations	23	(0.16)	43	(0.32)	52	(0.37)
Welding and cutting appliances	772	(5.52)	890	(6.62)	940	(6.72)
Blowlamps	NA		142	(1.06)	220	(1.57)
Electrical wiring	643	(4.60)	828	(6.16)	1 015	(7.26)
Other electrical equipment	1 243	(8.89)	1 423	(10.59)	1 555	(11.12)
Unknown	2 281	(16.32)	1 560	(11.61)	1 685	(12.05)
Other	5 242	(37.51)	4 265	(31.73)	3 906	(27.92)
Total	13 975	(100)	13 441	(100)	13 989	(100)

Table 4.3.4a COMPARISON OF MAIN SOURCES OF IGNITION WITH PREVIOUS YEARS 1964 AND 1973 EXPRESSED AS A PERCENTAGE

	1964	1973	1977	1978	1979
Children playing	11.5	7.5	5.8	6.9	6.4
Malicious	1.4	6.3	9.8	10.0	10.5
Smokers materials	10.5	6.8	5.4	7.1	7.4
Electrical apparatus	8.8	10.2	8.9	10.6	11.1
Electrical wiring	3.2	2.7	4.6	6.2	7.3

Note: Genuine comparison may be impossible due to changes in the methods of reporting during the 1970s

3. Tendency to stick to normal routes in buildings and to ignore fire escapes.

These all have one feature in common — a failure to cope with the unfamiliar.

Fires are not detected sufficiently early because people too easily find other explanations in more familiar terms for the early warning signals. An acrid smell is more likely to originate from an overheated oven than a fire. Even when a fire is detected, many people, perhaps through fear of making fools of themselves are reluctant to call the brigade at once, thus allowing bigger fire losses to be sustained. Constant reminders are therefore necessary for those detecting a fire to call the brigade promptly.

Failure to use an extinguisher properly may be partly due to excitement, but more often it is due to lack of practice and unfamiliarity. Thus, a man seeing a small fire in a workshop ran for a portable water (soda-acid type) extinguisher. He carried it over his shoulder upside down. The soda and acid mixed and expelled the water which formed a pool on the floor in which he slipped, falling and injuring his back. The fire was extinguished by someone else.

The tendency to use normal familiar routes rather than special escape routes will persist unless people have frequent fire practices in which the special routes are used, and are convinced that the (less familiar) escape route is the safer alternative. In many ways it would seem better to spend available funds in making normal existing routes safe and suitable for escape from fire rather than on special, but less familiar, fire escape routes.

The notion is commonly held that fire induces blind panic which causes many unnecessary deaths but recent work[2] suggests this may be exaggerated. There are many instances of irrational crowd behaviour where people have been squeezed or trampled to death in exits; this behaviour was often sparked off by some trivial incident quite unconnected with fire. Just as often, people die in vain rescue attempts through entering rooms and buildings in which smoke and heat are accumulating, without realising the hazard of being trapped by the smoke and being unable to find the exit. Fire always calls for swift but cool action with no half measures.

The first priority is to give warning, if none has been given.

The second is to evacuate the building, keeping all doors, windows and other openings through which air can enter and feed the fire closed.

The third, which may take priority over the second only when it is clearly feasible without jeopardising ones chances of escape, is to fight the fire. This must be done at source, using appropriate means and striking at the very heart of the fire. Fire fighting and rescue are skilled activities needing not only training but suitable equipment and clothing. A trained fireman using self-contained breathing apparatus with radio communication and protective clothing may survive and work in a situation where most unprotected people would perish.

The role of building design in fire protection is discussed in section 2.6, while fire hazards are pointed out in many other section, e.g. 3.10 (protective clothing), 5.1. (housekeeping), 5.6 and 5.7 (electrical ignition), and most sections of chapter 6.

The sub-sections which follow deal with types of fire, fire warning, evacuation, emergency fire fighting and finally protection of property both against fire itself and fire fighting operations.

4.3.1.1 Conditions essential to fire. The fire triangle or pyramid

Three conditions have long been regarded as essential components of any fire:

1. Fuel (i.e. the combustible material)
2. Oxygen (from the atmosphere)
3. Heat (essential to start the fire initially, but maintained by the fire itself once it has started)

These are familiar to fire fighters as the 'fire triangle' or pyramid (see *Figure 4.3.1*). If any one of these conditions is removed, the fire goes out. Methods of fire fighting thus depend on removing or shutting off the source of fuel, excluding oxygen or removing heat from the fire faster than it is liberated. A fourth condition is now recognised. Flames proceed chemically as branched chain reactions through the intermediary of free radicals which are constantly being formed and consumed. If the free radicals can be removed and prevented from continuing the chain reaction, the flame goes out.

Various chemicals used in dry powder and halogenated hydrocarbon extinguishers capture free radicals and put out the fire in this way. Potassium bicarbonate is more effective than sodium bicarbonate and free halogen radicals, especially bromine formed when a brominated

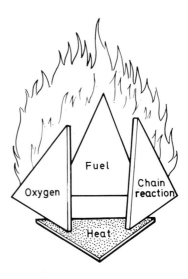

Figure 4.3.1   The Fire Pyramid
(National Safety Council, USA)

hydrocarbon meets a fire, are also effective. Thus the familiar fire triangle becomes a pyramid[3] and now includes the fourth condition:

4. Maintenance of chain reaction through free radicals.

4.3.2 Classes of fire[4]

Various classes of fire are recognised in order to rationalise the choice of extinguishing media and devices, and the precautions taken in fire protection and fire fighting.

Until 1970 three classes were recognised:

A. Fires of carbonaceous solids, wood, paper, cloth, etc.
B. Fires of flammable liquids, petrol, oils, greases, etc.
C. Electrical fires.

The third class differs from the first two in referring to the means of ignition rather than the combustible material, and applies only so long as the circuit causing the fire is alive. Once it is switched off the fire is of another class, perhaps A or B.

The old classification was superseded in 1970 by BS 4547 which lists four classes.

A. Solid materials corresponding to the old class A.
B. Flammable liquids corresponding to the old class B.
C. Gases and gas containers.
D. Metal fires.

4.3.2.1 Class A fires. Carbonaceous solids

The general method of extinguishing class A fires is by water jets which quench the fire and cool the material to below its ignition temperature.

Class A fires are often deep-rooted and well below the surface of the material, so that sufficient water must be applied to penetrate and cool the whole of the burning material to below its ignition temperature.

4.3.2.2 Class B fires. Flammable liquids

In dealing with flammable liquids two main hazards must be recognised:

1. If the liquid is lighter than water and does not mix with it, the use of water may actually spread the fire rather than extinguish it, since the liquid will float on the water and be carried into surrounding areas, cellars and drains.
2. If the liquid has a low flash point, its vapour will form an explosive mixture with air and this may spread and extend a considerable distance from the liquid itself. A source of ignition for instance a spark

or lighted match, anywhere in this area will cause a sheet of flame or flash-back which will set fire to liquid and any easily ignitable materials in its path.

Flammable liquids in general must be vapourised in order to burn, and it is the vapour not the liquid which burns.

Flash point. The flash point of a flammable liquid is the lowest temperature at which enough vapour is given off near the surface of the liquid to produce a flammable mixture with air; that is a mixture which may be ignited by a spark or other source of ignition, and which contains the proper ratio of vapour and air to support combustion. These ratios vary widely from liquid to liquid, but it should always be assumed that any flammable solvent above its flash point is in equilibrium with a flammable vapour-air mixture. Once a fire has started, its heat rapidly vapourises more liquid until the whole mass is aflame. In general, the lower the flash point of a liquid the more flammable is the material and the more violent the resulting fire.

The fact that a liquid is at a temperature below its flash point does not mean that it is safe. If a material such as kerosene with a flash point of 40°C is brought into contact with a source of intense heat — a welding torch, furnace or open fire — a small part of it could be heated above the flash point, give off vapour and burst into flame. The heat thus produced would heat the rest of the kerosene and the fire would spread. A mist of a high flash solvent is also almost as easily ignited as if it were a true mixture of air and vapour.

Nearly all flammable vapours are heavier than air, so that explosive mixtures of air and vapour will spread over the ground or floor when the air is still and flow into depressions in the ground, drains, trenches and cellars.

Explosive limits of vapour-air mixtures. The vapour of every flammable liquid has a minimum concentration in air below which it does not ignite when in contact with a source of ignition. There is also a maximum concentration of vapour above which flame is not propagated, although this is only found at room temperature if the flash point of the material is low. These limits are known as lower and upper explosive limits and they vary widely for different flammable liquids.

The flash points and explosive limits for several flammable liquids used in industry are given in *Table 4.3.5*.

General precautions for flammable liquids

Flammable liquids used in industrial buildings should be kept in safety cans which have a fire arrester in the spout and a spring closing cap so that they are always closed when not in use *(Figure 4.3.2)*. This obviates two of the main dangers; escape of flammable vapours from an open container, with risk of ignition and flash-back and ignition with explosive force of the residual vapour left in apparently empty containers.

Buildings in which flammable liquids are used should have good

Table 4.3.5 FLASH POINTS AND EXPLOSIVE LIMITS OF COMMON INDUSTRIAL LIQUIDS

Liquid	Flash point °C	Explosive Limits (per cent by volume) Lower	Upper
Acetone	−18	2.6	12.8
Benzene	−11	1.5	8.0*
Ethyl acetate	− 4	2.2	11.15*
Ethyl alcohol	13	3.5	19*
Ethyl ether	−45	1.0	50*
Hexane	−26	1.2	7.5
Kerosene	40	0.7	5*
Methanol	11	6.0	36*
Methyl ethyl ketone	− 1	1.8	10
Petrol (gasoline)	− 7	1.1	7.6
Toluene	4	1.3	7.0*
Xylene	17	1.0	6.0*

*At elevated temperature

ventilation and the quantities of liquid in a building should be kept to a minimum. No smoking should be the rule in a building where flammable liquids are used, and a careful check should be made regularly to eliminate all other possible sources of ignition. Empty containers of flammable liquids should not be kept in the building but returned to stores for refilling.

Many of the fires involving flammable liquids have occurred where the liquid was present as a paint, lacquer or rubber solution or paste solvent. The vapour from the drying article where the solvent was used or from the open paint or paste tin built up in concentration and spread to a point of ignition until a flash back occurred.

Where flammable liquids are used in plants for washing and dipping operations the tanks or containers should be provided with hinged covers which are automatically closed by a fusible link and a spring operated mechanism if the tank or container catches fire.

Water should only be used to extinguish a flammable liquid fire in certain limited and clearly defined circumstances:

1. For liquids heavier than water (e.g. carbon disulphide) and at temperatures lower than the boiling point of water.
2. For liquids readily soluble in water provided either that the quantity of diluted material is small enough to be contained or provided that the material can be washed away without causing an unacceptable pollution problem.

Alternatively the most suitable extinguishing agents are dry powder, foam, carbon dioxide or a suitable vapourising liquid provided they are used in conditions where they do not create a new hazard. If water has to be

Figure 4.3.2   Safety can (Walter Page Ltd.)

used to cool other equipment or plant (especially tanks) close to a vapourising liquid fire, care must be taken that the water does not flow into the burning liquid and spread the fire.

When a fire involving a flammable liquid has been extinguished there is often a danger that fresh vapour will form and mix with air creating an explosive vapour mixture which will re-ignite, often on hot or glowing material left from the earlier fire. While various measures may be taken to prevent this, there are some cases where it is better to let a vapourising liquid fire burn itself out, at the same time concentrating efforts on cooling objects exposed to the fire and preventing it from spreading.

Another hazard can occur if water is applied to a fire of a high flash point flammable liquid such as gas oil. If the liquid has been heated by the fire to a temperature above the boiling point of water (100°C), drops of water entering the liquid will vapourise rapidly with almost explosive force splattering the hot burning liquid over a wide area.

4.3.2.3   Class C fires. Gases

There is always a serious danger whenever a fire from a leaking or fractured gas main or container is extinguished, that the unignited gas continuing to escape will mix with air to form an explosive mixture. This when reignited may result in a serious explosion which causes more destruction and injury than the original fire would have done if left to burn itself out. The only safe way of estinguishing a gas fire is to shut off the supply of gas to the fire.

Cooling should, however, be applied to objects heated by the fire, particularly if they are combustible or contain flammable materials; water should be applied through a spray nozzle. Flammable materials, particularly compressed and liquefied gases in cylinders should be removed as quickly as possible from the neighbourhood of a fire. If a gas fire is extinguished accidentally by a water spray and the supply of gas cannot be instantly shut off, the area should be urgently evacuated.

Great care should be taken where electrical appliances and switches are exposed to a gas leak. It may seem logical to switch off any electric motors or other electrical equipment in the neighbourhood of such a leak. But unless the switch is flameproof the mere act of switching off a motor may create a sparking in the switch which could ignite an explosive gas-air mixture surrounding the switch. Electrical appliances exposed to such conditions should therefore be switched off remotely from a switch in a safe area and not from a switch exposed to the gas-mixture, unless the switch itself is known to be flameproof.

A high proportion of gas fires is caused by leaks from damaged or perished flexible rubber hose used with portable LPG cylinders for cutting and welding, etc. Besides taking the precautions listed in section 6.4, the use of rubber hose and portable cylinders inside buildings should be kept to the absolute minimum. Such equipment should be replaced as far as possible by fixed piping deriving its supply from the gas main or from a bank of cylinders secured in a safe place outside the building.

Most cylinders in the UK which contain flammable gases are not as yet fitted with pressure-relieving devices. Thus if a cylinder is involved in a fire, its internal pressure is likely to rise until it bursts, and its contents escape as a fireball. Unless the cylinder can be positively cooled, all fire fighters should withdraw to a safe distance.

If a gas cylinder starts to leak and the leak cannot be immediately stopped, it should be moved at once to the open air where the gas can disperse safely. If a cylinder has been involved in a fire, it should be emptied and tested in a proper cylinder testing station or returned to its makers properly marked for testing.

4.3.2.4 Class D fires. Metals

The most commonly encountered metal fires are those of magnesium and its alloys, although several powdered metals, notably aluminium, can form explosive dust clouds, whilst sodium and potassium react vigorously and catch fire in contact with water (see *Table 4.2.1*). The fumes from most metal fires are dangerous and some, e.g. those from cadmium, beryllium, and lead are extremely toxic.

4.3.2.5 Electrical fires

Fires may be caused by misuse of electrical equipment or by improper installation and maintenance. The resultant over-heating can then easily ignite insulation or other combustible materials and start a fire.

Wiring faults account for roughly one third of all electrical fires.

Overloaded circuits cause heating of cables which may result in direct fire hazards or more often in damage to insulation with the development of arcing and sparking. This may ignite vapours, dust or fluff or other combustible materials close at hand. Poor connections which have a high resistance cause overheating and a further deterioration of the connection until a temperature high enough to start a fire is reached.

Cables and wiring should be securely installed, properly supported and protected against mechanical damage, heat, vibration and the ingress of moisture and corrosive substances. They should be run so that they cannot accidentally come into contact with other services.

Transformers and other oil filled apparatus should be so installed that in the event of an oil leak and/or fire the oil is contained and cannot spread to other equipment.

With any fire involving electrical equipment the first step should be to switch off the current to disconnect the supply of electricity.

Provided it is safe to do so, personnel should attack the fire with suitable extinguishers containing carbon dioxide, dry powder or a vapourising liquid. Water or foam should not be used unless the supply of electricity has been cut off, and the fire is too large to be dealt with adequately by one of the recommended types of extinguisher. Water should not, however, be used if the fire involves oil and the fire should then be treated as a flammable liquid fire.

The fire brigade should be called to all electrical fires, and no attempt should be made to restart the equipment after the fire until all damage has been repaired and the equipment has been tested by a competent inspector.

4.3.2.6 Chemical fires

Fires involving chemicals require special treatment. There is such a wide variety of chemicals that in order for professional firemen to deal with them rapidly and effectively, a method of coding and labelling has become necessary. This tells the fireman at once how to handle the fire (or one which might involve the chemical) even if he is quite unfamiliar with the chemical or its properties.

The Hazchem system of coding and labelling is essentially an action code. It was started in the UK under the lead of the London Fire Brigade, originally for the marking of road tank cars which might be involved in an accident. This Hazchem code is essentially a transport code. It has been adopted on a voluntary basis as a labelling system by most UK oil and chemical companies, with the backing of the Chemical Industries Association,[5] and has been taken up by the Home Office and the Health and Safety Executive. It is expected to be obligatory for the marking of road tank cars in the UK very soon.

So successful has the code become for those dealing with transport accidents that it is now being extended to the labelling of chemicals and solvents in drums, bottles and cartons, and for their bulk storage. This again has been initiated by the London Fire Brigade, and is known as the London Hazchem Code.[12] The codes of most materials are the same on the

Hazchem transport code and on the London Hazchem Code which applies mainly to storage, but there are sometimes minor differences. In this case it is the London Hazchem Code which should appear on the label of the drum, carton, bottle or bulk storage container, but if a consignment of similar drums are put on a lorry for transport the transport code should appear on the outside of the lorry.[5]

4.3.2.7 The London Hazchem System[12]

The Hazchem code for any chemical consists of three parts, on an orange background:

(a) A number 1 to 4
(b) A capital letter P to T or W to Z
(c) An optional letter E

(a) The number 1 to 4 signifies the type of extinguishing medium to be used:

1 = Water jet.
2 = Water fog, produced by a high pressure atomising jet, *directed upwards*.
3 = Foam (normal protein foam as used by all fire brigades).
4 = Dry agent — commonly sand. Dry powders and vapourising liquids are also in this category.

Of these 1 and 2 are both water, but differ in form, whilst 3 is over 99% water, although its form and method of action are again different. Water jets which cool and quench are used mainly on carbonaceous materials.

Water fog (more akin to a 'Scotch Mist') is readily vapourised by a fire to form a mixture of steam and air which has a much lower oxygen content than air alone, and gives less support to a fire. Water fog may thus be used cautiously for dealing with large hydrocarbon fires where jets would be prohibited. It can also be used to wash and disperse clouds of toxic and flammable gases and vapours in the atmosphere, e.g. chlorine, ammonia and propane.

Foam has mainly a smothering or blanketing action in denying access of oxygen in the atmosphere to the fire. The water contained in it also has a quenching effect.

(b) The meaning of the letters P to T and W to Z are shown by *Table 4.3.6*.

Materials coded P to T are those which may in an emergency be diluted and flushed down the nearest drain. Materials coded W to Z may not be diluted and must be contained. Materials P and R, and W and X are those whose vapour or products of combustion are injurious to the human skin and for which full impervious protective clothing (generally nitrile rubber),

Table 4.3.6 HAZCHEM CODE LETTERS

Code	Reactivity	Personal protection	Treatment of spillage
P	Violent	Full Protection	
R			DILUTE
S	Violent	Breathing Apparatus	
T		Only	
W	Violent	Full Protection	
X			CONTAIN
Y	Violent	Breathing Apparatus	
Z		Only	

as well as self contained breathing apparatus with compressed air cylinder, must be provided. Materials S and T, and Y and Z are those for which breathing apparatus alone is necessary. Breathing apparatus is frequently necessary for work in smoke and oxygen deficient atmospheres even when the chemical itself presents no hazard. Materials P, S, W and Y are all capable of reacting violently either alone or with some other substance (e.g. water, air, metals or combustible materials) with which they may come into contact.

(c)   The final and optional letter E calls for evacuation of all persons other than fire brigade personnel to a minimum distance of 200 m from the incident, whether fire or spillage. This is used for materials which may be particularly hazardous, e.g. explosive, supertoxic, biologically active or radioactive.

The London Hazchem code is basically a code to enable firemen to determine instantly what action to take in the event of a fire in a storage area, works or laboratory. It is rather simpler than the Hazchem transport code which has to consider action to be taken by firemen in the event of spillage as well as fire. The Hazchem transport code requires the code letters S, T, Y and Z to be printed in orange inside a black box for substances for which breathing apparatus is required for fire only (i.e. not spillages), e.g.

The material coded S required the use of breathing apparatus in the event of fire only but not spillage. The London Hazchem code has no need to make such a distinction, and the black box device is omitted, the material being simply coded S.

Occasionally there may be other differences between the transport and the London codes for the same material, but these are not common and will probably disappear as the codings for different materials are revised.

Since the responsibility for dealing with any fire, whether chemical or not, rests with the fire brigade from the moment it arrives, it is important that all companies which carry and use chemicals and solvents should label all containers prominently with the London Hazchem code for the material. Their own staff, particularly those who may be involved in emergency fire fighting, should also be fully familiar with the codes and their meaning, so that they can use the appropriate extinguishing medium and employ personal protection, even where these fall short of those used by professional firemen. Personnel handling particular solvents and chemicals frequently, will generally be familiar with their properties and what fire extinguishing media may be employed, but the London Hazchem label displayed on the container should leave them in no doubt. If number 2 is displayed, it is clear that even small water jets should not be employed. If a fine water spray is not available, it will be necessary to use foam. If number 4 is displayed, a dry powder extinguisher or possibly sand is the most likely choice.

4.3.3 Fire warning systems

An excellent description of fire warning systems is given by Underdown.[6] Every works and factory should have both the necessary equipment and proper systems which people understand to warn of outbreaks of fire. Once a warning is given, the system should provide for the local fire brigade to be called at once. Should the warning turn out to be false, the brigade can be contacted by phone and the team and appliance diverted if necessary to a more urgent call by the fire station by radio.

It is better to err on the side of calling the brigade prematurely or unnecessarily rather than too late, since the first few minutes of a fire are vital. The times taken for the brigade to reach a fire from receipt of the signal are remarkably low in England and Wales in all but rural districts, see *Table 4.3.7*.

Table 4.3.7 ATTENDANCE TIME AND AREA IN WHICH FIRE OCCURS[7]

	Attendance time – per cent	
	4 minutes	*8 minutes*
Metropolitan borough	95.5	99.9
County borough	71.5	97.0
Municipal district	63.5	96.8
Urban district	41.1	88.0
Rural district	11.0	47.3

Methods of warning of fire range from the simplest — giving vocal warning — to sophisticated automatic electric warning systems.

Vocal warning. Vocal warning is usually unsatisfactory and dangerous since the very word 'fire', particularly when amplified over a public address system, is liable to cause panic. In some public buildings a code message, such as instructing a fictitious person to report to the manager, may be used to warn staff over the public address system. A second failing of vocal warning is that the person giving the warning has to remain in the building shouting until everyone has the message.

Fire triangles and hand-operated gongs. These suffer from the second defect of vocal warning in that someone has to remain sounding the warning until everyone has heard it.

Warning by internal telephone. This may involve dialling a special number which alerts the operator by a special signal — it has advantages and disadvantages. The main advantage is that it enables the person giving the warning to state the location and size of the fire, so that the brigade can be given this information and told which works entrance to use (if there are more than one). The disadvantages are:

1. It requires careful training of all personnel to ensure they remember which number to dial.
2. It requires special training of the telephone operators on the action they must take.
3. The switchboard may be unattended when the alarm is given.

It may still be possible to use such a system even when the works has a fully automatic switchboard. This should be discussed with the G.P.O., preferably before the switchboard is installed.

4.3.3.1 Electric fire alarm systems

The use of a separate specially installed electric fire alarm system is the preferred method. The alarm may be raised manually by a call point *(Figure 4.3.3)*, preferably one which only involves breaking the glass; this releases a spring-loaded button and initiates the alarm. Automatic fire detectors *(Figure 4.3.4)* may also be used with this type of system. There are four main types of fire detector:

1. Heat detectors.
2. Smoke detectors.
3. Flame detectors.
4. Combined heat and smoke detectors (calcium-arsenide infra-red beam).

Flame detectors where applicable, give the fastest warning. Care is needed in the choice of detector to avoid the likelihood of false alarms. The power supply to the system should be from an accumulator which is trickle-charged from the mains. Manual alarm points should be painted red and numbered, fixed at a height of 1.35 m from the floor, spaced at intervals of not more than 30 m, be easily accessible, well illuminated and conspicuous. They should all be actuated in the same way, preferably by smashing a light glass pane, by elbow, shoe or a special metal object provided. They should be inspected daily to ensure that they are not obstructed and they must be tested every three months, preferably by an outside electrical engineer under contract.

Four classes of fire alarm systems are recognised according to the size of the premises protected:

Figure 4.3.3   Manual fire alarm point, with hose reel (London Fire brigade)

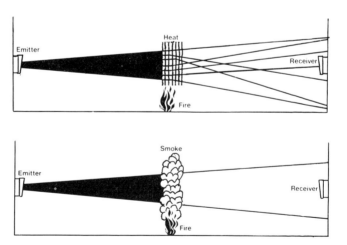

Figure 4.3.4   'Beam-master' combined smoke and heat detector system (Chubb Fire Engineering)

1. Small, single-storey premises with only few calling points connected to a single (or at most two) evacuation signalling points. There is probably no alarm receiving station or direct connection to the fire brigade; the fire brigade is warned indirectly, by telephone when the alarm rings.

2. Somewhat larger premises (two or more storeys) with a number of calling points connected to the evacuation bells, to an indicator panel to show where the alarm was first given, and usually by direct line to the fire brigade. The premises form a single fire compartment which would be completely evacuated when the alarm sounds.

3. Larger premises divided into several fire compartments so that only the compartment from which the warning is given is evacuated immediately on receipt of the warning. The fire call points are connected to a single indicating/control panel which usually has to be constantly manned by a trained person. He has to decide when to call the fire brigade and when to evacuate each compartment. He may also stop ventilating fans, close fire or smoke doors and shut valves by electrical signals.

4. Premises with a number of buildings which may be types 1, 2 or 3, (with their own sub control points) within the works area. These deliver a signal to a central control point, usually within the perimeter (see *Figure 4.3.5*). This control point would in a large works with large fire hazards be the same as the emergency control centre described in section 1.13. It would have an ex-directory telephone number

Figure 4.3.5   Fire control panel (GLC London Fire Brigade)

independent of the works switchboard, emergency lighting, toilet and canteen facilities and should be of fire, and, if necessary, explosion resistant construction.

The alarm signals are generally audible and are usually bells, sometimes supplemented by a visual signal. Care is needed to ensure that the alarm can be heard everywhere and by everyone in the compartment, even when noisy machines are in use. Sometimes a two-stage alarm is justified — an intermittent warning tone and a continuous tone which requires evacuation. The intermittent warning tone may be used as a signal to shut down hazardous processes, and for auxiliary fire fighters to proceed to assist in fighting the fire. No universal scheme suits all circumstances. Each one should be based on the particular circumstances. It is dangerous to have more than two such signals because of the danger of confusion, and even the use of two signals calls for careful training. The evacuation signal should be arranged to sound for at least five minutes before being automatically switched off.

Occasionally an additional 'all clear' signal is justified, i.e. in compartments adjacent to one with a fire where an intermittent warning signal only has been given.

4.3.4 Evacuation

The design of fire escape routes and doors was discussed in section 2.6. Requirements of such escape routes and doors are given in the Building Regulations. These insist that the individual should be able to escape from a fire by his own unaided efforts, and assume that collective action or rescue by the brigade cannot be relied on. The use of lifts to escape from a fire should generally be prohibited because of the danger of the power supply being cut off with persons trapped in the lift between floors. In tall buildings it may be necessary to specially protect certain lifts so that these can be used by the fire brigade.

Periodic fire drills under simulated fire conditions are necessary to ensure that every person is fully familiar with the alternative escape routes available to them, and is able to choose the most appropriate to the situation. The need to keep windows and doors on escape routes closed as far as possible to prevent them becoming filled with smoke or passages for air to feed the fire should be demonstrated, preferably by the use of smoke generators.

If a number of occupants in a building have to use an escape route containing doors or other constrictions, collective training in using the route without jamming the constrictions should be included. It is essential to avoid crowd pressure on exits which can arrest movement completely like an arch of granular solid or powder inside a hopper.

It is not enough for people to be able to use the escape route under fire and smoke free conditions with normal lighting. They may have to use it when it contains dense fumes, moving on hands and knees, feeling their way by the edge of the room, corridor or stairway, and keeping their mouths low

Firgure 4.3.6   Obstructed fire doors (GLC London Fire Brigade)

so as to breathe from a layer of cool and cleaner air adjacent to the floor.

Fire drill should also prepare people for survival in case the only exits are blocked and they are trapped by fire. Where possible they should then make for a room with a window which is free from fire and which has a door. Once inside all doors and openings between the person and the fire should be closed, and the bottom of the door sealed with any fabric available. The person should then open the window, stand by it and shout for help and learn to be patient, rather than risking his life by jumping from 9 m or more on to a hard surface. If doors are shut between him and the fire it may by-pass him completely.

Continuity in the colour, type of floor, wall, doors and lighting of an escape route, and clear and unambiguous marking of the escape route at every possible 'wrong turning' are important.

The time taken to evacuate a building by alternative escape routes once the warning has been given should be checked. It should not exceed 2½ minutes in a real fire situation.

In tall buildings containing eight floors or more, complete evacuation by stairways in the 2½ minutes specified is usually impossible. The solution likely to be adopted in the UK, following USA practice, is to provide fire resistant refuges at various elevations in the buildings to enable people to ride out the fire.

4.3.5 Emergency fire fighting

Fire appliances and extinguishing agents were referred to in section 4.3.2, in relation to the types of fire for which they are most suitable.

One needs to distinguish between large and specialised appliances, used by fire brigade and professional fire fighters for dealing with fires which have got out of control of the local works personnel, and smaller appliances used mainly by works personnel for dealing promptly with fires in their early stages. Incidentally, the old expression 'fire engine' is deprecated in BS 4422; Part 5, 1976, which recommends the use of the word 'fire appliance' to cover all equipment (including 'fire engines') provided for the purpose of detecting, recording or extinguishing a fire.

The larger appliances used exclusively by fire brigades which include mobile pumps, mobile turntables, platforms and extension ladders, rescue and demolition equipment are not dealt with in this section since they come under the control of a trained and experienced fire officer.

Equipment which is handled by the normal works personnel (and also in most cases by the works fire brigade or department) is dealt with here. It falls broadly into two categories:

1. Fixed appliances and hose reels;
2. Portable appliances.

4.3.5.1 Fire extinguishers and fixed appliances[8]

Fixed extinguishers may be designed for manual or automatic operation. They may also be classed as (1) those for general application and (2) those for use where there is a special risk.

Fixed installations using water and high expansion foam are most suitable for general protection while those using other extinguishing agents are intended for special risks such as oil or electrical fires. The choice and positioning of fixed installations should be considered when a works or building is being designed, since it is more expensive to install them once a building has been completed.

4.3.5.2 Hose reels

Hose reels are first-aid fire extinguishing equipment provided for the use of the occupants of a building or works and they may be installed instead of, or in addition to, portable water type extinguishers. When installed they will also be used on small fires by the brigade on arrival. This causes less water damage than the brigade's larger hoses.

A hose reel consists of a length (up to 36 m) of non-kinking rubber tubing with an internal diameter of 19-25 mm. A valve and nozzle are attached to the free end of the hose which is wound on a metal reel. The reel is usually supported by a wall braket and may be arranged to swing on a pivot. The reel has a hollow rotating shaft to the centre of which water is fed. The hose tubing is connected to an outlet on this rotating shaft. The shaft is permanently connected to a suitable water supply through special pipework.

With one type of hose reel, all that is necessary to obtain a jet of water is to grasp the nozzle, pull out the amount of hose needed to reach the fire and open the cock at the nozzle. The action of unwinding the reel or removing the nozzle from a special wall fitting turns on the water supply valve. With another type, a valve on the water inlet to the reel must be opened manually before the hose is run out.

A hose reel may be fixed or pivoted. The fixed type has guides fitted so that the hose can be pulled off the reel without kinking or jamming. The pivoted type swings to the direction in which the hose is pulled.

Nozzles for hose reels are available with internal diameters of 4.5-6.5 mm. The size chosen depends mainly on the pressure of water available. Nozzles should give a minimum flow of 0.38 litres per second. For a 4.8 mm bore nozzle, this requires a water pressure of 2.5 kg/cm^2 gauge at the nozzle. For a 6.4 mm bore nozzle this requires a water pressure of 0.8 kg/cm^2 gauge at the nozzle. The pressure loss caused by friction through 10 m of hose at a flow rate of 0.38 litres/second is 0.15 kg/cm^2 for a 19 mm bore hose and 0.035 kg/cm^2 for a 25 mm hose.

Hose reels may be supplied with fixed covers to protect them from dust, dirt and light which cause deterioration of the rubber tubing. They should be positioned so that no part of a building is more than 6 m from a nozzle when the hoses are fully extended, making due allowance for obstructions.

The flow of water through a hose reel with nozzle can be simply checked by measuring the maximum horizontal throw of the jet by directing it over a flat roof or open floor. A nozzle with a bore of 4.8 mm should give a maximum throw of at least 12 m and a nozzle with a bore of 6.5 mm should give a maximum throw of at least 18 m.

Hose reels require regular maintenance and checking at least once a year, in accordance with manufacturers' recommendations. Brief instructions for operating a hose reel should be displayed on or close to it. All employees should be trained to use hose reels, including how to pull the hose round obstacles.

4.3.5.3 Hydrants

Hydrants are arrangements of piping and outlets to which large diameter hose (64 mm or more) may be connected for use by the fire brigade or fully trained works firemen. Some hydrants inside buildings which are known as dry risers are kept empty until they are needed. These are used in cold climates where water in a wet riser might freeze, and also in very tall buildings where water will not reach the top of the riser until the fire pump is started.

Other hydrants, known as wet risers, are kept permanently full of water. If the water pressure in the fire main is not sufficient to deliver it to hydrant outlets at the top of tall buildings at sufficient pressure to operate the hoses, the risers inside the buildings must be supplied with water from a pump, usually a mobile one carried by a fire brigade. Wet risers are generally preferred to dry ones in situations where they can be used.

Hydrants may be fitted with foam inlets to which firemen attach a supply of foaming agent that mixes with water in the hydrant and hose.

4.3.5.4 Automatic sprinklers

These consist of a system of pipes, spray nozzles and heat operated valves by means of which a fire is automatically detected, the alarm given and water delivered to the fire. Sprinklers are useful for stores and other buildings containing combustible materials which are left unattended. The cost of the installation may be partly or wholly offset by the reduction in the fire insurance premium paid.

Similar systems may also be used on the outside of buildings and tanks to keep them cool if a fire develops near them and so to prevent the fire from spreading to them.

4.3.5.5 High expansion foam

The system consists of one or more foam-making machines fitted with short rigid ducts inside the roofs of single-storey buildings.

Multi-storey buildings are protected, with similar but larger ducts which run from roof to ground level. One or more floors may be flooded with a foam of very low density but sufficient stability not to collapse at once when exposed to a fire.

The foam may be filled with carbon dioxide instead of air. Its action is to smother and blanket a fire.

This is a relatively new system which has been mainly used to protect basements and tunnels to which access may be difficult if a fire develops there. It is finding wider application in warehouses and large buildings. The local fire brigade should be consulted for advice when the installation of such a system is considered.

4.3.5.6 High pressure water spray (fog)

The water is delivered at high pressure through special nozzles to form fine droplets. A high pressure water spray requires the use of a special booster pump (carried by most brigades) which gives a pressure of about 50 kg/cm^2. It rarely forms part of a fixed installation. Its main use is to protect against fires of flammable liquids and liquefied gases. The use of a very fine and carefully directed water spray removes the main disadvantages of water in dealing with such fires.

4.3.5.7 Medium expansion foam

Foam may be produced from a fixed foam solution vessel and carbon dioxide cylinder or from foam making equipment carried by the fire brigade. In either case it may be applied through systems of fixed pipework either to the seat of the fire or to the plant to be protected. Brigades normally carry supplies of normal protein foam only. This is mainly suitable for flammable liquid fires where the liquid is immiscible with water. Liquids such as alcohols which mix with water tend to break down the foam blanket. But for these conditions special compounds are available, which give stable foams, although they are more expensive than protein foam.

4.3.5.8 Carbon dioxide and other inert gases

These systems must be used with great caution indoors when people are present, due to their asphyxiating action in lowering the oxygen contant of the air. In at least one case a man fighting a fire in a basement with carbon dioxide extinguishers which were handed down to him from above collapsed and died as a result of oxygen deficiency in the atmosphere.

Gas extinguishing systems consist of a supply of the inert gas under pressure (usually in cylinders), a system of pipework and valves delivering the inert gas to the points of application and an automatic detection and initiating system which opens inert gas valves once the fire has been detected. It also, in many cases, closes doors and ventilation ducts. These systems operate by reducing the oxygen content of the atmosphere and/or by interrupting the chemical reaction in a flame.

Most of the gases used are suitable for electrical equipment and plants handling flammable liquids. This system is particularly suitable for protecting valuable equipment which is easily damaged by water and foam, such as computers. For these systems to operate most effectively, the fire fighting equipment should be housed in a gas-tight compartment which is closed to the atmosphere when a fire starts. Carbon dioxide installations need special care to avoid introducing risks of ignition by static electricity (see Section 5.6).

4.3.5.9 Dry powder

Dry powder is a term used for various free-flowing powders which when poured or otherwise discharged over a fire will extinguish it. The compositions of many fire extinguishing powders are not publicised by their makers for obvious reasons. They generally contain three principal ingredients each with a particular function.

1. *Sodium or potassium bicarbonate.* These liberate carbon dioxide when heated. Bicarbonates are thus a convenient means of applying carbon dioxide. They also react with and neutralise acids and some other reactive compounds and prevent damage from acids

released by a fire. Potassium bicarbonate is claimed to be more effective than sodium bicarbonate due to its greater chain terminating effect.

2. Certain finely powdered salts of metals which when present as a dust in the atmosphere strongly absorb radiant heat, thereby cooling and in some cases extinguishing flames.

3. A compound which prevents the powder particles from adhering to one another and forming lumps, thus preserving the free-flowing properties of the powder.

Dry powder installations comprise a dry powder container to which a gas cylinder (usually carbon dioxide) is coupled, and a system of piping and outlets which are located above the places where fires are likely to break out. They can be operated automatically or manually by opening a valve on the gas cylinder so that the gas drives the powder to the outlets.

These installations are suitable for flammable liquids and electrical equipment and for protecting some processes involving solids which are easily damaged by water or foam.

4.3.5.10 Portable appliances[9]

Portable fire extinguishers may be used to deliver water, dry powder, foam, carbon dioxide or a vapourising liquid to the seat of a fire. Their use should, as far as possible, be standardised and the minimum number of types necessary should be carried. Hose reels are generally preferable.

Supervisors should be able to identify the different classes of fires and should know which type of portable extinguisher to use. All personnel should know how to recognise and use the various types of extinguishers present. Practice sessions should be set up to ensure that all personnel act promptly and effectively in dealing with small fires.

4.3.5.11 Portable water discharging extinguishers *(Figure 4.3.7)*

Extinguishers which deliver water operate in various ways:

(a) *The water bucket.* This is the simplest of all; it must, however, be kept full of clean water and always in the place reserved for it. Some skill is required in directing the contents of a water bucket onto a fire — probably more so than closed portable extinguishers with nozzles.

(b) *Gas pressure applied from a cartridge.* A small cartridge of liquefied carbon dioxide is held inside the top of the cylindrical extinguisher. The cartridge has a brass cap which is pierced by a plunger passing through a gland in the top of the extinguisher. This is actuated by a sharp blow by the hand to the top of the plunger. Gas released in the extinguisher drives the water out through a discharge tube which extends to the bottom of the cylinder and is connected outside the cylinder to a nozzle via a short length of flexible hose.

Figure 4.3.7   Use of portable water extinguisher (Fire Protection Association)

(c) *Stored gas pressure.* The whole extinguisher is pressurised with gas at the time of charging with water. Water is discharged by opening a valve on the discharge tube.

(d) *Gas pressure formed by reaction between an acid and a carbonate within the extinguisher.* The extinguisher has an inner container filled with an acid solution (generally aluminium sulphate). The main body or outer container of the extinguisher is filled with a solution of sodium bicarbonate.

   The contents of the inner cylinder are released into the outer cylinder by inverting the cylinder and releasing a spring operated plunger. When the solutions mix, gas is formed which pressurises the extinguisher. A jet of water issues through a nozzle on the upper part of the extinguisher, so long as the extinguisher remains inverted. The flow of water stops when the extinguisher is turned the right way up and surplus gas escapes.

(e) *Hand pump inside the cylinder.* This is operated by a handle extending through a gland in the top of the cylinder.

The applications and limitations of water extinguishers have already been discussed. Water is best used for fires on solid materials which may re-ignite if not adequately cooled. It can readily penetrate to reach a deep seated fire.

Portable water extinguishers have capacities from 4 to 10 litres and an effective range of about 10 m. One or two extinguishers depending on their size are normally required for general protection per 220 m² of floor area.

When using a water-filled extinguisher, direct the jet at the base of the flame and keep it moving across the area of fire. A fire moving vertically should be attacked at its lowest point and followed up. Seek out any hot spots after the main fire is extinguished.

4.3.5.12 Portable foam extinguishers *(Figure 4.3.8)*

Foam extinguishers are of two types, mechanical and chemical. These correspond to water extinguishers in which the pressure is derived from a gas cartridge and from chemical reaction respectively. But the chemical foam extinguisher, unlike the soda-acid type of water extinguisher, is used in the normal upright position. Portable foam extinguishers have a capacity from 4 to 10 litres and a range of about 7 m. 10 litres of foam are normally required to extinguish 1 m² of burning liquid.

When a liquid fire has been extinguished by foam, the foam blanket left over the liquid remains in position thus preventing re-ignition and allowing the liquid to cool. Foam extinguishers should therefore be used for liquid fires where the liquid has been burning for some time and has become hot.

Foam is not effective on flowing liquids, whether the flow is horizontal or

Figure 4.3.8   Use of portable foam extinguisher (Fire Protection Association)

vertical. Foam conducts electricity and should not be used on live electrical fires. Most water miscible liquids break up ordinary foams.

When a liquid on fire is in a container, direct the jet at the far inside edge of the container, or at an adjoining vertical surface above the level of the burning liquid. This breaks up the jet and allows the foam to build up and flow across the surface of the liquid. When this is not possible, stand well back and direct the jet slightly upward so that the foam falls on to the surface of the liquid. Move the jet gently from side to side to cover the surface of the liquid. Do not direct the jet into the liquid because this will drive the foam beneath the surface and render it ineffective. It may also splash the burning liquid on to surrounding objects. *(Figure 4.3.8)*

4.3.5.13 Portable dry powder extinguishers[9] *(Figure 4.3.9)*

The use and composition of dry powder have already been discussed under fixed installations, (section 4.3.5). Portable dry powder extinguishers are made with capacities from 2 to 10 kg of powder. In operation and

Figure 4.3.9 Use of dry powder/vapourising liquid extinguisher (Fire Protection Association)

appearance they are like water extinguishers where the pressure is supplied from a gas cartridge. Their range is less than a water extinguisher, usually from 3-6 m.

These are the best type of extinguisher for dealing with fires of flammable liquids. They extinguish the flames over the liquid and thus act faster than foam. They can deal with larger areas of burning liquid than other extinguishers of the same size, and they are effective on fires of flowing liquid. Dry powder can be safely used on electric fires. The main limitation of dry powder is that it gives no protection against re-ignition after application ceases since it has poor quenching properties. It is less effective than foam on liquid fires where the liquid has become overheated (i.e. through prolonged burning).

Two kilograms of dry powder can normally extinguish a liquid fire covering an area of one square metre when properly applied.

4.3.5.14 Portable carbon dioxide extinguishers

Carbon dioxide extinguishers should only be used sparingly in buildings due to the dangers of asphyxiating personnel. A second hazard of carbon dioxide extinguishers is the formation of static electricity in the discharge which can ignite flammable vapours, sometimes with fatal consequences.

Carbon dioxide acts more rapidly than foam and is more suitable for dealing with fires which might spread to surrounding materials before a complete foam blanket could be formed over the burning liquid. Carbon dioxide extinguishers are suitable for dealing with small fires of liquids flowing over horizontal and vertical surfaces. They should be used where the main concern is to avoid damage or contamination by dry powder deposit or foam, for example to laboratory equipment or food preparation.

The cooling properties of carbon dioxide are limited and it gives no protection against re-ignition after application ceases. It is less effective than foam for very hot liquids burning in containers.

Carbon dioxide extinguishers contain the carbon dioxide under high pressure as a liquid in steel cylinders, with a valve leading via flexible hose to a horn shaped discharge tube. These extinguishers are normally used with the valve uppermost so that carbon dioxide is discharged as a gas. If they are inverted, a mixture resembling snow of carbon dioxide gas and solid carbon dioxide is discharged, provided the extinguisher is full and the ambient temperature is not excessive.

Portable carbon dioxide extinguishers have capacities ranging from 1 to 6 kg and a range from 1 to 3 m.

4.3.5.15 Portable vapourising liquid (Halon) extinguishers

Portable vapourising liquid extinguishers are now mainly restricted to the use of two compounds, Bromochlorodifluoro methane or BCF, and Bromotrifluoro methane or BTM.

These may be discharged either by gas cartridge (containing carbon dioxide) or by pressurising the container with nitrogen. They can be fitted with a control valve if desired, so that they can be discharged in short bursts, but once the seal has been broken they should be emptied, recharged and resealed. Their main action is by excluding oxygen from the flames. Since they do not conduct electricity they can be used on electrical fires. They have less static electricity risk than carbon dioxide, but they present the same asphyxiation hazard. In addition there is some risk of forming toxic decomposition products when their vapours are in contact with very hot metal, although this risk is far less than with older types of vapourising extinguisher which contained carbon tetrachloride, methyl bromide and other compounds which are little used now because of the toxic problem.

These extinguishers have a range of up to 6 m and 1 litre of liquid is sufficient to extinguish flames over an area of one square metre of burning liquid. The methods of using dry powder, carbon dioxide and vapourizing liquid extinguishers are essentially the same. *(Figure 4.3.9).*

On fires involving liquids, either in containers or on the ground, direct the jet or discharge horn towards the near edge of the fire and with a rapid sweeping motion drive the fire towards the far edge until all the flames are extinguished. On fires in falling liquids, direct the jet or horn at the base of the flames and sweep upwards.

When dealing with electrical equipment fires, first turn off the current. Then direct the jet or horn straight at the fire. When the equipment is enclosed, direct the jet or horn into any opening so that it penetrates the interior.

If the extinguisher has a control valve on the discharge, shut it when the fire appears to be extinguished, wait until the atmosphere clears and, if any flame is then visible, open the valve and discharge again.

4.3.5.16 Recharging extinguishers

All extinguishers should be recharged immediately after use, irrespective of whether they have been completely or only partly discharged. The safety or fire officer should arrange for books to be kept by supervisors to record every use of an extinguisher and when it was recharged.

4.3.5.17 Colour identification of portable fire extinguishers

The availability of many types of portable fire extinguishers for different types of fire have led to steps being taken to standardise their body colours for ease of identification. BS DD 48 1976 *Draft for development* proposes the following body colours for the different types of extinguishing agent:

Water	Signal red
Foam	Pale green
Powder (all types)	French blue
Carbon dioxide	Black
Halogenated Hydrocarbon	Emerald green.

4.3.6 Organisation, training and procedure[6,10]

Those whose duty it is to fight fires form three lines of defence:

1 *Personnel* working in the area or building where the fire starts. These form the first line of defence. Their weapons are hose reels, portable fire extinguishers and such other appliances as may be available, backed up by thorough training and practice.

2. *The works fire brigade.* Its size and composition and the equipment at its disposal will depend on the size of the works, the risks involved, and on the degree and promptness with which help may be expected from the public fire brigade. The works fire brigade will normally contain at least a nucleus of full time staff on every shift, sufficient in number to man the appliances available and drive them to any fires to which they are called within the works. It may be augmented by a number of volunteer firemen (frequently maintenance workers) who can be called upon instantly to assist the full time firemen when they are called to a fire. The works fire brigade will be under the command of an experienced fire officer.

3. *The public fire brigade.* In assessing what help may be expected from the public fire brigade, and how long it will take to arrive, account must be taken not only of the distance of the nearest public fire brigade from the works, but also other claims on its services and the equipment available.

Co-ordination of these three lines of defence and evacuation of personnel require good communications and a centre. Initial alerting will be mainly the responsibility of the works switchboard operator, but to ensure effective co-ordination of the actions that follow, a fire control centre should be arranged in advance that can be brought into operation quickly in an emergency. This may be either the works telephone switchroom or on the premises of the works fire brigade, or possibly the works manager's office or the safety specialist's office. The important point is that the whereabouts of the centre should be decided and known by all concerned in advance and that all reports, not only of the outbreak of fire but all subsequent developments, should be routed to it, thereby creating a position from which the situation can be controlled as it develops.

The training of the full and part-time members of the works fire brigade is obviously important, but so is that of the normal works personnel who form the first line of defence. The speed and effectiveness of their actions when a fire breaks out will determine the extent of the fire and the damage caused by it.

It is, therefore, strongly advised that all personnel be instructed in special sessions and that fire drills be held at least twice a year in which they can practise what they have been taught. The instruction sessions and fire drills should include the following points:

closing fire doors;
selecting and using portable extinguishers and hose reels;
directing firemen to a fire and closing valves on gas and fuel lines;
seeing that extinguishers are recharged;
use of gas masks and respirators.

4.3.6.1 Procedure in the event of fire

Although the various actions to be taken in the event of a fire have been covered in the foregoing sections, these have necessarily had to be broad and flexible to allow scope for various alternatives. No universal procedure can be laid down for every works or for each department within a works.

Nevertheless, it is strongly recommended that a firm procedure be established for each building or department within a works, so that precious time is not wasted in deciding which alternative to adopt. The responsibility of drawing up the procedure rests with management, advised where necessary by the fire and safety officers. The following points must be covered:

1. A senior member of staff is made responsible for safety in avoiding fire.
2. All personnel are issued with clear and comprehensive instructions.
3. Instructions are set out both for 'normal' (where applicable) and 'emergency' procedures.
4. Responsibilities in the event of fire are clearly allocated.
5. The design of the works or premises and the particular hazards of the materials and processes used are taken into account.
6. The alarm system for evacuating a building is clearly audible and distinguishable throughout the building.
7. A fire 'control centre' will come into operation in the event of fire with standing instructions as to the immediate actions to be taken including summoning the works fire brigade and notifying the management.
8. The evacuation procedure is efficient and all employees are familiar with the main and alternative escape routes from their places of work.
9. In the case of evacuation, provision is made to search every part of the building to ensure that no-one is left behind.
10. Practice fire drills are held at least twice a year.

4.3.7 Protecting against the effects of fire[11]

In all works it is advisable to plan the procedures to be taken in the event of a fire in order to reduce damage and speed up the return to normal operations. Besides considering direct damage to buildings and their contents from fire, the following points have to be taken into account.

1. Damage to machinery, stock and buildings from water used for extinguishing purposes.
2. The damage caused by the spread of smoke over a wide area.
3. Loss of production and business.

An organisation should be set up in every works to plan for fire rehabilitation, whether a fire has occurred or not. This may consist of the works manager, the chief engineer and his deputy, the fire officer, the insurance manager and specialist engineers in the maintenance department. The head of any department affected should be co-opted to the organisation if there is a fire. The fire brigade should be told in advance about particularly vital equipment or stock which could be damaged by fire.

Telephone numbers and addresses of people whose services may be needed to help renovate the building, provide temporary covering to damaged roofs, save equipment or materials or supply other equipment, should be kept by the organisation on an up to date basis. The list will include insurers, salvage experts, smoke removal experts, the architect of the building, consultants and contractors used by the firm and the emergency phone numbers of the suppliers of electrcity, gas and water.

From time to time the possible effects of fire in different buildings or departments should be considered and plans formulated to reduce the effects on production. These effects will include the following losses:

1. Power supplies;   4. Vital components;
2. Various machines;   5. Records and drawings;
3. Raw materials;   6. Premises.

Essential equipment, parts, stock or raw materials which would be very difficult to replace at short notice should be protected thoroughly against fire and where possible duplicated in well separated parts of the works.

4.3.7.1 Protecting against water damage

After some fires it has been found that the damage caused by water from fire hoses was even more costly than that caused by the fire itself.

Most of the safeguards that may be taken against water damage in the design of buildings were mentioned in section 2.6. These are summarised and extended below.

1. Waterproof floors above valuable equipment and goods.
2. Floor levels designed and sloped to direct flow of water from building.
3. Adequate drains clearly marked with manholes and scuppers.
4. Where drains cannot be provided, provide sumps with automatic pumps to collect and remove water.
5. Provide sills or ramps at door and other openings.
6. Provide drainage channels across large undrained floor areas to prevent lateral spread of water.

The contents of a building should be arranged with a view to minimising water damage, i.e.

1. Damageable stocks should be stored on non-combustible pallets to prevent water absorption from the floor.
2. Stacks should be of reasonable proportions and allow all round access so that they can be speedily covered with waterproof sheets.
3. Racks and shelves should not extend to ceilings, or butt directly on to walls. They should not retain water and should be non-combustible.
4. Vital parts of plant and equipment should be provided with waterproof covers such as polythene sheets for use in an emergency.
5. Valuable paper records and other vital items should be stored in fire-resisting cabinets or strong rooms. The most vital paper records and correspondence should be microfilmed and the film stored at a remote location.
6. Electrical switchgear should be mounted on battens and not directly on a wall so that it will not be affected by water running down the wall.
7. Electric motors should be mounted on plinths rather than at floor level.

4.3.7.2 Protecting against smoke damage

Precautions to limit the spread of smoke include smoke stop doors in corridors and at entrances to staircases and lobbies. Staircases should be ventilated by opening windows and skylights. Automatic dampers should be fitted at key points on conveyors and in air conditioning ducts.

Roof ventilators should be provided in large single-storey buildings. Basements should have smoke outlets provided with fitted covers that can be removed or broken to allow smoke to escape. Stores of materials which produce the most harmful or obnoxious smoke, cork, wool, jute and rubber should be segregated from other stores and departments.

4.3.7.3 Salvage during fire fighting

Much useful action may be taken by a salvage team to reduce fire damage from the moment a fire breaks out. Preventive work which should be carried out at the discretion of the officer in charge of the fire could include measures listed below.

1. Moving stock in danger.
2. Covering contents of buildings with waterproof sheets.
3. Preventing horizontal spread of water by making dams across large floor areas or at doorways.
4. Keeping drains clear.
5. Guarding adjacent buildings against water and smoke damage.
6. Ventilating premises to reduce smoke damage, but only as instructed by the senior fire officer.

4.3.7.4 Salvage after a fire

Efforts to salvage as much as possible and to restore the premises to normal as quickly as possible should continue after a fire. These steps include the following:

1. Informing the loss adjuster of the insurance company and getting his advice.
2. Searching debris carefully for anything that can be recovered, then clearing it away.
3. Pumping water out of flooded basements and pits, and drying out the premises, generally with mobile heaters. The work should be supervised to prevent further fire risk.
4. Drying and oiling machinery that has been sprayed with water or exposed to damp and covering it with plastic sheeting if it is exposed to the weather.
5. Covering damaged roofs as soon as possible.
6. Guarding premises against theft.

REFERENCES

1. Department of the Environment Building Research Establishment, *Annual Statistics of UK fire casulaties*, HMSO (Annually).
2. Canter, D., and Matthews, R., *The behaviour of people in fire situations: possibilities for research*. Building Research Station Current Paper CP11/76.
3. National Safety Council, *Accident prevention manual for industrial operations*, 7th ed., N.S.C. Chicago, 1341 (1974).
4. BS 4547:1972 and AMD 1658 (Jan 1975). British Standards Institution, London.
5. Chemical Industries Association, *Marking containers of hazardous chemicals*, CIA, London.
6. Underdown, G.W., *Practical fire precautions*, Gower Press, London (1971).
7. Langdon-Thomas, G.J., *Fire safety in buildings*, A & C Black, London, 21 (1972).
8. Fire Protection Association, *Fixed fire extinguishing equipment. The choice of a system*, Fire Safety Data Sheet F.S. 6004, 1970. FPA London.
9. Fire Protection Association, Fire Safety Data Sheets FS 6001, 1968; FS 6002, 1969 and FS 6003, 1969. Portable Fire Extinguishers, *How to choose, How to use, Siting care and maintenance*.
10. Fire Protection Association, *Planning fire safety in industry*. Technical Booklet No. 40, 1964.
11. Fire Protection Association, 'The aftermath of fire', *Fire Protection Journal Insert* 90, (1973).
12. GLC Fire Brigade, *Hazchem principles, interpretation and application*, GLC London.

5 COMMON INDUSTRIAL HAZARDS

5

Common industrial hazards

5.1 POOR HOUSEKEEPING[1,2,3]

Contents

Few children, as parents know, stay tidy for long and some grown-ups are the same. Poor housekeeping is an industrial hazard and a frequent contributory cause of accidents, often by masking other hazards. Housekeeping is largely a mental habit — best learnt young — as exemplified by the wearisome Victorian maxim, 'A Place for Everything and Everything in its Place'. It is a habit most of us are capable of forming, although both the very dull and men of genius often fail to acquire it. As good a description of bad housekeeping to be found anywhere is contained in a Prussian spy's report[4] on Karl Marx's home in exile in Soho in the 1850s:

'On entering Marx's room smoke and tobacco fumes make your eyes water to such an extent that for the first moment you seem to be groping about in a cavern, until you get used to it and manage to pick out certain objects in the haze. Everything is dirty, and covered with dust, and sitting down is quite a dangerous business. Here is a chair with only three legs, there another, which happens to be whole, on which the children are playing at cooking. That is the one which is offered to the visitor, but the children's cooking is not removed, and if you sit down you risk a pair of trousers'. (Perhaps this was only intended for spies).

This scene, exaggerated though the report probably was, would hardly have pleased Mrs Beeton.

Most safety specialists have a keen eye for good and bad housekeeping, typical features of the latter being poor lighting, dirt and untidiness, with materials, tools and scrap dumped on floor and benches, and no clear gangways between machines.

Poor housekeeping may constitute a tripping or falling hazard, a fire hazard or a dust explosion hazard. In the general disorder the greasy or damaged floor is camouflaged and defective guards, tools, machinery and electrical cables appear normal. It is not always an easy problem to overcome. Certain activities seem to lend themselves readily to good housekeeping, e.g. some oil refineries petro-chemical plants, where everything flows through closed pipes and vessels, cold wire drawing, and many other fully automated operations such as printing and packaging (at least so long as the machine does not run amok).

Other operations such as foundry work, demolition, oxy-propane cutting and welding, and even building construction and maintenance cause far more problems. Maintenance of plant and machinery which requires pipes, pumps and vessels to be opened up and sometimes entered leads to the frequent complaint that a bolt, spanner or bag has been left inside causing more trouble than the fault which led to the maintenance in the first place.

Good housekeeping requires more than good habits, regular cleaning, washing and maintaining walls, floors, doors and windows; in the first place it requires positive planning. This falls largely in the realm of production engineering. The flow of materials through a process must be studied and proper provision made for by-products (off-cuts, dust, turnings and packing materials) to be segregated, removed and disposed of. These are necessary but need not be unproductive activities.

By segregating scrap or off-cuts of particular materials in bins designated for them, it is often possible to find a use or market for them which pays for the effort of separation. Fabric off-cuts from a shoe factory found a market for making cuddly toys, and foamed plastic off-cuts from a mattress factory are used to fill cushions. Waste bins can be made inexpensively from metal drums and fitted with automatic self closing covers.

Proper planning is however essential; it may be difficult to find space for separate scrap containers in a congested factory or workshop, or there may be a temptation when business is brisk to install a new machine in the place previously occupied by the scrap bins. This is usually short sighted. Morale tends to fall off, standards and quality drop and visitors receive a poor impression before an accident with serious consequences often follows.

The hazards of poor housekeeping are specially acute when the materials left lying about are toxic, flammable or react violently with water. Some examples of these are:

Magnesium and aluminium dust and turnings.
Sodium metal (used as de-oxidant in certain castings).
Most chemicals and solvents.
Expanded polystyrene and polyurethane.
Insulation and packaging materials.
Rubber and rubber lattices.
Plastic wrappings and cartons of all sorts.

The subject of housekeeping is closely allied to cleaning. Many aspects of both of these are covered by the Factories Act 1961 which sets out the basic rules of factory cleanliness. These include:

Daily removal of refuse and dirt from floors and benches.
Weekly cleaning of workroom floors.
Inside ceilings, walls and partitions to be cleaned with hot water at least

every fourteen months and painted or varnished at least every seven years.

Adequate seating accommodation for all workers.

At least 400 ft³ (11.3 m³) of space per person employed, calculated at not more than 14 ft (4.2 m) from floor level.

It is often debated whether workers should be entirely responsible for cleaning their own workplace or whether reliance should be placed on special cleaning personnel. Much depends on the type of operation, whether shifts are worked and the method of payment, but as much responsibility as possible should be placed on the individual worker for his own housekeeping and cleaning.

One obvious psychological effect of untidiness and bad housekeeping is that other people seeing what appears to them a 'dump' will not hesitate to leave their own rubbish there — something they would be ashamed to do in a clean and tidy workplace.

The cleaning methods used deserve careful thought. Far too often reliance is still placed on the use of compressed air for removing dust from clothing, benches, sills, structures, cupboards and lighting fittings. This is nearly always short-sighted and often highly dangerous. The effect this has in creating explosive dust clouds and inhalation of dust which, if not actually toxic may lodge in the lungs thus impairing respiration, *should* be obvious to all.

While the dangers of inhaling asbestos fibres are now fairly well known it is not generally realised that nearly all short fibres inhaled into the lungs can cause similar damage. Proper vacuum cleaning equipment with tools for reaching into nooks and crannies and adequate and well maintained dust filters are a *must* for all operations where dust is present.

Brooms, brushes, waste for removing floor spillages, cleaning tools and detergents or other cleansing solutions should be provided for use by employees as the job demands.

Aisles and gangways must be clearly marked and everyone made aware of the necessity of keeping them clear. Areas where goods may be placed temporarily should also be marked. Often congestion of passageways can be relieved by a simple 'rule of the road' (keep left or right), with direction indicators where necessary.

Slipshod and ill-conceived lubrication methods can contribute seriously to bad housekeeping — either through oil spillages or through the discharge of a fine oil mist into the atmosphere. Specialist advice should be sought where this is a problem. Even the compressed air in many works contains small amounts of finely suspended oil from the compressor.

Supervisors and employees usually require only a little encouragement from management to maintain good housekeeping standards, so long as the work has been properly planned to facilitate good housekeeping without interfering with production and production bonuses. But even so, some positive incentives will often pay off. These may take the form of good housekeeping contests, either between different shifts within the same department, or between different departments. The judges of such a competition should if possible be fellow employees who may constitute a

PART ONE: Orderliness & Cleanliness

Check the following areas or objects to see if they are cluttered, dirty, or out of place.

		location	*comment*
1	Floors		
2	Gangways and storage areas		
3	Out of the way places (i.e. corners)		
4	Machines		
5	Work benches and surrounding areas		
6	Tools and tool storage cupboards or containers . . .		
7	Trucks, trailers, trolleys, conveyors and other handling equipment		
8	Adjacent foremen or supervisors offices		

PART TWO: Scrap & Rubbish

In many factories scrap material should be kept separate from rubbish. Two or more containers should therefore be supplied.

		location	*comment*
1	Removal method efficiency		
2	Condition of containers		
3	Use of containers (correct/incorrect)		

PART THREE: Tools

		location	*comment*
1	Suitability of tools		
2	Condition of tools		
3	Containers and storage facilities		

PART FOUR: Materials

Badly organised storage areas can cause accidents, damage valuable stocks and waste the time of storemen. Inspect the area and rate on the efficiency of the following.

		location	*comment*
1	Stacking		
2	Easy identification of different materials		
3	Stock taking facilities		

Figure 5.1.1   Housekeeping Check List (British Safety Council)

PART FIVE: General maintenance

	location	comment
1 Floors, stairs and steps		
2 Doors (including firedoors)		
3 Windows and walls		
4 Stationary machines and		
5 Machine accessories		
6 Trucks, cranes, conveyors and other moving machinery		
7 Tables, stands and benches		
8 Racks, trays, skids and platforms		
9 Ladders and other access equipment		

PART SIX: Power

Leakage or wastage of any of the following can be dangerous and expensive. All supply lines and apparatus must be checked.

	location	comment
1 Electricity		
2 Steam		
3 Gas		
4 Compressed air		

PART SEVEN: Light, Heat and Ventilation

	location	comment
1 Condition of fans, blowers or other cooling apparatus		
2 Quality of illumination (natural or artificial)		
3 Cleanliness of the above		
4 Efficiency of heating system (relate to normal working temperature)		

PART EIGHT: Welfare areas

Particular attention must be paid to the cleanliness of these areas.

	location	comment
1 Personal lockers		
2 Toilets and wash rooms		
3 Rest rooms and canteens		

Figure 5.1.1 (continued)

PART NINE: Safety

During a check on housekeeping conditions in a factory certain conditions may be found to be contrary to safety rules. These should be itemised below.

1 Are fire extinguishers, emergency exits and
   alarm systems readily available?   YES/NO

   (if no, detail) ....
   ....
   ....
   ....
   ....

2 Have any unsafe practises been observed
   during this check?   YES/NO

   (if yes, detail) ....
   ....
   ....
   ....

3 Are safety notices displayed effectively
   and have they been recently changed?   YES/NO

   (if no, detail recommendations) ....
   ....
   ....
   ....

PART TEN: Miscellaneous comment

....
....
....
....
....
....
....

Figure 5.1.1   (continued)

hazard spotting committee. A Housekeeping Check List as devised by the British Safety Council appears in *Figure 5.1.1* as an aid to judging such competitions.

Storage of raw materials, intermediate and final products, maintenance and ancillary materials poses many hazards, i.e. fire, hazards of handling, collapse of buildings and deterioration of stored materials. While some of these are dealt with in other chapters (see sections 2.6, 2.7, 4.2, 5.3 and 5.9), those which are best treated as housekeeping hazards are discussed in the following paragraphs.

5.1.1 Some hazards of storage

The commonest storage problem is that there is insufficient storage space to put everything. Space for storing supplies, tools, cleaning equipment and equipment which is only used occasionally is often forgotten when production departments are planned. This aggravates housekeeping problems by causing ladders, tools and machine parts to be left in working areas — particularly if the department is some distance from the main stores.

All materials stored whether on a temporary or permanent basis, should be arranged tidily. Storage of materials should be planned to minimise handling, especially man-handling.

Materials storage should be planned to ensure that fire alarms, lights and light switches, sprinkler controls, first aid equipment, fuse boxes and drains are not obstructed by goods in store. Exits and aisles should be clearly marked and must always be kept clear. Aisles carrying one way traffic should be at least 3 feet wider than the widest loaded vehicle using them. The turning radii of loading trucks need also to be considered when planning aisles.

The maximum permissible floor loading must be considered when planning storage and storage heights. This should take into account the weights of the goods soaked by water where a sprinkler system is employed. Many floors of buildings where there was a small fire have collapsed through the weight of water-logged materials rather than fire damage. There should always be a minimum clearance of 18 inches below sprinkler heads.

Gas cylinders containing liquefied or compressed inflammable or toxic gases, as well as inflammable or toxic liquids, most reactive chemicals, explosives and radioactive materials require to be stored in separate buildings.

Such materials should be clearly marked with the nature of the hazard, and a system established to ensure that they are stored at once in the proper place and that the date, quantity and a proper description of the material be recorded. There are limits to the quantities of hazardous materials which may be stored in any one place or building. The director responsible for safety must be aware of these and make arrangements to see they are observed.

Bins, racks, pallets or skids should be used where possible, with suitable

Figure 5.1.2   Safety refuse bin (Walter Page Ltd.)

mechanical handling equipment. Gas cylinders when piled on pallets or skids should be cross-tied.

Wire or banded boxes, cartons and bales should be stacked so that sharp ends of wire do not protrude into passageways. Bagged or other bulky materials such as skids of paper should be stacked in the form of a truncated pyramid, heights of the outer rows being limited to five feet or less. Bagged material should be cross-tied, with the mouths of the bags facing the inside of the pile.

Pipes, bars and other round materials should be stacked in layers with strips of wood or iron between the layers; the strips should be turned up at the end. Racks should be inclined towards the back so that pipes and round bars cannot roll out. Larger bar stock should be stored in racks with rollers to facilitate removal. These racks should be located in aisles which are not in use for other traffic.

Sheet metal, especially tin plate, is dangerous to handle. It should wherever possible be banded and handled mechanically. Strong leather gloves,

Figure 5.1.3   Storage with box pallet (H. & S. W. Booklet 47)

Figure 5.1.4 Typical pallet load, showing bonding (H. & S. W. Booklet 47)

preferably with metal inserts should be used when manhandling is necessary.

Material subject to spontaneous combustion (straw, sacking, charcoal) should only be stored in fire-resistant buildings fitted with sprinklers and dust-proof electric lights and equipment. Combustible materials must not be piled high and the interior of the piles should be ventilated.

Bins with hinged counterweight covers which contain combustible materials should have fusible links in the rope to close the cover, and the counterweights boxed in to prevent accidents if the link melts. Arching of material in the conical bases of bins can cause hazards. Vibrators attached

Figure 5.1.5 Carboy emptying by hand pump (Acfil Pumps Ltd)

to the base of the bin should if possible be used. Arches should not be cleared from below, nor should people enter bins to clear them.

Tanks containing hazardous liquids are preferably located out-of-doors at ground level. Traffic should not be allowed to pass underneath them. When a tank is located inside a pit, the atmosphere in the pit should be tested for oxygen content and any toxic or flammable vapours that could be present before anyone is allowed to enter without breathing apparatus, and then only with a lifeline and another person outside the pit.

The hazards of cleaning tanks must be recognised, and a proper written procedure set up suited to the contents of the tank (see section 6.1). Valves and pipework on filling and emptying lines from tanks should be identified by tags and colour coded (Appendix A).

Special precautions must be taken when a joint on a liquid filled line is broken to protect the fitter. Liquids in drums and carboys should not be poured. A syphon may be started by a rubber bulb or ejector but not by mouth suction. Emergency showers and eyewash fountains should be located nearby where acids, caustic or other dangerous liquids are handled or dispensed.

REFERENCES

1. National Safety Council, *Accident prevention manual for industrial operations*, 7th ed., National Safety Council, Chicago (1974).
2. Creber, F.L., *Safety for industry*, RoSPA (1967).
3. British Safety Council Leaflets, *Industrial cleaning and maintenance; Safety housekeeping; Housekeeping check list.*
4. Nicholaievsky, B., and Maenchen-Helfen, O., *Karl Marx man and fighter*, Pelican Books, London, 257 (1976).

5.2 THE SLIPPED DISC SYNDROME

Contents

Although this condition is often associated with handling and carrying heavy loads, it is so common in all walks of life — often disguised under other names — that a separate section has been devoted to it. It is only quite recently — largely due to the pioneering work of Dr. James Cyriax of St. Thomas' Hospital — that the slipped disc has been shown to lie at the root of a painful, often disabling, and almost universal complaint, known under many names, such as lumbago, sciatica or muscular rheumatism (which is not the same as rheumatic fever). Shakespeare refers to 'sciaticas' and 'loads of gravel i'th' back'.

Bone setters, osteopaths and bath attendants in Turkey treated the slipped disc with varying success whilst, in the UK, doctors were attributing some types of disc trouble to 'fibrositis' a disease of the muscles. Some authorities consider that cases of 'slipped disc' have been increasing in recent years, as *homo erectus* became a sedentary animal, although the mechanisation of so much heavy work, e.g. dockers', which claimed a higher than average proportion of victims, should have tended to reduce its incidence.

Whether we should class it as an injury or disease is debatable as is the question of to what extent it stems from either the working or the domestic environment. There now seems little doubt that it is largely preventable, although this involves breaking existing habits and thorough going changes in our education and upbringing. Human inertia may render it as intractable a problem as lung cancer caused by cigarette smoking.

The slipped or prolapsed invertebral disc is unusual among injuries in that the victim is frequently not aware of it at the time. The disc contains no nerves, hence nothing is felt at the moment of damage.

It is only some time later that pressure on the central spinal membrane or on the sleeve of the nerve-root causes symptoms, often with pain in quite a different part of the body. The injury may have been caused by a recent accident which led to no more than temporary soreness, but more commonly it has resulted from prolonged and unnecessary ill-treatment of the spine. This time lapse between injury and the first onset of symptoms makes nonsense of much of our ideas and legislation on industrial injury, insurance and compensation.

This point is unfortunately missed in the views expressed by some of our leading experts on industrial safety. An example exists in the chapter 'Ghost Accidents' in James Tye's book *Safety Uncensored*.[1]

This quotes among others the spokesman of a machinery company in Gateshead

> 'It pays a man to say that he injured himself at work, when in fact the injury took place at home. This is because he is £2.75 a week' (as the rate was then) 'better off drawing industrial accident benefit.'
>
> 'If a man strains his back at home over the weekend, goes to work on Monday and then complains of a sharp pain after bending at his job, or lifting something, then this is classed as an industrial accident. He will be able to get £7.25 a week. But if he stayed at home and had been completely honest, he could draw only £4.50. This situation might explain why industrial accident figures rise steadily year after year. The present system of State benefits needs revising.'

Most of the 'ghost accidents' referred to by James Tye are cases of slipped disc. One may therefore suspect that statements such as those quoted above reveal a certain over-simplification and even ignorance of the problem.

The sharp pain experienced by the worker on Monday morning may have been real enough, but to attribute it to a strained back incurred over the weekend is an over-simplification. The injury which caused a disc to rupture probably occurred many years ago. The pain on Monday morning was almost certainly not the first time it had caused trouble. The weekend activity which brought on the attack of pain on Monday morning may well have been sitting on the beach or in the sun on Sunday in a deck chair with his spine in a curved position for several hours. Had he gone to church instead and sat upright in a hard backed pew, he would probably not be suffering on Monday.

Between 80 and 90% of all adults at one time or another suffer from one or more disc lesions. The bent old man complaining about his rheumatics over a pint in the village local is often portrayed as a comic figure and impersonated by comedians. If the spectator was told the old man was suffering from one or more disc lesions, perhaps as the result of carrying two hundredweight sacks of corn in his younger days, he would perhaps be more concerned.

Some light is thrown on this aspect by an account[2] of the Parliamentary debate in 1956 on the proposal to limit the weight of sacks that might be lifted by farm workers. The maximum weight limit proposed was 130 lb for men and around 50-60 lb for women and young persons. It was suggested that some 13 million 2¼ cwt (252 lb) sacks, each at 6/- (now 30p) were in existence with a total value of £4 million and a life of 8-10 years, and that this represented too high a loss. A simple calculation showed that the regulation would be economically justified only if the cost of each accident exceeded £175.[2] While this was probably a gross underestimate of the cost to the community as a whole, the farmer did not have to pay this for the accident whereas he would have to pay for the sacks. Eventually the Agriculture (Lifting of Heavy Weights) Regulations 1959 (SI 1959/2120) (not implemented until 1965) limited the weight to 180 lb. This safeguard has, like some other safety measures, been, to some extent made redundant by technical change, the introduction of mechanical lifting devices and the automatic bagging of produce.

5.2.1 What is a slipped disc? [3,4]

Invertebral discs are the shock absorbers in our body. *Figure 5.2.1* shows front and side views of the spine, starting with the coccyx and sacrum at the base; above these are the lumbar vertebrae, the thoracic vertebrae and the cervical vertebrae, all strong bones. Between them lie the invertebral discs. Two lumbar vertebrae are shown in *Figures 5.2.2.* with the patient bending backwards and forwards. Each consists of a fibrocartilaginous ring attached to the edge of the bone with a softer pulpy centre. It is surrounded by a band which encapsulates it. The capsule is reinforced by ligaments front and rear. Between them they hold the discs in position. Behind the joint, a thick sheath called *dura mater* runs inside the bony canal, with the spinal cord within it, floating in a watery liquid to buffer it from shocks.

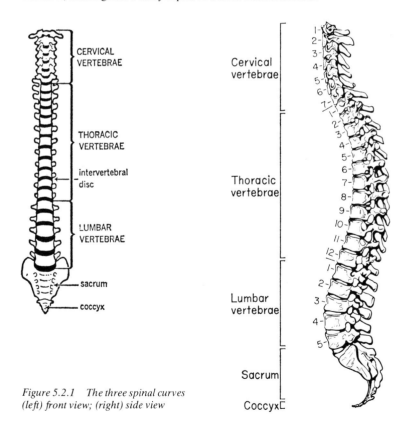

Figure 5.2.1   The three spinal curves (left) front view; (right) side view

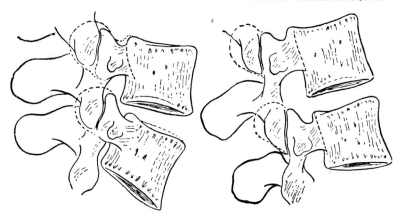

Figure 5.2.2 Lumbar vertebra with patient bending backward and forwards

Nerves pass out in pairs from the spinal cord opposite the invertebral discs and carry signals to and from the brain to all parts of the body. A normal spine shows two curves known as the lumbar and cervical lordoses (from the Greek lordos 'bent backwards'). Each joint is tilted at a slightly different angle. In slight lordosis the discs are subject to the least strain and

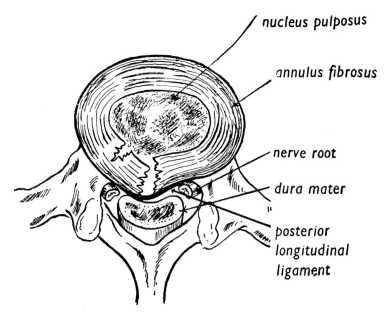

Figure 5.2.3   A cracked disc

held in proper alignment. When the spine is further bent or straightened the pressure at some part of the perimeter of the disc is increased, tending to move it beyond the edge of the bone and break it *(Figure 5.2.3)*. Increasing pressure coupled with movement tend to open the cracks, allowing part of the rim to protrude outside it.

Sometimes the pulp oozes out through the crack to form a nuclear protrusion. No symptoms appear unless the break is at the back of the spine, and a piece of disc or pulp is squeezed out backwards, protruding into the spinal canal and transmitting pressure through or to one side of the posterior ligament onto the dura mater or nerve-roots *(Figure 5.2.4)*. Now the disc has 'slipped'. An intact disc does not slip; this can only happen after

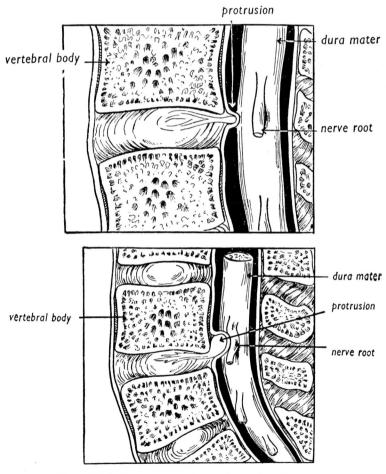

Figure 5.2.4   Pressure transmitted to dural tube and nerve root

a disc has fractured. Discs are composed of cartilage, which having no blood supply, cannot heal or reunite once they have broken. Not all human joints have cartilages lying between the bone ends. The joints of the ankle and heel contain no discs and are free of the troubles of the spine, despite their heavy loading and hard wear.

The lumbar spine did not develop in an evolutionary way as did the feet and hands of our ape forebears when they first learnt to stand upright. The spine had evolved to connect the pelvis to the chest, leaving room for the abdomen which it was well designed to support when our ancestors walked on all fours. It withstands backward strains well while held in its natural shape during evolution, but it was never intended to be arched or bent forward from the vertical position while supporting a load. To do this successfully, the whole spine would need to have been turned 180° on its axis. To have done this without man permanently facing backwards was too much even for the Almighty at the time of mans' creation, though perhaps orthopaedic surgeons of the future will complete His task.

Until then, we must live with and adjust to this inherent weakness, and avoid harmful bending exercises such as trying to touch our toes without bending our knees. Countless slipped discs owe their origin to this folly in our physical education.

Slipped discs are most common at the lumbar and cervical extent of the spine. Here attempts to bend the back forwards cause great distortion to the natural lordoses. They cause most trouble in middle age, but provide less trouble once the age of 55 is past. There are probably three reasons for this:

1. Bony outcrops called osteophytes grow at the edges of the vertebrae around the discs thereby surrounding the fragmented discs and holding the bits in place;
2. The pulpy centres of the discs degenerate and turn to cartilage at about this age;
3. By now at least the sufferer has generally discovered what brings on his pain and has altered his habits accordingly.

5.2.2 Some symptoms of the slipped disc

Only a few common features are given here. All those responsible for industrial safety and health — directors, safety officers and safety representatives — are strongly recommended to read at least Editha Hearn's popular book[3] *You are as Young as your Spine*, or better still, Cyriax's scholarly book[4] written for the layman, *The slipped disc*. This contains many important points which can only be appreciated by the man on the spot, and which cannot therefore be left entirely to the works doctor.

The first symptoms are usually slight, a dull ache in the lower back, between the shoulder blades or at the back of the neck. Sometimes the pain appears to come from a limb, calf, thigh, foot, forearm or even fingers, depending on where the nerve which is under pressure leads to. Treatment of the parts of the body where these pains are felt is ineffective. But one day, probably when the victim is bending forwards, the joint is forced open more than usual at the back, and the bulging disc is pressed harder against a nerve root. The pain may become intense, and other symptoms such as numbness, pins and needles in fingers or toes, and a need to urinate frequently or difficulty in urinating occur. It is then that the patient visits his doctor. If he is told he has a disc lesion he thinks the disc broke when the intense pain began, and does not connect it with the earlier symptoms.

Not all neck or back pains are the result of disc lesions, but it appears that the great majority are.

Dr. Cyriax has compiled a list of medical misnomers under which the slipped disc is variously passed off. These are reproduced in *Table 5.2.1*. Some are employed chiefly by doctors, others almost exclusively by bonesetters and osteopaths.

Table 5.2.1 MISNOMERS OF A SLIPPED DISC

Adhesions	Neuritis
Backstrain	Osteophytosis
Degenerate disc	Postural strain
Displaced hip	Pulled muscle
Displaced pelvis	Rotated facet joint
Displaced sacro-iliac joint	Sacro-iliac strain
Displaced vertebrae	Sciatica
Facet binding	Scoliosis
Fibrositis	Spinal arthritis
Lordosis	Spondylosis
Lumbago	Sprung back
Lumbar osteoarthritis	Strained lumbar ligament
Lumbo-sacral strain	Strained lumbar muscle
Muscle spasm	Torn muscle
Narrowed disc	Twisted pelvis

Treatment of a slipped disc may involve rest in bed, manipulation, traction, wearing a surgical corset, various sorts of injection and operation. A disc, once broken, does not heal. One modern method successfully introduced in the USA is to destroy the offending disc entirely by means of an enzyme, chymopapain.

5.2.3 Avoiding disc trouble

Misuse of the lower back leads eventually to disc lesions; it begins even in babyhood. Babies who move about while sitting, propelling themselves with one leg, instead of crawling, as well as those who fall hard and frequently on their buttocks while learning to walk, tend to develop precocious disc lesions in their teens. Walking machines which enable the baby to stay upright as he pushes himself along are recommended.

Old attitudes aimed at cultivating a graceful posture at school can be harmful. The key to a correct and natural spinal position is to stand and sit with the back hollowed, even though this leads to projecting buttocks and belly. Chairs must have proper back support, and children must be prevented from slouching forwards over their desks by ensuring that the slope and height of the working surface enables them to sit correctly.

Toe touching and forward 'back bending' exercises should be cut out. Hands on hip knee-bending exercises, with the back hollow are recommended, and weights should be lifted only by bending the knees and hip joints, not the back.

5.2.3.1 Beds and lying

Many people suffering from a slipped disc have been advised to sleep on a hard bed or even on the floor. Both are condemned, though a sagging bed is even worse. Hammocks are to be avoided.

A good bed contours the body curves and supports them in their proper position. A sagging bed can be improved by placing two boards laterally under the mattress, one below the small of the back and another below the thighs. Cushions should be used to support the lumbar curve when lying on a sofa.

5.2.3.2 Chairs and sitting

When sitting without back support, one is advised to lift the chest out of the abdomen, gently contracting the abdominal and buttock muscles and bringing the weight slightly forward. The forearm or elbow may be rested on one arm of the seat, the desk or one's thigh. When sitting in a chair with back support the latter should be practically upright. The back of the chair should support the lumbar curve on leaning back. If it does not, it should be padded there, or a cushion used in the small of the back, above the buttocks.

Most armchairs designed for the mass market are either too deep or slanted too much backward. If a chair is too deep the buttocks will not touch the back rest and any cushion for the lumbar curve will slide down behind the buttocks.

The back of a backward slanting chair should be long enough to support the head. Otherwise the buttocks tend to slide forward, the lumbar curve is obliterated and the neck is forced forward so that the cervical curve is lost too. Sitting with the knees straight and feet out-stretched, e.g. having breakfast in bed, or resting with feet on a stool, tends to open the lumbar joints at the back and should be discouraged. When sitting the knees must be allowed to bend.

Most chairs are too high in relation to the desk and table. The chair should be pulled well under the table to avoid the sitter bending too far forward *(Figure 5.2.5)*. Writing is best done on a sloping surface, as on a draughtsman's board.

To maintain correct posture when reading at a desk or typing, Dr. Cyriax recommends the use of reading stands shown in *Figures 5.2.6(a)* and *(b)*.

5.2.3.3 Standing and walking

The body should be upright, its centre of gravity above the ankles. Now the

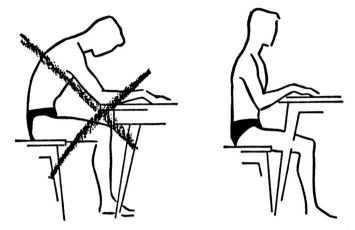

Figure 5.2.5   When working at a table the chair should be pulled well under the table to avoid bending too far forward. It is therefore important that the chair should not be too high

muscles contract to the minimum and pressure on the vertebral joints is the least possible.

The chest should be lifted out of the abdomen. This does not mean 'head up and shoulders back' which is most uncomfortable. To achieve correct head posture, one should think of oneself as suspended by a hair growing from the top of the head. The abdominal muscles should be slightly tensed, and the body held slightly forward so that the main weight is on the balls of the feet. When walking, the body should be held in this position, using the hips, knees, ankles and toes as flexibly as possible. At each step the body should be pushed forward by the toes, and the ankles and toes used properly in walking.

Overweight people tend to tilt the body backward, coming down heavily on the heels, which jars the vertebral discs. The shoe heels of a person who walks correctly are found to be worn evenly. Heavy wear at the back or on the side of the heels is a sign that the ankles and toes are not being properly used in walking.

When a person walks upstairs or up a slope the body should be held upright, he should place only the ball of the foot on each step.

5.2.3.4 Bending, lifting and carrying

To pick an object off the floor, the knees should be bent, not the back; if necessary the body should be supported by a hand on the thigh or any convenient object. If the knees are stiff and the object is light one can pivot from the hip joint with the weight on one foot, the other being raised behind as a counterweight without bending the back.

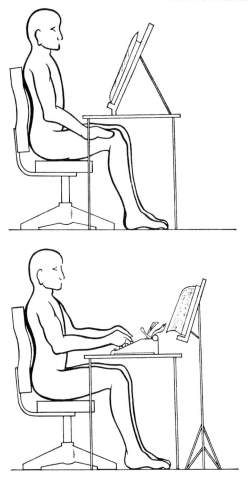

Figure 5.2.6   Good combinations of chair and desk
(above) Desk stand for holding papers
(below) Typing desk and copy reading stand

Lifting a large or fairly heavy object should be done by placing one foot on either side of it and slightly behind it (one foot slightly in front of the other), bending the knees and hip joint, not the back, and grasping the object with both hands. The object is lifted by straightening the knees and hip joint, still without bending the back.

When lifting a chair, rest it against the thigh or shin. When moving an object too heavy to lift, it should be pushed forwards rather than pulled. When lifting a suitcase at one's side, the knees should be bent, with the foot

Figure 5.2.7 St. Thomas's Hospital (London) posture chart. This card, showing how to avoid redisplacement in low lumbar disc trouble, is given to patients

furthest from the suitcase in front of the other; the body should not be bent over to one side. When carrying, the body should be kept as straight as possible. Where practical, the weight should be divided, e.g. carrying two medium sized cases, one at each side, rather than one heavy case on one side.

5.2.3.5 Sport and exercise

Most people, especially those who are office workers, require exercise for the general good of their health, as well as keeping their muscles (especially their back muscles) in reasonable trim. Swimming is the ideal exercise for the back, since the weight of the body is taken off the joints. In most strokes the spine naturally assumes its normal shape. Diving from the side or from a high board jars the spine at the moment of impact, and is best avoided. Company directors concerned with health and safety should do more to encourage their employees to swim regularly (but not in heavy seas!)

Games, exercises and dances to be avoided are those involving brusque and jerky movements and landing heavily on the heels. Squash and hurdling are hard on the spine. Opinions are divided on tennis and golf. Both are probably harmless and even beneficial in moderation, provided the soles of the shoes are well cushioned for tennis and one is careful how one handles a heavy bag of golf clubs, putts or picks the ball out of the hole. A powerful drive twists the trunk right round and provides a source of back trouble common among professionals.

5.2.3.6 The office

Sedentary workers are apt to bend all regions of the spine forward to read books or papers spread flat on the desk or table in front of them; this commonly leads to disc trouble. A desk stand which holds the papers in front of the readers' face like a music stand enables this bad posture to be avoided. A typist can also use a music stand placed behind and above her machine to good advantage.

Pressure inside the disc is greatest while sitting, so that proper chairs are essential. Work must be placed so that the worker can lean back while seated. A chair with a good back rest is of no advantage if its user is always leaning over the papers on his desk.

Library work which involves removing and replacing heavy volumes with arms outstretched on shelves at all levels frequently brings on back trouble.

5.2.3.7 The factory

Much has been said elsewhere in this book on lifting and carrying, but it is no less important to present objects to workers at the right height so that they do not have to stoop. Care should be taken that nothing has to be lifted with outstretched arms. This greatly increases leverage, for which the muscles have to accommodate by contracting much harder, thus inducing severe compression.

A wall poster showing right and wrong postures for various jobs is a constant reminder, and similar cards should be given to workers to take home. A man caught lifting the wrong way should be warned the first time and severely reprimanded the second time. He is jeopardising not only his own and his family's security, but rendering his employer liable to pay him wages while he is avoidably off work as a result.

Carrying weights resting on the hip should be encouraged, since the load on the spine is then transferred directly to the pelvis.

5.2.3.8 The garden

According to Dr. Cyriax, no one should dig the garden. Weeding should be done on all fours, preferably with a stool with a cushion on it to rest the chest on, so that the hands are free. If these ideas seem a little extreme, it is still worthwhile for those of us who are keen gardeners to ask ourselves honestly how much back trouble we suffer on account of it. It would be nice if we could combine swimming with gardening in some form of aquatic elysium, though that too would have its hazards.

5.2.3.9 Other disc hazards

Vibration as encountered in most forms of transport causes repeated compression strains. Most modern car seats slant too far to the rear; a wedge of wood or metal causing the seat to tilt forward and decrease the incline is recommended, so long as the head is clear of the roof. Usual back supports, provided as extras, are not recommended. A wedge-shaped cushion or back rest which supports the small of the back and holds the body upright is ideal. Cycling with uptilted handlebars is not harmful, but a racing bike can promote a bad back position.

Travelling over a bumpy road in any form of transport which imposes strains on the springs and shock absorbers is equally liable to cause spinal trouble. Coughing and sneezing contract the trunk muscles and can bring on symptoms of spinal trouble, especially if the individual is standing over a wash basin.

5.2.4 Economics of disc trouble

Disc trouble is the most common cause of avoidable absence from work. This is due partly to ignorance of its causes and partly to inadequate and antiquated methods of treatment which keep the patient off work far longer than necessary. Lay manipulators frequently produce rapid cures where traditional medical methods fail; the number of properly instructed orthopaedic physiotherapists is quite inadequate to care for all sufferers.

Data derived from a study[5] of 6.5 million periods of absence from work in the early 1960s are given in *Table 5.2.2*. Whilst the table may conceal various industrial diseases under other guises, it clearly demonstrates the enormous potential for savings within industry which a soundly directed campaign guided by an orthopaedic expert and backed by trained practitioners could achieve.

Table 5.2.2 CAUSES OF ABSENCE FROM WORK THROUGH ILLNESS IN THE UK

	Millions of days per year lost
Neurosis, phsychosis	52.0
Arthritis, rheumatism*	40.0
Influenza	35.4
Bronchitis	29.5
Accidents	16.5
Heart & blood vessel disease	16.3
Tuberculosis	15.4
Stomach trouble	12.1
Allergy	4.2
Hernia	2.6
Pneumonia	2.2
Cancer	0.6

*Most of this is attributed to slipped discs

REFERENCES

1. Tye, J., *Safety uncensored*, Corgi, London, 137 (1971).
2. Sinclair, T.C., *A cost effectiveness approach to industrial safety*, HMSO, London (1972).
3. Hearn, E., *You are as young as your spine*, Heinemann, London (1967).
4. Cyriax, J., *The slipped disc*, 2nd ed., Gower Press, Epping (1975).
5. Symposium on Clinical Problems of Practice, J. Coll. Med. Pract. Supplement 1, Vol XIII, No. 60 (1967).
6. *Lifting in industry*, Chartered Society of Physiotherapy.

(The illustrations in this section have been taken from 'You are as young as your spine (Hearn)[3] *and 'The Slipped disc' (Cyriax)*[4] *and are reproduced with the permission of the Publishers)*

5.3 MANUAL HANDLING

Contents

Some aspects of this subject have already been covered in section 4.5 'Ergonomics' and section 5.2 'The slipped disc syndrome'. This section amplifies these and deals with specific risks in manual handling.

Although many routine handling operations involving lifting and carrying heavy loads have been mechanised, every fit person is obliged both at home and at work to lift and carry various objects. 'Handling goods' is still by far the largest cause of industrial accidents in official statistics: thus out of 204 278 reported industrial accidents in Britain in 1975 under 'Factory Processes', 58 480 or 29% occurred while 'handling goods'. Back injuries and injuries to hands and fingers predominate.

One possible course to reduce these injuries would be to limit the weight and size of load that anyone be required to carry. This is not difficult where lifting and carrying objects of fixed weight and size are being done as a matter of routine; but it is just such routine handling operations which have been mechanised. The handling operations which it has not been possible to mechanise are often the most hazardous. The objects are often heavy but of unknown weight, irregular in size and have sharp edges. They are handled only occasionally and often for the first time as an incidental part of a person's activity.

5.3.1 Legal limitations on weights carried

At least one writer on safety has commented on the absence of official legislation or guidelines in the UK on how much a person should be expected to lift or carry.[2] Although such official guidance is scanty, it is not, however, entirely lacking.

The Factories Act, 1961, Part IV, section 72, tells us no more than that 'A person shall not be employed (a) to lift, carry or move any load so heavy as to be likely to cause injury (b) to him.'

Judgements based on (a) and (b) are given below:

(a) 'Where a person is employed to do work which involves the moving of an object, which to the employer's knowledge, is too heavy for him to move unaided, the employers are in breach of this provision if they do not instruct him not to move the object without assistance; it is immaterial that the employee could have obtained assistance had he wished for it.' (Brown v Allied Ironfounders Ltd. (1974) 2 All E.R. 135 (1974) I.W.L.R. 527. H.L.)

(b) In Kinsella v Harris Lebus (1963), 108 Sol. Jo. 14, the Court of Appeal held that a weight of 145 lb. was not likely to cause injury to a man of experience, so that there was no breach of this section in employing a man to lift that weight. It is relevant to note that by reg. 3 of the Agriculture (Lifting of Heavy Weights) Regulations, 1959 (SI 1959, No. 2120), the maximum weight of any load consisting of a sack or bag, together with its contents, which may be lifted or carried by an unaided worker exployed in agriculture, is 180 lbs.

A lower limit is given in 'The Woollen and Worsted Textiles (Lifting of Heavy Weights) Regulations, 1926', which state:

1. 'No person employed shall by himself lift by hand any material yarn, cloth, tool or appliance exceeding the maximum limits in weight set out in the Schedule to these Regulations.
2. 'No person employed shall engage, in conjunction with others, in lifting by hand any material, yarn, cloth, tool or appliance, if the weight thereof exceeds the lowest weight fixed by the Schedule for any of the persons engaged multiplied by the number of persons engaged.
3. 'A piece of cloth in the long cuttle or a sheet of loose material shall not be deemed to be a reasonably compact or rigid body for the purpose of these Regulations.'

Table 5.3.1 SCHEDULE

| *Person employed* | *Maximum weight where material, yarn, cloth, tool or appliance* | |
	is a reasonably compact or rigid body	*is not a reasonably compact or rigid body*
	(lbs)	(lbs)
(a) Man	150	120
(b) Woman of 18 years of age and over.	65	50
(c) Male young person over 16 and under 18 years of age.	65	50
(d) Female young person under 18 years of age.	50	40
(e) Male young person under 16 years of age.	50	40

New and more comprehensive guidelines now being prepared by the Health and Safety Executive are expected to be published at about the same time as this book. In Canada, a comprehensive survey written by J.R. Brown was published by the Ontario Ministry of Labour in 1972.[4]

It must be admitted that human variation and adaptability are such as to make it very difficult to lay down absolute rules on maximum loads to be carried. Athletes by undergoing special training whereby muscles are developed and proper techniques are learned, can safely lift loads which would cause serious injury to a sedentary worker whose muscles have atrophied through insufficient exercise.

5.3.2 Reducing handling injuries[5,6,7]

The first questions to be asked by managers and safety specialists faced with injuries caused through manual handling are:

1. Can manual handling be eliminated in the design of new plant or modification of existing plant?
2. What are the main causes and types of handling injuries? (e.g. cuts on sharp edges, abrasions by rough surfaces, crush injuries, back injuries, hernia, contact with injurious chemicals, dust exposure).
3. Can appropriate handling aids be provided — trucks, boxes, hooks and bars? (The handling aid must be carefully selected or designed so that it is right for the job and the worker.)
4. Will protective clothing or personal protective devices help?
5. Is there proper pre-employment examination and periodic re-examination to highlight persons with weaknesses who are pre-disposed to particular injuries?
6. What training are workers given in safe handling, and is it adequate?

These are merely preliminary questions to assist in a more detailed inquiry.

5.3.2.1 Safe habits in handling

Certain habits must be formed to avoid risk of injury. These may at first appear to 'slow the job down', although once acquired they eliminate expensive down time caused by accidents and confirm the adage 'More haste, less speed'. Some habits which must be learnt are:

(a) Inspect any object to be lifted. Has it sharp or jagged edges, burrs, splinters, slippery or rough surfaces? If so, suitable gloves should be worn.
(b) Wipe off grease, dirt or water from any object to be gripped or handled.
(c) Keep hands free from oil and grease.
(d) Grip the object firmly.
(e) Keep fingers away from pinch points.
(f) Keep hands away from the ends of long objects, e.g. pipes and bars to avoid risk of pinching them.

(g) Wear proper foot protection to protect toes and insteps in case object is dropped (see section 3.10).

(h) Wear eye protection as well as stout gloves when handling wire or metal bound boxes and bales, and take care that ends do not fly loose striking face or body.

(i) If material is dusty or toxic, wear a suitable dust mask or respirator (see section 3.10)

5.3.2.2 Lifting and carrying heavy or bulky objects by hand

Before a heavy or bulky object is lifted to be carried to another point, the employee should examine the floor or ground around the object and the route over which it is to be carried to ensure there are no obstructions or spillages where one could trip or slip, and that clearances are sufficient. If the route is unsatisfactory in any of these respects another route must be found. A rough estimate should be made of how long it will take a person to carry the load, whether stairs have to be negotiated and how and where the load can be put down *en route* if necessary. If there is danger of pinching his fingers when setting the object down, wooden blocks or cribs should be placed in position so that he can place the object safely on them.

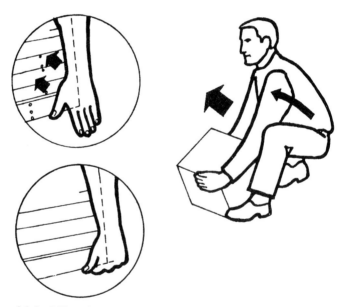

Figure 5.3.1   Lifting a heavy box. Grip the box with the palms to reduce finger strain. Keep back straight. Position feet as shown to maintain balance and to give a strong thrust forward and upwards off back foot. Arm close to sides and hands placed diagonally. The legs do the lifting. (From 'Safety for Industry' by F. L. Creber published by RoSPA)

The object should be examined to decide how it can be grasped safely, turning it over if necessary. If it cannot be grasped safely, some device such as a rope sling which enables it to be held safely should be used. The size, weight and shape of the object to be carried should be considered and checked by trial. If it cannot be lifted comfortably, help must be sought.

Feet should be set firmly on the ground, about 1ft apart, one foot slightly in front of the other, and as close to the load as possible. The knees and hip joint should be bent, with the back straight. The object must be gripped firmly. If the grip is unsatisfactory and has to be changed or adjusted, the object should first be set down. The object is then lifted by straightening the legs, tuckings in the chin to keep the spine straight and firm. (Figure 5.3.1.)

The employee must check that the load does not obstruct the vision and that the path is clear before starting his loaded journey. When turning to change direction or to set down the object, the whole body should be turned not just the waist. To lower the object to the floor, the knees and hips must again be bent. One corner should be put down first, so that the fingers can be removed from under the sides.

To place an object on a bench or table, the front edge should be set down first, pushing it far enough forward to ensure it will not fall off. After carefully releasing it, it should be slid forward into position by pushing.

Before an object is set down on a bench or some other support, the latter should be checked to see that it is strong enough and correctly placed to receive the load so that it cannot fall, tip over or roll off. Heavy objects are best stored at waist height.

5.3.2.3. Carrying by two or more men

The load should be carried so that it rides level and is equally divided. Test lifts should be made before carrying. Long objects such as pipes or bars should be carried on the same shoulder, the men walking in step. Shoulder pads should be used to prevent cutting or bruising the shoulder. Gang carrying should be directed by a foreman using special signals (e.g. a whistle.)

5.3.2.4 Special shapes

Boxes and cartons should be grasped at opposite corners, one corner being raised between the legs.

Sacks are lifted from the floor by grasping opposite corners. A light sack may be rested on one hip from the raised position before swinging onto the shoulder. To set it down it should be swung slowly from the shoulder to rest against the hip and belly, then if it has to be placed on the ground, lowered gradually, bending the knees and hip joint and keeping the back straight.

To handle *barrels and drums* safely requires special training, especially in up-ending and overturning full drums, when adequate room must be available *(Figure 5.3.2)*.

Figure 5.3.2 Handling drum. Use body weight to save muscular effort. To pull a drum over, first relax front knee, then put other foot well back to safeguard balance. The body weight acts as a counterbalance (From 'Safety for Industry')

One person should not be employed to up-end or overturn large full drums without lifting aid, and then only after training. A barrel or drum may be up-ended by two persons without an aid, standing on either side of the barrel. Both chimes are grasped near their high points, lifting one end and pressing down on the other. As the barrel is up-ended, balancing on the bottom chime, the grip of that chime is released.

Two persons may overturn a full barrel or drum in the open, both standing side by side, facing it. They grip the closest point of the top chime with both hands and with the palms of their hands on the side, push it until it balances on the lower chime. They then step forward, one on either side of the barrel, each releasing one hand to grip the lower chime as well. The barrel is then eased to a horizontal position.

Overturning and up-ending barrels and drums in confined spaces and against walls needs special care and should only be done by persons with training and experience. *Figure 5.3.3* shows a useful device for up-ending drums.

If a barrel or drum is to be rolled, it should be done by pushing with the hands, not the feet. To change direction a chime should be gripped with the hand.

Figure 5.3.3 Upending device for drums (Powell & Sons)

Barrels or drums should be lowered down a skid by sliding endways. Two persons are needed to roll a barrel or drum up a skid, outside the rails and not below the barrel or drum. It is, however, safer to control the motion of a barrel or drum on an incline by a rope, one end of which is secured to the platform at the top of the skid. The rope is passed round the drum or barrel and held firmly by a person on the platform.

Sheet metal and *window glass* should be handled with leather or other suitable gloves, and the feet should be protected. Those handling glass should wear aprons and have their arms and wrists protected by leather sleeves. Glass panes should never be carried under the arm, but with the palm turned outward holding the lower edge, while the upper edge is steadied by the other hand. The edges of large glass panes and plates should be marked by tape or grease crayon so that they are clearly visible. Special trucks and holders should be used where glass or sheet metal are frequently handled.

Long objects should be carried on a shoulder with the front end carried above head height to avoid striking others, and special care is needed at corners.

5.3.3 Aids and accessories for manual handling

Various mechanical aids which may be used when handling heavy objects are listed in the following paragraphs.

5.3.3.1 Handhooks, crowbars and rollers

These are frequently required, and workers should be trained to use them without danger to themselves and others. Hooks for handling wooden objects should be kept sharp and carried in a belt with the end covered. Crowbars should have sharp points; those using bars should be trained to position themselves to avoid falling or pinching their hands if the bar slips.

Rollers are often needed to move heavy objects. Rollers under a load should be moved (i.e. to change direction) with a bar or hammer (to avoid crushed fingers or toes).

5.3.3.2 Handtrucks and barrows

Two-wheel trucks and wheelbarrows should be equipped with knuckle guards *(Figure 5.3.4)* to protect hands from obstructions. Accidents have occurred when workers placed a foot on the wheel or axle of a handtruck to hold it, and handtrucks with brakes are therefore recommended. The wheels should be under the truck rather than outside; if they must be placed outside, they should have wheel guards. Axles should be kept well greased and trucks should be suitable for the loads they have to carry.

Workers must be trained not to leave handtrucks where they can be tripped over. Trucks with weighted tongues should be chosen if possible so that they stand naturally with the handles raised.

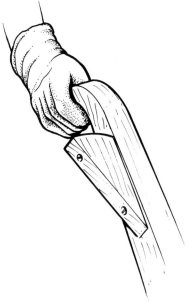

Figure 5.3.4   Knuckle guard for wheelbarrow or handtruck

The following points are important for users of hand trucks and barrows.

Heavy objects should be placed well forward near the bottom to keep the centre of gravity low and near the wheels so that the truck and not the worker carries the load.
Keep feet clear of the wheels.
Make sure load cannot slip, shift or fall or obstruct view.
Use leg muscles to raise truck to travelling position and keep back straight.
Never walk backwards with a handtruck.
Always keep truck in front except when going up an incline.
Move trucks at safe speed.
Provide a special area for parking trucks and ensure that they are kept there when not in use.

Precautions for four wheel trucks are similar to those for two wheel trucks and barrows. The main hazards for using trucks are:

Running wheels off bridge plates, planks or platforms;
Colliding with obstructions or other trucks, e.g. in doorways, and at corners;
Jamming hands between truck and other objects.

5.3.3.3 Manual lifting tackle

The use of jacks, chain blocks and rope block and tackle is a skilled task and should only be done by trained workers using protective hats and boots. Other workers must be effectively prevented from interfering accidentally or otherwise and they must be protected from falling loads. Warnings or barriers should be considered.

All lifting tackle must be regularly examined and tested and records kept of the results of the tests. Manually-operated lifting tackle, unless part of a fixed installation, should be kept in an appropriate store and only released to authorised persons on signature in a signed book.

When work is complete equipment should be checked by the borrower and returned to the storekeeper, and the book signed by the borrower to show whether the equipment was in safe working order on return.

Faulty equipment which cannot be repaired promptly and effectively and retested to satisfaction should be destroyed. Stretched links, fraying of ropes and undue wear are signs of danger.

Safe working loads must be clearly marked on all lifting equipment and not exceeded. Sharp corners and edges may damage lifting equipment, especially wire ropes, and wood or other packing must be used for protection.

Notes on using jacks

The footing should be sound and of sufficient area to distribute the load;

if not a steel plate should be used to distribute the load.
The head of the jack should have full face contact and precautions taken to prevent it slipping, i.e. using wood or cloth packing.
All lifts should be applied with the jack correctly centred.
As a load is jacked up, blocks should be placed under it on either side of the jack; if the load is left in the raised position it should be supported by the blocks as well as the jack.
Workers using jacks should wear safety shoes and instep protection.
Jacks must be carefully lubricated, maintained and inspected and inspection records kept.

5.3.3.4 Chain hoists

Chain hoists are stronger and more durable than fibre rope tackle and are to be preferred. There are three main types, spur-geared, differential and screw-geared. Spur-geared are generally preferred as the most efficient.

Screw-geared and differential hoists are self-locking and will automatically hold a load in position, but the spur-geared type is free running and is provided with a load brake similar to that of a crane.

Chain hoists should preferably be permanently hooked onto a mono-rail trolley designed for at least the maximum load of the hoist.

Chains should be of high quality welded steel with a safety factor of at least 5, and should have a capacity in excess of the normal working load.

Before attaching the load, the load chain should be examined to check that there is no twist, which easily occurs when blocks lift on two falls because the bottom block has been turned over.

A chain block should never be thrown or dropped from a height. The load chains should never be reeved round a load. The point of the hook should never be used for lifting. Lifting must always be done vertically or as near vertically as possible.

A chain block should never be used for purposes for which it was not intended, such as for towing vehicles.

REFERENCES

1. Health and Safety Executive, *Industry and services*, 1975. 81 (1976).
2. Halliday, J., Article on 'Manual handling' in *Industrial safety handbook*, edited Handley, 2nd ed., McGraw Hill, London, 224 (1977).
3. Fife, I., and Machin, E.A., *Redgrave's health and safety in factories,* Butterworths, London, 260 (1976).
4. Brown, J.R., *Manual lifting and related fields,* an Annotated Bibliography, Ontario Ministry of Labour (1972).
5. National Safety Council, *Accident prevention manual for industrial operations*, 6th ed., National Safety Council, Chicago, 503 *et seq* (1974). Chapter 19 'Principles of Materials Handling and Storing'.
6. Health and Safety Executive, *Booklet 2, Lifting and carrying*, HMSO, London.
7. Creber, F.L., *Safety for industry*, RoSPA (1967).

5.4 FALLS AND FALLING OBJECTS

Contents

'Falls of persons' rank in official statistics as one of the commonest types of injury-causing accidents. The chance of injury through being struck by a falling object is about a third of that of being injured by falling oneself.

Table 5.4.1 gives figures from 1975 to 1979 for injuries and fatalities for both kinds of accident in factories, docks and warehouses and in construction.[1] 'Falls of persons' include a higher proportion of fatalities than most other kinds of accidents and are the major cause of accidents in construction.

For factory processes, 'falls of persons' are broken down in more detail in *Table 5.4.2*. This shows that while falls on the same level comprise more than half the total number of accidents, falls from one level to another produce the great majority of fatalities. The hazards of working in high places are also demonstrated in *Table 5.4.3*.[1]

5.4.1 Anatomy of a fall

Depending on definition, it can be argued that falling, i.e. free fall, has never killed or injured anyone beyond perhaps the occasional split ear drum. It is not falling that causes the damage, but what happens when one stops falling. Icarus, the precocious aviator, fell a mighty long way, but his fall did not kill him. He drowned. A complete fall consists of three parts:

1. Release of body or object, caused by loss or failure of support, etc.
2. Free fall, when body accelerates downwards under the force of gravity. This is opposed by friction on the surface of the moving body by the air or other medium through which the body falls. After falling freely for a certain distance, which depends on the size, shape orientation, density and surface nature of the body, it reaches a constant or terminal velocity. For small light objects this distance is usually only a few feet, whilst for man it appears to be in the order of 100 ft.
3. The fall terminates when the body strikes an unyielding object, which causes rapid deceleration and brings it to rest. This is where the damage is done. The damage depends partly on the shape, hardness

Table 5.4.1 FALLS OF PERSONS AND TOTAL ACCIDENTS FOR 1975, 1977 AND 1979

		1975		1977		1979	
		Total	Fatal	Total	Fatal	Total	Fatal
1. Factory processes	Total accidents	204 278	231	207 875	205	189 281	189
	Falls of persons	32 049	39	34 141	42	37 125	28
	Struck by falling object	12 248	18	12 007	13	8 941	8
2. Docks & Wharves; quays & warehouses	Total accidents	4 702	15	4 693	21	3 513	17
	Falls of persons	1 013	3	1 307	4	916	2
	Struck by falling object	340	1	380	2	150	1
3. Construction	Total accidents	34 161	181	32 831	130	31 005	119
	Falls of persons	5 920	67	NA	73	9 020	56
	Falls of materials (other than burials in excavations & tunnelling)	2 743	23	NA	6	1 829	10
Total of 1, 2 and 3	Total accidents	243 140	427	245 399	356	223 799	315
	Falls of persons	38 982	109	—	119	47 061	86
	Struck by falling object	15 381	42	—	21	10 920	19

Table 5.4.2 BREAKDOWN OF 'FALLS OF PERSONS' IN FACTORY PROCESSES

Type of fall	1977		1978		1978	
	Fatal	Total	Fatal	Total	Fatal	Total
On or from stairs	1	4 503	1	3 442	2	3 015
On or from ladders	7	3 137	4	2 379	2	2 098
From one level to another	32	7 748	21	5 635	23	4 913
Falls on same level	2	18 753	4	26 403	1	27 099
Total 'falls of persons'	42	34 141	30	37 859	28	37 125

Table 5.4.3 INCIDENCE RATES OF FATAL ACCIDENTS PER 100 000 EMPLOYEES

Year	1973	1975	1977	1979
All Manufacturing Processes	4.5	3.7	3.4	2.9
Construction	23.1	17.7	13.1	11.7
Steel Erection	93.3	93.3	NA	NA

and elasticity of the stationary object, and partly on what part of the body hits it, and how well the body is cushioned by clothing or tissue.

People have survived some remarkable falls without serious injury. Ralph Nader quotes three falls of 45, 108 and 146 ft of this type.[2] In the one case 'A twenty-seven year old man jumped from the top of a fourteen-storey building falling 146 ft on to the top and rear of the deck of an automobile and landed in a semi-supine position. He suffered numerous fractures but did not loose consciousness and incurred no chest or head injuries. Two months later he was back at work'.

Most safety experts concentrate on trying to prevent falls from occurring in the first place; this is undoubtedly sound. For those much at risk from the hazard, e.g. air born commandos and high wire artists, we make their falls safer by providing parachutes or safety nets.

5.4.2. Falls of persons on the same level[3]

The usual causes for such falls are slipping and tripping. The common underlying causes, which are discussed elsewhere in this book, are listed in *Table 5.4.4*.

Few of the causes mentioned would cause a fall if the person were adequately aware or forewarned of the hazard. A warning however is useless unless the message gets through, and notices such as 'mind the step',

Table 5.4.4 COMMON UNDERLYING CAUSES OF PERSONAL FALLS AT SAME LEVEL[3]

Cause	Section where discussed
1. Floor surface poor or badly maintained	1.9, 2.6, 2.11
2. Obstacles on floor.	5.1., 5.3.
3. Poor lighting or sudden changes in lighting.	3.3
4. Worn or unsuitable footwear.	3.10
5. Poor eyesight or health.	1.7.
6. Running or jumping	1.7.
7. Influence of drugs, alcohol or fatigue.	1.8.

even where prominently displayed, too often do not penetrate the person's consciousness.

It also places an intolerable burden on a busy person to have to take in a number of warnings of unnecessary hazards in merely getting from A to B. The accent should, therefore, lie on removing the hazard from the journey. If this cannot be done, some guard or barrier should be placed round the hazard. Only as a last resort should reliance be placed on warning, and then in the full knowledge that it may be forgotten, ignored or not noticed.

5.4.3 Falls to a lower level

This type of fall can arise in several ways, the most common of which are:

1. Falls through floor openings, and into pits, vessels and trenches.
2. Falls from ladders and stairs.
3. Falls through roofs.
4. Falls from high working places.

5.4.3.1 Falls through floor openings and into pits, vessels and trenches

The most common feature of such falls is that the person is unaware of the opening until too late. Millions of us travel daily by train without stepping off the platform onto the track. This is because we all know there is a sharp drop at the edge of the platform which is distinctly painted and well lit. The unfamiliar and frequently ill-lit hole is dangerous.

Thus all openings in floors, including trap doors through which a person could fall onto a lower floor or into a pit or vessel, must be properly guarded with fixed barrier rails of adequate height, which are firmly supported and strong enough to withstand rough usage and occasional impact. This is a legal obligation if the pit or vessel contains any scalding, corrosive or poisonous liquid and its edge is less than 1 m above the highest ground or platform from which anyone could fall into it. (S.18, Factories Act 1961, Part II).[4]

Trenches should either be similarly guarded by rails, or covered. Covers for trenches and manholes must be strong enough to withstand the impact of the largest loads which could fall on them, and must be firmly secured and supported. Several fatal accidents have occurred when covers and floors were moved. This has happened with checker plates and reinforced concrete slabs supported on beams or angles on either side with only a small overlap. A blow on one side of the plate or slab caused it to slide off the support on one side, both the plate and persons on it falling.

When openings require covers, the design should specify metal covers with hand rings which fold flush with the top. The covers should fit snugly and be provided with lugs.

When a hole in the ground is covered by a metal sheet, a warning sign should also be displayed. Two men in a yard, seeing a corrugated sheet lying on the ground picked it up and started walking, without being aware of the hole it covered. The man behind fell in and was decapitated by the sheet.

Whenever a cover is lifted for access to the trench or hole beneath, an adequate barrier should be placed round it with suitable warning notices. Any trench or pit 1.2 m or more deep should be provided with ladders spaced not more than 15 m apart; these should extend from the bottom of the trench to at least 0.9 m above the surface of the ground.

5.4.3.2 Falls from ladders

Fixed stairs and fixed vertical ladders are treated in section 2.1.4 and only portable ladders are dealt with here.

Ladders must be well constructed of sound material and properly looked after. Wooden ladders must not be painted, as this hides defects, but may be protected with clear varnish. No ladder may be used which has any missing or defective rung or whose uprights show signs of splitting. Defective ladders should be destroyed or marked 'for repair' and locked away until repaired. Most of these requirements are obligatory under regulation 31 of the Construction (Working Places) Regulations[5] 1966, which adds:

'Every rung of a ladder shall be properly fixed to the stiles or sides. No ladder shall be used in which any rung depends for its support solely on nails, spikes or other similar fixing. Where in the case of a wooden ladder the tennon joints are not secured by wedge, reinforcing ties shall be used.

Wooden stiles or sides and wooden rungs of ladders shall have the grain running lengthways.'

(Somewhat different regulations apply to crawling ladders for use on sloping roofs.)

Ladders must be placed at the correct angle to the vertical (see *Figures 5.4.1* and *5.4.2*). The accepted angle in the UK is such that the vertical height from the ground to the top is four times the distance from the base of

Figure 5.4.1   Accepted ladder angle in the UK

the vertical height to the ladder base.[6] In the USA a slightly different formula is used:

> 'The base should be one fourth the ladder length from the vertical plane of the top support. Where the rails extend above the top landing, the ladder length to the top support only is considered.'[7]

This gives a slope which is slightly less steep than the British formula, and it is easier to apply since the length of the ladder and rung spacing is known, whereas the height of the top support is usually unknown.

Ladders must never be used in a horizontal position as gangways or scaffolding and should never be placed in front of a door that opens towards the ladder unless the door is locked or guarded.

Ladders must not be placed against a window pane or sash. Where it is necessary to support the top of a ladder against a window, a board should be securely fixed (not with nails) across the top of the ladder to give a bearing on either side of the window, or across the mullions or between window jambs.

It is recommended that portable ladders be equipped with non-slip bases. They must be placed so that both sides have secure footing. On soft ground a solid base must be provided, but this should not be done by packing with pieces of wood which may slip or split. The base must be substantial and level.

Ladders should only be set up against substantial buildings or secure and permanent objects — never against piles of loose boxes, barrels or similar objects.

When a ladder is used for access to a scaffold or high place, the top, and preferably the foot also, should be secured by lashing. The ladder side rails should extend at least 1 m above the upper landing.

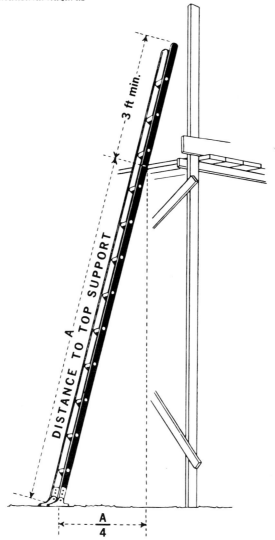

Figure 5.4.2   Accepted ladder angle in the USA

When first using a ladder after placing it in a new position, the bottom should be held by another person until the ladder has been lashed or its security is beyond all doubt.

Ladders must not be placed close to live electric wiring or against any operational piping which could be damaged, and should not be used in a strong wind except in an emergency.

Erected ladders should not be left unattended unless they have been anchored at both ends. Short ladders must not be spliced together, as they are designed for use in their original lengths, and are not strong enough for use in longer lengths.

Ladders should be inspected before use and never used if faulty. They must be kept clean and free from dirt and grease. Care is needed when ladders are used in busy places, especially if power trucks are operating. A protective barrier and warning notices should then be placed round the footing and a look-out man posted at the bottom. Both hands should be used when climbing and descending, and no loads should be carried in the hands.

Ladders should be stored in a well ventilated place where they are neither exposed to the weather nor to heat, e.g. from stoves, radiators or steam pipes.

They may be hung on brackets against a wall, with enough supports to prevent warping, or placed on edge on racks or rollers (to facilitate removal) rather than stored flat. Ladders should be clearly marked to identify them and the department to which they belong.

5.4.3.3 Ladder inspection

Ladders should be inspected regularly — every three months if they are used frequently, and the results of the inspection recorded. A printed inspection record with an appropriate checklist is recommended.

For straight portable ladders, the main points to be examined are:

Loose steps or rungs;
Loose nails, screws or metal parts;
Cracked, split or broken uprights, steps or rungs;
Splinters;
Damage to non-slip bases.

In addition, stepladders must be carefully inspected for steadiness, loose or bent hinges and hinge spreaders, broken, split or worn steps. Extension ladders should be additionally checked for loose, broken or defective locks and for deterioration of rope.

Trestle ladders, trolley ladders and sectional ladders all have special features which should be checked by regular inspection.

5.4.3.4 Falls from sloping roofs and through fragile roofs

Many deaths have occurred through men sliding down and off sloping roofs and falling through fragile roofs of asbestos cement and light plastic sheeting which break under the load of a man. Warning notices should be securely fixed on all approaches to fragile roofs. Where people are liable to walk on a platform or solid roof adjacent to a fragile roof, guard rails should be placed to prevent them stepping onto the fragile roof.

Work on sloping roofs is covered by sections 33 to 36 of the Construction (Working Places) Regulations 1966,[5] and by DOE Advisory Leaflet 67.[6]

For normal work on pitched roofs, and whenever the slope exceeds 30°, there must be either an effective barrier at the eaves to stop a man who slides down from falling off, or the work must be done from a platform, at least 450 mm wide, with guard rails and toe boards on the open side.

But if the work is not extensive, such as replacing one or two tiles, crawling boards or ladders are allowed *(Figure 5.4.3.)*. These must be used and fixed so that one cannot slide down with somebody on it. The crawling board or ladder must never rely on the eaves gutter or ridge tiles for support, although a secure anchorage may be obtained by a strong bracket with a surface which bears on the far roof surface over the ridge.

Crawling boards must always be used on fragile roof surfaces.

No one should walk on any roof without first checking that it is safe for his weight.

5.4.3.5 Falls from high working places and platforms

Many accidental falls occur from unfenced platforms. Often the workers are self-employed and working on a piece rate system. By using perhaps

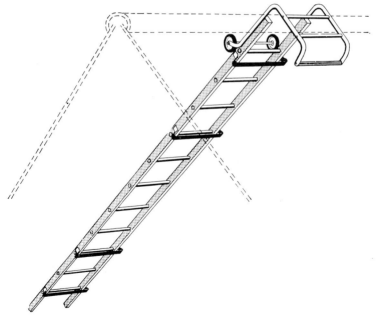

Figure 5.4.3   Crawling ladder for use on pitched roof. The ladder is fitted with wheels to assist placing in position (H. & S. W. Booklet 6b)

only a single plank, without any barrier or toe boards, they are taking the speculative risk of loss of life and limb for the chance of higher earnings. We can easily delude ourselves into thinking that by ignoring safety precautions workers only take pure risks when, in fact, the risks are often speculative. Now, however, under HASAWA (1974), the owner of the building may be liable for injuries caused in this way.

The open sides of a working platform must always have guard rails and toe boards, as well as barriers to prevent materials stacked on the platform from falling off *(Figure 5.4.4)*. Legal responsibilities for this are contained in the Construction (Working Places) Regulations No. 94, regulations 27, 28, 29 and 30.[5]

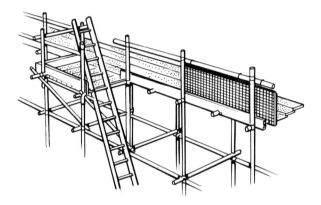

*Figure 5.4.4. Working platform with guard rails and toe-boards for elevated work
(DOE Advisory leaflet 67)*

Fixed vertical ladders providing access to platforms must be provided with cages to prevent anyone on them from falling off backwards.

Where work has to be done near machinery which, because of its position is not fully guarded, the motor must be switched off, the fuse removed and the fusebox locked with a warning notice on it. If work has to be done near the track of an overhead crane, positive measures must be taken to prevent the crane coming nearer than 6 m to the work.

In cases where it is necessary to work aloft without a proper platform, safety nets or sheets or belts with proper anchorage must be used.[9] These are covered by Regulation 38 of the Construction Regulations referred to above, and by CP 153, 1969.[10] The subject is too specialised to be discussed here. A good discussion of safety belts is given in the National Safety Council's Accident Prevention Manual for Industrial Operations.[8] People required to work in this way must be especially selected, trained and subjected to regular medical examinations. Since April 1st 1975, when Section 4 of HASAWA (1974) came into effect, owners of premises are responsible for ensuring the safety of persons such as window cleaners working on their premises who are not their regular employees.

5.4.4 Falling objects

Protection against injury from falling objects is partly provided by protective clothing, hard hats, and safety shoes, which are discussed in section 4.6. Other important rules are:

Never walk under loads suspended from cranes or other machines, and move out of the way if it appears that a load is about to pass over you.
Never throw any things down from aloft. Get them lowered or carry them where it is safe to do so.
Place tools and materials in stable positions away from ledges where they might fall. It is recommended to use a box or paint pot secured to a ladder or platform by a hook.
Use mechanical lifting equipment when available, but avoid unsafe improvisations.
Take care in stacking materials and ensure that the stack is stable.

Many injuries are caused to men in tunnels and excavations by falling objects and by collapse of the sides and roofs. Hand dug holes (such as graves) require shoring below a certain depth to protect the diggers, and owners of such holes are now obliged to protect the diggers under Sections 2 and 4 of HASAWA (1974).[11]

REFERENCES

1. HM Chief Inspector of Factories, *Annual Report 1974*, HMSO, London (1975). Health and Safety Executive, *Industry and Services 1975* and *1976*, HMSO, London.
2. Nader, R., *Unsafe at any speed*, Grossman, New York (1965).
3. Creber, F.L., *Safety for industry*, RoSPA (1967).
4. Fife, I., and Machin, E.A., *Redgrave's Health and Safety in factories*, Butterworths, London, 82(1976).
5. ibid. 869.
6. Department of The Environment, *Advisory leaflet 67. Building without accidents*.
7. National Safety Council, *Accident prevention manual for industrial operations*, 7th ed., Chicago, 416(1974).
8. National Safety Council, *Accident prevention manual for industrial operations*, 7th ed., Chicago, 509(1974).
9. BS 1397:1967. *Industrial safety belts and harnesses*, British Standards Institution.
10. CP 153:1969. 'Windows and rooflights. Part 1.' *Cleaning and safety*, British Standards Institute.
11. Health and Safety Executive, *Health and safety industry and services (1975)*, HMSO, London, 20 (1976).

5.5 MACHINE HAZARDS

Contents

The wide variety of machines makes it impossible to treat the hazards of particular ones in detail. Many important aspects of the problem are discussed elsewhere in this book. In this section the emphasis is on hazards to people, although the same hazards are also responsible for most property damage accidents.

A new responsibility for supplying safe machinery for use in the UK was placed on manufacturers, suppliers, vendors and importers by section 6B of HASAWA (1971)[1] which states:

'It shall be the duty of any person who designs, manufactures, imports or supplies any article for use at work to ensure, so far as is reasonably practicable, that the article is so designed and constructed as to be safe and without risks to health when properly used . . .'

The inference from this is that it is no longer acceptable to supply bare and unguarded machinery, leaving it to the user to design and improvise his own safeguards to be fitted later. The hazard must 'as far as is reasonably practical' be removed from the job in the design stage. It is far better, more economic and more effective to incorporate guards when the machine is being designed than to add them as appendages later, when they are just as easily removed.

Machine hazards are classified and identified in BS 5304:1975,[2] which includes a comprehensive section on means of eliminating these hazards by design.

It must not be overlooked, however, that the time and effort required to remove the hazard from the job can often be better spent in removing the human from the job and fully automating the machine or assembly line. The question is discussed in section 1.6. It leads logically to the further question of what to do with the worker displaced, but that fortunately lies outside the scope of this book. Machine hazards are themselves often compounded or ameliorated by the human element and by the environment:[3, 4, 5]

Has the operator been properly selected and trained for the job in hand (section 1.7)?

Is his alertness likely to be diminished by long hours of working, (excessive overtime), diurnal cycling (i.e. shift working), or domestic influences (section 1.8)?

Has the machinery been well laid out, within delineated areas, so that the operator has his working space where he can work without interference, or is he literally working in a gangway?

Are the protective systems appropriate, adequate and functional?

Is the machinery to be regularly inspected and properly maintained (section 2.11)?

What of the working environment?

Is the lighting satisfactory and are noise levels tolerable (sections 3.3 and 3.4)?

Is the air clean and free from toxic fumes and is the temperature conducive to good work (sections 3.2 and 3.6)?

Is the worker provided with a suitable chair or stool, and can he work in a comfortable posture which does not induce back or other strains (sections 3.7 and 5.2)?

Are his actions natural and in accordance with his likely reflexes, and has his biology been considered in positioning controls and planning his tasks (section 3.7)?

Has he been provided with appropriate protective clothing and devices, and can he wear and use them comfortably (section 3.10)?

Are there adequate containers for waste and scrap, and is the factory housekeeping of good standard (section 5.1)?

These numerous questions, whilst mostly peripheral to the machine and its own hazards, play a vital part in determining the safety of the man-machine relationship.

5.5.1 Automation

The human hazards associated with the manual placing and positioning of work in machines and transferring it from one machine to another are eliminated by automation. Most hand and finger injuries in light engineering occur in this way, and can often be avoided by the use of mechanical indexing fingers, tongs and turning devices.

In heavier engineering, e.g. steel rolling mills, similar but larger devices have reduced burns, back injuries, hernias and foot injuries. But whereas automation has eliminated or reduced hazards to the operator, it has often increased the hazard exposure of the maintenance engineer and cleaner. This is largely because of the need for rapid repair on an item in a production line which has broken down, and the frequent temptation to start such work before the equipment is de-energised and isolated.

Sometimes, particularly in the case of high temperature processing operations, it is only possible to carry out the cleaning or repair job while the equipment is hot and moving, but every effort should be made to avoid this. Such operations require special care and training and a careful and thorough analysis of the hazards to which the worker is exposed.

5.5.2 Types and examples of machine hazards

Machine hazards may be characterised by the type of machine, the industry wherein it is used, or by the mechanical motion. The last method is now preferred, since it enables the hazard to be recognised at once in different industries and machinery and appropriate measures adopted. The four main groups of motion are:

1. Rotary
2. Reciprocating and/or sliding
3. Oscillating
4. Complex

5.5.2.1 Rotary motion

Several types of hazard can arise from rotary motion — shafts and other rotating parts, in-running nips where two or more rotating parts rotate parallel to one another in opposite directions, in-running nips of the belt and pulley type, and screw or worm mechanisms.

Many revolving shafts appear smooth and harmless. This is nearly always deceptive. They may be smooth or rough, large or small, revolve slow or fast, or have projections such as key heads, set screws or cotter pins (*Figure 5.5.1*). They can catch loose or flapping clothing such as ties, tapes and torn overalls, hair, shoe laces, or string projecting from a pocket. Revolving shafts cause air currents which can assist in trapping the loose material. When this has been lapped once round the shaft it winds up the material rapidly, injuring the person attached to it. Examples are drill spindles, lathes and drill chucks, boring bars and shafts of all types.

All these are dangerous unless enclosed. But since there are many operations where complete enclosure is impossible, turning, drilling, boring, grinding, etc it should be a rule that the only person allowed access

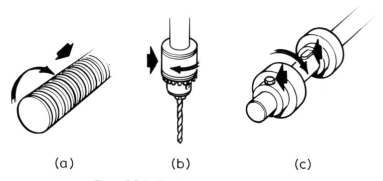

(a)   (b)   (c)

Figure 5.5.1   Rotating parts operating alone

to these rotating parts should be the fully-trained operator who is using them. They should never be allowed to revolve when exposed and unattended.

More dangerous rotating parts are those with arms and discontinuities, open arm pulleys, fan blades, spoked gear and flywheels, as well as centrifuges, grinding wheels, circular saws, milling cutters, circular shears and planing machines. These too should be totally enclosed except for the working edges of tools, which must only be used by a trained operator.

Other dangerous rotating machinery includes revolving drums with projections or spikes, beaters used in paper making, carding machines and cotton openers.

An in-running nip point exists where two or more parallel shafts, rolls or wheels rotate parallel to one another in opposite directions (*Figure 5.5.2*). They may be in close contact or some distance apart. There is less danger when the shafts rotate in the same direction. These are found on steel rolling mills and callenders in the rubber and paper industries, and rotary printing machines.

Another type of in-running nip is that between a belt, chain or moving fabric and a pulley wheel, sprocket or roll. Nip points are also found between moving and stationary objects. Employees must be protected from the dangers of trapped fingers and hands in nip points by proper guards. Wearing gloves in the proximity of a nip point increase the danger. A man with bare hands whose fingers are trapped at a nip point may lose one or more fingers, but if he is wearing strong gloves his whole hand and arm is likely to be drawn in. Guarding nip points presents special problems when the machine (e.g. a rotary printing machine) has to be cleaned whilst in operation.

Screw and worm mechanisms pose a hazard in the shearing motion between the moving screw and the fixed parts of the machine. Examples are screw conveyors, food choppers, mincing machines and helical blade mixers.

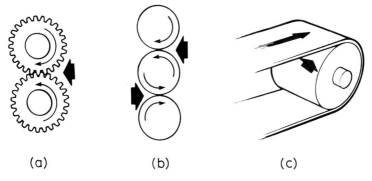

(a)   (b)   (c)

Figure 5.5.2   In-running nips

5.5.2.2 Reciprocating motion

Reciprocating machinery includes hydraulic, pneumatic and hand presses, drop and relief stamps, guillotines, trimmers, perforators, corner cutters, the transverse carriages of metal planing machines, and riveting machines *(Figure 5.5.3)*.

The danger lies where the moving part of the machine crosses a fixed part. The guillotine and shear in which a moving knife crosses a stationary knife are specially dangerous.

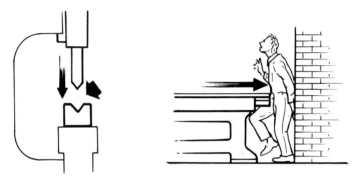

Figure 5.5.3   Reciprocating or sliding motion

5.5.2.3 Oscillating and complex motions

Cam-operated mechanisms combine a sliding and rotary motion. Others employ an oscillating movement such as a pendulum or crankshaft.

These compound motions are not always predictable and they may be more dangerous than the simple ones from which they have been derived.

5.5.3 Guarding devices

A wide range of guarding devices and principles have been developed. These include fixed guards, interlocking guards, automatic guards, trip devices, mechanical restraint devices, feeding and take-off devices. Many are highly ingenious and sophisticated, depending on light beams and photo cells and measurement of electrical capacity, with electronic circuits, electromagnets and electromechanical interlocks. One critical question which must be asked of these is 'Do they "fail safe" in case of power or circuit failure?'

Sometimes a combination of two or more devices may be required. For instance, an interlocking guard protecting moving parts which have considerable inertia may be actuated by the current passing through the drive motor. The interlocking guard would be de-activated when the

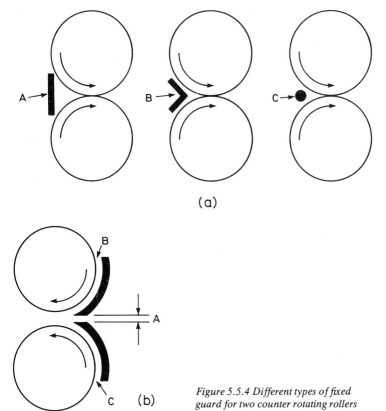

(a)

(b)

Figure 5.5.4 Different types of fixed guard for two counter rotating rollers

motor was switched off, and before the moving parts had come to rest. This might call for an 'overrun' device in addition, to keep the guard in position until the machine had stopped.

Wherever possible, fixed guards should be used, which prevent access to dangerous parts of the machine at all times. Where these have been designed as an integral part of the machine, the designer can usually arrange that the machine will not operate without the guard in position (see *Figure 5.5.4*).

To enable work of sizeable cross section to be fed to a machine, fixed guards are sometimes used some distance from the machine, with larger openings than would be employed if the guard were closer to the machine. Recommended maximum widths of openings for various distances between the opening and the danger point are given in *Table 5.5.1*.

When choosing a guard system, simplicity, reliability and ease of maintenance are of utmost importance. If part of a complex guard system fails and it takes six months to get a spare part, trouble is certain.

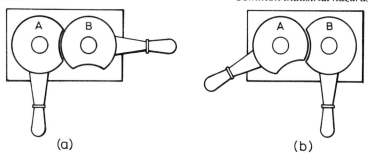

Figure 5.5.5 Interlocking guard using hydraulic valves

Table 5.5.1 SAFE OPENINGS IN FIXED MACHINE GUARDS

Distance of opening to danger point, mm	Maximum width of opening, mm
0-40	6
40-65	9
65-90	12
90-140	15
140-165	18
165-190	22
Over 190	30

A successful machine guard is one which allows operators to increase production by removing the fear of injury from their minds. The problem of preventing guards being removed from operating machinery only arises when the guard is one which retards production or is constantly having to be removed to clean the machine or remove defective work.

Detailed descriptions of the principles, modes of operation and selection of different types of guard have been published elsewhere and should be studied by those faced with this problem.

5.5.4 Training and work permits for use of guarded machinery

Operators of guarded machinery must have thorough training on the hazards of the machine and the reasons for the guard before being allowed to operate it.

Positive procedures must be established to ensure that the following safe practices are observed and respected:

1. No guard may be adjusted or removed without the written authority of the responsible supervisor, and then only by the person responsible for maintaining the machine.

2. No machine may be started unless the guard is in position and functioning properly.
3. Before any maintenance is done on the machine the operating supervisor must ensure that the power is turned off and the main switch locked and a sign attached to it. This must be entered and signed on the appropriate isolation certificate/request for maintenance form.
4. Employees working on or near mechanical equipment must be appropriately dressed, with no loose hair, clothing, pendants or other loose articles which could get caught by a moving part.

5.5.5 Lubrication, maintenance and inspection

A proper system of lubrication appropriate to all machinery must be established and implemented, with clearly laid down responsibilities and procedures, including the signature of the person who actually does the lubrication, giving date and time and any abnormalities found.

Inspection and maintenance are covered in section 2.11. The subjects are referred to here to ensure that the importance of inspection and maintenance of the machine guards themselves is not overlooked.

5.5.6 Special machine hazards

Some special machine hazards are discussed later in this book in sections 6.5, 6.6, 6.7 and 6.8.

REFERENCES

1. Fife, I., and Machin, E.A., *Redgrave's health and safety in factories,* Butterworths, London (1976).
2. BS 5304:1975. 'Code of practice for safeguarding of machinery'.
3. National Safety Council, *Accident prevention manual for industrial operations,* 7th ed., Chicago (1974).
4. Hammer, W., *Handbook of system and product safety,* Prentice Hall, New Jersey (1972).
5. Health and Safety Executive Booklet 43, *Safety in mechanical handling,* HMSO, London.

(The diagrams in this section are reproduced from DD32:1974 by permission of the British Standards Institution. It is emphasized that this document is not a British Standard, but a draft for development, and should not be treated as a British Standard. Copies of the document can be obtained from BSI, 2 Park Street, London W1A 2BS.

5.6 STATIC ELECTRICITY

Contents

Static electricity is always present in the industrial environment. It is generated whenever two different materials come into contact and are separated, or when they are rubbed together. Some typical situations capable of producing static electricity are shown in *Figure 5.6.1*. Eventually an electric field may be built up which is sufficient to break down the resistance of the air gap separating the charged object from another which is earthed or at a different potential. The resultant sparks often have sufficient energy to ignite flammable mixtures of gases, vapours, mists or even powders with air and which happen to be present in their path. They can also cause explosives and other unstable substances (such as peroxides) to ignite, explode or decompose rapidly.

Although ignition is the main hazard caused by static electricity, it can produce unexpected shocks in humans, false readings in sensitive instruments, and cause dusts to be deposited in unwanted places. It can also cause bagging and agglomeration of powders thus preventing free flowing in hoppers, shutes and silos.

Though a person is unlikely to be seriously harmed by a shock produced when, after acquiring a static charge, he touches some earthed metal object, the shock may cause some involuntary movement on his part which could result in an accident.

Static electricity is artifically generated for use in several processes — removal of dusts from gases, and some paint spraying and modern printing processes. The hazards are discussed briefly at the end of this section, but first we deal with the hazards of natural static electricity formed as a by-product of the industrial operation itself.

The discharge of liquid droplets and solid particles at high pressure is always accompanied by static; if the liquid is flammable and the discharge results from a leak, say a blown joint, it is sometimes possible for a spark to be produced which ignites the material. Static charges on small particles, particularly non-conductors can also be discharged through air without a spark by a 'corona discharge'. This may take place spontaneously or be induced by applying a voltage to sharp pointed wires. Unlike sparks, a

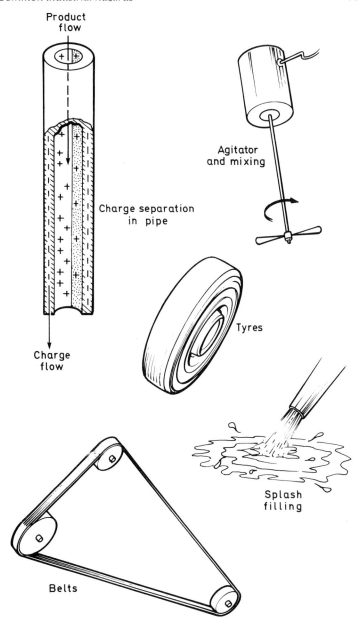

Figure 5.6.1 Typical static producing situations

corona discharge is continuous and depends on the formation of ionized gas molecules which form a conducting path. A corona discharge is seldom capable of igniting a flammable mixture, although it is visible and usually produces ozone. (This method is used for preventing the build up of static, e.g. on a fabric passing over rollers, or on a plastic powder conveyed in an air stream).

The build up and discharge of static electricity in the microclimate of an operating theatre, ship's tank or industrial building is, of course, analogous to the build up of charges in clouds and their release as lightning.[20] A large number of fires and explosions have been attributed to static electricity.

5.6.1 Some accidents caused by static

Prior to 1960 there were, on average, five explosions a year in hospital operating theatres of flammable mixtures of anaesthetics and air.[1,2,3] Special precautions to prevent the build up of static electricity in operating theatres were introduced in 1960. Since then explosions in operating theatres have been less frequent.

Some of the fires and explosions which have destroyed oil tankers were certainly sparked off by static electricity — probably by the sparks from clouds of charged water droplets formed during tank cleaning using high pressure water jets.[4],[10]

Another explosion in an oil tanker which killed four and injured seven was attributed to static electricity generated by carbon dioxide.[5] This was being discharged from cylinders into a nearby empty tank containing naphtha in order to produce an inert atmosphere.

Static electricity has been suspected of causing fires and explosions in plants handling combustible powders and dusts in driers and pneumatic conveyors, as well as several explosions in explosives factories. It was given as the 'official cause' of the Hindenberg disaster by the inquiry which followed,[6] though subsequent confessions have thrown doubt on this. Fires caused when resin powders were being poured into a stirred vessel containing xylene solvent to dissolve them were attributed to static electricity.[7] Moving belts of plastic film, fabrics and paper passing over rollers can present ignition hazards if flammable vapours are present (e.g. from coating materials).

5.6.2 Approximate electrical quantities

A great deal of research has been carried out on the generation and elimination of static electricity in industrial and other processes. This is a rather specialised subject, and the following treatment gives only a rough outline of the main points. For more detailed information, references 8 to 12 should be read.

The rate of charge build up or current flowing in systems where static electricity is continuously being formed by the flow or movement of material is low — between 1 and 1000 picoamps (10^{-9} amps) for liquids flowing in pipelines and between 10 and 100 000 picoamps for powders leaving a grinder. But potentials of 10 to 40 kV can easily built up unless the charge is conducted away from the object as fast as it is formed.

The electrical energy of a conducting object carrying a charge is given by:

$$E = 10^{-3} \frac{CV^2}{2}$$

where
E = energy, microjoules ($= 10^{-6}$ joules)
C = capacity of object, picofarads ($= 10^{-12}$ farads)
V = potential, kilovolts ($= 10^3$ volts)

A man has a capacitance of about 250 pf and a tin lid about 10 pf. Thus a tin lid charged to a potential of 20 kV has an energy of about 2 μJ.

The minimum energy of a spark required to ignite various materials depends both on the material and on the method of test. Mixtures of flammable vapours in air require from 0.1 to 1 μJ and flammable dust clouds from 5 μJ upwards. Sensitive explosives can be ignited with energies of far less than 0.1 μJ. Thus a small metal object on an insulating mounting and rubbing against a plastic belt can readily acquire sufficient electrical energy to produce a spark capable of igniting the vapour of a flammable solvent which may be present. Poor conductors such as plastic sheets, can in certain circumstances, produce sparks with equal energy, although this happens less easily as most of the charge has first to flow through the poor conductor.

If a stream of non-conducting but charged powder flows into a earthed metal container, the electrical energy will probably be released as a harmless corona discharge as it enters the container, and the same applies to a stream of non-conducting liquid entering a tank. But if the powder strikes a metal object separated from the bin or tank by some non-conducting material, the object will build up sufficient potential energy to give an intermittent spark discharge. The sparks may have enough energy to ignite any flammable vapour present, or even the powder itself.

Thus we see that three conditions must be present for static electricity to produce a fire or explosion:

1. A flammable vapour/air or powder/air mixture must be present;
2. An electric charge must have built up, generally on a conducting object, insulated from its neighbours with sufficient potential to discharge as a spark to a neighbouring, usually earthed (grounded) object;
3. The spark must have sufficient energy to ignite the surrounding flammable mixture.

Safety precautions against ignition by static aim at eliminating one or more of these conditions.

5.6.3 Protection against static

The principal methods used to prevent the build up of electrostatic charges to dangerous levels are:

    (a) Earthing and bonding of stationary conductive equipment *(Figures 5.6.2 and 5.6.3)*;
    (b) Increasing the conductance of floors, footwear, wheels and tyres for personnel and moving equipment;
    (c) Increasing the conductivity of non-conductors by incorporation of conductive additives, surface layers and films and humidification of the atmosphere;
    (d) Increasing the conductivity of the atmosphere by ionisation.

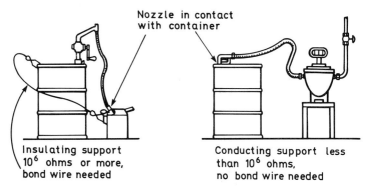

Figure 5.6.2   Examples of bonding and earthing

5.6.3.1 Earthing and bonding

Bonding eliminates a difference in potential between objects whereas earthing eliminates a difference in potential between an object and the ground. Earthing would be clearly impossible in a flying aircraft. Neither bonding nor earthing are effective where the objects are themselves non-conductors. Earthing is generally preferred to bonding, although bonding may also be used as an added precaution. Earthing may also be needed as protection against lightning and electrical circuits. Examples of earthing and bonding are shown in *Figure 5.6.2*.

    The earth connections used for protection against electrical circuits are usually adequate for electrostatic earth connections, although they should not be entirely relied on since they may be disconnected when the circuit is moved or changed. The following information applies only to earthing for static protection.

The first need is to identify all conducting equipment which may be separated by insulating materials where the build up of static electricity could have dangerous consequences. Even metal shafts revolving in bearings can build up electrostatic charges since they are usually separated from the bearing by a non-conductive oil film. A resistance to earth of 1 megohm is usually sufficient to prevent the build-up of dangerous static charges. But a lower resistance is necessary when sensitive explosives and unstable chemicals are handled.

The National Fire Protection Association in the USA has published standards on electrostatic protection.[13] While the normal construction of a plant generally provides adequate earthing, special earthing wires are sometimes needed, e.g. for equipment mounted on rubber anti-vibration mountings. Electrostatic earthing should be planned, positive, recorded and tested. When equipment is maintained, moved or painted, the earthing should be checked afterwards to ensure it has not been removed or impaired.

A special problem arises on equipment which is insulated from the earth and its surroundings as a result of the use of cathodic protection. A satisfactory compromise can generally be found which gives sufficient insulation for effective cathodic protection without excessive consumption of power or anodes, yet provides adequate conductance for electrostatic earthing. An expert should be consulted in such cases.

5.6.3.2 Protecting personnel and moving equipment

People are conductors but their footwear, clothing, hair and the floors they walk on are often non-conductors. Two types of protective footwear and flooring are available:

1. Antistatic flooring and footwear of moderate resistance are available which give adequate protection against static hazards other than those involved in handling explosives, where a higher degree of protection is required. The resistance of anti-static footwear is still sufficiently high to protect the wearer from shocks arising through contact with live wires or caused by failure of insulation of mains circuits.
2. Conductive flooring and footwear of much lower resistance should be used by personnel handling sensitive explosives. It does not protect against contact with live wires, etc., but prmises where such operations are carried out should have such high standards of circuit protection as to rule out this possibility.

Trolleys and vehicles which in the past were mainly earthed by trailing wires and chains are now usually adequately protected by tyres of rubber which include a conducting additive.

5.6.3.3 Increasing the conductivity of non-conductive materials

These materials may be broadly sub-divided into two classes — organic and inorganic. The organic materials include natural products, wood, natural fibres, fur and leather as well as synthetic polymeric plastics, fibres and rubbers. The synthetic materials have the highest resistances, especially plastics such as polypropylene and polyethylene. They are now, however, available with anti-static additives. Articles such as tyres, wheels, belting, hosepipes, sheets and footwear containing such additives are described in BS 2050:1961.[15]

Most natural products are sufficiently conducting to prevent the build up of electrostatic charges providing the atmosphere is not very dry.

The inorganic materials of main concern are ceramics and glass, which normally have very high resistances. Permanent surface conducting films can be applied to both. In the case of ceramics, a semi-conductive glaze can be applied which contains a high proportion of metal oxides.[16] This can only be done during manufacture.

A transparent conducting film of stannic oxide can be applied to glass articles after manufacture. This provides a useful means of shielding delicate instruments from electrical fields as well as preventing the local build up of electrostatic fields on glass windows, covers and shields.[16]

Less permanent conducting films can be applied to organic and inorganic materials by applying various anti-static agents, mostly based on condensates of ethylene oxide. They are applied to nylon, silk and other fibres to facilitate spinning and weaving, and are afterwards washed off. In the same way a temporary conducting film can be applied to a non-conducting surface such as glass or plastic.[16]

Humidification of the atmosphere to maintain a permanent relative humidity of at least 65% produces a very thin conductive moisture film on most surfaces. This method is best incoporated as part of a complete air conditioning system, and is usually expensive.

5.6.3.4 Increasing the conductivity of the atmosphere

The conductivity of the atmosphere may be increased by producing electrically charged ions in it. This can be accomplished by the use of

1. Powered static bars;
2. Induction needle bars;
3. Nuclear static bars, which employ a radioactive source.

Figure 5.6.4 illustrates the principles of needle bars and nuclear static bars. Powered and nuclear devices both introduce potential new hazards of their own and should only be installed with caution after alternative methods have been carefully considered.

Figure 5.6.3   Fully bonded solvent can from container (Walter Page Ltd.)

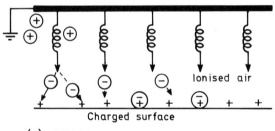

Ionised air

Charged surface

(a) INDUCTION NEEDLE BAR

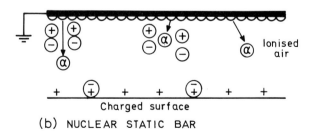

Ionised air

Charged surface

(b) NUCLEAR STATIC BAR

Figure 5.6.4   Principles of needle bars or nuclear static bars

5.6.4 Electrostatic charge control

The absence of electrostatic charges can only be verified by measurement.[16] A non-contacting static voltmeter is required. The electrical capacitances of objects liable to acquire a charge can be measured using a capacitance bridge. Periodic surveys by a person competent in this field are recommended in situations where static electricity poses a hazard or processing problem.

5.6.5 Special ignition hazards and precautions

In operations involving the handling of gases, liquids and solids where static charges could lead to ignition either of the material itself or some other one in the vicinity, several situations have been identified where special hazards exist and which require suitable precautions. These are summarised in the following paragraphs. The explosion hazards inherent in these situations are described further in sections 4.2.

5.6.5.1 Gas discharges

Gases discharged at high velocity through jets under conditions where neither liquid droplets nor solid particles are present, seldom acquire sufficient static charge to ignite. When they contain liquid droplets or solid particles, or when these are formed during the discharge, these can acquire sufficient charge to ignite flammable vapours present. Several fires and explosions have been caused in this way, including two mentioned earlier in section 5.6.1.

The release of carbon dioxide gas from cylinders in which it is stored as a liquid under pressure, is often accompanied by cooling, when charged solid particles of carbon dioxide and ice may be formed. Carbon dioxide is thus not suitable for the rapid blanketing or 'inerting' of flammable areas.

Discharge of liquified petroleum gases is also usually accompanied by charged liquid droplets, and the same applies to the discharge of many other liquefied gases.

The discharge of steam into the atmosphere, especially in the form of steam 'curtains' for the dispersion of flammable gas or vapour escapes in oil refineries and petrochemical plants, can produce charged water droplets which might cause ignition.

The main safeguard lies in earthing all electrical conductors near the escaping gas stream which could become charged. Techniques which are intended to be applied for the 'inerting' of flammable atmospheres should be thoroughly examined before they are used to ensure that they cannot themselves cause ignition.

5.6.5.2 Flammable liquid[1, 7, 8,]

When liquids flow in pipelines electrical charges are generated at a rate which increases rapidly with liquid velocity. Constrictions such as valves and filters increase the rate of charge generation. If a second liquid phase or gas bubbles are also present, the rate of charge generation is much increased. Liquids falling freely into a tank can also acquire considerable charges. The main precautions employed are:

1. *Earthing and bonding.* Above ground tanks used for the storage of flammable liquids should be earthed, preferably with an uninsulated ground wire which is easily inspected for mechanical damage. The transfer system should also be earthed and bonded, as well as the road and rail tankers, or any drums or cans into which the liquid is to be transferred. Earthing clips should be spring loaded and capable of making metal contact through any surface layer of rust or paint; they should be fixed before the hose is connected and not removed until after it has been disconnected.
2. *Vessel inlet points.* Flammable liquids should enter a tank or vessel through a bottom inlet or dipleg to avoid free fall.
3. *Flow velocity.* The liquid velocity in the transfer line should be restricted if the resistivity of the liquid is high. This applies specially if droplets of a second liquid phase are present. Maximum flow rates for particular circumstances may be found from published guides.[17]
4. *Restrictions.* Valves and filters which cause restrictions and high rates of charge generation should be as far as possible from the entrance to the tank or vessel receiving the liquid.
5. *Anti-static additives.* Special additives are available for reducing the resistivity of very high resistance liquids and are effective at low concentrations (2-10 ppm). They are normal ingredients of aviation turbine fuel and some other flammable hydrocarbon liquids.
   A minimum relaxation time may be required after a tank has been filled to enable any charge on the liquid surface to dissipate before it is sampled or dipped, and personnel performing these tasks should be earthed. Non-conducting plastic containers larger than 5 litres capacity should not be used for flammable liquids with flash points below 30°C. Flammable gas atmospheres inside tanks can be avoided by the use of floating roofs or blanketing with an inert gas atmosphere. There are several instances where the latter may be necessary, though not all are associated with static.
6. *Mixing and stirring.* Mixing of flammable liquids in vessels containing an explosive atmosphere should be carried out with low speed agitators in the lower part of the vessel and fully submersed. Special advice may be needed.

5.6.5.3 Combustible powder handling[9, 1J]

Most powder handling operations — sieving, pouring, conveying and grinding, cause the build up of static charges. Although dust clouds are in general less easy to ignite than vapour-air mixtures, this varies enormously and depends on the particle size and moisture content as well as the chemical composition.

Most industrial systems handling combustible powders in air require explosion vents of large cross section and often also an explosion suppression system which rapidly releases an inerting gas into the system when the start of a pressure rise, which would develop into an explosion, is detected.

In some cases the use of an inert gas in a closed circulating system is justified, although this can be expensive. Often it is found that if the plant is well protected by explosion vents, the occasional fire is less expensive than the use of inert gas.

The plant itself should be well grounded, and very low conductivity materials should not be employed. The charges on flowing powders can often be effectively removed by sharp pointed corona needles connected to earth.[18] Footwear and floors used by personnel should be anti-static or conducting.[21]

5.6.5.4 Explosives and unstable compounds

Special advice should be sought on the hazards of static initiated ignition or explosion.

5.6.6 Electrostatic paint spraying

In electrostatic paint sprayers[19] a voltage of approximately 60 000 V is applied to the head of the gun. Current-limiting devices must be incorporated into the supply circuit to protect personnel from shocks and prevent the formation of sparks with sufficient energy to cause ignition of the paint spray. The floor and the soles of operators' shoes must be conducting. The gun, its operator and all persons and conducting objects within 3 m of the charged head must be earthed to prevent the possible build up of induced charges. Solvents used in the paint composition and for cleaning the gun should have a flash point of at least 23°C in the UK — preferably 30°C. Solvent containers must be earthed and conducting and the voltage supply must be switched off before the gun is cleaned.

5.6.7 Electrostatic dust removal and printing

The apparatus should be supplied and installed only by specialist firms who are fully cognisant of the hazards, and used only in accordance with their recommendations.

REFERENCES

1. Redding, R.J., *Intrinsic safety*, 1st ed., 162 (1971).
2. 'Notable fires', *F.P.A. Journal*, 109, No. 27 (Oct 1954).
3. 'Operating room fires', *Fire Journal*, 23 and 26 (March 1970).
4. 'Million dollar explosion on tanker', *Fire Prevention*, 30-32 (No. 103, May 1974).
5. 'Static spark from CO_2 discharge causes second explosion on Alva Cape', *F.P.A. Journal*, 121-123 (No. 75, July 1967).
6. Walker, J., *Disasters*, Studio Vista, London, 71 (1973).
7. Vos B, van de Douwes, C., Ramackers, L., and van de Weerd, J.M., 'Electrostatic charging of suspensions during agitation', *Proc. of 1st Int. Loss Prevention Symposium*, Delft, The European Federation of Chemical Engineering, (Elsevier, Amsterdam) 38, 1 (May, 1974).
8. Klinkenberg, A., and Van Der Minne, L.J., *Electrostatics in the petroleum industry*, Elsevier, Amsterdam, 1st ed. (1958).
9. Gibson, N., 'Safety problems associated with electrostatically charged solids', 2nd *Int. Conf. on Static Electricity*. DECHEMA. Monograph 1370-1409 Frankfurt, Germany.
10. International Chamber of Shipping, 'First and Second Reports on Explosions in very large tanks London (1971).
11. Hay, D.M., and Napier, D.H., 'Minimum ignition energy of dust suspensions', Proceedings of Symposium. Chemical Process Hazards VI, 73 (April 1977).
12. Gibson, N., 'Static electricity' in *Industrial Safety Handbook*, edited Handley, 2nd ed., McGraw Hill, London, 132 (1977).
13. National Fire Protection Association, 'Occupancy Standards and Process Hazards', 9, Standard 77, N.F.P.A., Boston, USA (1961).
14. *BS 5451:1977 Specifications for electrically conducting and anti-static rubber footwear*. British Standards Institution.
15. BS 2050:1961 *Electrical resistance of conductive and antistatic products made from flexible polymeric material*, British Standards Institution.
16. Jowett, C.E., 'Control of static electricity', *Fire Prevention Science and Technology*, Fire Protection Association, London, 4 (No. 15, August 1976).
17. Shell Chemical Co. Ltd., 'Safety in fuel handling', Shell Chemical Co., London (1963).
18. van de Weerde, 'Generation and prevention of electrostatic charges in pneumatic transport of plastic powder'. *Proc. of 1st Int. Loss Prevention Symposium*. Delft, The European Federation of Chemical Engineering, (Elsevier, Amsterdam), 71 (May 1974).
19. 'Electrostatic Paint Spraying', *Accidents* (A Quarterly Journal produced by HM Factory Inspectorate), London No. 93 (December 1972).
20. Golde, R. H.. *Lightning protection*, Edward Arnold London (1973).
21. BS 3187: 1959 *Electrically conductive rubber flooring*, British Standards Institution.

5.7 PURPOSE GENERATED ELECTRICITY

Contents

The previous section dealt with the hazards of electric charges produced as an unwanted side effect of some normal operation. This section deals with the hazards of electricity generated and used for a purpose,[1,30,31.]

The hazards fall into three groups:

1. Those leading directly to human injury — electric shock and associated flashes;
2. Those leading to the ignition or decomposition of combustible, flammable and unstable materials;
3. *Miscellaneous hazards*. These include many in which electricity plays a somewhat less direct role, e.g. the hazard of unintentional starting and stopping of electrically powered equipment, or the failure of a critical equipment item to operate when needed.

5.7.1 Electric shock and flash burns

An electric current flowing through the human body causes shock and injury; the severity of the injury depends on:

1. The magnitude of the current flowing;
2. Whether the current is direct or alternating;
3. If a.c., whether the frequency is low or high;
4. Whether the current passes near nerve centres and vital organs; and
5. The time during which the current flows.

Serious and sometimes fatal shocks can be obtained on contact with the normal mains voltage.

In most electrical accidents the current flows from hands to feet through the body near the heart. The results are, therefore, serious. The internal resistance of the human body is quite low, though the skin if dry has a high resistance. This is much reduced if the skin is wet, and most fatal accidents occur then. The effects of currents of various magnitudes flowing through a man are given in *Table 5.7.1*.[1]

A woman's body is more susceptible to electric currents than a man's and the same effects are said to be produced with about 60% the current required in a man.[1]

Table 5.7.1[1] EFFECTS OF ELECTRIC CURRENT ON A MAN'S BODY

Current in milliamps			Effect
a.c. 50 Hz	d.c.	a.c. 10 000 Hz	
0-1	0-5	0-9	No sensation
1-8	6-55	10-55	Mild shock
9-15	60-80	60-80	Painful shock
16-20	80-100	80-100	Some loss of muscular control
20-45	100-350		Severe shock and loss of muscular control
50-100	400-800		Possible heart failure (ventricular fibrillation)
Over 100	Over 800		Usually fatal

Table 5.7.1. shows that the effects of mains frequency alternating current are more serious than either direct current or high frequency alternating current. A man receiving a shock caused by grabbing a bare conductor, can generally release himself when the current passing through his body is less than 15 mA. But with currents above 20 mA muscular control is generally lost, the muscles contract, and the victim can no longer release himself. The electrical resistance of the human body is given in *Table 5.7.2*.[1]

Table 5.7.2 HUMAN RESISTANCE TO ELECTRICAL CURRENT

Body area	Resistance, ohms
Dry skin	100 000-600 000
Wet skin	about 1000
Internal, hand to foot	400 to 600
Ear to ear	about 100

The high resistance of a dry skin is rapidly broken down by a high voltage. High voltage mains frequency current causes violent muscular contraction, often so severe that the victim is thrown clear. The less violent contraction caused by a lower voltage may, however, prevent the victim from freeing himself, so that the effect is no less dangerous.

An electric shock may injure the human body in several ways:[1]

1. Contraction of chest muscles causing interference with breathing, leading eventually to asphyxiation.
2. Temporary paralysis of the nerve centre leading to breathing failure. This failure may continue for a considerable time after the current has been removed. During this time the victim must be kept alive by artificial respiration.
3. Irregular movement of heart muscles — known as ventricular fibrillation — which causes blood circulation to fail. The heart cannot spontaneously recover, and death usually follows.
4. Suspension of heart action by muscular contraction, from which the heart may recover when the flow of current ceases.
5. Haemorrhage and destruction of nerves, tissues and muscle, caused mainly by heat.

Flash burns can be caused by opening switches, removing fuses from energised circuits or by shorting cables. These burns may be deep and slow to heal. Two fatal burn accidents occurred in 1974 caused by arcing on switchgear in sub-stations on which men were working. The confined space in the sub-stations and the use of inward opening doors (which in one case was closed by the blast of the arc, trapping the victim inside) were considered major contributory causes of the deaths.[29] Welding flashes cause eye pain to personnel exposed to them even at considerable distances. The pain, due to ultra-violet light, is often not immediate but is felt for two or three days afterwards. Electric welding operations should be screened so that the eyes of personnel not involved with them are not exposed to the flashes (see section 6.4).

There is today a great deal of specialised electrical equipment which can introduce special operating hazards. This includes electric welding equipment, X-ray machines, ultra-violet and infra-red light sources and high-frequency heating installations. They should be selected, installed and protected only by professionally trained and qualified electrical engineers. Some of their hazards are discussed in other sections (e.g. section 6.9 'Radiation hazards'). High frequency currents (200 kHz upwards) flow in a thin surface layer on any conductor. A person coming into contact with high frequency power naturally pulls away from it. This sets up an arc, which can cause painful burns.

High-frequency generators, transmission lines and heaters or other equipment using h.f. power thus need special shielding and safety interlocks to prevent injury to personnel.

The overall responsibility for safe electrical installations must lie with the professional electrical engineer. The following remarks are intended only to warn the layman of some common hazards, but not to interfere with the work of professionals. Such interference may be the greatest hazard of all.

5.7.1.1 Safety colour code for 3-core flex, single phase a.c. mains supply

The colour code for 3-core cables used for domestic, commercial and light industrial applications in the UK was changed by law on 1st July, 1969, to bring British Standards into line with those commonest in Europe.[2] The old and new colours are given in *Table 5.7.3*.

Table 5.7.3 OLD AND NEW COLOURS FOR 3-CORE CABLES

Contact	Colour	
	pre 1969	*post 1969*
Earth	Green	Green/Yellow
Live	Red	Brown
Neutral	Black	Blue

Since many installations are still wired with the old colours, some confusion can unfortunately arise.

5.7.1.2 Temporary wiring

Temporary electrical wiring should only be allowed in very few situations. It is best controlled by the issue of a permit which limits the period of use and contains a signed undertaking that it will be replaced by conduit or other permanent wiring or removed altogether before the permit expires. The main danger of temporary wiring is that it may become semi-permanent, sometimes with festoons of flex covered with dust and grime.

5.7.1.3 Defects and damage to electrical apparatus

All personnel need to be instructed to report immediately any defects which they observe in any electrical apparatus or installation. They should be alerted to look out for cracked or perished insulation, loose joints in conduits, damaged fuse boxes and switch covers, damaged plugs and loose pins, faulty sockets and detached earth wires.

These faults must be recorded and a maintenance form made out and a proper system implemented whereby they are replaced or repaired promptly by the electrical department.

Faulty electrical fittings and appliances must not be allowed to become familiar sights.

5.7.1.4 Portable electric tools[4]

Three different and alternate types of protection against electrocution are available to users of portable electric tools:

1. The use of an earthed conductor which gives a low-resistance path to earth. This earths all parts of a tool which might be held or touched, so that in case of an insulation failure, the holder receives at worst a very mild shock.
2. The use of low voltage power to reduce the shock potential.
3. The use of double insulation.

5.7.1.5 Earthed electric tools (220-260 V)

Earthed electric tools, operating off the 220-240 V single phase a.c. supply have long been in use in the UK. These are manufactured to Class 1 of BS 2769:1964.[3] The tool is earthed via the yellow or green earth core of the 3-core flex through the plug top and socket outlet. The earthing wire and connections must be tested regularly and a proper signed record of these tests kept in a book.

   Care is needed in the use and storage of portable tools to ensure that the cable and plug top are not subjected to tension — both in use and when the tool is removed from store. Serious accidents have occurred when the earth

Figure 5.7.1   Electric shock to user of portable tool caused by damaged plug (Reproduced by permission of the Controller of HMSO)

connection in a plug top broke under tension and came into contact with a live lead, and when cables were cut or damaged e.g. by a drum rolled over them.

Trailing cables to portable tools should be as short as possible, and suspended from structural members or hand rails. They must not be allowed to rest on floors, particularly those of gangways, or in pools of oil or other liquids. Where trailing cables cross gangways, they should run above them and be properly secured and marked so that they cannot be hit and damaged by traffic. They must not be passed round sharp objects or close to sources of heat such as steam pipes.

5.7.1.6 Low voltage tools

A range of low voltage tools is available which operate at 110 and 50 V (Classes I and II, 110 V and Class III, 50 V to BS 2769:1964).[3] They must be fitted with special plug tops to distinguish their voltage and make it impossible to plug them into a normal mains socket outlet.

They are supplied from step-down transformers which may be fixed or portable. The output from a fixed step down transformer is distributed through a low voltage distribution system with fixed socket-outlets. The secondary winding of the transformer must be earthed in the middle so that the maximum shock voltage that could be accidentally received is 55 V.

When a 110 V a.c. supply from a transformer with a centre earthed secondary winding is available, great care must be taken only to use it with circuits and equipment with which it is compatible. Accidents have been reported through the use of such a supply with incompatible control circuits when the control failed to function when required.[29]

Another main hazard lies in incorrect wiring. Some experts regard a 25 V d.c. system as the most they would allow as a 'safe system'.

5.7.1.7 Double insulation.

With the advent of strong non-conductive plastic materials for gears, tool handles and holders, a new and better standard of insulation for portable tools became available. These are known as Class II 'double insulated' tools, because they have two barriers of protective insulation between the circuit and the tool holder. They are available for 240 or 110 V a.c. supply and do not require an earth lead. The elimination of the earth lead has eliminated the hazard in earthed tools of the earth lead becoming disconnected in the plug top and coming in contact with a live lead.

Class II double-insulated tools were legalised in 1968[5] for use in all premises subject to the Factories Act (1961) and their safety record has been excellent.

Portable electric tools are tested by the British Standard Institution and approved tools are marked with their seal. The international sign for double insulation is shown in *Figure 5.7.2*.

Figure 5.7.2   International sign for double insulation

5.7.1.8 Further protection against shock

Portable electric tools are protected by a fuse in the plug top. It is essential that the correct fuses be fitted, and that the cause be established after a fuse has blown before another is fitted.

Insulating rubber footwear complying with BS 5451:1977[6] gives good protection to the wearers against shock from a line to earth short from normal mains voltage, whilst at the same time providing sufficient conductance to allow the rapid discharge of static electricity to earth. The construction of electrical equipment for shock protection is covered in BS 2754:1976.[8]

Several types of sensitive ground fault interrupters are available which rapidly break any circuit in the event of quite small currents leaking to earth. These will protect an individual involved in a line-to-earth short, but not in a line-to-line contact.[9] The circuit for an earth-leakage circuit breaker is shown in *Figure 5.7.3*.

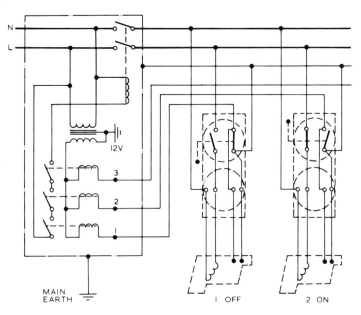

Figure 5.7.3   Earth-leakage circuit breaker equipment for portable tools

Electrical equipment suffers from the effects of dirt, dust, oil, sometimes water and overheating, and requires protection from them in design, construction and operation.

5.7.1.9 Prohibited buildings and areas

Most works have buildings or areas which contain especially hazardous electrical equipment, i.e. transformers, rectifiers, high-voltage switchgear and live conductors. Warning notices must be displayed outside forbidding entry to all except specially trained specialist electrical staff.

The buildings or gates on the area enclosures should be kept locked, and only these specialist staff should be issued with keys.

5.7.1.10 Hazards in construction and maintenance work

Exposed and energised high voltage lines and trolley bars are serious hazards to men carrying out construction or maintenance work, and they should as far as possible be banned in new works and replaced by insulated, sheathed and buried power cables and enclosed trolley bars.

An adequate number of connection points to power lines should be provided with isolating switches so that it is not necessary to make live connections to them. Where such hazards exist, every effort should be made to arrange that construction, maintenance and other work in their proximity is done at a time when the power can be switched off and the exposed power lines, etc. have been earthed. This should be covered by a permit of agreed duration. Steps must be taken to ensure that power cannot be switched on again until the work has been completed.

Special precautions have been published for use in the construction industry when there is no alternative to working in proximity to live lines.[10]

Underground cables are usually protected by a layer of tiles, and their positions shown at ground level by cable markers which show the run of the cable and all bends or changes in direction. The positions and depth of such cables must be marked on all site drawings.

Before beginning any excavation or erection work the exact position of all underground cables in the area must be checked. If there are any doubts, the position should be found by careful hand digging to expose the cable tiles. Special instructions for construction work near buried cables are given in the reference cited.[10]

5.7.1.11 Rescue of victims of electrocution

Anyone who has been electrocuted and remains in contact with a live electrical conductor has to be rescued very quickly. But rescuers face a serious risk of electrocution themselves.

The first step is to switch off the power supply to the conductor.

Supervisors and leading hands must therefore know the position of all switches in their area so that they can go at once to the right one and switch it off. This may create considerable dislocation and other hazards, e.g. light failure and motor stoppage. In buildings and areas where even a remote danger of electrocution exists, a proper safety drill should be worked out and rehearsed with full cooperation between the production manager, the safety officer and the electrical engineer.

Physical rescue of a victim in contact with a live conductor should only be attempted if, because of the remote location of the switch, it is impossible to reach it quickly. If the conductor is at a voltage greater than 500 V rescue should only be attempted by a skilled electrician who is fully familiar with the circuit and its hazards.

If the voltage is below 500 V, the rescuer should not touch the victim with his bare hands, but either push him away from the conductor with a piece of dry wood or plastic, drag him away by his clothing (providing it is dry), throw a piece of material such as an overall round him and drag him clear with that, or use rubber gloves without holes or tears to grasp him and drag him away. Electrocution victims are often not breathing and they should be given immediate artificial respiration, such as the mouth to mouth method as describe in Appendix B. This should be continued until the victim revives or he is formally pronounced dead by a doctor.

Victims of electrocution who survive usually make a rapid and complete recovery.

5.7.2 Hazards of ignition of flammable gases, vapours, liquids and powders by purpose-made electricity

Ordinary industrial electrical installations are not designed to eliminate sparking or local hot spots which might ignite an explosive mixture of a flammable gas, vapour or powder and air. Despite the fact that the Health and Safety at Work, etc Act 1974 section 4, enjoins employers to do their best to prevent emissions of noxious substances into the atmosphere, this is easier said than done.

Arising out of bitter experience, including the Senghenydd Colliery disaster in 1913 (where methane was thought to have been ignited by a spark from a signalling bell), and the needs of various industries, a number of methods, systems and standards have been evolved which allow specially safeguarded electrical equipment to be employed in so-called 'hazardous areas'. A 'hazardous area' is one where a 'flammable atmosphere' may be present. A 'flammable atmosphere' may occur in the open or inside buildings if a flammable gas, vapour, liquid or dust present in a container (tank, vessel, equipment, etc.) is released. It can also occur inside the container if air is present there.

The use of electrical equipment in 'hazardous areas' poses two difficult questions which need to be answered before deciding whether any electrical equipment may be allowed within the area at all, and if so what degree of safeguarding is required.

The questions are:

1. How easy is it to ignite the flammable material?
2. How far does the hazard extend?
3. How much hazardous material is liable to be ignited, and how big will the resulting bang be?

The second and third questions are of course inter-dependent. In practice the policy adopted is to safeguard against the effects of relatively small escapes by allowing no possible source of ignition within say 7.5 m of likely sources of emission, whilst accepting the impractability of safeguarding against a really large emission[11].

Large emissions in an industrial complex, such as that which caused the Flixborough disaster, are almost certain to reach a source of ignition before being diluted by the air to a harmless concentration (i.e. below the lower flammability limit). At sea or in open country, there is a better chance that a large emission will disperse harmlessly.

The subject then is treated in three parts:

1. Flammable materials, viewed from their ease of ignition;
2. Extent of hazardous areas with varying degrees of probability of hazard;
3. Means of safeguarding and systems and methods available.

The whole subject is exceedingly complex, and only the main points can be covered here. But it must be said that the end result of any such analysis can at best be doubtful, and may fail to take one or more important factors inherent in any situation into account. So the precautions and methods of safeguarding reached from even the most thorough formal analysis should be treated as minimum ones. The question should always be asked, 'What else should be taken into account?'

If this seems somewhat laboured, it is because the boundaries of 'hazardous areas' which have been determined by the application of rules which in the last resort are arbitrary, can assume an importance which is wholly unwarranted. They have, for instance, been widely used in fixing the separation between one hazardous chemical plant from another, without much consideration of the different degrees of process hazard or the size of fire or explosion which could result. If one asks why this was done, the short answer is that these arbitrary rules are or were the only mandatory rules available. Unfortunately flammable gases and vapours can never learn the rules.

In case it is questioned why so much trouble is taken to differentiate between various degrees of ignition hazard and varying probabilities of the hazard being present, the answer is economic. Normal industrial electrical equipment is very much cheaper and more compact than 'flameproof' equipment, as well as being available in a wider range and at shorter delivery. Most satisfactory methods of protection against electrical ignition are costly.

5.7.2.1 Potentially explosive atmospheres and their ease of ignition. Codes and standards

Flammable gases and vapours which may produce a potentially explosive atmosphere when mixed with air between lower and upper flammability limits have been classed by their ease of ignition in several standards and codes. The principal British Standards are:

BS 5501 Electrical apparatus for potentially explosive atmospheres. This is in seven parts, all of which are European Standards, and accepted by the members of 'CENELEC'. These members comprise the electrotechnical committees of most EEC countries including the UK. The seven parts are:

Part 1. General requirements   Part 5. Flameproof enclosure 'd'
Part 2. Oil immersion 'o'   Part 6. Increased safety 'e'
Part 3. Pressurised apparatus 'p'   Part 7. Intrinsic safety 'i'
Part 4. Powder filling 'q'

BS 5345 Code of practice for the selection, installation and maintenance of electrical apparatus for use in potentially explosive atmospheres (other than mining applications or explosive processing and manufacture) This code of practice will consist of twelve parts. Six parts are currently available, i.e:

Part 1: 1976 Basic requirements for all Parts of the code
Part 3: 1979 Installation and maintenance requirements for electrical apparatus with type of protection 'd', flameproof enclosure.
Part 4: 1977 Installation and maintenance requirements for electrical apparatus with type of protection 'i', Intrinsically safe apparatus and systems.
Part 6: 1978 Installation and maintenance requirements for electrical apparatus with type of protection 'e', Increased safety.
Part 7: 1979 Installation and maintenance requirements for electrical apparatus with type of protection N.
Part 8: 1980 Installation and maintenance requirements for electrical apparatus with type of protection S. Special protection.

BS 4683 Electrical apparatus for explosive atmospheres. Four parts have been published:

Part 1: 1971. Classification of maximum surface temperature
Part 2: 1971. The construction and testing of flameproof enclosures for electrical apparatus
Part 3: 1972. Type of protection N
Part 4: 1973. Type of protection 'e'

Earlier standards and codes of practice, now withdrawn, though often referred to, include BS 229: 1957 and CP 1003 in three parts dated 1964, 1966 and 1967.

Table 5.7.4 PROPERTIES OF SOME FLAMMABLE GASES, VAPOURS AND LIQUIDS AND RELATED T CLASS AND APPARATUS SUB-GROUP

Compound	Flash Point °C	Vapour density (air=1)	Flammability Limits in Air,%v Lower	Upper	Ignition Temp °C	T Class	Apparatus Sub-Group
Acetaldehyde	−38	1.52	4	57	140	T4	IIA
Acetone	−19	2.0	2.15	13	535	T1	IIA
Acetylene		0.9	1.5	100	305	T2	*
Ammonia		0.59	1.5	28	630	T1	IIA
Benzene	−11	2.7	1.2	8	560	T1	IIA
Butadiene		1.87	2.1	12.5	430	T2	IIB
Butane	−60	2.05	1.5	8.5	365	T2	IIA
Butanol	29	2.55	1.7	9.0	340	T2	IIA
Carbon disulphide	−20	2.64	1.0	60	100	T5	*
Carbon monoxide		0.97	12.5	74.2	605	T1	IIB
Cyclohexane	−18	2.9	1.2	7.8	259	T3	IIA
Cyclohexanol	68	3.45	1.2		300	T2	IIA
Diethylether	−20	2.55	1.7	36	170	T4	IIB
Ethylene		0.97	2.7	34	425	T2	IIB
Ethylene oxide		1.52	3.7	100	440	T2	IIB
Hydrogen		0.07	4.0	75.6	560	T1	IIC
Hydrogen sulphide		1.19	4.3	45.5	270	T3	IIB
Methane (firedamp)		0.55	5	15	595	T1	I
Methanol	11	1.11	6.7	36	455	T1	IIA
Nitrobenzene	88	4.25	1.8		480	T1	IIA
Nitroethane	27	2.58			410	T2	IIB
Phenol	75	3.24			605	T1	IIA
Propane		1.56	2.0	9.5	470	T1	IIA

Note* Flameproof equipment is not specified but intrinsically safe equipment of Group IIC may be used.

BS 5501 classifies electrical apparatus for use in potentially explosive atmospheres in two ways: By grouping, and by temperature class. Two main groups of apparatus are recognised:

Group I   for mines susceptible to firedamp (methane).
Group II   for other places where a potentially explosive atmosphere may occur.

Apparatus for Group II is sub-divided (sub-divisions A, B and C) according to the nature of the potentially explosive atmosphere for which it is intended. This sub-division is based on the maximum experimental safe gap (MESG) for flameproof enclosures or the minimum ignition current (MIC) for intrinsically safe apparatus.

Group IIA applies to most flammable gases and vapours encountered in industry. Group IIB applies to several more flammable gases and vapours, e.g. ethylene, ethylene oxide, and coke oven gas. Group IIC applies to acetylene, carbon disulphide, ethyl nitrate hydrogen and water gas.

The temperature class gives the maximum allowable surface temperature of the apparatus in zones where the vapours of particular flammable compounds may be present. This depends on the minimum ignition temperature of the material, although this is somewhat arbitrary since the ignition temperature depends on many factors including the time of contact and the nature of the surface (including any catalytic effect it may have on combustion). Maximum surface temperatures for various temperature classes for Group II electrical apparatus are as follows:

Temperature Class	T_1	T_2	T_3	T_4	T_5	T_6
Max. surface temp. °C	450	300	200	135	100	85

The properties of some flammable gases, vapours and liquids and related temperature class and apparatus sub-groups are given in Table 5.7.4.

5.7.2.2 Area classification (Division or Zone numbers)

The idea of classifying areas according to the frequency or probability of a flammable gas or vapour being present originated in the American Oil Industry, with the publication of code API. RP 500 A, *Recommended pratice for classification of areas for electrical installations in petroleum refineries*.[17]

In the early days of oil refining the flammable materials present all had much the same degree of hazard, and fell into Group II of BS 229. The recent widespread use of hydrogen in oil refining (hydrogen is in Group IV) has increased the risk of ignition. American practice has been followed by the oil refining industry in the UK, with the publication of the Institute of Petroleum's *Model code of safe practice in the petroleum industry, Part 1. Electrical safety code*.[18]

It has been accepted by the British Standards Institution in BS 5345, and by the International Electrotechnical Commission in IEC 79, *Electrical*

Table 5.7.5 HAZARDOUS AREA CLASSIFICATION

I.E.C. Publication 79-10 and BS 5345

Zone O. A zone in which an explosive gas-air mixture is continuously present or present for long periods.

Zone 1. A zone in which an explosive gas-air mixture is likely to occur in normal operation.

Zone 2. A zone in which an explosive gas-air mixture is not likely to occur and if it occurs it will exist only for a short time.

Non-hazardous. An area in which an explosive gas-air mixture is not expected to be present in quantities such as to require special precautions for the construction and use of electrical apparatus.

apparatus for explosive gas atmospheres — Part 10. Classification of hazardous areas. [9] It has been developed for use in the chemical industry and elsewhere in the RoSPA-ICI code *Electrical installations in flammable atmospheres.* [21]

Areas in a plant containing flammable materials (of whatever class or[21] group) are assigned Zone numbers which correspond to the probability of explosive concentrations of gas or vapour being present at any time. The use of the term 'Division', found in the earlier codes of practice has now generally disappeared in favour of the term 'Zone'. Zone numbers are given with their definitions in Table 5.7.5.

The four types of Zones or Divisions centre around *sources of hazard* these being defined as points from which a *flammable material* which creates a *flammable atmosphere* may be released. Three grades of source of hazard are recognised:

Source of hazard 0 (Continuous)., A source of hazard from which the release is for long periods (i.e. more than 1000 hours in a year for equipment in continuous use).

Source of hazard 1 (Primary). A source of hazard from which the release is frequent (i.e. for a total period of between 10 and 1000 hours in a year for equipment in continuous use).

Source of hazard 2 (Secondary). A source of hazard from which the release is infrequent and of short duration (i.e. for a total period of less than 10 hours in a year for equipment in continuous use).

Any hazard source of grade '0' will be surrounded by an inner Zone 1 and an outer Zone 2. Likewise any hazard source of grade I will be surrounded by a Zone 2. The areas outside all Zone 2 areas count as 'non hazardous' or 'safe' areas, where normal industrial electrical equipment and fittings may be used.

Typical sources of hazard include:

Pumps.	Roof tanks.
Compressors.	Pressure storage vessles.
Pipe joints.	Low pressure refrigerated storage tanks.
Valves.	Open-topped oil/water separators.
Meters and associated filters.	Petrol filling stations.
Road or rail tanker loading.	Open drain channels.

Division or Zone 0 areas are generally those inside tanks, vessels and other equipment containing a flammable liquid.

The zone 1 areas around a Zone 0 source are relatively small; not more than a metre in any direction from the source. But the extent of Zone 2 is considerably larger; ranging from 6 to 30 m horizontally from the source. The vertical extents of these areas depend on whether the escaping gas or vapour is lighter or heavier than air. Mostly it is heavier. Complex sets of rules have been devised whereby the extent of these hazardous areas can be

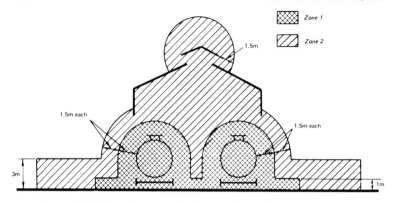

Figure 5.7.4. Hazardous area classification for road or rail tanker loading (ICI-RoSPA)

delineated in the vicinity of an oil, gas or chemical plant or storage area or depot.

A few examples from the oil industry, are given below.

Open situations. The most common situation is one where there is a source of emission in the open air which may create a 'flammable atmosphere' only under abnormal conditions. This might be a pump with a mechanical seal handling light naphtha, where the seal could leak. The hazardous area would then be classified as Zone 2. The classified area extends vertically 7.5 m above the source of hazard and 15 m horizontally in all directions from the source. Beyond 7.5 m from the source of hazard in the horizontal plane, the vertical extent of the hazard may be reduced to 4.5 m.

In the case of a pipeline, where a well maintained valve or meter is installed in a well ventilated situation or a pit, the extent of the Zone 2 area above ground may be reduced to 3 m in all directions from the source of hazard, or the edge of the pit. The pit itself should be classified as Zone 1 area.

But if the source of emission described above gives rise to a dangerous atmosphere under normal operating conditions, the area defined above would be classified as Zone 1, and would be surrounded by a Zone 2 area of the same vertical and horizontal distance.

Enclosed premises and surrounding areas. Suppose a source of emission which can give rise to a flammable atmosphere only under abnormal conditions is located within a building. The whole of the inside of the building is then classified as Division 1, as rapid dispersion of the flammable atmosphere cannot be expected due to lack of ventilation.

The area round the building is then classified Zone 2, taking the openings in the building as the source of the hazard. For a flammable

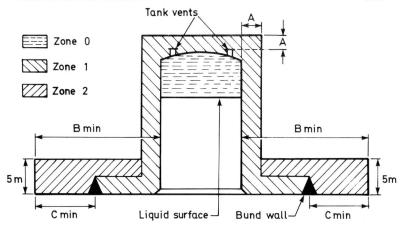

Figure 5.7.5 Classification for fixed-roof tank

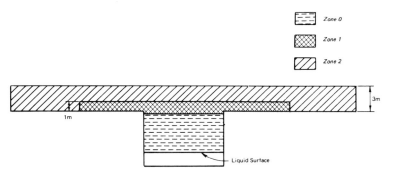

Figure 5.7.6 Hazardous area classification for open-topped oil/water separator (ICI-RoSPA)

vapour heavier than air, Zone 2 area extends vertically 7.5 m above the source of hazard and horizontally 15 m in all directions from the source. Beyond 7.5 m from the source in the horizontal plane, the vertical extent may be reduced to 4.5 m.

Despite the somewhat arbitrary nature of these rules, they afford useful guidelines and disciplines in the design of plant containing flammable materials and in the selection and protection of electrical equipment. These rules are considerably refined in the RoSPA-ICI code, to take account of variations in the volatility of the material escaping and the type of source. This refinement can be taken further by taking more factors into account and by the use of a computer for rapid evaluation of the extent of hazardous areas. An important practical feature of the RoSPA-ICI code is the recommendation to check the frequency and extent of hazardous emissions by the use of flammable gas detectors at key positions. Illustrations of hazardous area classification are given in *Figures 5.7.4* to *5.7.6*.

5.7.2.3 Methods of safeguarding

Electrical apparatus even when properly safeguarded should not be employed in a 'hazardous area' when it is practicable and economic to site it elsewhere. For quite apart from its higher initial cost, it will require more frequent inspection and maintenance.

All electrical apparatus designed and manufactured for use in 'hazardous areas' is certified by BASEEFA, whose certification standards should be referred to, as well as BS 4683 and BS 1003.

Switchgear and control instruments of plants where there are risks of emissions of flammable materials are wherever possible located in safe areas at safe distances from the plant. The following methods of safeguarding are employed:

1. Segregation.
2. Flameproof enclosures.
3. Intrinsically safe systems.
4. Pressurising and purging.
5. 'Approved' and specially protected apparatus (Protection types 'N' and 'e').
6. Non sparking and totally enclosed apparatus.

5.7.2.4 Segregation

Segregation makes use of fire-resistant impermeable barriers to create a lower risk zone or *non-hazardous area* in which electrical apparatus appropriate to the less stringent classification may be used. Examples are:

1. A commutator motor in a *non-hazardous area* drives a pump in a Zone 1, the shaft passing through a gas-tight gland in a wall separating the two areas.
2. Industrial lighting fittings in a non-hazardous area illuminate a Zone 0 enclosure through suitably designed translucent panels.

5.7.2.5 Flameproof enclosure

This is defined in BS 4683, Pt 2, as:

'An Enclosure for electrical apparatus that will withstand an internal explosion of the flammable gas or vapour which may enter it without suffering damage and without communicating the internal flammation to the external flammable gas or vapour for which it is designed, through any joints or structural openings in the enclosure.'

Since it is designed to stand the pressure of an internal explosion, no openings for pressure relief are needed. But products of an explosion may escape through joints or openings for shafts or spindles. These must be sufficiently long and narrow to cool the products of the explosion to prevent ignition of a flammable atmosphere outside.

Flameproof enclosures may not be modified without the written permission of the certifying authority. They are not necessarily capable of withstanding the effects of an internal electrical fault. The maximum permissible dimensions of gaps for various enclosures are given in BS 4683, Pt 2, and BS 229.

Enclosures are marked with the appropriate British Standard and group of gases for which they are designed. A flameproof enclosure designed for a more hazardous gas may be used for a less hazardous one, but not vice versa, i.e. a Group IIB enclosure may be used in place of a Group IIA enclosure.

A limited range of flameproof enclosures is now available for use with hydrogen, but none are available for use with acetylene or carbon disulphide.

Flameproof enclosures are mainly intended for use in Zone 1 gas and vapour risks. They may also be used, but with no relaxation of standards in Zone 2 risks, but they are not allowed in Zone 0. Flameproof enclosures exposed to the weather must be specified and designed as 'flame' proof/weatherproof' enclosures.

Flameproof enclosures are characterised by wide metal-to-metal flanges between covers, and labyrinth paths round spindles and shafts. They are heavy metal boxes which while tolerable for motors, transformers and simple motor controls, represent an intolerable limitation on small items such as lamps, instruments and telephones.

In dust and combined dust and gas/vapour conditions, flameproof enclosures may be used providing some additional precautions are taken.

All cable entries to a flameproof enclosure must be machine cut by the manufacturer; this may not be done 'on site'.

5.7.2.6 Intrinsically safe systems

These are defined in BASEEFA Certification Standard SFA 3012[16] as:

'Systems comprising apparatus and interconnecting wiring in which any spark or thermal effect in any part of the system for use in the hazardous area is incapable, under prescribed conditions, of causing ignition of a given gas or vapour. Intrinsically safe systems are treated in depth by Redding in his book *Intrinsic safety*.'[17]

Because the minimum ignition energy is very small, usually below 1 mJ, the method of safeguarding can only be applied to low power systems, e.g. controls and instruments.

Instrinsically-safe systems may comprise an intrinsically safe apparatus in a hazardous area (e.g. a switch or electric thermometer) connected to safe area apparatus (power supply, recording instrument, etc.) in a safe area. They may alternately comprise a complete intrinsically safe system with battery as power source. The latter may be used in portable instruments (e.g. gas detectors) employed in hazardous areas.

Intrinsically safe systems must be approved either by BASEEFA or other recognised approvals body. They must also be approved for the groups of gases and the Zone for which they are to be used.

Special precautions, i.e. clearance and creepage distances, insulation and earthing must be observed with intrinsically safe systems to ensure that they are not 'invaded' by other circuits. Some intrinsically safe systems are now approved for use in Zone 0 risks, whereas all approved systems are suitable for Zones 1 and 2.

An intrinsically safe system may generate a high enough temperature at some point to cause an ignition of some vapour present whereas its spark is too weak to cause ignition. This applies particularly to flammable gas detectors, where the detector head is enclosed in a flame trap.

Intrinsically safe systems are, therefore, designated by temperature class (T.1 to T.6) and should not be installed where flammable materials are used which have ignition temperatures below the maximum for that class.

Intrinsically safe systems should not be used where there are flammable dust risks without first establishing that the minimum ignition temperature and energy of the dust cannot be exceeded. The presence of a dust layer on a piece of intrinsically safe apparatus may impair heat dissipation and increase the temperature of the apparatus.

5.7.2.7 Pressurising and purging

Pressurising is a method of safeguarding whereby air or inert gas in a room or enclosure is maintained at a pressure high enough to prevent the ingress of the surrounding atmosphere which might be flammable.

Purging is a method of safeguarding whereby a flow of air or inert gas is maintained in a room or enclosure in sufficient quantity to reduce or prevent any hazard which could arise in the absence of the purge.

A combination of both methods can also be used.

As an example of pressurising, a control room in a hazardous area would be sealed and fitted with double doors to form an air lock. The atmosphere in it would then be maintained at a higher pressure than that outside by a fan and ducting system drawing air from a non-hazardous area.

As an example of purging, a hopper handling a flammable dust is fitted with an extraction hood to remove airborn dust and so prevent its accumulation in the surrounding area.

Special attention must be given to the design, construction, use and monitoring of pressurising and purging systems. Useful information on this point is given in the RoSPA-ICI code[21] from which the above definitions have been taken.

5.7.2.8 'Approved' and specially-protected apparatus

Several types of apparatus approved by HMFI before 30 September 1969 for use in flammable atmospheres are listed on HMFI form F. 931.[22] These include certain battery handlamps. Others have been individually approved by HMFI by letter prior to that date.

Other apparatus with special protection has since been approved by BASEEFA for use in Zones 1 and 2 for particular atmospheres and subject to limitations of ignition temperature.[23] They are designated 'Type of Protection s' and include a factory sealed fluorescent handlamp with flexible cable.

Type of Protection N covers other apparatus certified by BASEEFA[24] as complying with BS 4683, Pt 3, wherein it is defined as:

'A type of protection applied to electrical apparatus such that, in normal operation, it is not capable of igniting a surrounding explosive atmosphere and a fault capable of causing ignition is not likely to occur.'

The main feature of such apparatus is that any spark occurs in an enclosed break device, a hermetically sealed device, or has insufficient energy to cause ignition of a flammable atmosphere. Type N protected devices are only suitable for Zone 2 gas and vapour risks, for gases and vapours whose ignition temperature exceeds the temperature rating of the device.

There are, in addition, similar devices individually approved by HMFI by letter.

5.7.2.9 Non-sparking and totally-enclosed apparatus

Non-sparking apparatus is that which will not produce arcs or sparks except on failure. Totally-enclosed apparatus is so constructed that the risk of entry of a flammable atmosphere surrounding it for a short time is very small. Neither of these are certified.

A solenoid valve may be classed as a non-sparking apparatus and a specially made lighting fitting as a totally-enclosed apparatus.

Both types of apparatus should only be employed in low Zone 2 risks by an experienced engineer after careful study of all aspects, both of their design and construction and of the risks involved.

5.7.2.10 Special applications

Several cases arise where special electrical apparatus may be required to operate in hazardous areas.

1. *Battery-operated vehicles.* Battery-operated vehicles for use in hazardous areas are certified to BASEEFA Certification Standard SFA 3006.[25] The methods of safeguarding usually comprise

flameproof enclosures, intrinsically safe systems and type of protection 'e'.

2. *Portable and transportable apparatus.* This should not be used in a 'hazardous area' unless it has been approved by BASEEFA or H & SE for the particular risk of the area, i.e. gas or vapour group and Zone number. Non-approved portable and transportable apparatus should under no circumstances be used in hazardous areas, nor should apparatus approved for use only in a less hazardous area be employed in one of greater hazard.

3. *Electrically-driven canned pumps.* Special precautions are needed when canned pumps (which allow the pumped liquid to enter the motor) are used to pump a flammable liquid because of ignition risks in the motor. Precautions are given in EEUA Handbook No. 26.[26] some of which may be required even when the pump has a flameproof certificate.

4. *Risks with light metals and alloys.* These risks are not inherent in the electrical circuits but in the use of aluminium and similar light metals for electrical enclosures and paint. These can spark and burn fiercely under certain conditions. Portable and transportable electric apparatus with unprotected light metal enclosures should not be taken into hazardous areas unless protected against frictional sparking, e.g. by coating with an abrasion resistant material. Rigidly mounted electrical apparatus with light metal enclosures and aluminium sheathed cable should not be used in Zone o areas.

5. *Personal electric apparatus.* Hearing aids, key ring torches, transistor radios, walkie-talkies, cameras with electronic light meters or flash attachments, 'bleepers' and pocket calculators must be certified or approved by a competent authority before being taken into a hazardous area. Warnings against inadvertently taking such uncertified equipment which might cause ignition into a hazardous area should be prominently displayed. The risk with electronic wrist watches is, however, negligible and these may be worn without certification.

6. *Use of apparatus to foreign standards.* Apparatus certified for use in a hazardous area to a foreign standard may be used with discretion provided the engineer responsible has studied the appropriate foreign standards and/or code of practice carefully making sure that it covers the hazard in question. The certificate must be checked to ensure that it applies to the apparatus concerned. In case of doubt advice should be sought from BASEEFA.

5.7.2.11 Dust risks

In processes and industries where there are risks of dust coming into contact with electrical equipment, the characteristics of the dust need careful examination, and the following questions asked:

Is it combustible?
Does it decompose or melt readily? If so at what temperature?
Does it conduct electricity?
Does it form a thermally insulating layer when it deposits on any surface?

The usual methods of safeguarding electrical apparatus against dust risks are:

1. Enclosing it so as to limit the amount of dust which can reach sources of ignition;
2. Ensuring that the surface temperature of the enclosures does not exceed the ignition temperature of the flammable dust in cloud or layer form.

Two standards are given in IEC Publications 144[27] and 34-5.[28] Two types of protection referred to, IP 54 and IP 65 are suitable for use in dust risks:

IP 54. This is met by most standard totally enclosed weather proof or flameproof apparatus.
IP 65. This is met by enclosures meeting the requirements of BS 3807.

The ERA are normally prepared to carry out the tests required. These may be required in cases where the dust is conducting and may thus short circuit any insulators on which it may settle.

Motors are especially susceptible to dust hazards, and totally-enclosed fan-cooled motors should be used where the problem arises. Dust must be removed by regular cleaning, using vacuum cleaning and a hand bellows. Where compressed air has to be used (e.g. to remove dust from windings) the pressure should not exceed 3 bars g the air should be clean and dry, and there should be a means of extracting the dust laden air.

Oil filler caps must be kept closed and dust seals and gaskets kept in good condition. Where oil and dust have been allowed to accumulate, e.g. on motor windings, the deposit should be removed with a non-flammable solvent which does not damage the insulation. The cleaned windings should be dried and given a coat of insulating varnish.

REFERENCES

1. Dalziel, C.F., 'Effects of electric current on man', *Electrical engineering* (Feb 1941).
2. BS 4410:1969. *Connection of flexible cables and cords to appliance*, British Standards Institution.
3. BS 2769:1964. *Portable electric motor-operated tools*, British Standards Institution.
4. Bland, J.G.V., 'Portable electric tools: safe working' in *Industrial Safety Handbook* edited Handley, 2nd ed., McGraw Hill, London, 162 (1977).
5. *Electricity Regulations, 1908 (Portable Apparatus Exemption) Order 1968*, HMSO, London.
6. BS 5451: 1977 *Conducting and antistatic rubber factories*. British Standards Institution.

7. BS 3398:1961. *Antistatic rubber flooring*, British Standards Institution.
8. BS 2754:1976. *Construction of electrical equipment for protection against electric shock*, British Standards Institution.
9. Say, M.G., *Electrical engineers' reference book*, 13th ed., Newnes-Butterworths, London (1973).
10. National Federation of Building Trade Employers (London), *Construction Safety* (1969 — amended yearly).
11. King, R.W., 'Plant hazards', *Engineering* (April 1976).
12. BS 5345. *Code of Practice for the selection, installation and maintenance of electrical apparatus for use in potentially explosive atmospheres*. (12 parts). Part 1. 1976, B.S.I
13. BS 1259:1958. *Intrinsically safe electrical apparatus and circuits for use in explosive atmospheres*, British Standards Institution.
14. BS 4683, Pts 1-4. *Electrical apparatus for explosive atmospheres*, British Standards Institution.
15. BS 5501. *Electrical apparatus for potentially explosive atmospheres* (7 parts) 1977. BSI.
16. BASEEFA *Intrinsic safety*, S.F.A. 3012 H & SE, Buxton, Derbyshire
17. American Petroleum Institute, 'Recommended practice for classification of areas for electrical installations in petroleum refineries', A.P.I. RP 500 A.
18. Institute of Petroleum, 'Model code of safe practice in the petroleum industry', Part 1: Electrical safety code.
19. International Electrotechnical Commission, 'Electrical apparatus for explosive gas atmospheres'. Part 10 'Classification of hazardous areas IEC 79
20. Redding, R.J., *Intrinsic safety*, McGraw Hill, London (1971).
21. RoSPA-ICI Ltd., 'Electrical installations in flammable atmospheres', *Engineering Codes and Regulations* (1972) RoSPA Birmingham
22. H.M.F.I., 'Intrinsically safe and 'approved' electrical apparatus for use in certain specified atmospheres. List of certificates issued by H.M. Chief Inspector of Factories'. Form F. 931.
23. BASEEFA, *Special Protection*, S.F.A. 3009.
24. BASEEFA, *Type N control gear*, S.F.A. 3011.
25. BASEEFA, *Battery-operated vehicles*, S.F.A. 3006.
26. Engineering Equipment Users Association, 'Electrically-driven glandless pumps', Handbook No. 26.
27. International Electrotechnical Commission, 'Degrees of protection of enclosures for low-voltage switch gear and control gear', IEC. 144.
28. International Electrotechnical Commission, 'Rotating machines Part 5 — degree of protection of enclosures for electrical machines', IEC 34-5.
29. Health and Safety Executive, *Industry and services, 1975*. HMSO, London (1976).
30. Burkett, J., and Plump, P., *How to find out in electrical engineering. A guide to sources of information*, Pergamon Press (1967).
31. Fordham Cooper, W., *Electrical safety engineering*, Newnes-Butterworths, London (1978)

5.8 ROPES, LIFTING TACKLE AND CRANES

Contents

Ropes and lifting tackle, i.e. slings, lifting beams, tongs and electromagnets are required with cranes and hoists of all descriptions whether they are mechanically or manually powered. A high proportion of crane accidents is caused by the incorrect use of ropes and lifting tackle, and they are therefore treated first.

Reported accidents involving lifting equipment in construction processes in the UK in 1975/1979/ were as follows:[1]

	1975	1977	1978	1979
Fatalities	22	15	11	17
Total lifting equipment accidents	635	—	473	300

The subject is specialised and only some of the main points can be given here. References should be studied for detailed information. It is subject to various legislation[2] in Britain, the principal being

1. The Factories Act, 1961, Part II, S.26.
2. The Construction (Lifting Operations) Regulations, 1961, S.34 to 41.
3. The Shipbuilding and Ship Repairing Regulations, 1960, Part IV, Raising and Lowering, Etc. S31 to 47.
4. The Shipbuilding (Reports on Lifting Appliances) Order, 1961, (SI 1961, No. 433).

Points covered by (1) above include:

Definition of lifting tackle.

567

Mandatory examination of all chains, ropes and lifting tackle every 6 months by a competent person.

Keeping a register of all chains, ropes and lifting tackle except fibre rope slings.

Table of safe working loads of slings at different angles to be displayed.

Mandatory annealing of chains every 14 months.

5.8.1 Ropes[3,4,5]

Ropes are classed as wire ropes (usually steel) and fibre ropes, the latter being natural or synthetic. Fibre ropes are more flexible than wire ropes, but more easily damaged by heat; natural fibre ropes shrink when wet and are susceptible to rotting on exposure to damp atmospheres, and acid or alkaline fumes. They are widely used for slings, cordage and to some extent for haulage. In general, natural fibre ropes are being replaced by synthetics.

Synthetic fibre ropes are generally stronger and more resistant to moisture and chemicals than natural fibre ropes, but their properties depend very much on the type of synthetic.

Wire rope is generally used because of its greater strength/weight ratio, its constant strength and length whether wet or dry, and its great durability. British specifications for all types of rope are quoted in references 11, 12, 13 and 14.

5.8.1.1 Fibre ropes

The main natural fibre ropes are manila, sisal and hemp.

Manila rope is the strongest and is firm and pliant when new, white or light yellow and lustrous. It begins to char at only 150°C.

Sisal rope has the same colour but less gloss. The fibres are stiff and tend to splinter and the rope is harsh to handle.

The main synthetic fibre ropes are nylon, polyester and polypropylene and these are replacing the natural fibres. Nylon rope is over twice as strong as manila and four times as elastic, has high abrasion resistance and only loses 14% of its strength when wet. It resists rot and mildew, and many oils and chemicals, but it is attacked by strong acids. It loses strength above 150°C and melts at about 250°C. The elasticity of nylon rope makes it useful for shock loading, e.g. for safety lines, but it also increases danger on breakage, as the ends whip violently and can cause injury.

Polyester rope is good for general purposes and critical applications. It has about half the elasticity of nylon, retains its strength when wet, and is very resistant to rot, sunshine, chemicals and abrasion.

Polypropylene ropes are intermediate in strength between manila and nylon for the same size. They are lighter than water, and quite unaffected by it, as well as being rot proof and resistant to most chemicals. But they swell with hydrocarbons and soften at a rather low temperature. These ropes have rather poor abrasion resistance, especially when one rubs in tension on another, and they tend to slip easily as their coefficient of friction is low.

The breaking strength, weight and average safety factors for several synthetic and natural fibre ropes is given in *Table 5.8.1*.

Table 5.8.1 MINIMUM BREAKING LOAD (KG) AND WEIGHTS OF THREE-STRAND HAWSER LAID FIBRE ROPES

Diameter mm	Manila grade 1 (kg)	Sisal kg	Nylon kg	Poly-ester kg	Poly-propy-lene kg
7	370	330	1020	770	740
8	540	480	1350	1020	960
10	710	635	2080	1590	1425
12	1070	950	3000	2270	2030
16	2030	1780	5300	4100	3500
18	2440	2130	6700	5100	4450
20	3250	2840	8300	6300	5370
24	4570	4060	12000	9100	7600
Weight per coil kg/220m					
7	8.0	8.0	7.1	8.7	5.0
12	23.0	23.0	20.6	25.5	14.5
20	61.0	61.0	57.0	70.0	40.0
24	88.0	88.0	82.0	101.0	57.0
Recommended safety factors					
	5	5	9	9	6

Natural fibre ropes are damaged when loaded to 50% of their breaking strength.

New rope should be thoroughly examined for irregularities which indicate possible degradation, before being used. Rope in service should be inspected frequently; at least once a month if used regularly. The whole length should be examined, inch-by-inch for wear, broken fibres, powdered fibre between strands, displaced strands or yarn, variation in size or roundness of strands, discolouration and rotting. The rope should be untwisted in several places to examine inner yarns which should be spotless. The strength of fibres may be checked roughly by scratching with a finger nail.

The diameters of pulleys used with fibre ropes should be at least eight times the rope diameter. Ropes should not be dragged, particularly through sand, dirt, oil, water and grease. Kinking should be avoided. Wet rope made from natural fibre should be dried out promptly by hanging in a dry place in a loose coil, but should on no account be allowed to freeze.

All personnel working with ropes should be properly trained in their use, including how to coil and unwind whilst maintaining the correct twist, how to avoid kinks, and methods of jointing and knotting.

Rope should be carefully stored in a dry clean place, free from fumes, where air can circulate through it. Rope too large to be hung up should be laid on grids so that air can circulate. Dirty rope should be cleaned before being stored.

5.8.1.2 Wire ropes

Wire ropes are used on every type of crane and hoist. The size and type of rope to be used are determined by the maximum load and the safety factor, which in turn is governed by the working conditions. The correct rope is specified by the crane manufacturer.

The most frequent safety factor for running ropes is 5, although lower safety factors are used for some standing ropes or guys. A detailed schedule of recommended safety factors is given in National Standards such as reference 4. The breaking load of the rope is determined by testing a sample to destruction. Purchasers of wire ropes receive copies of the manufacturers test certificates which give the breaking load, the date of test and the testing authority.

Wire ropes consist of strands each containing the same number of wires and a core. There are many types of construction, each intended for a particular use. The main points to consider in rope construction are:

1. The number of wires in each strand.
2. The number of strands in the rope.
3. The lay of the rope or the direction in which the strands spiral.
4. The type of lay which depends on whether the wires spiral in the same direction or the opposite direction to the strands.
5. The type of core, particularly whether fibre rope or wire rope.

Typical rope cross sections are shown in *Figure 5.8.1*. The strands are generally built up of layers of wire, with one wire in the middle. Each strand generally contains 4, 7, 12, 36 or 41 wires, depending on the number of layers.

The flexibility of the rope is governed by the number of wires per strand. Guy ropes which require little flexibility generally have 7 wires per strand, standard hoisting rope has 19 wires per strand. Most ropes contain six strands and a core, though some contain 8 strands.

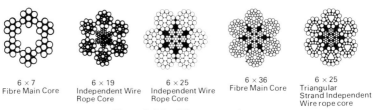

6 × 7
Fibre Main Core

6 × 19
Independent Wire
Rope Core

6 × 25
Independent Wire
Rope Core

6 × 36
Fibre Main Core

6 × 25
Triangular
Strand Independent
Wire rope core

Figure 5.8.1   Wire rope sections
(British Ropes, Ltd.)

Ordinary Lay   Lang's Lay

Figure 5.8.2   Lay of wire ropes

Ropes are normally made in right-hand lay, that is the strands spiral to the right, but they can be supplied in left-hand lay.

Ropes are supplied preformed unless otherwise specified. In a preformed rope the strands and wires have been given the helix they take up in the completed rope. In a non-preformed rope the wires and strands are held forcibly in position, and immediately fly apart when the rope is cut.

There are three types of lay:

1. *Ordinary lay.* The wires and strands spiral in opposite directions. Thus in right hand ordinary lay where the strands spiral to the right, the wires spiral to the left *(Figures 5.8.2)*. The rope is easily handled and does not tend to untwist when a load is freely suspended by it. But it tends to wear as only a few crown wires are in contact with the pulley at any one time.

2. *Lang's lay.* The wires in the strand are laid in the same direction as the strands are laid in the rope. Lang's lay ropes offer a better wearing surface when in use, and in many cases can be expected to last longer. Lang's lay ropes should not be used when one end of the rope is free to rotate.

3. *Multi-strand rope (sometimes called 'non-rotatory').* A stranded rope in which two or more layers of strand are spun helically around the main core. The layers of strands are so disposed that, when under tension, the tendency to rotate will be a minimum.

Where high abrasion resistance is needed, 'triangular strand' construction may be used. Here the outer strands have a triangular section. It is always made lang's lay.

Fibre cores are usually saturated with lubricant. Wire ropes with fibre cores cannot be used when any part of the rope may be subjected to temperatures above 100°C, steel cores must then be used.

In general steel core ropes are to be preferred for many applications.

Wires and strands are lubricated during manufacture, and the finished rope is protected with an external coating.

This protection is normally adequate to prevent deterioration due to corrosion during shipment and storage. For optimum life in use, however, the rope must be lubricated. This in-service, lubrication is termed 'dressing.'

Wire for ropemaking should be cold drawn from plain carbon basic steel according to BS 2763:1968. For corrosive conditions, galvanised wire and stainless steel wire are used. The size of the wire rope in the UK is always given by its diameter measured in mm.

STEP 1.

APPLY FIRST CLIP — one base width from dead end
of wire rope — U-Bolt over dead end — live end
rests in clip saddle. Tighten nuts evenly to recommended torque.

STEP 2.

APPLY SECOND CLIP — nearest loop as
possible — U-Bolt over dead end — turn on nuts
firm but DO NOT TIGHTEN.

STEP 3.

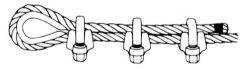

ALL OTHER CLIPS — Space equally between first
two — this should be no more than approx. 6-rope
diameters apart — turn on nuts — take up rope
slack — TIGHTEN ALL NUTS EVENLY ON ALL CLIPS
to recommended torque.

Figure 5.8.3   Method of clip installation (National Safety Council, USA)

5.8.1.3 Handling wire ropes

Gloves, preferably leather, should always be worn when handling wire rope. Personnel handling wire rope must be properly trained in order to avoid kinks, keep the right balance and join ropes by splicing, never knotting. Wire ropes should only be laid out on a clean dry surface, and in a straight line with no slack.

A coil of small rope of short length — say up to 100 m, may be unrolled along the ground but should always be kept under control. Never pull a rope away from a stationary coil. If larger sizes or long lengths of rope have to be handled, these should be supplied on a reel. The reel must be mounted on a stand and supported by a shaft. The reel and stand must be placed to limit the fleet angle of installation to 1°30′ max. (i.e. 1 m width between flange to 40 m of horizontal distance). If space allows for a longer run then it can be used with advantage.

A large or long length of rope on a reel has a high inertia. Make sure the reel is braked to prevent overrun which can lead to loops forming. A simple brake consisting of a wooden plank acting as a lever against the reel flange may be used. If, for any reason, a loop does form, ensure that this does not tighten to cause a kink. (A kink in a six strand rope can severely affect its strength, in a multi-strand rope can lead to distortion which could lead to its immediate discard). Wherever possible the reel stand should be mounted so as not to create a reverse bend during reeving (i.e. for a winch drum with an underlap rope — take the rope off the reel underlap).

Multi-strand ropes are normally supplied with fixed ends unless otherwise specified. If, for any reason, it is necessary to cut a multi-strand rope on site, at least three tight servings of soft seizing wire should be applied on each side before cutting. It is important to maintain the manufactured condition of the rope. Do not use a swivel and if replacing with the aid of an old rope use a short strop of fibre rope between the old and new ropes. Preferably install with a fibre rope as a pilot.

5.8.1.4 Securing ends of wire ropes[6]

If skilled labour is available the rope may be spliced or socketed but if not the ends of a rope may be secured by the use of wire rope grips and a thimble as shown in *Figure 5.8.3*. Other methods of fixing are shown in *Figures 5.8.4 and 5.8.5*. The minimum number of grips for ropes of different diameters are as follows:

Rope dia in mm	*Min. No. of grips*
Up to and including 22	3
Over 22 and up to and including 32	4
Over 32 and up to and including 38	5
Over 38 and up to and including 51	6
Over 51	7

The wire rope grips should be fitted with the U bolts on the short end of the rope and spaced at a distance of six rope diameters, the first grip being fitted as close to the thimble as possible.

Figure 5.8.4   End fittings for wire ropes (top) Swaged sockets; (bottom) zinc poured sockets (Courtesy John A. Roebling's Sons Corp., Subsidiary of the Colorado Fuel & Iron Corp.)

When grips are fitted correctly, the connection should hold up to 85%of the breaking strength of the rope.

After one or two loadings, the grips should be inspected and the nuts tightened as required.

5.8.1.5 Maintenance of wire ropes

Ropes need to be cleaned and lubricated regularly. Cleaning fluids should not be used because they remove lubricant. The best method of cleaning is with a wire brush. The rope should be dry before new lubricant is applied. The lubricant may be applied by hand or as a spray, or by passing the rope through an oil bath.

Ropes must be inspected frequently by a specially trained inspector, and a routine established for frequency of inspection when the machine or sling is taken out of service.

Within the UK, regulations governing the inspection and removal from service criteria of wire ropes are still under discussion at the time of writing. Those directly involved in this aspect are advised to refer to the appropriate regulations under the Health & Safety at Work, etc. Act, as they appear.

Broken wires must not be cut off but should be bent backwards and forwards with a pair of pliers until they break. Fittings such as anchor points and grips need frequent inspection for serviceability and signs of deterioration.

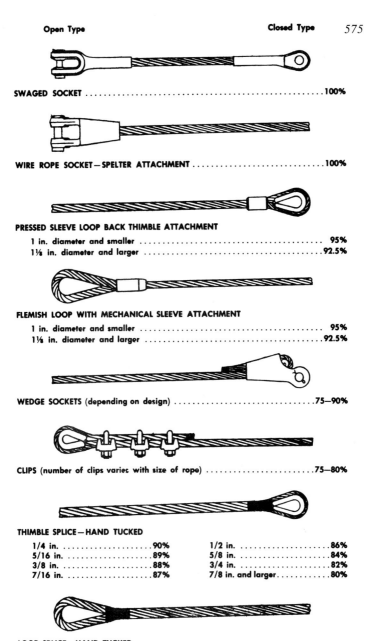

SWAGED SOCKET 100%

WIRE ROPE SOCKET – SPELTER ATTACHMENT 100%

PRESSED SLEEVE LOOP BACK THIMBLE ATTACHMENT
   1 in. diameter and smaller 95%
   1½ in. diameter and larger 92.5%

FLEMISH LOOP WITH MECHANICAL SLEEVE ATTACHMENT
   1 in. diameter and smaller 95%
   1½ in. diameter and larger 92.5%

WEDGE SOCKETS (depending on design) 75–90%

CLIPS (number of clips varies with size of rope) 75–80%

THIMBLE SPLICE – HAND TUCKED

1/4 in.	....................90%	1/2 in.	....................86%
5/16 in.	....................89%	5/8 in.	....................84%
3/8 in.	....................88%	3/4 in.	....................82%
7/16 in.	....................87%	7/8 in. and larger	............80%

LOOP SPLICE – HAND TUCKED
   Efficiencies of loop splice are the same as those given for thimble splice.

Figure 5.8.5   Efficiencies of attaching wire rope fittings in percentages of strength of rope

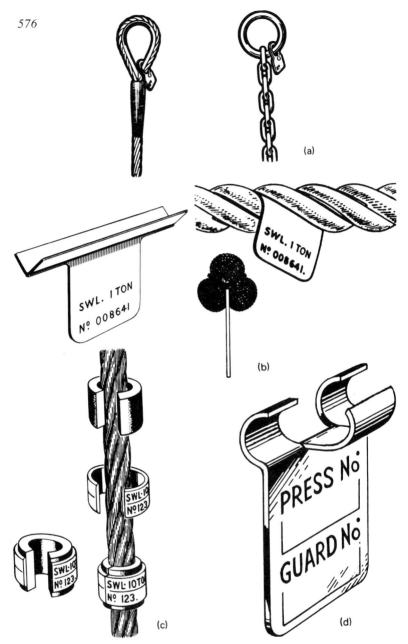

Figure 5.8.6   Identification labels (not to scale)
(a) Chaintabs (c) Cabletabs
(b) Ropetabs (d) Guardtabs
(From 'Principles of planned maintenance' by Clifton)

All wire ropes and slings must be clearly marked with an identification number and the safe working load (*Figure 5.8.6*).

5.8.1.6 Storage of wire ropes

Before storing a wire rope it should be inspected and rope dressing applied as required; it should then be wrapped. Ropes should be stored in a dry place with an even temperature to prevent condensation of water. They should be kept clear of the ground during storage, i.e. on pallets or on a frame if they are on drums. Stored ropes should be covered, inspected regularly and coated with dressing as needed.

5.8.1.7 Damage to wire ropes and slings

The main causes of deterioration of wire ropes are:

Wear. Particularly on the crown (outside) wires.

Corrosion. Generally on interior wires, which is difficult to detect and dangerous. Corrosion on outer wires is more obvious.

Kinks. Often formed by faulty installation of new rope.

Fatigue. Indicated by square wire breaks with crystalline fracture — caused mainly by bending stresses.

Lack of lubricant. Loss can occur rapidly at high working temperatures.

Overloading. Both static and dynamic.

Incorrect coiling. Causing crushing of inner layer against drum.

Mishandling. By dragging ropes or cutting wires.

Wire ropes, particularly in the form of slings, are easily damaged. Typical types of damage and their causes are illustrated in *Figure 5.8.7*

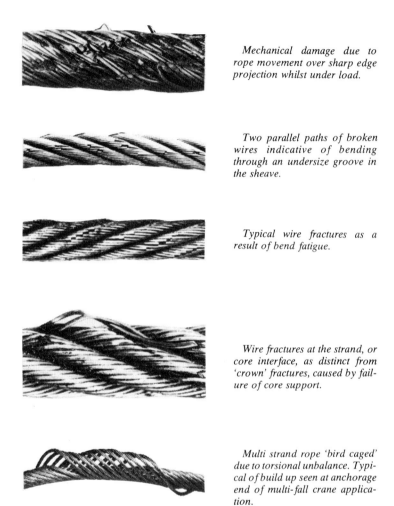

Mechanical damage due to rope movement over sharp edge projection whilst under load.

Two parallel paths of broken wires indicative of bending through an undersize groove in the sheave.

Typical wire fractures as a result of bend fatigue.

Wire fractures at the strand, or core interface, as distinct from 'crown' fractures, caused by failure of core support.

Multi strand rope 'bird caged' due to torsional unbalance. Typical of build up seen at anchorage end of multi-fall crane application.

Figure 5.8.7   Typical examples of wire rope deterioration (Reproduced by kind permission of British Ropes Ltd.).

5.8.2 Lifting tackle[3,4,5,15]

This includes chain and rope slings and their fittings and special tackle which includes lifting beams, tongs and electromagnets. Accidents involving failure of lifting tackle are the commonest ones with cranes. Special care is needed in training slingers in the selection, use, inspection, maintenance and storage of slings and their fittings.

5.8.2.1 Types of sling

Two basic types are used, chain slings and rope slings. These are available in various materials. Every sling must be clearly marked with its number and the safe working load which should be stamped on the upper terminal fitting. Deep indentations which could damage the fitting must be avoided. Only one working load should be shown. The range of slings should be standardised to avoid confusion, and the identification numbers should conform to the range. Safe working loads should, if possible, be given in whole numbers of tonnes to avoid possible confusion caused by a decimal point.

The angle between the legs of a two-leg sling is most important because the safe working load decreases as the angle increases. Although the Law calls for the use of wall charts giving the safe working load for any sling at various included angles, this often leads to confusion. It tends to be ignored and the slinger relies on his own judgement.

A simpler and safer practice is to ensure that no sling is used with an included angle between any two legs greater than 90°. The safe working load marked on the sling will then correspond to an included angle of 90°. The included angle for a three leg sling is the maximum angle between two adjacent legs, and the included angle for a four leg sling is the angle between two diagonally opposite legs.

Table 5.8.2 SAFE WORKING LOADS OF SINGLE AND MULTIPLE LEG SLINGS

Type of sling	% of S.W.L. for 2 leg sling
Single leg	80
Three leg	128
Four Leg	160

Note. These factors allow for the effect of tilt and unequal distribution of the load between the legs.

Safe working loads for single and multiple leg slings are given in *Table 5.8.2.* as percentages of the safe working load for a two leg sling (as defined previously) when all legs are identical in construction and have the same breaking load.

5.8.2.2 Safe use of slings

The following points apply to all types of sling.

1. Shock loads must be avoided, as they can amount to several times the weight of the load. The commonest causes of shock loads are: slipping of the sling; the sudden application of a crane brake; or the snatching of a load.
2. Slings must not be passed round sharp corners without adequate packing to prevent damage.
3. A sling should never be dragged from under a load unless it is quite free.
4. Slings should never be crossed, twisted, kinked or knotted.
5. If there is any risk of personal injury, shock loading, heavy wear or exposure to high temperature the actual load should be well below the safe working load.
6. Crane hooks must always be properly attached to slings, generally via a ring.

5.8.2.3 Inspection of slings

All slings should be examined and tested by a competent and preferably independent inspector (i.e. an insurance surveyor) at intervals of not less than six months, and the results entered in a register in which each sling is clearly identified. The inspection will include proof loading of some or all of the slings. Each component of each leg is to be subject to a proof load equal to twice the safe working load of a single leg. The upper terminal fittings should be subject to a proof load equal to twice the safe working load of the entire sling.

5.8.2.4 Storage of slings

Slings must be stored away from heat and damp in a place where they cannot be damaged by other objects placed on top of them. A special sling store with a place marked for each sling is recommended. Chain slings may be placed by the crane in marked vertical pipes slightly smaller than the diameter of the ring which projects above the top of the pipe.

Rope slings may be looped in coils and placed on pegs or hooks mounted on a vertical board, one sling per peg *(Figure 5.8.8)*.

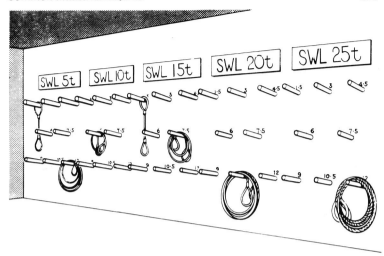

Figure 5.8.8   Storage of slings. Rope slings are stored on a wall-rack marked with the safe-working load and length of the sling

5.8.2.5 Rope slings

Fibre rope slings are particularly suitable for handling loads which might be damaged by contact with wire ropes, and where flexibility is important. But they should not be used if they are liable to be exposed to high or freezing temperatures, or the action of chemicals and acid fumes. Only a high quality fibre rope (preferably manila) should be used.

Wire rope slings are best made from preformed wire rope of top grade wire with an independent wire rope core. For rope up to 1⅛in diameter type 6 by 19 is usual for flexibility. Braided wire rope slings are preferred for lifting heavy round objects. They are flexible, resistant to kinking and have little tendency to twist under load.

Wire rope slings have special end fittings which form a permanent eye and allow a link, shackle or other fitting to be connected. They can be formed by manual splicing, by squeezing a ferrule on to the rope in a hydraulic press, or by placing the end of the rope in a socket and filling it with molten zinc after removing the core and treating the wires. This is a skilled technique which should only be performed by trained and experienced men. Only correct fittings of the right size may be used. Bulldog grips should not be used for slings as they are easily caught on obstructions and are troublesome to handle. The identification number and safe working load must be stamped lightly on the ferrule. All slings should be examined and proof tested before use. Properly made end connections can have 100% of the strength of the rope.

Wire rope slings require less lubrication than lifting ropes. Just sufficient should be applied to prevent corrosion. Excess lubricant should be carefully removed to prevent sand and dirt adhering to the surface.

5.8.2.6 Chain slings

High tensile and alloy steel chain slings are now in general use. They have good shock absorbing properties and require no heat treatment. Wrought iron slings, once common, require skilled periodic heat treatment and should not be used.

Chain slings should be watched for signs of damage or deterioration such as stretched links, undue wear, corrosion and pitting. If damage is serious, the sling should be scrapped. A chain should never be hammered to straighten a link or force one into position. A chain should never be dropped from a height. A broken chain must never be joined by a nut and bolt.

The maximum permissible wear of a chain link is 10% of its nominal diameter, provided the rate of wear is not rapid and the chain is not subjected to shock load. But a chain sling should be scrapped earlier if wear affects smooth working of links or if the metal has flowed over the edges of the worn part *(Figure 5.8.9)*.

Figure 5.8.9   Wear on bearing surfaces

5.8.2.7 Sling fittings

Many different types of end fittings are available. They include eye bolts, shackles, (Both 'die' and 'bow' shape), rings, collars and hooks of various sizes.

When using an eye bolt, the load must be applied in the plane of the eye so that both sides of the eye bolt are in line with the sling. Shackles and not hooks must be used with these eyes. The shackle pin must always be screwed fully into the bow or nut.

Crane and hoist hooks and hooks used to suspend cradles and working platforms are a prolific source of accidents. No more than one sling or sling leg should be suspended directly from a hook, but where two or more slings have to be suspended they should be attached to a ring which is hung on the hook.

Whenever people's lives depend on the support of a load (usually including themselves) from a hook, a safety hook or a housed hook should be used. Plain hooks should only be used for vertical lifts when there is no danger of the rope whipping or the attachment slipping or jumping from the hook, and the ring or other means of attaching the load to the hook must sit snugly in the bight of the hook.

5.8.2.8 Special lifting tackle

Special types of tackle such as beams are required for long objects which cannot be supported by a sling. Other tackle such as electromagnets and tongs may be used for speed and convenience provided special precautions are taken. All such equipment must be inspected regularly and a test certificate issued.

5.8.2.9 Lifting beams

Lifting beams and spreader bars should be purpose designed for a particular task and for use with a particular crane. Each beam should be clearly marked with its identiflcation number, safe working load and the weight of the beam. Care must be taken that the combined weight of the beam and load does not exceed the safe working load of the crane.

5.8.2.10 Electromagnets

Electromagnets are widely used for moving steel, but they have many hazards. The principal one is that a power cut or blown fuse may cause a load to fall. Loads should not be carried near passageways or where people are working. No one should be allowed to stand near a magnet in use. Crane

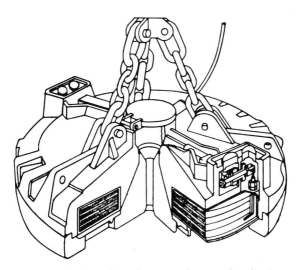

Figure 5.8.10   Typical circular magnet for general applications

drivers should be specially instructed and tested in the use of magnets. The following points must be emphasised:

1. Power to magnet must always be off except when in contact with load.
2. Only wooden poles with no metal attachments or fastenings should be used to guide magnet.
3. No one should be allowed in a wagon or lorry where material is being loaded or unloaded by magnet.
4. The magnet cable must never foul any obstruction.
5. Magnets should never be used near iron manhole covers, checker plate flooring, steel containers or loose magnetic objects.
6. Magnets must not be used for hot loads which could damage insulation. Check temperature limitation of magnet.
7. No shunting of wagons by magnet may be allowed unless rigidly controlled.
8. The weight of magnet plus load must be well below the safe working load of crane and rope to allow for snatching load when magnet is energised.

A typical circular lifting magnet is shown diagrammatically in *Figure 5.8.10.*

5.8.2.11 Lifting tongs

Like lifting beams, these should only be used for particular loads for which they are designed, and with a particular crane. They must be checked before use to ensure that they work correctly.

5.8.3 Cranes[4,5,7,16]

Cranes, hoists and lifts[10] are of so many types and have so many hazards that only those using them who are prepared to study the subject in depth can hope to avoid accidents. The company director responsible for safety and the works safety manager will not normally possess this knowledge unless crane manufacture and operation form part of their business.

If the firm has a production building or workshop with an overhead travelling crane, it will gain familiarity with many aspects of crane construction and operation, but if it only hires a crane occasionally for construction, piling or demolition, it may find itself responsible for the safety of personnel and machinery of which its knowledge is only superficial.

If possible the firm should have a site engineer with the necessary training and experience. One important reason for this is that it is often the hirer rather than the owner of the equipment who is responsible for its safety and to see that it complies with legal requirements. The main regulations which at present apply are:[2]

The Factories Act, 1961, Sections 22 to 27 and Sections 48, 80, 122, 125 and 163.
The Construction (Lifting Operations) Regulations, 1961.

The Shipbuilding and Ship Repairing Regulations, 1961, especially Part IV, Raising and Lowering, etc.

These may be superseded at anytime by new regulations under HASAWA 1974. The most detailed and authoritative guide today is CP 3010, 1972 *Safe use of cranes (mobile cranes, tower cranes and derrick cranes)* which should be studied by those concerned with crane safety. 'A little learning is a dangerous thing', and the limited information given here cannot substitute for such detailed study.

The hazards of cranes and other lifting apparatus may be roughly split into four groups and interactions between them.

1. Those inherent in the apparatus itself, caused by design, construction or subsequent deterioration or mis-erection.
2. Those inherent in the load or the duty, for which the apparatus may be inadequate or unsuitable.
3. Environmental hazards, including sudden wind gusts and storms and unstable anchorage.
4. Human limitations. This is not always the same thing as error. The driver may suddenly develop muscular cramp or a fit of sneezing while lowering a load.

Two accidents demonstrate these points.[1] A rail-mounted tower crane in a siding adjacent to the main Cardiff-London line collapsed under load, falling across the track a few minutes before a train passed. The train was derailed and a number of people injured.

The following points emerged from the subsequent investigation:

1. The weight of the load was twice that marked on it;
2. The slinger and driver were both new to the crane and the job. Had they been experienced they would probably have developed that sixth sense which would have told them that the load was not what it appeared to be;
3. There was no telephone at the siding nor other means of warning British Rail of the accident;
4. The crane collapsed through metal failure. Had the broken part been new or in good condition it would probably have withstood the extra weight.

Whilst the 'proximate cause' appeared to be the abnormal load, the limitations of the persons involved and of the crane itself clearly contributed.

In another accident, when a fire started in a building with a lift, all the occupants except three escaped unharmed by the stairs and fire escape. The three however attempted to descend by lift which stuck between floors when the fire caused failure of the power supply. All three died.

Hazards in the man-crane interaction can be considerably reduced by redesigning the crane cab and controls. Edholm[8] gives an example where the cabs of overhead travelling cranes used in steel mills were vastly improved to give better visibility for the driver, protection from high temperatures by air conditioning, and a seat which allowed the driver to use foot pedals as well as hand controls instead of standing and using hand controls only as before.

Another development of overhead travelling cranes is the use of unmanned cranes which work entirely from radio signals from the ground. Their use however appears limited since the crane driver in his cab generally has the best overall view.

5.8.3.1 Uses of cranes, hoists and lifts

Besides merely lifting things into position (as in construction) and loading and unloading ships, rail and road cars, cranes have many other uses. Overhead travelling cranes and gantry cranes are used mainly for horizontally conveying machinery and materials in process from one end of a factory building or yard to another.

Cranes are used with electric or pneumatic pile drivers for piling, and with demolition balls or 'skull breakers' for demoliton and breaking up scrap. Cranes with grabs and drag lines are used in excavation.

5.8.3.2 Types of crane

The main types considered here are:

1. Electrically operated overhead travelling cranes.[7]
2. Mobile cranes, generally self propelled, and containing their own power source — a diesel or petrol engine.
3. Tower cranes, which are of many types and which may be mobile or stationary.
4. Derrick cranes, including Scotch derricks and guy derricks.

5.8.4 Electrically-operated overhead travelling cranes

The following paragraphs refer specifically to electrically-operated overhead travelling cranes.

5.8.4.1 Operator selection and training

Operators of EOT cranes (as of all other types of cranes) need to be specially selected. They must be able to read and write and be able to understand visual and verbal messages quickly. Operators must be over 18 years old and must have at least an elementary understanding of engineering principles, and be able to judge speed, distance and perspective correctly. Their disposition is all important; they need to be self reliant and suited to working in comparative isolation. At the same time they must be dependable and practical. They should be medically examined before selection to check eye-sight, reaction time and general fitness, and re-examined periodically. An assessment of their trainability for their job should also be made if possible as a valuable aid in selection.

Training should include competent instruction in a classroom in basic structural, mechanical and electrical engineering principles, the use of slings, estimation of weights and angles, stacking and the use of electromagnets. Signalling codes, works safety rules, work permit systems and emergency procedures in case of breakdown, accident or fire should be taught. The training must also include practical work on a crane involving working with slingers, riggers and production and maintenance personnel. This should increase in complexity using different loads at varying speeds under various conditions. On satisfactorily completing the course and passing a practical and oral test, the driver should be given a provisional licence and assigned to work with an experienced driver.

Slingers, riggers and look-out men must also be trained and be thoroughly familiar with hand signals *(Figure 5.8.11)*. The look-out men are required in a busy works such as a steel mill, to warn crane drivers and anyone who might be struck by the crane or load. They should be supplied with a distinctive coloured jacket and flags or bells or both, and should have no other duties. They should be men with a good record of conscientious service and receive proper instruction in their duties which include:

Warning the driver when the crane is approaching a low speed area;
Warning those in the danger area of the approach of a crane or its load to ensure they move to a safe place and remain there while the crane is passing;
Warning all concerned not to enter a danger area when there is any danger of objects falling from a crane or the track.

5.8.4.2 Safe working with EOT cranes

Rules and procedures are necessary for driving, slinging, maintenance and inspection which state clearly who does what, how and when. A register containing details of inspections must be kept.

Great care must be taken to prevent anyone entering or leaving a crane from coming into contact with bare electrical conductors. If any conductors are unavoidably within reach, they should be guarded by permanently fixed screens. Access to cranes should be via stairs and platforms of open tread plates on which dust cannot settle, so designed and located to minimise the possibility of anyone being struck by a crane.

All moving machinery and shafts including the long travel shaft of the crane must be securely and permanently guarded. Guards which need frequent removal must be robust and easily handled, and some means of securing them to the crane when they are disconnected for maintenance must be provided

Special escape equipment, preferably a portable extension, light metal ladder, should be fitted in every crane cab to enable the driver to escape if

Hoist

Lower

Slew left

Slew right

Jib up

Jib down

Travel to me

Travel from me

Signal with both hands

Trolley to me

Trolley from me

Stop

Emergency Stop

Figure 5.8.11   Hand signals for overhead travelling cranes

trapped by fire, or enter and leave the crane if, because of power failure, he could not use the normal access route provided.

5.8.4.3 Maintenance and construction hazards

Maintenance and construction personnel working on or near cranes are subject to various hazards which must be explained to them:

1. Being struck by a crane or its load;
2. Being trapped in places of limited clearance;
3. Being hit by, or dropping, tools or equipment;
4. Falling from an EOT crane;
5. Touching live conductors.

Written procedures must cover the inspection and testing of a crane in motion, and under whose instructions this is performed.

Suitable low voltage plug points should be provided on the crane itself and near the tracks for portable tools used for maintenance.

When more than one crane uses the same tracks, they should not be allowed to approach closer than 6 m to each other. Beyond this there should be a low speed area somewhat longer than the braking distance of the crane when travelling at full speed with its maximum load. Both areas must be clearly marked.

5.8.5 Mobile cranes, tower cranes and derrick cranes[5,9]

The code of practice, CP 3010:1972 should be carefully studied. This gives guidance for the safe use of cranes and includes general recommendations for testing, maintenance, erection, dismantling procedures, etc.

REFERENCES

1. Health and Safety Executive, *Industry and Services, 1975*, HMSO, London (1976).
2. Fife, I., and Machin, E.A., *Redgrave's Health and Safety in Factories*, Butterworths, London (1976).
3. Baumeister, T., *Mark's Standard Handbook for Mechanical Engineers*, 7th ed., McGraw Hill, New York (1967).
4. National Safety Council, *Accident Prevention Manual for Industrial Operations*, 6th ed., Chicago (1973).
5. CP 3010:1972. *Safe use of cranes (mobile cranes, tower cranes and derrick cranes)*. British Standards Institution.
6. BS 462 *Bulldog grips, Pt 2, Metric Units*, 1969.
   BS 463 *Sockets for wire ropes, Pt 2, Metric Units*, 1970.
   BS 464:1958 *Thimbles for wire ropes*, British Standards Institution.
7. BS 466:1960. *Overhead travelling cranes for general use in factories, workshops and warehouses*, British Standards Institution.

8. Edholm, O.G., *The biology of work*, Weidenfeld and Nicolson, London, 166 (1967).
9. Health and Safety Executive, Guidance Notes PM 3 *Erection and dismantling of tower cranes,* and PM 9 *Access to tower cranes.*
10. Heath and Safety Executive. Technical Data Notes, 51 *Periodic thorough examination and testing of lifts.*
11. BS 2052:1965, Specification for ropes made from coir, hemp, manila & sisal.
12. BS 4928, specification for man-made fibre ropes. Part 1:1973 Polypropylene ropes. Part 2:1974 Polyamide, polyester and polyethylene filament ropes.
13. BS 302:1968. Specification for wire ropes for cranes, excavators and general engineering purposes.
14. BS 2763:1968, Specification for round steel wire for ropes.
15. Dickie, D. E. (UK reviser: D. Short), *Lifting tackle*, Butterworths (1981).
16. Dickie, D. E. (UK reviser: D. Short), *Crane handbook*, Butterworths (1981).

5.9 POWERED WHEELED TRANSPORT WITHIN WORKS AND FACTORIES

Contents

Three different types of wheeled transport are included within this category. They comprise:

1. *Rail transport.* This may be confined to a works siding linked into the country's rail network, or it may consist of an extensive internal works rail network.
2. *Road transport.* This refers to the use of vehicles which pass in and out of the works to pick up and deliver goods and services within the works. Many of the drivers will be unfamiliar with the works and its internal road network and traffic regulations.
3. Industrial powered trucks predominantly owned by the firm and used by it for internal transport; these are mostly slow moving and include lift trucks, tractors and trailers and crane trucks; some may work entirely inside buildings.

Fatal accidents in the UK on factory premises for these types of transport are given in *Table 5.9.1.*[1] A somewhat different breakdown, giving all reportable accidents, is given in *Table 5.9.2.*[1],[2] These figures show that vehicle accidents in factories are a serious problem.

Table 5.9.1 FATAL TRANSPORT ACCIDENTS IN UK FACTORY PROCESSES

Year	1971	1973	1975	1979
Rail transport	10	12	6	7
Non-rail transport total	40	54	38	47
Overturning of vehicle	5	9	4	
Collision between vehicles	5	13	2	
Collision between vehicle & fixed structure	6	7	7	not known
Run over	15	14	15	
Other non-rail accidents	9	11	10	

Table 5.9.2 FATAL AND REPORTABLE ACCIDENTS IN FACTORY PROCESSES

Year	1973		1975		1979	
	Fatal	*Total*	*Fatal*	*Total*	*Fatal*	*Total*
Rail transport	12	590	6	489	7	303
Non-rail transport total	60	17 144	42	14 118	47	13 492
Vehicle in motion not moved by power	1	5 657	1	4 200	} 41	} 7 461
Vehicle moved by power	48	6 307	33	5 282		
Vehicle stationary	11	5 180	8	4 636	6	6 031

Factory law appears to be mainly concerned with rail transport in the *Locomotive and Waggons (used on lines and sidings) Regulations 1906*, which have stood the test of time. The Construction (general provisions) Regulations 1961, Part IX, deals with other types of transport including mechanically propelled vehicles and trailers, but these presumably only apply to factories under construction. Neither the Factories Act 1961 nor HASAWA 1974[3] has much to say about transport within factories. HASAWA includes any 'vehicle, vessel, aircraft or hovercraft' under the term 'premises', which should be noted by their owners.

5.9.1 Rail transport within works

The *Locomotive and Waggons (used on lines and sidings) Regulations* attempt to obviate several expected hazards and one or two less obvious ones. They include these admonitions:

Point rods in accessible positions should be covered.
Proper sprags and scotches are to be fitted to the wheels of stationary wagons.
No person should crawl under a train or waggon to cross a railway line.
No person should ride on the bumpers or couplings.
Locos and waggons should be uncoupled only by suitable appliances which do not require anyone to place any part of his body between the bumpers. (*Note.* In spite of this, serious and frequently fatal accidents, constantly recur from this cause.)
When more than one waggon is pushed by a loco, someone should accompany or precede the first waggon if anyone could be endangered.
The engine driver should give an efficient sound warning on approaching a level crossing or blind curve.
Moving rail traffic should be well lit at night and in fog.
All drivers should be at least 18 years old.

A less expected hazard is revealed in the regulation: 'waggons should not be moved by a pole nor by towing by a rope or chain from a loco *on an adjacent line of rails!*'

Locos and trains have two hazards which are unusual among wheeled transport:

1. They are heavy and their braking distance at any speed is much greater than that of a car or lorry.
2. In most locomotives, the driver's position is behind the body of the loco, so that he does not have clear vision of the track immediately ahead. The position is even worse when he is pushing a number of waggons.

Railway systems within works usually differ from public railways in that there are many unguarded crossing points, and although train speeds are low, these three factors together appear to be responsible for many rail accidents within works and factories. Another danger point lies in the approach and joining up of a slowing moving train or loco with stationary waggons on which men are working. At the point of impact they are thrown off balance and are liable to fall or be hit by falling or moving objects.

5.9.2 Road vehicles within works

These are a main cause of industrial injuries and fatalities. The subject is dealt with in H & SE booklet 43,[4] in the Health and Safety at Work Series, in H & SE Technical Data Note 44[5] and in other publications.[6]

The road system within a works is usually quite different from the public system in the following respects:

1. In an old works particularly, there may be many blind intersections, sharp corners and concealed exits;
2. Roads are used by road vehicles, internal factory transport trucks and pedestrians alike. The road transport vehicle is often not the main road user.

In view of the above the following recommendations are made:

1. The use of road transport vehicles within works should be restricted. Car and vehicle parks should be provided outside the works perimeter for visitors. A clear need must be demonstrated before a vehicle is allowed inside a works.

2. Special passes should be issued for drivers and vehicles before they are allowed to enter a works. These should contain essential safety rules (such as observation of a fixed speed limit), and should require the drivers signature to a statement that he has read and understood the rules and agrees to abide by them. Where possible a simple road plan of the works with indications of special danger points, i.e. restricted head room, sharp corners, dead ends and roads where only one way traffic is permitted should be printed on the rear of the pass. The works patrol man on the gate should explain these carefully to new pass holders who are unfamiliar with the

works. If there are sharp corners or narrow roads which cannot easily be negotiated by long or wide vehicles restricted passes should be used for the drivers of such vehicles which preclude them from using these roads, and appropriate warning signs must be provided at the entrances to those roads.

3. Roads inside works should be provided with standard markings and road signs, particularly speed limits, identical to those used on public roads; buildings and departments should also be clearly marked.

It must not be forgotten that a driver who has just left a public road where a higher speed limit applies will not immediately adjust to a much lower limit within a works. Ridges of standard dimensions should be constructed across the roads at strategic points to act as speed retarders and reminders to drivers.

4. Expert advice should be sought before planning new roads, particularly those to be used by long and articulated vehicles.[7]

5. Separate pedestrian routes should be provided with designated road crossings. Suitable pedestrian barriers should be provided at entrances to and exits from buildings, and at the corners of a building.

6. Special attention should be paid to blind intersections where road vehicles may cross internal factory traffic; fixed wide angle mirrors should be used in conjunction with appropriate road signs.

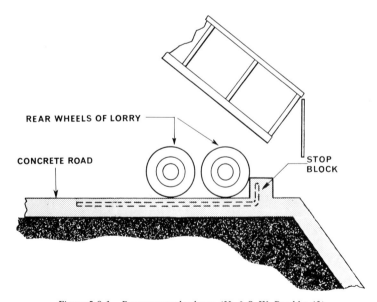

Figure 5.9.1   Permanent wheel stop (H. & S. W. Booklet 43)

7. Special attention must be paid to the dangers of reversing vehicles. First the works road system should be planned to minimise the need to reverse vehicles, particularly long ones. Second, trained assistance must be provided for the drivers of vehicles who need to reverse. These people should be provided with traffic wardens' armbands and be trained not merely in the correct signals, but also to stand in safe positions where they are clearly visible to the driver and have clear vision of the rear of the vehicle and what lies behind it. Where these traffic duties are only part of the person's job, they should nevertheless take priority and the job should be one which makes this possible.

8. Special bays should be provided for loading and unloading so that vehicles placed there do not interfere with through traffic. The road surface there should be firm and level. Wheel chocks and earthing connections should be provided where necessary.

9. Where a road borders on a pit, tip or hopper, into which material is to be tipped or shovelled, safety stops should be provided across the road to prevent the wheels of the vehicle going over the edge. These must be carefully chosen and installed to ensure they are suited to the tipping vehicles *(Figure 5.9.1)*.

10. Roads should be properly constructed of suitable material with even surfaces, and be properly drained. They must be properly maintained and swept, and kept clear of snow in winter.

11. All factory roads must be adequately lit.

12. Warning must be given to road users at places where visibility may be unavoidably curtailed by mist, e.g. from steam trap discharges.

13. Care must be taken in designing plant and roads to ensure that process pipes valves and other equipment are protected by their position from being hit by road vehicles.

14. Loading and unloading must be effectively supervised to ensure that nobody falls or gets hit by a falling object. Proper means of access to the vehicles for those working on it must be provided. When loading, care must be taken that a vehicle is evenly loaded, not overloaded, and that the load is secure, cannot slide and does not overhang. When unloading, care must be taken that the load does not become unstable when ropes, tarpaulins or part of the load itself is removed.

15. Special attention must be given to the loading and unloading of detached trailers. Chocks are generally needed under the rear set of trailer wheels, and additional support to that of the landing gear, i.e. blocks or jacks, are frequently needed under the front.

16. All vehicles must be parked with the brakes on. When a vehicle is parked on a gradient, it should also be left in gear and chocks applied.

The maintenance of road vehicles has its special hazards which are discussed in section 3.4. Two only are mentioned here:

1. A blow out of a large tyre during inflation can cause serious injury. Although it should rarely occur, it is recommended that a special cage be provided for inflation of tyres where this is carried out regularly.
2. When work has to be carried out beneath vehicles supported on jacks, or under the raised body of a tipping lorry, there is always some risk that the lifting mechanism may fail to hold or be de-activated as a result of movement or action on the part of those working on the vehicle. Supporting blocks should be used for added safety.

Driver selection and training has not so far been mentioned, although many road vehicle accidents result from driver fault. One source of error and confusion which firms owning their own vehicle fleets should take pains to avoid is that which comes through drivers changing from different vehicles, with different lengths and different types and patterns of gear and other controls. Adaptable though humans are, it takes time to get accustomed to the length of one's vehicle, the gear pattern and the layout of controls and instruments. Vehicle fleet owners should take pains to standardise their fleets as far as possible, and should not require their drivers to change frequently from one type to another.

5.9.3 Powered trucks

Many types of powered truck are used for movement and stacking of materials in works and factories. They may be classified by the way they are powered, by the position of the driver, and by the type of load carried and the mode of attachment. This is shown in *Table 5.9.3*.

Vehicles may have pneumatic or cushion tyres, and be designed for rough or smooth ground, low or high gradients. It is therefore most important that the right type of truck be selected for the load and type of ground to be traversed, as many accidents have arisen through the use of unsuitable trucks. The speeds are usually limited to about 6 m.p.h. Each type of powered truck has its own hazards and fields of application.

5.9.3.1 Electric trucks

Electric storage batteries are heavy and have to be recharged frequently and/or changed. Employees, particularly their eyes and clothing, must be protected from acid spills and splashes when batteries are changed and recharged. Battery charging should only be carried out in a special area by trained personnel provided with goggles, rubber gloves, boots and aprons.

Battery charging rooms should be well ventilated and smoking and open flames prohibited in them to guard against hydrogen explosion. They should be provided with rubber mats and means of flushing and neutralising

Table 5.9.3 CLASSIFICATION OF POWERED TRUCKS

Classed by	Main classification	Sub classification or attachment	Application
Power source	Electric battery		Indoor use
	Internal combustion	LPG	Indoors, well ventilated and outdoors.
		Petrol	Outdoor
		Diesel	Outdoor
Driver position	Riding		Largest loads and greatest distances
	Walking		Smaller loads and short distances
	Remote		Within single building
Load attachment	Tractor	Trailers	Longest distances, rough ground, varied load.
	Low lift	Fork	Palletted materials
	High lift, fixed mast	Shovel	Loose material
	High lift, tilting mast	Ram	Coils and open centre loads
	High lift, telescopic mast	Suction	Light rolls, e.g. paper.
		Clamp	Pipes, lumber
	Straddle carriers	Hoist	Large heavy loads
	Crane trucks	Boom and jib	

spillages. Charging apparatus and batteries must be guarded against being bumped by trucks.

An overhead hoist with insulated hooks and yokes should be provided for lifting batteries, and a proper truck bay provided for trucks with leads to each truck from the charging equipment. Electric trucks should be used in most buildings in preference to trucks with internal combustion engines.

5.9.3.2 LPG trucks

LPG-fueled trucks give a cleaner exhaust with a lower carbon monoxide content than gasoline and diesel trucks and may be used in well ventilated sheds and buildings subject to satisfactory air sampling and testing under a range of atmospheric conditions when an LPG truck is in use. The carbon monoxide concentration should never exceed 50 p.p.m.

LPG trucks should be refuelled from an LPG storage cylinder outdoors in a special fenced enclosure. This should be a safe distance from buildings and sources of ignition and a special parking and off loading area should be provided with the necessary valves and pipework for it to be refilled from an

LPG road tank car. The fire and explosion hazards of handling LPG are discussed in section 4.2. The symptoms and treatment of carbon monoxide poisoning are discussed in Appendix B.

If an LPG truck catches fire in a building through leakage of LPG, it should be allowed to burn to prevent an explosive gas mixture accumulating. Fire fighting efforts should be restricted to preventing the fire from spreading.

5.9.3.3 Petrol and diesel engine trucks

Petrol (gasoline) and diesel oil should be handled by a normal filling station located in the open air and designed in accordance with the appropriate regulations and codes. Engines must be stopped and no smoking while filling with petrol. The petrol tank cap should be replaced and any spills allowed to evaporate before the engine is restarted.

Petrol and diesel oil fueled trucks are unsuitable for regular use in buildings, although they may start and finish their journeys in a building provided there is free flow of air through it (open sides or large doors), and the carbon monoxide content is monitored.

Petrol engined trucks should carry a suitable fire extinguisher.

5.9.3.4 Tractors and trailers

Tractors should not normally be allowed to tow more than one laden or two unladen trailers. The couplings should be such that they do not come unhitched on curves or permit the trailer to veer from side to side. The driver must be trained and licenced and no passengers may be carried.

5.9.3.5 Trucks: low and high lift

A variety of different types of trucks which pick up the load, carry it in front of the truck, and deposit it sometimes at a height well above the driver, is available. These trucks possess varying degrees of hazard and require differing degrees of skill. While the choice will depend largely on the application, certain safety aspects should be checked.

Trucks which can lift loads higher than the driver, or which operate in areas where objects might fall, should have a strong guard over the operator to protect him without interfering with his vision or access to either side of his driving position *(Figure 5.9.2)*.

All exposed moving parts should be guarded, and tyres should have guards to prevent stones, etc. being thrown at the driver. The driver's seat should be such that he can readily turn his head round to get a clear view behind the truck when reversing.

Every powered truck should carry a plate showing its weight and its rated capacity. It should also have a horn or warning device, loud and distinctive

Figure 5.9.2   Overhead protection for driver of fork-lift truck (H. & S. W Booklet 43)

enough to be heard above other noises in the area. It should be painted a distinctive warning colour.

Many accidents that occur with fork lift trucks result from inadequate training and supervision of their drivers. The most frequent cause of accidents is driving the truck with the load raised, when the centre of gravity of the truck is high. Overturning can then readily occur, e.g. by centrifugal force when turning corners or on driving across a slope, when the centre of gravity is outside the wheel base. Once wheels leave the ground, the overturning moment increases and it is usually too late to prevent the truck and load from falling over and injuring the driver *(Figure 5.9.3)*.

Fork lift truck drivers should be selected, trained, tested and licenced in the same way as drivers of overhead travelling cranes. Since there are several different types of truck with different characteristics, the licence should apply only to the model of truck on which the driver was trained. If required to drive a different type, each driver should undergo further training and pass a test on that type before his licence is extended. Licences should be valid for a limited period, e.g. three years, and drivers should be retested at the end of that period before their licence is renewed.

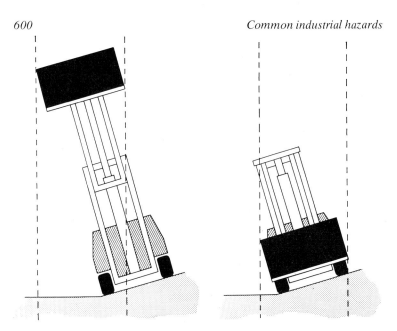

Figure 5.9.3 The effect of tilting equipment (H & S.W. Booklet 43)

The skills, hazards and training of fork lift truck drivers are described in detail elsewhere,[4,6,8] only the more important points are listed here:

Tyre pressures should be checked and adjusted where necessary when taking over a fork lift truck;

Drivers should find out the weights of all heavy objects before attempting to lift them, checking first that they do not exceed the maximum load of the truck; this is a matter where management can help by insisting that the weights of all such articles be clearly marked on them;

Loads must be carried as low as is practicable without risking contact with the floor or ground if this is uneven;

Forks should be kept tilted back except when raising or depositing loads;

Trucks must be driven at a safe speed with consideration for others;

Horns should be sounded at bends with restricted vision;

Care must be taken when materials or stacks have been left in a gangway, and this should be reported at once to the supervisor, and the culprit reprimanded;

Trucks should never be operated on floors inadequate for the load. If necessary, such floors should be clearly marked at entrances with signs prohibiting the use of fork lift trucks. Sudden braking with a load should only be done in an emergency;

The carrying of passengers should be forbidden;

The width between forks should be adjusted to suit the load;

If a load is so bulky as to obstruct forward vision, the truck should be driven in reverse;

When carrying a load down a slope, the truck should be driven in reverse;

When carrying a load up a slope the truck should be driven forwards;

Special care should be taken when going through doorways or travelling over rough or slippery ground and whenever overhead clearance is restricted;

The handbrake should always be kept on when the mast is tilted;

When unloading, the load should not be tilted forward until it is in position over the loading place;

No one should be allowed under the forks when elevated;

Forks should be lowered and the truck immobilised when a truck is parked;

Trucks should, as far as possible, be parked only in special parking areas reserved for them.

Lift trucks are different in many ways from other vehicles, e.g. steering by the rear wheels, tendency to oversteer, frequent movements in reverse, complications caused by the variable height of the forks and tilt of the masts. Hence it should not be assumed that a competent driver of another type of vehicle will quickly and automatically master the handling of a fork lift truck. A five-day course of full time training under a competent instructor is normally considered necessary, although it has been claimed[8] that, with fully experienced instructors and carefully selected trainees, the necessary skills can be acquired in only half this time.

5.9.3.6 *Other powered trucks*

Powered handtrucks should be equipped with brakes which are applied automatically when the handle is released. The hands of the operator should be protected from coming into contact with obstacles, by guards where necessary. If the operator is intended to ride, i.e. by standing on a platform, his feet and legs should be protected, but if he is not intended to ride on the truck, it should be so constructed as to make it impossible to do so without considerable discomfort.

Automatic and remotely controlled trucks should operate only over clearly defined and marked floor areas which should where possible be fenced off to a height of 1 m, with sufficient clearance inside to prevent anyone being trapped between the fence and the truck. The main hazard of these trucks is that they are quiet and give no warning of their approach. They should have a flexible trip device or probe in front which will trip the power and apply brakes if it strikes any person or object in the path of the truck. Their sides and wheels should be designed to protect employees' legs, feet and toes.

Straddle carriers should have horns, head and tail lights, safe access ladders, wheel guards and chain guards, and a properly protected cabin for the driver. They should only be employed on routes when the ground or

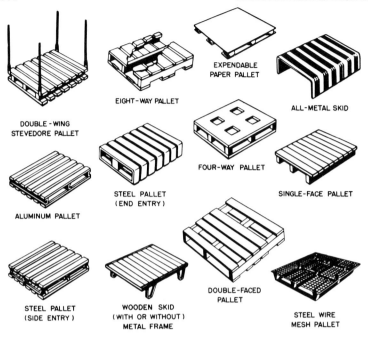

DOUBLE - WING
STEVEDORE PALLET

EIGHT - WAY PALLET

EXPENDABLE
PAPER PALLET

ALL - METAL SKID

ALUMINUM PALLET

STEEL PALLET
(END ENTRY)

FOUR - WAY PALLET

SINGLE - FACE PALLET

STEEL PALLET
(SIDE ENTRY)

WOODEN SKID
(WITH OR WITHOUT)
METAL FRAME

DOUBLE - FACED
PALLET

STEEL WIRE
MESH PALLET

*Figure 5.9.4   Types of pallet and skids used in UK and USA (National Safety
Council & Industrial Truck Assn. USA)*

road surface and the overhead clearance have been surveyed and found
suitable.

Crane trucks come under the heading of mobile cranes and CP 3010[10]
should be consulted on matters of safety.

5.9.4 Pallets

Fork lift trucks, pallet trucks and stackers are generally used with palletised
loads which may form a free standing stack or be placed in storage racks,
which though more expensive, have the advantage that they allow any
pallet and its load to be removed without disturbing the others.

Specifications for the construction of pallets are given in BS 2629, Pt 1,
1967, and for shelving in BS 826:1965. Their safe use is dealt with in Health
and Safety at Work booklet No. 47.[9]

The main types of pallet are flat pallets, post pallets and box pallets. Flat
pallets are usually of wood and are used with loads strong enough to allow
further similar palletised loads to be stacked on top, up to four or five high.
They have two parts, the decks or surfaces and the stringers or spacers,
which separate the upper and lower decks. The deck boards are at right

angles to the stringers. They may be intended to be used a number of times, or they may be expendible, and used for one journey only. The type of wood, its moisture content and the type of fastening are important factors in the strength and durability of wooden pallets. Wooden pallets are liable to warp and develop strains in a dry atmosphere. Several types of skid and pallet are shown in *Figure 5.9.4*.

Post pallets have metal posts at each corner, generally square or circular, each with a foot at the bottom shaped so as to fit securely on top of the corner post below. The decks are attached to the four posts just above the feet. Post pallets generally have open sides. They are used to store machinery, drums and other large objects where one unit conveniently fits within the space bounded by the posts and deck.

Box pallets are similar to post pallets, but have sides, usually of a metal rod grid, which are attached to the posts. They are used to carry loose objects or cartons which might fall through the open sides of a post pallet.

Storing materials in high stacks on pallets, either free standing or in racks, has many potential hazards — fire, collapse under load, or disturbance when a neighbouring stack is being loaded or removed, as well as through the leakage of toxic or other hazardous materials. It requires careful planning and selection of building, storage system and choice of handling equipment.

REFERENCES

1. Health and Safety Executive, *Industry and services* 1975, HMSO, London (1976).
2. H.M. Chief Inspector of Factories *Annual Report* 1974, HMSO, London (1975).
3. Fife, I., and Machin, E.A., *Redgrave's health and safety in factories*, Butterworths, London (1976).
4. Health and Safety Executive, *Booklet 43, Safety in mechanical handling*, HMSO, London.
5. Health and Safety Executive, Technical Data Note 44, *Road Transport in Factories*, HMSO, London.
6. National Safety Council, *Accident prevention manual for industrial operations*, 7th ed., National Safety Council, Chicago, 738 (1974).
7. Henderson, A., and Cole, M., 'Design vehicle criteria and geometric design', *Traffic Engineering and Control*, **9** (9), 431-435 (1968).
8. Partridge, H., 'How to minimise fork truck accidents and reduce operation costs', Chapter 20 in Handley's *Industrial Safety Handbook,* 2nd ed., McGraw Hill, 260 (1977).
9. Health and Safety Executive, Booklet 47, *Safety in the stacking of materials*, HMSO, London.
10. CP 3010:1972, *Safe use of cranes (mobile cranes, tower cranes & derrick cranes)*, British Standards Institution.

6 SPECIAL INDUSTRIAL HAZARDS

6

Special Industrial Hazards

6.1 ENCLOSED SPACES. TANKS, PITS, VESSELS AND DRUMS

Contents

Work, both inside and outside closed spaces frequently results in accidents, particularly where it involves the use of heat (welding and cutting torches, soldering) which may ignite flammable materials present. Such accidents may happen even when apparently thorough precautions have been taken to avoid them.

Work on or inside enclosed spaces is beset by many pitfalls; proper preparatory steps must always be taken to ensure that the work can be done safely. The steps required vary from one situation to another. They must be planned on a sound understanding of the chemical nature and behaviour of the materials originally present in the enclosure, the materials of the enclosure itself and their behaviour when in contact with liquids used in cleaning and when heated. Safe procedures are unfortunately sometimes long and expensive.

The problems were poignantly illustrated by the explosion which occurred in a storage tank at Dugeons Wharf, London in which a subcontractor who was using an oxy-propane cutter and five members of the London Fire Brigade were killed.[1] The tank contained gummy residues formed during the storage of myrcene, a compound rather similar to turpentine. A manager at the depot remarked after the explosion 'I can't understand it. I steamed the bastard for three days . . .' Unfortunately steaming had little effect on the gum, but at the higher temperature of a flame it decomposed, giving off flammable vapour which formed an explosive mixture with air.

The gums or residues left after the evaporation of many industrial organic liquids not only give off flammable vapours when heated; they are likely to contain peroxides formed by reaction between the organic liquid and oxygen in the air. Some of these peroxides are explosive; the peroxide formed from common (di-ethyl) ether is notorious for this — while many others decompose vigorously once a certain temperature is reached and without further application of heat, thereby forming flammable vapours. These hazards are far too seldom appreciated, even by qualified chemists.

Time can ultimately be saved and much danger avoided when careful thought is given to the cleaning, maintenance, repair and even the scrapping of the tank, vessel, drum, etc when it is being designed. But even this foresight can be negated when a container designed for one material and a specific purpose is pressed into service with a different material or for a different purpose from that for which it was designed.

The subject is wide and has many hazards to life. It is conveniently considered under two headings:

1. Work on enclosures from the outside, which involves heating the wall of the enclosure at some point. This includes repair, modification or demolition. The main hazard is that of explosion.
2. Work inside enclosures which requires someone to enter the enclosure. The main hazards are asphyxiation and gas poisoning. Accidents are frequently fatal.

The first heading is dealt with in section 31(4) of the Factories Act 1961[2] which states:

'No plant, tank or vessel which contains or has ever contained any explosive or inflammable substance shall be subjected (a) to any welding, brazing or soldering operation, (b) to any cutting operation which involves the application of heat, or (c) to any operation involving the application of heat for the purpose of taking apart or removing the plant, tank or vessel or part of it, until *all practicable steps have been taken to remove the substance and any fumes arising from it, or to render them non-explosive or non-flammable*; and if any plant, tank or vessel has been subjected to any such operation no explosive or inflammable substance shall be allowed to enter the plant, tank or vessel until the metal is cooled sufficiently to prevent any risk of igniting the substance.'

The second heading is dealt with in section 30 of the Factories Act 1961, which reads:[2]

'30. *Dangerous fumes and lack of oxygen* —
1. The provision of sub-sections (2) to (8) of this section shall have effect where work in any factory has to be done inside any chamber, tank,

vat, pit, pipe, flue or similar confined space in which dangerous fumes are liable to be present to such an extent as to involve risk of persons being overcome thereby.

2. The confined space shall, unless there is other adequate means of egress, be provided with a manhole, which may be rectangular, oval or circular in shape, and shall not be less than eighteen inches long and sixteen inches wide or (if circular) not less than eighteen inches in diameter, or in the case of tank wagons and other mobile plant not less than sixteen inches long and fourteen inches wide or (if circular) not less than sixteen inches in diameter.

3. Subject to subsection (4) of this section, no person shall enter or remain in the confined space for any purpose unless he is wearing a suitable breathing apparatus, and has been authorised to enter by a responsible person, and, where practicable he is wearing a belt with a rope securely attached and a person keeping watch outside and capable of pulling him out is holding the free end of the rope.

4. Where the confined space has been certified by a responsible person as being, for a specified period, safe for entry without breathing apparatus and the period so specified has not expired, subsection (3) of this section shall not apply, but no person shall enter or remain in the space unless he has been warned when that period will expire.

5. A confined space shall not be certified under subsection (4) of this section unless:

(a) effective steps have been taken to prevent any ingress of dangerous fumes; and

(b) any sludge or other deposit liable to give off dangerous fumes has been removed and the space contains no other material liable to give off dangerous fumes; and

(c) the space has been adequately ventilated and tested for dangerous fumes and has a supply of air adequate for respiration;

— but no account shall be taken for the purpose of paragraph (b) of this subsection of any deposit or other material liable to give off dangerous fumes in insignificant quantities only.

6. There shall be provided and kept readily available a sufficient supply of breathing apparatus, of a type approved by the chief inspector, of belts and ropes, and of a suitable reviving apparatus and oxygen, and the apparatus, belts and ropes shall be maintained and shall be thoroughly examined at least once a month or at such other intervals as may be prescribed, by a competent person; and a report on every such examination, signed by the person making the examination and containing the prescribed particulars shall be kept available for inspection.

7. A sufficient number of the persons employed shall be trained and practiced in the use of the apparatus mentioned in subsection (6) of this section and in a method of restoring respiration.

8. The chief inspector may by certificate, grant subject to any conditions specified in the certificate, exemption from compliance with any of the requirements of the foregoing provisions of this section in any case where he is satisfied that compliance with those requirements is unnecessary or impracticable.

9. No person shall enter or remain in any confined space in which the proportion of oxygen in the air is liable to have been substantially reduced unless either:
(a) he is wearing a suitable breathing apparatus;
(b) the space has been and remains adequately ventilated and a responsible person has tested and certified it as safe for entry without breathing apparatus.
10. No work shall be permitted in any boiler-furnace or boiler-flue until it has been sufficiently cooled by ventilation or otherwise to make work safe for the persons employed.'

6.1.1 Hot work on drums and small tanks

Precautions necessary to prevent fires and explosions during the hot repair of drums and small tanks (especially petrol and fuel oil tanks) are discussed in H and SE booklet 32.[3] The first important point made is the following warning:

> '*If facilities are not available for ensuring the removal of flammable material either in the works or elsewhere, repairs or taking apart by methods involving heat should not be attempted.*'

The only possible exception to this is where the work can be carried out with the exclusion of air or oxygen from the tank or drum. This is considered in section 6.1.1.5. The booklet goes on to mention the possibility of cold repairs of small leaks, e.g. in vehicle petrol tanks, by cold curing resins (best reinforced by glass fibre). This suggestion, while sound, raises the possibility that someone may later attempt to repair a further leak by welding in a tank which has already been repaired cold by a resin or other plastic material. This can present a very real hazard on which more is said later.

The drums and small tanks discussed here have all one feature in common. They are readily disconnected from the pipes and frame or structure to which they are attached and it is assumed that this is done before they are cleaned, made safe and repaired. One is, therefore, starting with a tank or drum with perhaps one filling connection and cap between 25 and 60 mm diameter, a smaller emptying connection at the lowest point of the tank or drum, and perhaps another connection for a level indicator or gauge.

All external pipes and valves have been removed; most of the fuel or other combustible material has been drained already. The drum or tank has been detached from its supporting frame, and is fairly readily man-handled. It will probably still contain a thin film of flammable or combustible material on the inside. This is quite sufficient to cause an explosion if the drum or tank is heated. Three points must be made clear:

1. As little as 15 cc of a volatile liquid in a 200 litre drum will form an explosive atmosphere with the air inside the drum.

2. The ignition temperature of many such liquids is sufficiently low for the explosive atmosphere to be ignited by a hot soldering iron.

3. If the tank or drum contains small amounts of a combustible but non-volatile material such as diesel oil, lubricating oil, wax, linseed oil or resin, and is heated, the combustible material will be cracked or decomposed by the heat, forming a flammable vapour and hence an explosive atmosphere.

The steps required and which may be used to make the drum or tank safe for hot work depend on:

1. The materials and method of construction.
2. The previous contents.
3. The past history.
4. The internal configuration (presence of baffles, etc).

The main methods of cleaning available are:

(a) *Washing or filling with cold water.* This is particularly useful where the material to be used is readily soluble in water. It is preferable to fill the tank or drum completely with water, wetting all surfaces, and then allows the contents to drain completely. It may also be carried out to displace gross amounts of liquids which are insoluble in water and cannot be completely drained (such as lubricating oil), as a preliminary to steaming out or hot washing.

(b) *Washing with high pressure water jet.* This is sometimes used to remove sludges and very viscous materials, usually as a preliminary to steaming. It is more applicable to large tanks than small ones. The operator should be guarded (eyes, body) against splashback or losing grip on the hose, and the possibility of static sparking, though remote, should be considered.

(c) *Air blowing.* Air blowing is only applicable to the removal of very volatile liquids which leave no residue. A fan or source of compressed air is required, and the air must be directed well into the interior of the drum or tank, and must sweep out all compartments. The air nozzle and the tank must be grounded to prevent static sparking. Air blowing must be carried out in the open air or in an open-sided building with no sources of ignition, and must always be followed by inspection and testing for flammable vapour. The amount of air and the length of time required to clear a tank or drum of a flammable liquid by air blowing are often far more than expected.

(d) *Steaming out.* This is generally the preferred method (see *Figure 6.1.1*). The tank or drum is best placed over a fixed low pressure steam jet which is inserted through the filling opening so that the condensed water drains freely. Steam hoses can also be used. Both the steam jet and container should be grounded, and the cleaning operations carried out in an area where the vapour evolved causes no nuisance problem or ignition hazard. If

Figure 6.1.1   Steaming out an aircraft fuel tank before repairing (Courtesy British Airways)

the material to be removed is oily, it is possible to check its presence in the condensate draining from the container from time to time. The flow of steam must be sufficient to heat up the whole of the container to 100°C, after which further steaming for half an hour is usually sufficient. Some pitches, heavy residues, polymers, resins and most plastics resist steaming.

(e) *Washing with hot detergent solution (e.g. sodium silicate or sodium phosphate), followed by water wash and/or steaming or air blowing.* Some of the materials resistant to steaming can be removed by washing with a hot detergent solution. Cleaning units are available for drum cleaning with a circulating pump, detergent tank, heater and jets.

(f) *Washing with hot alkaline solution (caustic soda) will remove some materials not shifted by (e).* Workers must wear eye protection, gloves and aprons to protect them from caustic splashes.

The use of organic solvents is not recommended except as a last resort. Where used, they should be followed by one or more of the methods (a) to (e) as appropriate to remove all solvent.

The atmosphere in the drum or tank which has been cleaned, and gas freed for hot work, should be tested using a portable flammable gas or vapour detector or explosimeter of approved type. This itself requires periodic checking and calibration, and replacement of spent batteries.

Internal visual inspection requires good illumination with a flameproof lamp. The use of an endoscope for illumination and viewing is strongly recommended.

6.1.1.1 Materials and method of construction

Unlined steel tanks and drums of all welded construction may be cleaned by any of the above methods.

Soldered steel tanks and drums; tin-plated or galvanised tanks and drums; aluminium or aluminium alloy tanks and drums; copper and magnesium alloy tanks and drums. These may be cleaned by methods (a), (b), (c) and (d) but not by method (f). Method (e) is generally applicable, but the solution to be used and the temperature at which it is used should first be checked with the suppliers of the tank or drum.

Tanks and drums lined with resins, plastics, lacquers, rubber, and tanks or drums which have been repaired with resin or plastic compounds are generally unsuitable and unsafe for welding and cutting and usually also for soldering. Only cold methods of repair should be considered.

6.1.1.2 Previous contents

Water soluble organic liquids, which leave no gum or residue on evaporation (examples are alcohol, cold acetone and acetic acid). Methods (a) or (d) are recommended. Sometimes method (a) alone is sufficient, i.e. the container is completely filled with water, wetting all surfaces, drained, and the process repeated. This, however, should be checked by testing for flammable gases and visual inspection. Method (b) is sometimes applicable, provided a rigorous procedure is laid down and followed.

Volatile flammable liquids insoluble in water, which leave no residue on evaporation (e.g. benzene, toluene, dichlorethylene and light solvent naphtha). Method (a) may be used if required to remove gross quantities of flammable liquid, but must be followed by method (c) or (d), either of which may be used alone provided full precautions are taken. For carbon disulphide, which has a very low ignition temperature, method (d) is not considered safe, and a combination of methods (a) and (c) is recommended.

Liquids of medium solubility and medium to low volatility, and volatile liquids with a less volatile residue. Method (d) is recommended. In addition to testing the atmosphere in the container for flammable vapour the inner surface of the container after steaming should be inspected for non-volatile residue. If there is even a thin film or residue, it is usually unsafe to apply hot methods of working. Some of the residue may be removed by scraping and its behaviour on heating tested (in a laboratory). If it gives off flammable vapours which burn on heating this may be

Figure 6.1.2   Portable inflammable gas detector being used to test atmosphere inside vessel on chemical plant prior to maintenance work (Detection Instruments Ltd)

regarded as confirming that it is unsafe to carry out hot work on the tank or drum.

Sludges, gums, greases, tars, polymers and heavy oils. Methods (b) and/or (e) or (f) will probably have to be applied, followed probably by (d). Inspection and testing are again important. Unless all organic material can be removed, hot work should not be allowed on the tank or drum.

6.1.1.3 Previous history.

The previous history of the tank or drum should be checked to see whether the tank contains an insoluble organic residue or has been repaired or coated internally with a resin which cannot be removed by any of the methods recommended.

If so, no hot work should be allowed unless it can be clearly shown that the heated area is well removed from the area where resin has been applied and that the latter area will remain cool when the hot work is in progress.

6.1.1.4 Internal configuration

Many small tanks contain baffles which makes it difficult to inspect some parts of the interior visually (see section 2.10.3.5). The only evidence that these are free of organic materials which could evolve flammable vapours on heating is that gained from knowledge of their previous contents or manufacture. If it is considered necessary to carry out hot work on such a baffled tank, this should be only done after authorisation in writing by a responsible person after the tank has been cleaned and gas freed.

Some fuel tanks installed on military vehicles are now filled internally with lightweight expanded metal mesh, which greatly reduces the hazards of hot cutting or welding. It seems unlikely however that these will be widely adopted on the fuel tanks of civilian vehicles.

6.1.1.5 Hot work on small tanks and drums with air excluded

In cases where all flammable or combustible material cannot be completely removed from a small tank or drum it may be possible to exclude air from it while hot work is done on the outside by passing steam through it, keeping it full of water or filling it with carbon dioxide or nitrogen gas.

Carbon dioxide gas is available as 'dry ice' which should always be handled with gloves to prevent cold burns, and in a well ventilated place. About 1 kg of crushed dry ice is required per 400 litres of tank or drum volume. Carbon dioxide is heavier than air and will flow out of any openings on the lower side of the tank or drum. When carbon dioxide is used, all openings in the tank or drum should be closed except one on the upper side which should be loosely plugged or fitted with a simple non-return valve to prevent air re-entering.

One difficulty with hot work on a tank or drum filled with carbon dioxide, nitrogen, etc. is that once the tank or drum is cut, the inert gas will escape and be replaced by air.

A promising technique which largely obviates this problem has been pioneered in the UK by British Oxygen Company. *(Details are available from BOC–NOWSCO Ltd, Units G2 and G3, Boundary Road, Harkeys Industrial Estate, Great Yarmouth, Norfolk NR31 0LY).* This is Nitrogen Foam Inerting which is now available as a commercial service (see *Figures 6.1.3* and *6.1.4*). The air or other gas or vapour in the tank is displaced by a low density but stable foam containing only nitrogen as gas. The foam is introduced at the bottom. Foams of various stability are available to suit the work in hand. It is possible to use a stiff and heat resistant foam which remains intact close to welding and cutting operations. Nevertheless the foam inside a tank should be continuously topped up as work proceeds.

Other possible hazards are that the residue left in the tank or drum is an unstable peroxide which may explode or decompose vigorously when heat is applied, or that it gives off toxic vapours which affect those working on the tank or drum. Hence hot work on tanks or drums containing combustible materials can be dangerous even when all air or oxygen is excluded, and such work must be carefully supervised.

616

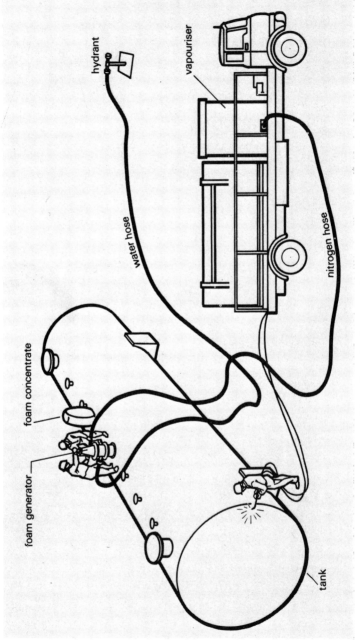

Figure 6.1.3   Welding of vessel with nitrogen foam inerting (British Oxygen Co.)

Figure 6.1.4 Nitrogen foam inerting. Vessel being dismantled after inerting. Final section of vessel removed after third cut (British Oxygen Co.)

6.1.1.6 Scrapping of small tanks and drums

Unless the tank or drum to be scrapped can be readily cleaned and made gas free, it is generally uneconomic to cut it by a flame for scrap. If it has to be cut, a cold method is preferred. Usually it is better to flatten it by a tup or flattening rolls.

6.1.1.7 Standard procedure for hot work on drums and small tanks

In works where hot work on drums and small tanks is frequently called for, it is advisable to work out a standard written procedure, with the necessary permits to work, based on the circumstances prevailing and in the light of the considerations discussed above.

6.1.2 Isolating, cleaning and gas freeing fixed vessels and tanks and confined spaces[4,5]

Fixed vessels and tanks must be cleaned and freed of flammable gases before hot work is done on them from the outside, and must in addition be freed of toxic gases and vapours, contain only air and be well ventilated before any person is allowed to enter them for work inside unless that person wears breathing apparatus and any other necessary protection.

Such isolation, cleaning and gas freeing is critically important when the vessel or other confined space has to be entered, and many lives have been lost through neglect in these operations. A works which has neither proper cleaning equipment nor personnel informed on tank cleaning and gas freeing should consult the supplier of the liquids contained in the tanks which require cleaning. There are several firms of tank cleaning contractors who have all the necessary equipment as well as experienced crews.

The operations discussed below should be carried out under the supervision of a competent person who is familiar with the hazards and with fire accident prevention and first aid measures. Before starting operations, the supervisor should check what equipment is needed at each stage and ensure that he has everything needed. This may include beryllium copper non-sparking tools, blank flanges and stub ends and jointing material, water, steam and compressed air hoses and jets, portable air blowers with flameproof motors and long lengths of suction and delivery hose, fresh air hoses masks with hose and blower, self-contained breathing apparatus (for rescue) suitable clothing, safety belts, safety lines, ladders, ropes and lifting tackle and oxygen resuscitation equipment.

6.1.2.1 Vessels and tanks above ground

This is the most common case where cleaning is required. The steps required are set out below in their logical order.

1. *Normal emptying or pump out.* The inventory of flammable, combustible or toxic process liquids are reduced as far as possible by normal process procedures, floating out hydrocarbons remaining below the emptying lines of tanks and vessels with water introduced through fixed connections. Stirrers and other motors are then stopped and valves shut.

2. *Removal of ignition sources, erection of barriers and notices.* All ignition sources — matches, open flames, smoking, gas engines, welding, exposed electrical wiring and equipment — are removed to an appropriate distance from the vicinity of the tank or vessel, bearing in mind that flammable vapours are likely to be released from the tank or vessel to atmosphere during cleaning. Barriers are then erected around the tank or vessel at a suitable distance with notices prohibiting entry to all but authorised personnel, prohibiting the use of flames and ignition sources within the area, and warning that this is a hazardous area in which flammable vapours will be released.

3. *Disconnecting and blanking all lines.* Liquid remaining in the tank or vessel should first be drained as far as possible through drain valves or lines, running the drainings where required into drums, though often this will run into oily or chemical drains of the works.

All process lines into and out of the vessel or tank should then be disconnected at flanges on the vessel or tank, where possible removing a stool piece or elbow so that the tank or vessel is open to the atmosphere. Blank flanges should be inserted between the vessel and the nearest valve to it on all process lines that are not disconnected while work is in progress. Internal valves in the vessel (e.g. excess flow valves and 'mushroom' valves) should be removed. Shutes and lines for entry or discharge of solids should also be removed or blanked.

All service lines should be similarly disconnected or blanked except for any which will be required during the subsequent cleaning and gas freeing steps (e.g. steam, water, air). These must be isolated after use. Relief lines which discharge into a common discharge header which is in use at the time the tank or vessel is out of service should also be blanked — the position of the blank being carefully selected to minimise possible hazards.

Valves may only be considered for isolation in these circumstances where there are two closed block valves in the line with an open bleeder valve between them. The block valves must then be locked closed and the bleeder locked open. This, however, is more hazardous than the installation of blank flanges, and the practice should only be allowed in approved cases.

The isolator switches of all motors attached to the vessel and in the area isolated should be locked in the OFF position.

4. All *manholes* should be opened and the tank or vessel allowed to air, unless the tank is liable to contain highly toxic or pyrophoric material — iron sulphide on the walls, finely divided carbon, or unless the contents are themselves pyrophoric. In this case the tank should be given a preliminary steaming or other treatment to remove the pyrophoric materials before the manholes are opened.

5. *Sludge and material* which may not have been completely drained should be removed as far as possible from outside the tank, using high pressure water hoses, grounded to prevent possible static build up. Suitable breathing apparatus (for the fumes from the tank or vessel) should be worn and flameproof lamps or torches used. If the tank or vessel contains an agitator it may be preferable to run the agitator whilst removing sludge. The exact procedure for removing sludges which are difficult to handle should be worked out with the co-operation of a chemist who understands the properties and hazards of the materials.

6. *The wind conditions and the disperal of vapour* during steaming or air blowing should be considered. Only when these conditions are deemed safe should steaming or air blowing be started. This is continued for a number of hours, depending on the size and flow of air or steam. Air blowing should not be considered when the tank contained a volatile liquid with no 'tail'. If the contents of the tank or vessel were highly toxic, it may be necessary to

connect a special condenser and receiver to the tank or vessel during steaming.

If steam is used, the tank or vessel should be cooled and ventilated after steaming. If all openings are closed, the tank or vessel will probably collapse as the steam condenses inside it.

7. Test the atmosphere in the tank with a flammable gas indicator which has recently been calibrated using a long sampling probe manipulated from outside. If the instrument shows more than 10% of the lower explosive limit at any point, there are likely to be pockets of higher concentrations and steaming or air blowing should be continued until a lower level is found.

6.1.2.2 Vessels and tanks below ground in pits, cellars and basements.

While the principles employed for vessels and tanks above ground still apply, the problems are more difficult because of several complicating factors:

1. The pit itself is a trap or enclosure for heavy vapours, and a vessel or tank in a pit is an enclosure within an enclosure. Any person working in a pit containing a tank or vessel with flammable or toxic material in it should wear an air line breathing apparatus, safety belt and lifeline which is attached to a simple lifting apparatus above the pit operated by a trained person who is able to extricate the worker if required. This applies even when the atmosphere inside the pit itself is safe.

2. It is more difficult to isolate buried lines connected to a pit than to a vessel above ground. This applies especially to drains from the pit, through which flammable toxic vapours can sometimes back up into the pit. All drains should as far as possible have water seals, although even those cannot entirely be relied on. A surge of gas pressure in the drain can blow the seal and quickly fill the pit with toxic gas.

3. When a vessel within a pit is being drained into the pit, it is necessary to provide such drainage facilities in the pit as may be required, including sump pumps with flameproof electric or pneumatic motors, and air extraction fans and exhaust ducting extending to the bottom of the pit to remove heavy vapours. Such fans should be sparkproof.

4. If a man collapses in a pit without a lifeline by which he can be rescued, nobody should risk their own life attempting to rescue him unless the rescuer is wearing a self-contained breathing apparatus, has a lifeline attached to him and sufficient trained support above the pit to pull both him as well as the victim out. In practice this may mean that the man who has collapsed inside the pit is beyond rescue. Whilst this may sound callous, there have been too many multiple casualties in pits and sumps through urgent impromptu rescue attempts by unprepared persons for any other advice to be given.

The fact that a man has collapsed in a pit is a near certain indication that the atmosphere in the pit, harmless though it may appear, is lethal and a sure death trap to anyone else entering it without an independent air or oxygen supply. It also suggests that the safety measures where this happens are inadequate.

6.1.3 Work inside confined spaces where the atmosphere is dangerous

The previous paragraphs relating to pits are also relevant to this section. Details of protective breathing apparatus and clothing to be worn are given in section 3.10.

Nobody should be allowed to enter a tank containing toxic vapours or an oxygen deficient atmosphere unless it is absolutely necessary. An air line breathing apparatus supplied by a blower outside the tank where the atmosphere is quite safe is generally most suitable in the circumstances.

The fit of the facepiece should be carefully checked before the man enters the tank. If the tank or vessel contained an atmosphere that is dangerously irritant or corrosive to the skin, a complete attire of impervious clothing should be worn which is supplied with clean air inside by the air line. Persons with burst eardrums should not be allowed to work in vapour-laden atmospheres.

An important and sometimes forgotten point is that the person entering the tank or vessel must be able to enter through the manhole wearing the breathing apparatus and clothing without displacing the face mask or tearing the clothing. *This requires forethought.* It is even more difficult for a rescuer wearing a self-contained breathing apparatus with air or oxygen cylinders on his back to enter through a standard tank manhole.

The person working in the tank or vessel must wear a safety belt or harness and lifeline and must be supported by someone outside who has been trained with him and is able to rescue him from outside the tank. If entry is through a top manhole, suitable davits and lifting tackle must be positioned for rescue.

6.1.4 Work inside confined spaces, where the atmosphere is safe

An atmosphere which shows an adequately low reading on a flammable gas detector is not necessarily safe. It may not even contain any oxygen at all or it may contain pure oxygen. So the oxygen content should also be checked, by a competent person using approved apparatus.

Next, one must consider what toxic gases may be present in the atmosphere initially or may be formed there while the man is present. This problem was discussed in section 3.2, which gives a short list of common toxic gases and their TLVs (*Table 3.2.3*) and a list of booklets published by the Health and Safety Executive on methods of tests for many of these. Firms supplying respiratory protective equipment also supply simple and rapid portable test equipment[6] for many gases, in which a known volume of gas is sucked through a glass tube containing an absorbent material. This shows a colour change which is specific for a particular compound and which gives an approximate measure of the concentration present. A selection of tubes may be carried so that the atmosphere may be tested quantitatively for a number of possible contaminants in a matter of minutes.

It must however be stressed that such simple analytical methods as these are at best very approximate. The National Safety Council of Chicago has drawn attention to a number of pitfalls in their use.

Before any work is started inside a tank or other confined space it is essential that the atmosphere be thoroughly tested and the results entered on a confined entry certificate before anyone is allowed to enter. One form of confined space entry certificate which is suitable for many works is shown in *Figure 6.1.5*.

Next, the questions of ventilation and generation of toxic gases within the confined space have to be considered. Unfortunately, one of the most dangerous confined spaces still continues to be the night watchman's hut, with carbon monoxide the lethal gas, generated by a simple oil or calor gas burner. Water heating boilers installed in small rooms or basements with inadequate ventilation are another common and easily overlooked trap.

Tanks and vessels inside which operatives have to work must be well ventilated when work is in progress, and a specially close watch must be kept when hot work involving flames and/or torches, grinding or work with

---

FACTORIES ACTS 1961, SECTION 30.

CHEMICAL WORKS REGULATIONS, 1922—REGULATION 7.

Certificate No. **A   98714**

Certified that the under-noted place which requires to be entered and in which there is reason to apprehend the presence of dangerous gas or fume:

(a) is isolated and sealed from every source of such gas or fume and has a supply of air adequate for respiration and is free from danger.

(b) is not so isolated and sealed from every source of such gas or fume and free from danger and therefore must not be entered by any person unless he is wearing a 'breathing apparatus' and (where there are no cross-stays or obstructions likely to cause entanglement) a 'life-belt' the free end of the rope attached to which shall be left with a man outside whose sole duty shall be to keep watch and to draw out the wearer of the life-belt if he appears to be affected by gas or fume.

(Either (a) or (b) must be crossed out in full).

Particulars of place referred to:

Certificate valid until...................................(Time)   ....(Date)

Signed.. (Time)   ....(Date)
            Signature of responsible person appointed for the purpose.)

Serial Nos. of Clearance Certificates issued under this Section 30/Regulation 7 Certificate:

COMPLETION OF JOB: This Section 30/Regulation 7 Certificate is hereby cancelled.

Date and Time..   Signature...

Figure 6.1.5   Confined space entry permit

chemicals or solvents (e.g. lining with cold cure resins) has to be carried out inside the vessel. In such cases fan assisted ventilation should be employed, and appropriate additional respiratory and eye protection should be considered, provided where appropriate and made compulsory wearing where necessary.

The importance of the formal entry certificate must not be overlooked or forgotten. Signature of the certificate by a person responsible for any one of the various precautions is a salutory reminder that he will be liable to serious penalties if he has skimped or neglected his responsibilities. The certificate also draws the attention of everyone concerned to all the precautions considered necessary, and thus enables everyone to make a mental check that he himself is satisfied that all necessary precautions have been taken.

6.1.5 Repair and demolition of large storage tanks for flammable liquids[6]

Large tanks which have been used for storing flammable and combustible liquids present some special problems:

1. Because of their size, it is often impossible to steam them out. There just is not enough steam available.
2. Tanks often contain residues — oily, tarry or gummy — on the bottom, sometimes on the side of the tank and occasionally inside the roof of the tank which are not properly removed by steaming.

Where steaming cannot be used or is ineffective, the use of high pressure hot or cold water mechanical washing, with or without a detergent is recommended. Suitable machines are available which can be lowered into a tank through the top manhole, and automatically spray the walls, roof and bottom of the tank in the same way as a garden sprinkler. There are several specialist firms who engage in this work.

It is not, however, always possible even with such equipment to remove all residues and flammable vapours from a tank, whilst hand cleaning of residues by men with breathing apparatus is arduous, dirty and potentially hazardous. The following cases are considered, using other methods:

1. Repair of storage tanks holding flammable liquids (flash point below 38°C.)
2. Demolition of storage tanks holding flammable liquids (flash point below 38°C).
3. Repair of storage tanks holding combustible liquids (flash point above 38°C).
4. Demolition of storage tanks holding combustible liquids (flash point above 38°C).

6.1.5.1 Repair of large storage tanks holding flammable liquids

It is assumed that the tank has been either steamed or mechanically washed and aired but still contains residues.

If the roof and side walls are clean and the residue is entirely on the bottom, and it is heavier than water, it may be covered with a few inches of water. The tank is then blown through with air using a spark free ventilation fan and the atmosphere tested for combustible gases. Cutting or welding may start on the side when the atmosphere is free of combustible gases, but ventilation should continue while work is in progress.

If the residue is lighter than water, it may still be possible to blanket it with high protein foam, then proceed as before. But in this case a continuous supply of foam may be needed to make up for foam which collapses.

Where these methods are not applicable (e.g. there is a gummy residue adhering to the sides of the tank) then it may be necessary to consider inerting the tank by filling with nitrogen, carbon dioxide or products of combustion from an inert gas generator. Purging and inerting must be carried out thoroughly and the atmosphere in the tank tested at several levels and positions before hot work starts. At no point should the oxygen content exceed 10% before hot work is started. The use of nitrogen foam discussed in section 6.1.1.5 will often provide a safe and economic solution to these problems.

Welders and other workers must be protected (usually by air line breathing apparatus) against asphyxiation by the inert gas once they have made a hole in the side of the tank. The supply of inerting gas must be maintained to make up for this loss.

6.1.5.2 Demolition of large storage tanks holding low flash point liquids

After steaming or mechanical washing, it may be possible to cover any residues with water or protein foam, gas free the tank by ventilation, and proceed to demolish from the top downwards.

If this cannot be done, the possibility of first filling the tank with water, then making a substantial opening at a high point with a cold tool (e.g. pneumatic chisel) should be considered. The top sides can then be demolished working downwards, keeping the tank full of water.

6.1.5.3 Repair of storage tanks containing less combustible liquids

As an alternative to the methods given in 6.1.5.1., the use of high expansion foam is suggested, after gas freeing the tank. The foam is generated using a portable foam generator and injected through a bottom manhole and allowed to flow out over the top while work is in progress. The advantage of this method is that it obviates the need for preliminary cleaning of the tank. If the tank contained a material of low flash point it may be possible to employ nitrogen foam. (see section 6.1.1.5).

6.1.5.4 Demolition of storage tanks containing less volatile liquids

If there are non-volatile residues on the side walls and roof, after atmospheric testing the roof can be cut away by cold methods without the need to fill with water.

Sometimes the roof can be demolished using a crane and skull-breaker or other heavy weight.

6.1.5.5 General

It is clear from the many different conditions that can arise that the cleaning and preparation of large tanks for repair or demolition have many complications. In case of doubt specialist advice should be sought from the Health and Safety Executive or from firms or consultants specialising in the subject.

REFERENCES

1. Tye, J., *Safety uncensored*, Corgi, London, 119 (1971).
2. Fife, I., and Machin, E.A., *Redgrave's health and safety in factories,* Butterworths, London (1976).
3. Health and Safety Executive, *Repair of drums and small tanks: explosion and fire risk*, Booklet 32, HMSO, London (1976).
4. Wareing, T.H., 'Entry into confined spaces', Chapter 6 of *Industrial Safety Handbook*, edited Handley, W., 2nd ed., McGraw Hill, Maidenhead (1977).
5. National Safety Council, *Accident prevention manual for industrial operations*, 7th ed., National Safety Council, Chicago, 1314-1318 (1974).
6. Health and Safety Executive, *The safe cleaning, repair and demoliton of large tanks for storing flammable liquids*, Technical Data Note 18 (rev) H & SE.

6.2 CHEMICAL HAZARDS

Contents

Chemicals are used in many industries and it is inevitable that chemical hazards are discussed in several different contexts and places in this book. Since all matter has its own chemical composition, a great many materials used or formed in industry may be classed broadly as chemicals. This chapter is mainly concerned with substances which are generally recognised as 'chemicals' and their use both in laboratories and industrial processes.

Hazards can arise leading to serious results. One example was the mass poisoning of livestock and humans (through eating poisoned meat) in Michigan, USA. This was caused by the mistaken incorporation into animal food of a toxic chemical manufactured as a fire retardant for plastics.

The principal references to chemical hazards which have already been discussed in other sections are given below:

Section 1.4 *Safety responsibility and organisation*
The responsibilities of manufacturers and suppliers to warn customers of hazards in the handling and use of their products is discussed.

Section 1.9 *Safety inspections and audits*
The audit programme given here was specifically devised for use in the Chemical Industry.

Section 1.12 *Secrecy*
A hypothetical example is given of how a serious accident in chemical manufacture could result from commercially justifiable secrecy which could well prevent the hazard coming to light.

Section 1.13 *Planning for Major Emergencies*
This relates particularly to major hazards in the oil and chemical industries.

Section 2.8 *Oil and chemical plant*
The rating of fire and explosion hazards in chemical processes and their influence on plant layout, design and protective features is discussed.

Figure 6.2.1 Washing away lead from streets contaminated over a period from a nearby factory (Popperfoto)

Section 2.9 *Hazard analysis and plant and equipment design*
The example given to illustrate quantitative hazard analysis is a hydrocarbon-water separator such as is employed in the chemical industry.

Section 3.2 *Air and breathing*
Air contaminants and their toxicity and control are discussed in relation to respiratory disease. Threshold limit values for concentrations of toxic chemicals in air are discussed, values given for a number of common ones, and references given for methods of determining actual concentrations in factory atmospheres.

Section 3.9 *Industrial hygiene and toxicology*
The health hazards of many different types of chemicals are discussed.

Section 3.10 *Personal protective clothing and devices*
Toxic chemicals feature prominently among the hazards from which personnel have to be protected.

Section 3.11 *Locker, dressing and washing rooms*
The subject is related to the chemical contamination to which the worker is exposed.

Section 4.2 *Explosion hazards*
Chemical explosions are treated at some length.

Section 4.3 *Fire hazards*
Chemical fires and suitable extinguishers for a number of chemicals are
discussed.

Section 5.1 *Poor housekeeping*
The storage and safe handling of gases in cylinders and liquids in drums and
carboys are discussed briefly.

Section 6.1 *Enclosed spaces, tanks, pits, vessels and drums*
The cleaning and gas-freeing of tanks, vessels and drums which have
contained toxic chemicals are discussed.

6.2.1 Chemical hazards in laboratories

A useful starting point in this discussion is the attitude of the scientist or
chemist who handles the chemicals in a laboratory. Some remarkable
changes have occurred in the last fifty years in the attitudes of scientists to
the hazards of the chemicals which they handle.

The traditional image of the absent-minded scientist working away far
into the night surrounded by clouds of smoke or fumes from his
experiments was often not far from the truth. Chemistry has been and
probably still is, a vocation for its most devout adherents, demanding
sacrifice of time and where necessary of health in order to advance the
frontiers of knowledge in the service of mankind. The occupational cancer
suffered by Madame Curie as the result of her researches in radio-activity is
regarded as a noble and inspiring example to other scientists.

Nobel suffered severely from inhaling the vapour of nitroglycerine.
J.B.S. Haldane the biologist was more than once the willing human guinea
pig of his own experiments. Dr. C.H. Barlow another martyr in the cause of
science innoculated himself in 1944 with the bilharzia worm which
produced 30 000 eggs a day in his liver. After nearly dying he was cured
after a year and a half by treatment as painful as medieval torture. The
frontiers of knowledge will always find volunteers ready to sacrifice
themselves in its cause.

The absent-mindedness of the scientist is usually associated with the
character of the person pursuing this field. Absent-mindedness is, however,
just as likely to have been the effect of one or more toxins which the
scientists handles on his central nervous system and brain. The vapours of
mercury, hydrogen sulphide, benzene and many carcinogenic chemicals
were commonly present in the atmosphere of many a teaching and research
laboratory even quite recently.

Such sacrifice can only be justified if it reduces the exposure of others to
toxic hazards. Often the reverse seems to be the case. A recent survey of
causes of death in members of the American Chemical Society between
1943 and 1967 revealed that deaths from cancer of the pancreas and

malignant lymphomas were significantly higher than among the general population.[2] Members of the American Chemical Society (or indeed any other Chemical Society) probably regard working conditions which they readily put up with as quite good enough for their laboratory assistants. Thus it is not long before they may expect this of industrial workers and even the general public.

It has taken several tragic shocks involving members of the public to even make a dent on this traditional attitude of the chemical scientist. These shocks include the thalidomide tragedy, the death and disablement of many Japanese from eating shell fish containing high levels of mercury from Minamata Bay, the Seveso disaster when a few kilograms of a 'super toxic chemical' formed as a by-product of a process for making weed-killers were released to the atmosphere, and the large-scale Michigan food poisoning referred to earlier.

At the same time, and possibly partly as a result of these shocks, the following improvements are taking root:

1. Proper health precautions in chemical laboratories, including strict housekeeping standards, adequate numbers of fume cupboards, glove boxes and remote handling methods for potentially hazardous chemicals, and adequate personal protection.
2. Experimental evaluation using test animals of the toxic hazards of all industrial chemicals (particularly those used in drugs and as food additives) and of new compounds hitherto only synthesised in chemical laboratories.
3. Improvements in hygiene and toxicological control in industry.
4. Withdrawal of some toxic materials from use, and restrictions on the use of many others with known or suspected toxicological effects.

The research chemists' activities are bound to continue unless a complete and internationally agreed halt is called to all scientific research involving the synthesis and utilisation of new compounds. At present it seems that new compounds are still being synthesised several times faster than the rate at which they can be tested, and the number of suspected toxins is steadily increasing.

A scheme under which all UK manufacturers and importers will be required to notify the Health and Safety Executive of all new substances introduced in quantities greater than 1 tonne/year, *and of their likely hazards*[13] was circulated by the H & SE for discussion in 1977.

6.2.2 Chemical hazards in industry

The first requirement is adequate and systematic information about the hazards of the materials used. One useful standard form for compilation of this information is the *Material Data Safety Sheet* issued by the US Department of Labour (*Figure 6.2.2*). This was originally drawn up in response to a legal need for employers in Ship Repairing Ship Building and Shipbreaking to have such information at hand in order to enable

U.S. DEPARTMENT OF LABOR
Occupational Safety and Health Administration

Form Approved
OMB No. 44-R1387

MATERIAL SAFETY DATA SHEET

Required under USDL Safety and Health Regulations for Ship Repairing,
Shipbuilding, and Shipbreaking (29 CFR 1915, 1916, 1917)

SECTION I

MANUFACTURER'S NAME	EMERGENCY TELEPHONE NO.

ADDRESS *(Number, Street, City, State, and ZIP Code)*

CHEMICAL NAME AND SYNONYMS	TRADE NAME AND SYNONYMS

CHEMICAL FAMILY	FORMULA

SECTION II - HAZARDOUS INGREDIENTS

PAINTS, PRESERVATIVES, & SOLVENTS	%	TLV (Units)	ALLOYS AND METALLIC COATINGS	%	TLV (Units)
PIGMENTS			BASE METAL		
CATALYST			ALLOYS		
VEHICLE			METALLIC COATINGS		
SOLVENTS			FILLER METAL PLUS COATING OR CORE FLUX		
ADDITIVES			OTHERS		
OTHERS					

HAZARDOUS MIXTURES OF OTHER LIQUIDS, SOLIDS, OR GASES	%	TLV (Units)

SECTION III - PHYSICAL DATA

BOILING POINT (°F.)		SPECIFIC GRAVITY (H_2O=1)	
VAPOR PRESSURE (mm Hg.)		PERCENT, VOLATILE BY VOLUME (%)	
VAPOR DENSITY (AIR=1)		EVAPORATION RATE (_____ =1)	
SOLUBILITY IN WATER			

APPEARANCE AND ODOR

SECTION IV - FIRE AND EXPLOSION HAZARD DATA

FLASH POINT (Method used)	FLAMMABLE LIMITS	Lel	Uel

EXTINGUISHING MEDIA

SPECIAL FIRE FIGHTING PROCEDURES

UNUSUAL FIRE AND EXPLOSION HAZARDS

Form OSHA-20

SECTION V - HEALTH HAZARD DATA

THRESHOLD LIMIT VALUE

EFFECTS OF OVEREXPOSURE

EMERGENCY AND FIRST AID PROCEDURES

SECTION VI - REACTIVITY DATA

| STABILITY | UNSTABLE | | CONDITIONS TO AVOID | |
| | STABLE | | | |

INCOMPATABILITY *(Materials to avoid)*

HAZARDOUS DECOMPOSITION PRODUCTS

| HAZARDOUS POLYMERIZATION | MAY OCCUR | | CONDITIONS TO AVOID |
| | WILL NOT OCCUR | | |

SECTION VII - SPILL OR LEAK PROCEDURES

STEPS TO BE TAKEN IN CASE MATERIAL IS RELEASED OR SPILLED

WASTE DISPOSAL METHOD

SECTION VIII - SPECIAL PROTECTION INFORMATION

RESPIRATORY PROTECTION *(Specify type)*

| VENTILATION | LOCAL EXHAUST | | SPECIAL |
| | MECHANICAL *(General)* | | OTHER |

| PROTECTIVE GLOVES | EYE PROTECTION |

OTHER PROTECTIVE EQUIPMENT

SECTION IX - SPECIAL PRECAUTIONS

PRECAUTIONS TO BE TAKEN IN HANDLING AND STORING

OTHER PRECAUTIONS

Figure 6.2.2   Material data safety sheet

employees to be protected from material hazards. It is now widely used in the USA by manufacturers to inform their customers of hazards of their products and to train employees. Rather similar information sheets have been recommended for use in the UK.

Section I of the sheet includes the emergency telephone number of the manufacturer and the chemical name, chemical family and formula of the material as well as its Trade Name. (Many hazardous materials have been 'passed off' under trade names which give no clue as to their composition.)

Section II gives a detailed breakdown of the hazardous ingredients of the material with the per cent and Threshold Limit Value of each.

Section III gives important physical data which show whether the vapour will rise or sink in air, whether the liquid or solid will float, sink or dissolve in water, and how volatile the material is.

Section IV gives fire and explosion data including recommended media and procedures for extinguishing fires, and any unusual hazards.

Section V gives health hazard data which includes emergency procedures.

Section VI gives reactivity data including materials with which the substance is incompatible, and which must be kept well away from it.

The other sections are straightforward.

The proper use of such sheets represents a major step in reducing chemical hazards. This involves:

1. Careful preparation of the sheets for every material to be handled. Sources of information include:
   A. *Supplier's product information. Under HASAWA '74 all the information called for on the sheet should be provided by the supplier.*
   B. *The Handbook of Chemistry and Physics published by the Chemical Rubber Publishing Company, Cleveland, Ohio.*
   C. *Dangerous Properties of Industrial Materials by N. Irving Sax.*
   D. *The Chemical Industries Association, London.*
   E. *Kirk and Othmer, Encyclopaedia of Chemical Technology published by John Wiley.*
   F. *Materials and Technology (8 volumes) published by Longmans.*
   G. *The Toxic and Hazardous Materials Group of the Atomic Energy Research Establishment, Harwell.*
   H. *The Health and Safety Executive.*
2. Distribution of copies of the data sheets to all likely to be involved in handling and using the materials, including fire fighting personnel, and
3. Discussion, training and demonstration sessions to ensure maximum practical understanding of the material hazards.

For some purposes, particularly for dealing with transport accidents and fire fighting, the information given on these sheets if not already known to the people concerned, is too much to take in when a sudden emergency arises. To cope with these emergencies, a number of labelling and identification systems have been devised, and in some cases their use is obligatory. The main fault with these systems is that there are too many of them. This causes confusion, especially to exporters and shippers of chemicals. Only some of the more important are discussed here.

6.2.2.1 Labelling systems[3]

1. *The United Nations Labelling System*. This is based on a hazard class number and appropriate labels with suitable wording and pictures some of which may be selected or deleted for each hazard class. The hazard class numbers are shown in *Table 6.2.1*.[4] Some of these labels are illustrated in the colour plates at the end of this book.

Table 6.2.1 UN. HAZARDOUS CLASS NUMBERS

Hazard Class Number and description

1. *Explosives* — Class A, B and C.
2. *Non-flammable and flammable gases*
3. *Flammable liquids*
4. *Flammable solids* (Readily combustible)
   Spontaneously combustible substances
   Water-reactive substances
5. *Oxidising materials* (and/or organic peroxides)
6. *Poisonous materials* (Class A, B and C poisonous or toxic substances)
7. *Radioactive materials* — White I, Yellow II or Yellow III
8. *Corrosive materials* — acids, corrosive liquids or solids and alkaline caustic liquids
9. *Miscellaneous hazardous materials.* These are materials which during transport present a danger not covered by other classes.

Each chemical is assumed to belong to only one class and to have a characteristic number, but if it also has secondary hazards characteristic of another class, a second label for this class may also be added without the number. The system tries to overcome language barriers by pictures, but conveys a minimum of information.

2. *The UK 'Hazchem' System*[5,6,7] The Chemical Industries Association in the UK have introduced a voluntary system for marking road and rail tank vehicles, which is widely used by their member companies. This is based on the 'Hazchem' code which was explained in section 4.3.2.7 and is summarised in *Figure 6.2.3*. CIA members employ a composite sign which includes the Hazchem code, the UN number of the chemical, a coloured hazard warning diamond, the name of the manufacturer and the telephone number for specialist advice in an emergency. An example is shown in *Figure 6.2.4*. The coloured hazard warnings are given with explanations in Appendix A.

back

Notes for Guidance

FOG

In the absence of fog equipment a fine spray may be used.

DRY AGENT

Water **must not** be allowed to come into contact with the substance at risk.

V
Can be violently or even explosively reactive.

FULL
Full body protective clothing with BA.

BA
Breathing apparatus plus protective gloves.

DILUTE
May be washed to drain with large quantities of water.

CONTAIN
Prevent, by any means available, spillage from entering drains or water course.

front

Hazchem card

Hazchem Scale

Issue No 1

FOR FIRE OR SPILLAGE

Hazchem	◇
UN No	

1 JETS
2 FOG
3 FOAM
4 DRY AGENT

P V	FULL	DILUTE
R	BA	
S V	BA for FIRE only	
S	BA	
T	BA for FIRE only	
T		
W V	FULL	CONTAIN
X	BA	
Y V	BA for FIRE only	
Y	BA	
Z	BA for FIRE only	
Z		
E	CONSIDER EVACUATION	

Figure 6.2.3 Hazchem card

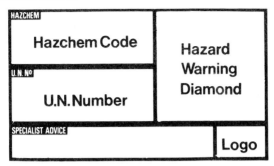

Figure 6.2.4   Composite Hazchem sign

3. *The European 'Tremcard' System.*[6][,7] The Transport Emergency Card System has been developed by the European Council of Chemical Manufacturers Federation.

One copy of the card is to be displayed on the outside of the vehicle or load and another in the driver's cab. *Figure 6.2.5* shows the Tremcard which gives the name and description of the chemical, the nature of the hazard, protective devices required when handling it, and emergency action to be taken.

4. *The National Fire Protection Association System for the Identification of the Fire Hazards of Materials (NFPA 704 M).*[7] The NFPA system is intended to alert fire fighting personnel to the hazards of materials in transport containers, storage tanks and process vessels which may already be involved in a fire or be in danger of leaking, rupturing or combustion as a result of a fire. This depends on the use of an easily read diamond shaped label which contains four squares. It is illustrated in *Figure 6.2.6*. Three of the four squares contain numbers ranging from 0 to 4 which indicate the toxicity, flammability and reactivity of the material. The fourth square contains a marking to show whether the material is radio-active or not and whether or not water may be used to deal with a spillage or fire of the material. NFPA 704 M contains a comprehensive list of chemicals with the numerical grading of their different hazard ratings. The scheme is used quite widely outside as well as inside the USA — particularly by American companies.

5. *Other systems.* The International Air Transport Association (IATA) publishes a detailed handbook of regulations for the international air transport of hazardous materials.

6.2.2.2 Practical handling problems

Many accidents occur during the emptying of chemicals from road or rail tankers, drums or other transport containers into fixed storage tanks or

TRANSPORT EMERGENCY CARD (Road)

CEFIC TEC(R)-9a
May 1971 Rev. 1
Class V ADR
Marg. 2501 A, (a), 2⁰ (a)

Cargo

NITRIC ACID (above 70%) fuming

Colourless liquid giving off yellowish-brown vapour with perceptible odour
Completely miscible with water

Nature of Hazard

The vapour poisons by inhalation
Causes severe damage to eyes, skin and air passages
Corrosive
Attacks clothing
May react with combustible substances creating fire or explosion hazard and formation of toxic fumes: nitric oxide

Protective Devices

Suitable respiratory protective device
Goggles giving complete protection to eyes
Plastic gloves, boots, suit and hood giving complete protection to head, face and neck
Eyewash bottle with clean water

EMERGENCY ACTION — Notify police and fire brigade immediately

- Stop the engine
- Mark roads and warn other road users
- Keep public away from danger area
- Keep upwind
- Put on protective clothing

Spillage

- Contain leaking liquid with sand or earth, consult an expert
- Do not absorb in sawdust or other combustible materials
- If substance has entered a water course or sewer or contaminated soil or vegetation, advise police
- Use waterspray to "knock down" vapour

Fire

- Keep containers cool by spraying with water if exposed to fire

First aid

- If the substance has got into the eyes, immediately wash out with plenty of water for several minutes
- Remove contaminated clothing immediately and wash affected skin with plenty of water
- Due to delayed effect of poisoning, persons who have inhaled the fumes must lie down and keep quite still. Patient should be kept under medical treatment for at least 48 hours
- Seek medical treatment when anyone has symptoms apparently due to inhalation or contact with skin or eyes
- Even if there are no symptoms send to a doctor and show him this card
- Keep patient warm
- Do not apply artificial respiration if patient is breathing

Additional information provided by manufacturer or sender

TELEPHONE

Prepared by CEFIC (CONSEIL EUROPEEN DES FEDERATIONS DE L'INDUSTRIE CHIMIQUE, EUROPEAN COUNCIL OF CHEMICAL MANUFACTURERS' FEDERATIONS) Zurich, from the best knowledge available; no responsibility is accepted that the information is sufficient or correct in all cases
Obtainable from NORPRINT LIMITED, BOSTON, LINCOLNSHIRE
Acknowledgment is made to V.N.C.I. and E.V.O. of the Netherlands for their help in the preparation of this card

Applies only during road transport   English

Figure 6.2.5   Transport emergency card (TREMCARD)

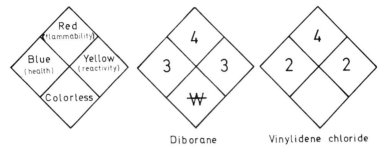

Diborane   Vinylidene chloride

*Figure 6.2.6   Fire hazards of materials. NFPA label 704 Reproduced by permission
from NFPA No.704, 1975 (The Fire Hazards of Materials), Copyright 1975,
National Fire Protection Association, Boston, MA*

directly into a process. Similar accidents happen when dissolving reactive solids such as caustic soda, to make up solutions.

In nearly every case the supplier will have studied the problems of the safe handling of his products and worked out safe methods. The supplier should be consulted when the handling systems are being designed — not afterwards. Some neat, safe and effective methods are often to be found in supplier's literature. Thus to dissolve the content (solidified cake) of a 200 litre drum of solid caustic soda, the drum is placed vertically on a grid above an open top tank to receive the solution, a hole is knocked in the side of the drum close to the base, the lid in the centre of the top of the drum is opened, and a water hose with a rose is placed inside. The water supply is then turned on gently. The cake will have contracted in the drum when it solidified leaving a gap between the wall and the cake. This provides a passage for the solution formed to run down,[8] (see *Figure 6.2.7*).

Accidents often occur with sulphuric acid by adventitious mixing with water. Mild steel is not appreciably corroded by sulphuric acid at concentrations above 85% but at concentrations between 5 and 65% rapid corrosion occurs. The conical roof of a 200 ton sulphuric acid tank was being washed down with a hose without closing the small hole on the roof used for 'dipping' and sampling. Water ran down inside the wall of the tank and formed a light layer on top of the acid. Next morning the manager was surprised to find the tank had been cut in half far more cleanly than any oxy-acetylene torch could have done.

Particular care is needed with the flexible tubes and hoses sometimes used to empty drums of corrosive toxic chemicals and flammable liquids. The correct method of opening and emptying a drum is as follows.[9]

Set the drum upright with emptying bung on top under a ventilating hood. Unscrew bung slowly to release any internal pressure. Fit valve in place of top bung and make sure it is tight. Tip drum on one side with side bung up and fit another valve in its place, making sure both connections are leak tight. Support the drum in a drum cradle with one valve uppermost.

Connect the upper valve to a vent pipe leading to the open air. Connect the other valve to a closed but vented storage vessel or processing unit. Look out for leaking connections and rectify at once before proceeding

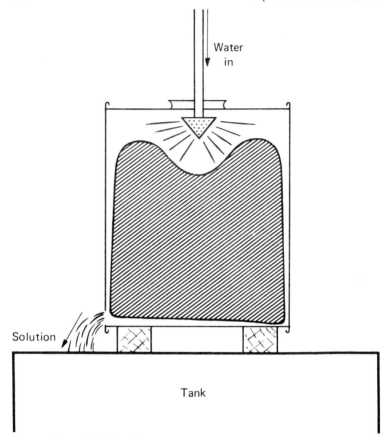

Figure 6.2.7 Dissolving solid caustic soda in a tank (ICI Limited)

further. If liquid spills, use dry clay or other absorbent material and ensure sweepings are safely disposed of.

Empty drums must be decontaminated before being scrapped or used for anything else. Methods of doing this are described in section 6.1.

Drums should never be emptied by pressurising with compressed air, nitrogen, etc., unless they have been clearly designed for pressure, even then proper safety precautions must be taken to prevent the drum being overpressurised. Another cause of accidents in handling drums and carboys is that the cap or bung is not securely closed when the drum is handled.

Quite a few accidents have been caused by the fact that a drum was under slight internal pressure when the screw cap was removed—caused often by the fact that the drum was filled at a different temperature and atmospheric pressure. The cap should always be slackened first a turn or two to allow air pressure inside and outside to equalise through the thread before the cap is finally opened.

6.2.2.3 Some basic chemical plant hazards

Technical literature abounds with papers and reports of hazards and accidents that arise on chemical plant, and only a few basic hazards can be mentioned here. The following examples are illustrative only:[10]

A. Hazards may arise when impure or contaminated chemicals are used.
B. A temperature measuring instrument may give a false reading (caused by breakage of a thermocouple wire) so that the actual temperature is considerably higher than believed by the operator. This may lead to a run-away reaction and over-pressure.
C. By-products may accumulate at relatively high concentrations in parts of the plant and cause unexpected effects. A serious explosion in a butadiene recovery plant was caused by the localised concentration of vinyl acetylene (a by-product) in one part of a distillation column.
D. The course of a reaction may be affected by a different material of construction to that previously used.
E. Strongly exothermic reactions carried out in batch type reactors may 'run away' if the temperature rises more than 10 to 20°C above the design temperature.
F. A poor distribution of reactants within a reactor may lead to hot spots accumulations of unreacted materials or undesirable side reactions.
G. If a stirrer is stopped or slowed down during a reaction, reactants may accumulate. Later starting the stirrer or speeding it up may cause the reaction to go out of control.
H. Air may be sucked into a centrifuge containing flammable liquids by 'windage' in spite of the supply of inert atmosphere to the centrifuge, resulting in an explosion inside the centrifuge housing.

6.2.2.4 Monitoring of chemical plant hazards

It has long been the practice in the manufacture of explosives to employ skilled technical personnel in a purely monitoring role, quite divorced from production, with power and authority to have the plant shut down if they have justified reasons to feel that conditions are not safe or that some new or unexpected hazard has arisen. The need for such monitoring activity depends very largely on the hazard potential of the chemical plant and this can vary enormously.

A plant using the solar evaporation of sea water to make salt has little hazard and probably requires a minimum of monitoring. A plant containing substantial quantities of low boiling liquid hydrocarbon under pressure and at a temperature considerably in excess of its atmospheric boiling point is a potential bomb, with a hazard comparable to that of an explosives plant. Such a plant can only be operated safely when there is a fully effective system of hazard monitoring.

The danger is greatest when a dominant manager wholly committed to production occupies a very senior position.[11] The normal balance between production and effective hazard monitoring then easily breaks down and a situation such as that described in section 1.12 readily arises.

6.2.2.5 Legal regulations[12]

The main regulations which apply specifically to chemical works are the Chemical Works Regulations 1922, as amended by SI 1961 No. 2435 and SI 1973 No. 36. These contain some useful general regulations regarding the provision and use of breathing apparatus, protective clothing, wash rooms, exhaust ventilation, life-belts, and non-metallic implements for cleaning out residues from stills or chambers which have contained acid.

The regulations also contain a number of specific regulations which apply to particular processes in use in 1922, but very little which is specific to more modern processes. Furthermore it seems unlikely under the spirit of the Robens Committee Report and HASAWA (1974) that much in the way of new regulations specific to modern processes may be expected. There seems, therefore, to be little point in discussing here regulations which today have only limited application in the Chemical Industry.

REFERENCES

1. Taylor, G.R., *The Doomsday book*, Thames and Hudson, London, 91 (1970).
2. Weisburger, E.K., 'Industrial cancer risks', Chapter 8 in Sax, N.I., *Dangerous properties of industrial materials*, 4th ed., Van Nostrand, New York, 280 (1975).
3. Lewis, R.J., *Labelling and identification of hazardous materials*, in Sax, N.I., loc cit, pages 326 and 330.
4. Lewis, R.J., loc cit, page 343.
5. Chemical Industries Association, *Marking containers of hazardous chemicals*, CIA, London.
6. Chemical Industries Association, *Road transport of hazardous chemicals*, CIA, London.
7. National Fire Protection Association, Standard 704 M, *Identification systems for fire hazards of materials*, NFPA, Boston, Mass., USA.
8. Imperial Chemical Industries Ltd., *Dissolving caustic soda and handling its solutions*, Technical Brochure, ICI, London.
9. Sax, N.I., Industrial Fire Protection, Sax, loc cit, page 263.
10. Factory Mutual System, Loss Prevention Data, 7-43, *Loss prevention in chemical plants* and Loss Prevention Data 747 *Physical precautions in chemical plants*, FMS, 1151 Boston Providence Turnpike, Norwood, Mass, USA.
11. Council for Science and Society, *Superstar technologies*, CFSS, London.
12. Fife, I., Machin, E. A., *Redgrave's health and safety in factories*. Butterworths, London, 728 (1976).
13. Health and Safety Commission. Discussion document. *Proposed scheme for the notification of the toxic properties of substances*, H & SC (1977).

6.3 METAL CASTING

Contents

Casting of metals is carried out in foundries (from the French 'fondre' — to melt) and in special machines such as those used for die casting small objects in large quantities. The casting of bronze vessels and images is of great antiquity and probably predates written language. The casting of iron which requires considerably higher temperatures is more recent, having been developed in Europe in the late middle ages, mainly for the manufacture of cannon.

Most of the metal objects in use today have been through at least one casting process in the course of their manufacture. A great range and variety of casting methods are used. The main casting methods are given in *Table 6.3.1*[1] which shows the materials for which they are used and remarks on their application.

Almost as varied as the methods of casting are those of melting the metals for casting. The main types of furnace use are listed against the metals cast in *Table 6.3.2*.

All metal casting involves at least four basic steps:

1. Preparation and assembly of the mould or die, and any core required for hollow castings. In most cases this requires the prior fabrication of a pattern, which has the same shape as the casting but is slightly larger to allow for shrinkage of the casting on cooling.
2. Melting of the metal to be cast.
3. Pouring or compression of the molten metal into the mould or die, and solidification of the casting.
4. Removal of the casting from the mould or die and removal of surplus material including the sprue or gate used to fill the mould or die.

Table 6.3.1[1] BASIC CASTING METHODS IN USE

Method	Metals cast	Remarks
Sand casting	Most ferrous and non-ferrous	Most widely used casting method
Shell-mold casting	Most materials except low carbon steels	Good low cost method
Permanent mold casting	Restricted — brass, bronze, aluminium, some gray iron	Economic for large numbers
Centrifugal casting	Stainless steel and aluminium alloys, cast iron and tin/lead alloys	Mainly used for pipe and heavy wall cylinders
Plaster-mold casting	Narrow range — brass, bronze, aluminium	Suitable for complex objects. Requires little finishing
Investment casting (lost wax process)	Wide range	Best for parts too complicated for other casting methods
Die casting	Narrow range — zinc, aluminium, brass, magnesium	Most economic for large numbers where applicable

Table 6.3.2[1] TYPES OF MELTING FURNACE USED FOR CASTINGS

Metal	Types of furnace used
Steel	Electric — direct arc — acid and basic Open hearth, acid and basic Bessemer, electric induction (high frequency)
Cast iron	Cupola, reverberatory, electric (direct arc)
Malleable iron	Cupola and/or reverberatory, rotary, open hearth (acid)
Brass & bronze	Crucible, electric (indirect-arc, induction, both low and high frequency), air furnace
Aluminium	Crucible furnace with metal or refractory crucibles, induction (low frequency)
Magnesium	Crucible furnace with steel crucibles, induction

The term 'mould' is generally used for castings poured by gravity and the term 'die' for castings made under pressure.

Whilst most casting is carried out in specialised foundries and factories, many large works and factories producing non-engineering products have a small foundry, employing the sand casting method as an adjunct to their maintenance workshop. Metal casting presents a wide spectrum of hazards to workers. Some constitute immediate risks of personal injury whilst others present long term and delayed risks to health.

The principal hazards, not necessarily in order of importance, fall under the following headings:

1. Burns caused by bodily contact with hot or molten metal or slag.
2. Explosions caused by water coming into contact with molten metal and injuries resulting from them.
3. Respiratory diseases caused by the cumulative inhalation of fine airborn particles in the 1 to 5 μ range — notably silica and lead particles.
4. Carbon monoxide poisoning caused by dangerous concentrations in gases from the melting operation, cupolas and open-hearth furnaces.
5. Eye injuries resulting both from prolonged exposure to radiation from white hot objects and from physical injuries from flying particles, often hot.
6. Impairment of hearing through prolonged exposure to high noise levels (e.g. the roar of a cupola furnace).
7. Material handling injuries, often compounded by the fact that the materials handled are hot.
8. (Mainly with magnesium casting). Fires and dust explosions.

It is clear from the foregoing that foundry operations are too varied and specialised to allow detailed discussion or even reference to all of the hazards that can arise. Some of these hazards are only faced by a limited number of experienced personnel for whom the safe accomplishment of their tasks is an essential part of their skills. As examples one might pose the questions:

How would you tap an acid hearth furnace without danger of getting splashed by molten metal?
What would you do if the tool used to tap the furnace stuck in the hole?
What precautions would you take when placing a bott in the tapping hole to stop the flow of molten metal when the ladle was nearly full?
How could water be used to cool and chill a continuous ingot of aluminium made by continuous casting without risking an explosion?
How would you drop the bottom doors of a cupola containing perhaps a foot of molten iron in order to take it out of service for re-lining?

The hazardous operations listed above have one feature in common; the accidents to be guarded against would occur quickly, and it would hence be surprising if those affected had not given a good deal of thought to the questions and found methods of working which minimised their risks. However, the delayed effects of breathing apparently clean air containing very fine silica or other toxic dust in suspension are more insidious and less likely to have been considered seriously by the persons exposed — partly because they may be largely unaware of the hazard at the time of first exposure.

6.3.1 Legislation, safety studies and training

A number of UK regulations apply to casting metals,[2] notably:

The Foundries (Parting Materials) Special Regulations, 1950.
The Iron and Steel Foundries Regulations 1953; and
The Non-ferrous Metals (Melting and Founding) Regulations, 1962.

Other regulations which affect foundries include:

The Blasting (Casting and Other Articles) Special Regulations, 1949
(See section 6.5)
The Grinding of Metals (Miscellaneous Industries) Regultions, 1925
(See section 6.5)
The Protection of Eyes Regulations, 1950 (See section 6.5)
The Foundries (Protective Footwear and Gaiters) Regulations, 1962
(See section 3.10)
The Abrasive Wheels Regulations, 1970 (See section 6.5)

6.3.1.1 The Foundry (Parting Materials) Special Regulations, 1950

These prohibit the use, as a parting medium, of materials containing more than 3% of silica, free or combined, calculated on the dry basis, with the following exceptions.

1. Zirconium silicate (Zircon).
2. Calcined china clay.
3. Calcined aluminous fireclay.
4. Sillimanite.
5. Calcined or fused alumina.
6. Olivine.

The prohibition includes the use of dust or other matter deposited from a fettling or blasting process as a parting material.

These regulations prohibit the use of the highly dangerous silica flour as a parting material, although it is still permitted as a necessary ingredient of some moulding sands.

6.3.1.2 The Iron and Steel Foundries Regulations, 1953

These Regulations cover the following main points:

Work near cupolas and furnaces. No one is allowed within 3.7m (12ft) from a vertical line passing through the delivery end of a spout used for delivering metal on any cupola or furnace or within 2.5m (8ft) from a vertical line from the nearest part of a ladle positioned at the end of the spout.

Gangways and pouring aisles. The surface number and minimum width of gangways and pouring aisles are specified.

Dust and fumes. The use of open coal, coke or wood fires are restricted and virtually prohibited in workrooms. Mould stoves, core stoves and annealing furnaces shall not emit offensive or injurious fumes into any work room.

Knockout operations shall as far as possible be carried out in a separate room of the foundry provided with local exhaust ventilation and a high standard of general ventilation. Dressing and fettling operations shall be similarly carried out in a separate room and provided with similar ventilation.

Protective equipment. Suitable gloves and approved respirators shall be provided for persons exposed. Bathing facilities and clothing accommodation are to be provided as discussed in sections 3.10 and 3.11.

6.3.1.3 The Non-ferrous Metals (Melting and Founding) Regulations, 1962

These are more comprehensive than the Iron and Steel Foundries Regulations, 1953 and cover the following main points:

Cleanliness of floors. All accessible parts of floors of rooms where casting and allied processes are performed shall be cleaned at least once a day except for sand floors, which shall be kept in good order.

Construction of floors. Floors of rooms where casting and allied processes are performed shall have an even surface of hard material. Sand floors are only permitted where the work done makes them necessary; where used, they shall be maintained in a firm and even condition.

Manual operations involving molten metal. Adequate space must be provided and kept free of obstructions for all manual operations involving molten metal.

Disposal of dross and skimmings. These should be placed in a suitable receptacle immediately after removal from the molten metal or furnace.

Arrangement and storage. All movable items used in casting must be arranged and placed in an orderly manner so as to minimise risks at work. Scrap metal, sand and other loose materials when stored indoors shall be kept in suitable bins, bunkers or other receptacles.

Gangways and pouring aisles. Minimum widths for several circumstances are specified.

Dust and fumes. Similar though perhaps somewhat stricter precautions are specified to those noted in 6.3.1.2.

Protective equipment. Gloves and approved respirators must be supplied to exposed workers, and suitable screens to protect against flying particles and drops of molten metal shall be provided where necessary. All employees shall use this equipment properly.

Room temperature. The room temperature where any approved process is being carried out shall after the first hour of work not be less than 10°C or if the outside temperature falls below freezing point, not less than 11°C above the outside temperature.

Washing facilities and clothing accommodation. The essentials of these have been given in section 3.11.

Facilities for meals. Proper facilities for heated meals must be provided for foundry workers and main meals may not be taken in rooms where foundry processes are carried out.

Cleanliness of indoor workplaces. All walls of rooms in which foundry processes are carried out must be cleaned to a height of at least fourteen feet not less than once every fourteen months.

Dressing operations. These operations must be done inside a building.

Disposal of waste. All waste products from shell mouldings including burnt sand must be disposed of as soon as practicable after the castings have been knocked out.

Material and equipment left out of doors. Such material and equipment must be placed and arranged safely with safe means of access by roads or pathways with firm and even surfaces. These must be properly maintained and kept free of obstruction.

6.3.1.4 Safety studies in foundry operations

Health and safety in foundries is kept under continuous review by the Joint Standing Committee on Health, Safety and Welfare in Foundries. This comprises representatives of employers, trade unions, foundry equipment manufacturers and suppliers, the Health and Safety Executive (including Medical Advisers and Engineering Inspectors) and others.

Various sub-committees have been appointed to study particular aspects of health and safety in foundries. Two of these have been concerned with Machinery Safety,[3] and Continuous Casting and High Speed Melting[4] respectively; they have published a number of reports giving detailed recommendations resulting from their studies. Those dealing with Machinery Safety[3] have covered the following topics:

Guarding moulding machines.
Use of mechanical restraint in guarding.
The guarding of automatic foundries.

Those dealing with continuous casting and high-speed melting[4] have dealt with:

Vertical semi-continuous and continuous casting of aluminium operational safety and causes and prevention of break out.
Warning and control for continuous casting of copper alloys.

Other studies based on the experience of HM Factory Inspectors deal particularly with the foundry environment.[5]

6.3.1.5 Training in foundry operations

Training in the theory and practice of foundry operations is carried out at a large number of technical colleges and universities. Industrial training is catered for by the Foundry Industry Training Committee, Industrial Training Board, 50 Charlotte Street, London, W.1.

The FITC provide a 48-week molder-coremaker course and a 1-week foundry appreciation course at their Midland Training Centre in West Bromwich near Birmingham. Safety is treated as an integral part of these training courses. A short course in Hazard Recognition and Identification for foundry managers is in preparation.

6.3.2 Toxic dusts, fume and gas hazards[5]

Air-borne particles, particularly silica, are probably the main health hazards in foundry work, and considerable effort is needed to control the problem. This is discussed in the Department of Employment booklet *Improving the Foundry Environment* and the main points are summarised below.

Besides silica, foundry atmospheres are liable to contain a number of other hazardous dusts, vapours and gases. A list of the main hazardous materials with their TLVs, origins and short notes is given in *Table 6.3.3.* The toxic dusts and gases most likely to be present vary considerably from one type of foundry or foundry operation to another.

The solution to the problems of toxic dusts and gases in foundries requires a threefold approach.

1. Identification and monitoring of the hazard,
2. Elimination of the toxic dust, fume or gas at source (where possible),
3. Control of the hazard.

Table 6.3.3 HAZARDOUS MATERIALS IN THE FOUNDRY ATMOSPHERE
(OXIDES INCLUDED WITH ELEMENTS)

Material	TLV mg/m³	Origin in foundry atmosphere	Notes
Acrolein	0.25	Thermal decomposition of core oils	Irritant and health hazard
Aluminium		Cleaning castings with grinders or sanders	Fire and explosion hazard in dust collecting systems
Antimony (Oxide)	0.5	Present in some lead & copper alloys	
Beryllium (Oxide)	0.002	From some copper alloys	Highly toxic
Cadmium (Oxide) fume	0.05	From some copper alloys	Highly toxic
Carbon Monoxide	55	Leaks from flues & charging doors of cupolas	Fixed and portable carbon monoxide monitors should be used to measure concentrations & warn personnel
Chromium (Oxide)	1.0	In stainless steel casting. Released during thermal cutting and grinding	
Fluorides	2.5 as F	From magnesium and ductile steel casting	
Hexa methylene tetramine		From phenol formaldehyde resins used in shell moulding	Skin irritant. Dust explosion hazard
Iron (Oxide)	10	Melting, burning, oxygen lancing, pouring, grinding, welding, machining	High concentrations common in foundry atmosphere
Lead (Oxide)	0.15	Oxide formed in melting, pouring & welding Elemental dust formed in cleaning and machining	Serious hazard in non-ferrous foundries
Magnesium (Oxide)	10	Cleaning & machining magnesium castings	Fire and explosion hazard. Metal fume fever
Organic Solvents		Cold coating sand with resin	Fire and vapour explosion hazard
Phosphorus (Oxides)	1	From phosphor bronze melting, pouring & cleaning	
Silica		Silica flour in sand handling systems, knock-out, dressing & cleaning castings	Cause of silicosis
Silicones		Mould release agents in shell moulding	Some types are toxic. Skin and eye irritants
Sulphur dioxide	13	Oxidation of sulphur in castings, especially magnesium	Mainly irritant

6.3.2.1 Dusts

Dust particles vary greatly in size. The larger ones with diameters of 10 microns and more settle fairly rapidly. They are more of a nuisance than a health problem since they are mostly trapped by the hairs and cilia in the upper respiratory system — the nose, throat, larynx and trachea. They are swept up and collected in the mucus and brought to the mouth where they can be spat out or swallowed. It is the very fine dusts with diameters of less than 5 microns which reach the bronchioles and alveoli, of the lungs and cause the pneumoconioses such as silicosis. Some larger dust particles such as pollens with diameters of approximately 20 microns cause allergies and diseases of the upper respiratory system.

The fine and dangerous inorganic dusts are invisible in ordinary lighting conditions, although their presence is often inferred through that of coarser dusts. The settling rates of fine silica dusts in still air are given in *Table 6.3.4*.

Table 6.3.4 SETTLING RATES FOR SILICA PARTICLES IN STILL AIR

Size, microns	*Settling velocity, mm per second*
0.5	0.03
1.0	0.09
2.0	0.30
5.0	2.03

Several methods have been developed for collecting, measuring and analysing fine dusts in the foundry atmosphere. One of the simplest yet effective methods of collecting and measuring dust is the use of a set of Hexlet filters. A known volume of the atmosphere is sucked through a series of filters each of which retains only particles within a certain size range. The concentrations of dust of several size ranges in the atmosphere are given by the gain in weight of the filters, and the dusts analysed chemically. For crystalline silica the X-ray diffraction method is satisfactory.

Portable samplers (*Figure 6.3.1*) are now available to be worn or carried by workers who are liable to be exposed to harmful dusts, particularly silica. These can collect and continuously filter air taken from the breathing zone of the worker.

Besides sampling and analysing the atmosphere for toxic dusts, it is desirable to have methods for visually detecting the very fine dusts in the atmosphere and making an approximate visual appraisal of their concentrations under various conditions. These should show where fine dusts are most prevalent, and how effective are the means used to suppress their formation or eliminate them once they are present in the atmosphere. Optical and photographic methods which involve the light scattering by small particles at angles between 5 and 15° to the direction of the light beam have been in use since 1951[2], and can be used in foundries for special surveys.

Figure 6.3.1   Portable air sampler for toxic dusts (Rotheroe & Mitchell Ltd.)

6.3.2.2 Dangerous gases and fumes

Carbon monoxide is produced in fired melting furnaces and mould and core stoves by incomplete combustion. Its presence in the foundry atmosphere can normally be eliminated by attention to burners and by proper design of flues. One place where its emission is difficult to control is the charging door of cupola furnaces.

Sulphur dioxide is also formed in foundries where sulphur containing fuels are used, and also from sulphur compounds contained in some metals which are being melted and cast. It tends to be an irritant nuisance as much as a health problem, and is best avoided by the use of sulphur free fuels and improved ventilation.

Metallurgical fumes formed in melting are mostly very small particles, about 0.2 microns in diameter, the commonest being those of iron, zinc and lead. Special precautions should be taken whenever alloys containing lead, cadmium and beryllium are being cast.

Organic vapours and fumes are formed from some of the materials used from fluxing and degassing non-ferrous alloys, and from resins used in making shell mouldings. This occurs both in pattern spraying and from the hot mould during casting as a result of thermal decomposition.

A rapid but approximate method of sampling and testing the atmosphere for many of these toxic vapours is the use of an aspirator and gas detector tube as specified in BS 5343:1976 (see section 6.1).

6.3.2.3 Elimination of toxic dusts, fumes and gases

Foundry processes can be broadly split between those carried out in:
    1. The moulding shop.
    2. The dressing shop.

Toxic dusts, etc are formed in both sets of operations. The extent to which they are formed in the dressing shop depends very largely on the amount of adherent sand and surplus metal that has to be removed. This in turn depends on the moulding techniques and practices employed. Careful attention to sand practice, the moisture content of green sands, core and mould paints and heading, gating and casting techniques is therefore important.

In sand practice, it is important to limit the extent and duration in which sand is handled dry, since wet sand can be handled without causing any significant amount of dust. In green sand mould casting, the sand is dry only from the time of casting to the time when water is added after the knockout. This time should be kept as short as possible, and moisture added to the sand immediately after knockout instead of conveying dry sand from knockout back to the sand preparation.

In core making, the use of binders, especially organic ones, should be reduced to a minimum, and the proportion of sand and clay fines used also kept to a minimum. The sand should be cool before mixing to prevent evaporation of water and the need for excessive amounts of binder. Where oil bonded cores have to be employed, care should be taken in the technical control of their production, particularly to ensure that they are adequately baked. If persistant organic fumes arise as a result of some oil employed, efforts should be made to find alternative oils which give off less fume.

Moulding methods should be studied and where possible modified to eliminate the need for easing or slackening of large ferrous castings which causes a great deal of dust. The use of collapsible cores and the incorporation of wedges which can be withdrawn after casting can lead to great improvements.

Dust can often be greatly reduced in the stripping of heavy castings by building up the mould in pieces which remain bonded after casting, and can be removed intact. The use of compressed air for cleaning moulds is the cause of much unnecessary dust, and a fixed vacuum installation is greatly to be preferred.

Care is needed in the choice of fuel and in the ventilation of portable mould driers to avoid the production of carbon monoxide and sulphur dioxide in the foundry atmosphere. Hard cupola coke in the size range 50 to 75 mm is preferable to gas coke as fuel.

The use of silica flour as a sand filler in steel foundries should be restricted and where possible eliminated. It should be bought in the moist state and stored and used with at least 6% moisture to reduce dust formation. When choosing a moulding process for any operation the relative liability of the various processes to produce dust and fumes should be carefully considered.

The use of excessive temperatures and oxidising conditions during melting seriously aggravate fume production. Melting and casting should be carried out as quickly and expeditiously as possible to limit the formation of fumes from molten metal, and the distance over which fuming ladles have to be carried from furnace to mould should be as short and direct as possible.

In green sand moulding, the box should be knocked out as soon as possible after casting to reduce the amound of dry sand produced. The amount of hand dressing of castings should be kept to a minimum and should be preceded by some form of mechanical cleaning. Where annealing of castings is required this should where possible be done before dressing. Wet cleaning methods such as the use of wet decoring bars should where possible be used instead of dry methods.

6.3.2.4 Control of dust and fumes

Local exhaust ventilation employing high velocity low volume air streams should be used as far as possible to control dusts and fumes. It is particularly required when conveying dry sand, and in all other operations where dry sand is employed. The conveyor or operation should be as completely enclosed as possible with an exhaust air connection to the hood of the conveyor or other equipment.

Mills in which resin and sand are mixed for shell moulding should be fitted with special covers so that the resin can be added without removing the cover. Shell moulding processes usually have several stages where detailed attention to ventilation can make a vast improvement in the amount of dust and fume formed.

Furnaces which produce large amounts of metallurgical fume such as arc furnaces with the use of oxygen should be provided with adequate hoods and local exhaust ventilation, which may completely surround the opening from which the fumes would emerge.

Side-draught local exhaust ventilation which acts on a rising column of dust and fume and continuously removes it before it reaches the breathing zone of the operators is generally preferred to updraught or downdraught local ventilation. Downdraught ventilation can only be applied to small castings where the thermal air currents produced are of low velocity. Updraught ventilation does not offer much protection to workers who have to put their heads between the source of the dust and the ventilation hood.

Figure 6.3.2 Effect of exhaust rig for non-ferrous furnace
(H. & S.W. Booklet 17)

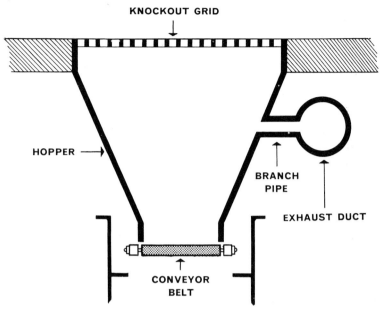

KNOCKOUT GRID

HOPPER →

BRANCH
PIPE

EXHAUST DUCT

CONVEYOR
BELT

Figure 6.3.2 Effect of exhaust rig for non-ferrous furnace (H. & S. W. Booklet 17)

Figure 6.3.2 shows the effect of an exhaust rig for non-ferrous furnaces.
Figure 6.3.3 shows local exhaust ventilation.

6.3.3. Heat in foundries

Conditions in many foundries are characterised by excessive workroom
temperatures, as well as by low thermal efficiency and excessive fuel
consumption. The problem is really best tackled at the design stage.

The use of small melting furnaces with high melting rates which reduce or
eliminate the need to keep reserves of molten metal can greatly improve the
situation. Careful attention to furnace efficiencies and to insulation can also
reduce the temperature in foundry rooms and save fuel costs.

6.3.4 Physical explosions in foundries

Serious explosions occur from time to time in foundries and smelters where
molten metals are handled, caused by the sudden contact between water
and hot or molten metal. Water expands by over 1000 times when it
vapourises, and the pressure generated can be enormous. Many
opportunities for such accidents arise in foundries, and great care is needed.

Typical examples of how they can arise are:

1. Charging wet ingots or scrap fuel or other material to the melting furnace.
2. Presence of water in ladles used to transfer molten metal from the melting furnace to the mould.
3. Presence of water in the moulds, dies or pits.
4. Leakage of cooling water (used to cool some essential part of a furnace) into the furnace or ladle through a fracture in the part cooled.
5. Overfilling of a mould, die or ladle with molten metal causing it to spill over into a pool on the floor.

In some casting processes, particularly the continuous casting of aluminium, the metal is actually run through a die and into and through a pit of water. This apparently hazardous operation can be carried out safely when special precautions are taken. Among others, the water in the pit should be at least ten feet deep.

Precautions must be taken to check that all mould surfaces are dry and free from cracks before casting. The surface of iron moulds should be free from rust (which undergoes a thermic reaction with aluminium), and preferably painted with a special paint which causes water to run off the surface. When an ingot or casting sticks in a mould, it should never be forced or hammered out until the pipe is quite solid. Further precautions are given in references 4 and 6.

6.3.5 Guarding foundry machinery

The more mechanised foundry operations such as shell moulding, permanent mould casting and die casting, make use of a great deal of machinery and some are set up as fully mechanised and automated production lines, which require adequate guarding to protect operators. Difficulties sometimes arise when attempting to fit guards around existing lines because of restricted space. Guards should be robust, they should facilitate maintenance and they should sustain or improve production. Fixed guards are preferred where possible, other methods such as mechanical restraint, interlocking, a key exchange system, a plug and socket system and automatic conveyor safety control all have their uses.[3]

'Mechanised restraint' of some moving parts is sometimes needed to protect operators from traps between a stationary and a moving object — caused by the unexpected movement of a power-operated part — in turn caused by the malfunction of a control mechanism. The restraining device usually contains pawls and ratchets which permit unidirectional movement only and provide a second line of defence against the unwanted movement.

The key exchange system employs identical locks for the power supply and an access door to the equipment. Only when the power supply is locked in the off position can the key be removed and used to open the access door. A spare key of course defeats the system.

The plug and socket system allows a door or panel to be removed only after the power circuit has been broken. This suffers the disadvantage that the power circuit may be broken while the machine is running. The automatic conveyor safety control system prevents a conveyor from moving when an operator or tool is in a certain position, or unless the person or thing is in some safe position.

6.3.6 Other safety points[6]

Foundry work calls for detailed consideration of safe working methods and equipment for the wide range and variety of jobs done. The points mentioned below represent only a short selection.

6.3.6.1 Housekeeping

Housekeeping which was discussed in section 5.1 requires particular attention in foundry work because of the natural handicaps to good housekeeping which the work presents. Each employee should be responsible for the housekeeping in his own work area, and should be allowed sufficient time for keeping everything clean and in order. All necessary equipment and cans and rubbish disposal bins should be available and emptied regularly.

Floors beneath and immediately surrounding melting units should be pitched away from the unit to provide drainage and floors and pits must be kept free from water. Where water is needed to hold down dust, the minimum quantity should be used. Pits and floor openings should be protected by a cover or solid guard rail.

Die moulds and receiving stations for surplus metal from ladles should be located clear of passageways and at least 300 mm (1 ft) above floor level. Any door which may have to be kept closed to prevent draughts should have an adequate window at eye level to permit a view beyond it.

6.3.6.2 Illumination

The nature of foundry operations makes good illumination difficult to achieve. This applies particularly where overhead cranes are used which necessitate placing light fittings at a considerable height. The visible radiation from molten metal further complicates the problem. Expert advice should be called in where necessary to help solve some of these difficult problems.

6.3.6.3 Ladles

Ladles should be provided with suitable covers and the ceramic rim or lip of lined ladles should provide between 12 and 25 mm cover over the metal shell (see *Figure 6.3.4*).

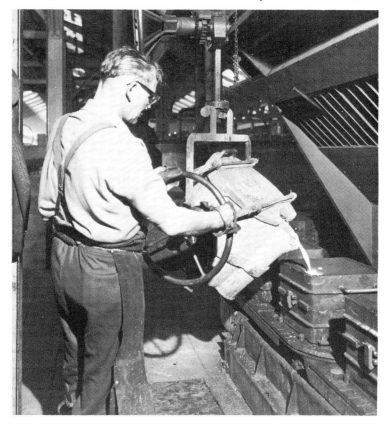

Figure 6.3.4   Tilting ladle with safety device (Morgan Crucible Ltd)

Ladles mounted on trucks or handled by overhead cranes or monorail should have safety locks to prevent tilting when casting is not being carried out, and they should be equipped with warning bells or sirens to be used when molten metal is being transported. Ladles should be thoroughly dried before being used, if necessary in a separate drying shed outside the foundry.

6.3.6.4 Scrap

All equipment used to break up scrap (shears, demolition balls) must be adequately guarded to protect operators and passers-by from flying metal.

6.3.6.5 Storage

Adequate storage on level and substantial foundations must be provided for all foundry materials and equipment when not in use. Patterns should be stored on substantial racks and shelves in a special building.

6.3.6.6 Slag disposal

Furnaces and pits should have sufficient removable receptacles or pots into which slag may flow to allow it to solidify before it is dumped.

6.3.6.7 Refractory crucibles

Refractory crucibles in which metal is melted and then transported present a serious hazard if they break when full of molten metal. They must be thoroughly inspected for cracks and flaws by a competent inspector, stored in a warm dry place and protected from moist air.

Special care is needed in annealing crucibles and heating them for the first few runs. Only tongs of the proper shape and size which fit well below and around the bilge should be used for moving crucibles, and these should never be driven home tight with a tool.

Figure 6.3.5 shows a crucible being lifted out of a furnace.

6.3.6.8 Magnesium grinding

The hazards of fires and dust explosions in magnesium grinding led to the Magnesium (Grinding of Castings and other Articles) Special Regulations,

Figure 6.3.5   Crucible with lifting device (Morgan Crucible Ltd.)

1946. These make it obligatory to collect the dust in wet scrubbers situated not more than 4.6 m (10 ft) from the grinding wheel or polishing mop. The dust should be washed into a sludge pit in which it is kept under water until dispersed.

There is a further danger in the use of wet collecting systems with magnesium and aluminium dusts and powders. These react slowly with water to form hydrogen which can form an explosive mixture with air. To avoid this the sludge should be removed frequently from the collector, and the scrubbing system must be well maintained.

6.3.6.9 Inspection

Frequent inspection of foundry equipment is required, especially of metal parts which are exposed both to heat and mechanical stresses. A proper formal inspection is required to operate in conjunction with a well planned preventative maintenance programme.

Castings should be subjected to a carefully selected system of non-destructive testing using a combination of the methods described in section 2.10.

6.3.7 Personnel welfare and protection

A programme for protecting the health of foundry workers should include appropriate medical examination and chest X-ray before starting work, periodic examination and chest X-rays during subsequent employment and adequate first aid facilities with properly trained first-aiders.

Adequate washing and changing facilities are needed, particularly to protect core-room workers from dermatitis, and in non-ferrous foundries where lead and other toxic fumes may be produced (see section 3.11) Protective clothing, discussed in section 3.10 is of critical importance to foundary workers.

REFERENCES

1. Loper, C.R., 'Foundry practice and equipment' from *Marks standard handbook for mechanical engineers*, 7th ed., McGraw Hill, London, 13-2 (1967).
2. Fife, I., and Machin, E.A., *Redgrave's health and safety in factories*, Butterworths London (1976).
3. Joint Standing Committee on Health Safety and Welfare in Foundries, *Reports of the subcommittee on machinery safety* Reports 1 to 5, Guarding foundry machinery, HMSO, London (1971 to 1977).
4. Joint Standing Committee on Health Safety and Welfare in Foundries, *Reports of the subcommittee on continuous casting and high-speed melting*, Reports 1 to 3, HMSO, London (1972 to 1976).
5. Department of Employment, *Improving the foundry environment*, Health and Safety at Work Booklet 17, HMSO, London (1974).
6. National Safety Council, *Accident Prevention Manual for Industrial Operations*, 7th ed., NSC Chicago, Chapter 33 (1974).

6.4 WELDING AND THERMAL CUTTING OF METALS

Contents

A weld is defined as a local coalescence of metal wherein coalescence is produced by heating to suitable temperatures, with or without the application of pressure, and with or without the use of filler metal. The filler metal may have a melting point the same as the base metals (as in arc or gas welding) or it may have a lower melting point but above 427°C (800°F). This definition includes brazing, but excludes soldering. It distinguishes welding from mechanical joining and adhesive bonding.

Thermal cutting processes sever or remove metal by local melting or by the reaction of the metal with oxygen sometimes with fluxes at an elevated temperature, or by a combination of both.

Welding and thermal cutting of metals are similar operations, often carried out by the same personnel using the same or similar equipment. Most welding and cutting operations carried out are variants of one of three main methods:

Gas cutting and welding.
Arc cutting and welding.
Resistance welding.

Less commonly used methods include electron beam welding, friction welding, ultrasonic welding, explosive welding and laser beam welding. Only the three main methods are discussed here, the main emphasis being placed on manual and semi-automatic welding.

The safety of these operations depends largely on the hazard awareness and training of the welders and their supervisors and managers. The same hazards can of course exist in automatic welding operations, although with these it is generally possible to eliminate them in the design and layout of the welding machines and allied operations.

Considering the numerous hazards of almost any welding operation it seems a little surprising that the subject has received so little attention in

British legislation. The main legal references to the subject are in The Protection of Eyes Regulations, 1974, Schedule 1, Parts II, III and IV, these were made under the powers of Part IV of the Factories Act 1961, and were discussed in section 3.10.2 of this book and in section 31 of the Factories Act 1961 Part II which deals with the welding of tanks which contained flammable vapours. This is discussed in section 6.1 of this book.[2]

Here the distinguishing features of the main welding and cutting methods and their principal variants are first discussed briefly. This is followed by a discussion of the hazards common to all these operations. Finally the special hazards common to particular methods of welding are discussed.

6.4.1 Gas welding and cutting of metals[1][3]

This covers the following processes:

1. *Manual oxy-acetylene welding of steel and many other metals*, usually with a welding rod (filler) of similar composition to the parts being joined. Oxy-acetylene welding of steel is carried out without a flux, whereas fluxes are necessary for most other metals.

2. *Oxy-hydrogen welding of metals with low melting points* such as lead and thin aluminium sheet.

3. *Oxy-acetylene and oxy-propane or natural gas braze welding and brazing* using a variety of filler rods (principally copper and silver alloys) which have melting points lower than those of the metals being joined (steel, cast iron, copper, etc.).

4. *Mechanised oxy-acetylene welding*, used generally in the manufacture of steel barrels, tubes and pipes, usually without filler metal.

5. *Oxy-acetylene flame spraying* using filler rods, wire or powder of special alloys to form a protective surface to resist abrasion, corrosion, heat, impact or a combination of these factors. This is also often used to build up worn surfaces.

6. *Oxygen cutting* using a torch with a central oxygen jet and an oxy-fuel gas heating flame, burning acetylene, propane or natural gas. This depends on the rapid oxidation of iron when heated to 815°C (1500°F) in the presence of oxygen, and is used with ferrous sheet, plate, bars, castings, etc. for severing, trimming, edge preparation for welding, metal removal and gouging. This can also be used underwater.

7. *Oxygen-powder cutting* of stainless steel. Because stainless steel is not cut by an oxygen jet due to the resistance of chromium oxide, special

techniques have been developed in which iron powder or a chemical flux is fed with oxygen to the jet in order to remove the chromium oxide.

6.4.2 Arc welding and cutting[1,3]

Arc welding comprises a group of welding processes wherein coalescence is produced by heating with an electric arc or arcs with or without the application of pressure and with or without the use of filler metal. The arc is maintained between an electrode and the work. Either direct or alternating current may be used, depending on the process, the filler metal, the type of shielding, the base metal and other factors.

Arc welding and cutting processes include:

1. *Shielded-metal-arc welding*, which aims to exclude atmospheric oxygen and nitrogen from the work. Shielding may be accomplished by a flux cover applied to the electrode, powdered flux heaped on the weld area, or a flow of inert or active gases projected round the arc and molten metal. The process is used on mild steel, alloy steels, stainless steels and on some non-ferrous metals.

2. *Submerged arc welding* is usually an automatic process in which a mechanically fed bare metal electrode is used. The weld and arc are shielded by a blanket of powdered 'flux'. Either the work or the arc may be moved. Welding is performed in the flat position. The process is used on mild steel, alloy steels, stainless steels, copper and nickel and their alloys. Welding quality is high. A manual version of the process can also be used.

3. *Gas tungsten arc welding* employs a single tungsten electrode and a shield of argon or helium gas projected round it. A filler metal can be used. This is introduced separately into the arc, either manually as a rod or mechanically as a wire, fed from a reel. The process is used for welding aluminium, magnesium, nickel, titanium and stainless steels without flux. The tungsten electrode is consumed very slowly and is not deposited in the weld. The process may be manual, semi-automatic or fully automatic.

4. *Carbon arc welding* employs a carbon or graphite electrode. Only d.c. power is employed. The welding rod is held in one hand, fed into the arc and fused into the joint while the electrode holder is held with the other hand.

5. *Arc spot welding* is used to produce small circular welds in lap joints using a manual tool employing the gas tungsten-arc-welding process. Filler metal is not usually employed.

6. *Arc cutting processes* rely on the heat of an electric arc to melt a path through metal, and are capable of cutting non-ferrous as well as ferrous metals. Some means of flushing the molten metal from the cut has to be provided.

The most successful arc cutting processes are:

Air-carbon arc cutting employing a jet of compressed air attached to the carbon electrode holder to blow away molten metal.

Plasma arc cutting employing a tungsten cathode in a tubular water cooled nozzle through which air or a special gas mixture is passed as a high speed jet at a very high temperature.

The hot gas makes a rapid narrow cut in the metal and flushes away the melt. Mixtures of hydrogen and argon or nitrogen are used for stainless steels and non-ferrous alloys, air or oxygen is used for mild steel.

6.4.3 Resistance welding

Coalescence in resistance welding is produced by the heat developed by the resistance — at the points of contact between two parts to be welded — to a current flowing from one part to the other, when the parts are held in contact by external pressure. Resistance welding processes include:

1. *Resistance spot welding* takes place between two parts which are lapped and held together by localised pressure between water-cooled copper electrodes.

2. *Resistance seam welding* takes place in a lapped seam which is moved between two copper rollers which form the electrodes. The weld may be a series of closely spaced spot welds, or overlapping spot welds, or a continuous weld nugget.

3. *Projection welding* is a form of spot welding where the localised points of contact between the parts are formed by stamping or machining the parts before they are brought together. The welds may be made singly or in multiples. Projection welding can be used with most metals except copper and red brasses. Galvanised iron, terneplate and tinplate are handled, and dissimilar metals such as steel and aluminium can be welded by special techniques.

4. *Upset welding* is used to join two members of roughly equal cross section. An electrode is attached to each of the two members which are brought together under pressure, creating a forge weld of symmetrical shape.

5. *Flash welding* is a development of upset welding. The parts are brought together lightly with the current flowing, then separated slightly which creates a flashing action at the surface which generates most of the heat and melts the metal. Heavy pressure is then applied which forces out the molten metal and makes the weld in the plastic metal behind it. Flash welding is faster than plain upset welding, and enables dissimilar metals of widely different melting points to be welded.

Nearly all types of resistance welding are done by machines which may be semi or fully automatic. With some materials (hardened steels) it is necessary to pre-heat before welding; others require immediate post-weld heat treatment.

6.4.4 Common welding hazards

The welding processes described above have a number of hazards in common, the extent of which varies from process to process, in addition to their own particular hazards. The common hazards are light rays, fire, toxic gases and fumes, general material handling hazards, noise and ionising radiation. Protection is discussed in conjunction with the more serious hazards.

6.4.4.1 Light rays

Arc and gas welding and cutting operations produce infra-red and ultra violet radiation which can harm the eyes and skin. Permanent damage to the lens of the eye may result from looking at a powerful ultra-violet source without eye protection. Ultra-violet light can also produce burns and tumours on the skin which occasionally turn malignant. Ultra-violet light produces ozone from the oxygen in the air up to several feet from the welding or cutting arc. This light can also decompose chlorinated hydrocarbons such as trichloro-ethylene and perchloro-ethylene to form highly toxic substances, even at a considerable distance from the arc. Argon arc welding is a particularly powerful source of ultra-violet radiation.

The only serious effect of infra-red radiation is that of heating the tissues with which it comes into contact. Thus if there is no skin burn there is unlikely to be any harm done.

The main precautions required, which apply particularly to inert gas shielded arc welding, are:

1. *Eye protection.* Tinted goggles or spectacles with side shields, helmets and shields should be worn by welders and their helpers *(figures 6.4.1 (a), (b) and (c))*. Arc welding operations should be isolated from other workers by booths or screens painted on the welding side with dark non-reflecting paint so that other workers are not exposed to either direct or reflected rays.

2. *Skin Protection.* All parts of the body exposed to strong ultra-violet radiation should be covered with dark clothing which as well as protecting the skin is needed to reduce reflection to the operator's face underneath the helmet. Woollen clothing is preferable to cotton, as it is less easily ignited and more resistant to ultra-violet rays.[3] Clothing liable to generate static sparks, e.g. nylon, should not be worn. The clothing should be thick enough to prevent the radiation from penetrating it. The use of protective skin cream should also be considered.

3. *Degreasing* operations and other work using chlorinated solvents should

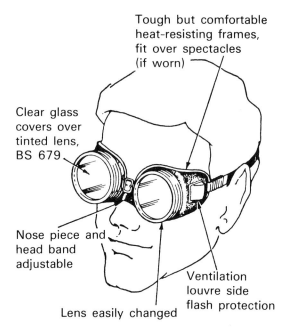

Tough but comfortable heat-resisting frames, fit over spectacles (if worn)

Clear glass covers over tinted lens, BS 679

Nose piece and head band adjustable

Ventilation louvre side flash protection

Lens easily changed

Figure 6.4.1(a)   BS approved safety goggles for welding

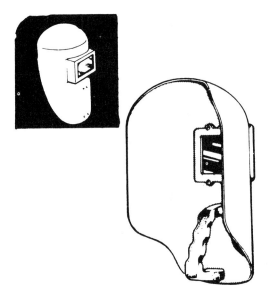

Figure 6.4.1(b)   Hand held screen

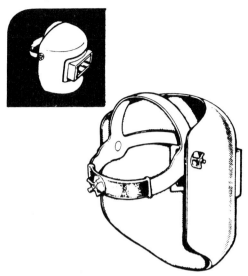

Figure 6.4.1(c)   Adjustable helmet

be so located that no solvent vapour can enter the welding or cutting area. The work in particular must be dry and free of these solvents before being brought into the welding area.

4. In areas where inert gas shielded arc welding or plasma arc cutting is carried out, *local exhaust ventilation* is likely to be needed to prevent hazardous concentrations of ozone from building up. Ventilation may in addition be needed for other reasons, e.g. for removal of fumes, nitric oxides, products of combustion and heat.

6.4.4.2 Fire

Special precautions must be taken to prevent sparks or hot slag from welding and cutting from reaching combustible floors and other materials. Sparks from welding can travel 10 m or more! The more important precautions are listed below:

1. When welding or cutting has sometimes to be done in hazardous areas where flammable materials are present, a 'hot work permit' system should operate.
2. When welding or cutting has to be done near combustible materials and neither the work nor the combustibles can be moved, the latter should be covered by fibre glass curtains or metal sheets. Spray booths and ducts should be cleaned to remove combustible deposits.

3. Where welding or cutting has to be done over a wooden floor, the floor should be cleaned and those parts where sparks or hot metal may fall should be covered by metal sheet or some non-combustible material. Sometimes the floor may be wetted, providing that electric welders can be protected against any additional shock hazard.

4. Cracks or holes in floors, walls and other openings through which hot metal or slag could pass or fall, thereby creating a hazard, should be covered with sheet metal or other non-combustible material.

5. Where it is necessary to weld or cut near to wood or other combustible material which cannot be removed or protected, a fire watcher equipped with a hose or suitable extinguisher should be posted near the operation, and the fire watch maintained for half an hour after the job is completed to check that there are no smouldering fires.

6. When welding or cutting has to be carried out on tanks, drums etc. that have contained combustible materials, the precautions given in section 6.1. should be followed.

6.4.4.3 Welding fumes and toxic hazards[3,4,5]

Metal fumes evolved during welding, brazing and metal cutting are a constant hazard. The condition known as metal fume fever is quite common among welders. The attacks have flue-like symptoms with raised temperature, aches and pains and sometimes respiratory difficulties.[4] More serious illnesses are experienced by welders and cutters exposed to fumes containing toxic elements present in brazing and welding rods, in the parts being cut and welded and in paint protecting them.

Examples of such poisoning are:

Coppersmiths using brazing rods containing up to 20% of cadmium have suffered liver and kidney damage,[5]
A number of demolition cutters removing iron structures heavily coated with lead paint were admitted to hospital with lead poisoning,[6]
Cobalt used as an alloying element in steels causes lung inflammation and fibrositis and possibly lung cancers, and has caused a number of deaths,[7]
Beryllium, used particularly as an alloying element in copper, is highly toxic causing chronic lung inflammation which is frequently fatal.[8]

Threshold limit values of some metallic elements commonly encountered in welding and cutting are given in *Table 6.4.1*.[9] These TLVs are mostly extremely low, and it is reasonably certain that such concentrations are often greatly exceeded for short periods in the air breathed by some welders and cutters in spite of the protection given by hoods. In addition to these metal fumes welders are exposed to oxides of nitrogen formed by the combination of oxygen and nitrogen at high temperatures and ozone formed by the action of ultra violet light on oxygen in the atmosphere. The fluxes also may contain toxic elements such as fluorine and boron which can form dangerous fumes in air. Welders and cutters must be protected against these hazards where they arise.

Gas welding of mild steel in the open air is unlikely to warrant respiratory protection, although gas cutting of an old heavily painted steel structure would certainly warrant it. Thus it is necessary for safety specialists to make a critical assessment of the respiratory protection needed in various welding operations, if possible in conjunction with an industrial hygienist. Local exhaust ventilation is usually the preferred solution. Whenever welders complain of tightness of the chest, dizziness and fever it is generally a sign that respiratory protection is inadequate.

Table 6.4.1[9] THRESHOLD LIMIT VALUES OF ELEMENTS COMMONLY FOUND IN WELDING AND CUTTING FUMES

Element of substance	TLV mg/m³	Element of substance	TLV mg/m³
Beryllium	0.002	Molybdenum	5.0
Cadmium oxide fume	0.05	Nickel and compounds	1.0
Cobalt metal fume and dust	0.1	Platinum (soluble salts)	0.002
Copper fume	0.2	Rhodium (metal fume)	0.1
Ferrovanadium dust	1.0	Silver metal	0.01
Manganese and compounds	5.0	Tellurium	0.1
Lead fumes and dusts	0.15	Vanadium fume	0.05
		Zinc oxide fume	5.0

6.4.4.4 Other common hazards

Welders are constantly faced with the hazards of lifting and carrying (discussed in section 5.3) usually under handicap. The articles to be handled are often hot or have sharp edges. The welder is wearing tinted goggles and/or a helmet, which reduces visibility especially in artificial light, he is also encumbered by heavy personal protective equipment. Some welding and cutting operations are also extremely noisy. Housekeeping is also a difficult problem when slag and hot offcuts are being produced at a considerable rate. These problems call for understanding on the part of safety specialists and discussions with welders, cutters and management as to how they can best be solved. Radiation exposure from gamma rays and X-rays used in weld inspection is another hazard to which welders may be exposed. Education and careful layout and planning of the work are essential to avoid this.

6.4.4.5 General clothing requirements (see section 3.10)

Eye and skin protection from strong ultra-violet sources have been considered in sub-section 6.4.4.1 above. Most welders require flame-resistant gauntlet gloves, aprons of fire resistant material, safety shoes with protected tops and head protection against falling objects. Those engaged in heavy work will require high boots or fire resistant leggings.

For overhead work, leather or fire resistant capes or shoulder covers are required, and insulating flame resisting skull caps to be worn under helmets to prevent head burns. Ear protectors may also be needed.

Outer clothing should be kept free from oil and grease as far as possible; sleeves and collars should be kept buttoned. Clothing should be designed to minimise places where sparks and hot metal particles could lodge, i.e. no pockets in outer clothing and no turn-ups on trouser legs or cuffs.

6.4.5 Gas welding and cutting hazards[10]

In addition to the hazards common to all methods of welding discussed in section 6.4.4, each group of welding processes has its own special hazards.[3]

Gas welding and cutting involves the use of oxygen, acetylene and other gaseous fuels, cylinders, hoses, reducers, torches and intense flames. These are considered in the following paragraphs.

In the UK oxygen is supplied in steel cylinders as a gas, now at 172 bar or 2500 psig, or sometimes as a liquid at low temperatures to be gasified on the consumer's premises.

Combustible materials burn more rapidly in oxygen than in air and any slightly smouldering material, such as a cigarette end or pipe will burst into flames in a stream of oxygen. Oxygen also forms highly explosive mixtures with acetylene and all fuel gases. Oil and grease may ignite and burn in the presence of pure oxygen, and must never be applied to oxygen cylinders and fittings.

Acetylene is supplied in cylinders which are completely filled with an approved porous material impregnated with acetone in which acetylene is soluble, generally under a pressure of 17.2 bar or 250 psig at 21°C (70°F). Acetylene itself is explosive and cannot be handled safely on its own at pressures above 15 psig since it is liable to detonate, forming carbon and hydrogen, in the complete absence of air or oxygen. This restricts its use in under water welding to very shallow depths. It also makes it important that the porous material in acetylene cylinders should have no voids of appreciable size.

Acetylene is sometimes generated as required in special apparatus by the reaction between water and calcium carbide, which is stored in watertight tins or drums. The main danger in this lies in damage to the tins or drums in handling. Should water come into contact with carbide, there is a danger of ignition and explosion.

Hydrogen is generally supplied in cylinders under a pressure of about 2000 psig or 138 bar. In the UK and most other countries, the threads on oxygen cylinder valves and regulators are right hand, and those on acetylene and fuel gas cylinders and regulators are left hand. Acetylene cylinders are painted maroon, hydrogen red and oxygen black.

6.4.5.1 Gas cylinders

The handling of compressed and liquefied gases in cylinders presents a number of hazards. The following points should be made regarding the use of cylinders for gas welding and cutting.

1. Cylinders should be used in an upright position and secured against being knocked over. This may be done by the use of a cylinder trolley, by cylinder stands, or by standing them against a wall with a length of chain to secure them.
   Empty cylinders should be stored and handled as if they were full as they can be equally as dangerous.
   Cylinders should never be brought inside tanks, vessels or other confined spaces, but should be secured vertically outside with their reducers, so that only low-pressure hose and the torches are inside. For work aloft it is often safer to place the cylinders on ground level and run low pressure hoses to the work.
2. Open cylinder valves slowly using only the proper key. Before making a connection to a cylinder valve outlet, crack the valve open for a second to clear any particles of dirt in the opening, pointing the valve and opening away from the body and not towards anyone else. Fuel gas cylinders should be well away from sources of ignition when this is done.
3. Never use gas cylinders without a pressure reducing regulator attached to the cylinder valve, except where cylinders are attached to a manifold, when the regulator will be attached to the manifold header. Pressure regulators must have two pressure gauges, one for cylinder contents and one for the outlet regulated pressure, and these pressure gauges must be frequently checked against gauges of known accuracy. Regulators and gauges should only be used with

Figure 6.4.2   Acetylene cylinders should be stored in a cool dry place away from heat and direct sunlight without switches or lamps.

gases for which they are designed and intended. Defective cylinders, valves, regulators, etc. should always be returned to the manufacturer for repair.

4. Before attempting to stop a leak between a regulator and a cylinder by tightening the union nut, check that the cylinder valve is tightly shut.

5. If a leak appears on a fuel gas cylinder, take it to a safe place or out of doors well away from sources of ignition, and attach a suitable label to the cylinder. A leak though a valve seat may be temporarily stopped by attaching a regulator to the valve. If the leak cannot be stopped in this way, post warnings against smoking and naked lights near it, open the valve slightly and allow the contents to escape slowly. A responsible person should be present until the cylinder is depressurised. Notify the supplier promptly and return the cylinder to him, following any instructions he may give about its return.

6. Take care never to allow sparks, molten metal, electric currents, excess heat or flames to come into contact with the cylinder or attachments.

7. Never use oxygen as a substitute for compressed air in any operation for which compressed air is intended to be used, e.g. pneumatic tools, dusting clothing or starting engines.

8. Never bring cylinders into tanks or unventilated rooms or other closed quarters. Wherever possible they should be placed out of doors on a north wall.

9. Before removing a regulator from a cylinder valve, close the cylinder valve and release the gas from the regulator.

10. Never fill gas cylinders except with the agreement of the owner and then only in accordance with a recognised safe procedure. Never mix different gases in a cylinder.

6.4.5.2 High pressure manifolds and regulators

Gas cylinders are frequently connected together by manifolds to provide a continuous gas supply or one at a higher rate than a single cylinder can give. They are best obtained from a reliable supplier and installed in accordance with his instructions. There are basically two types — small portable manifolds to connect up to five cylinders together and stationary manifolds to connect a large number of cylinders to a pipe distribution system. Special care is needed with oxygen manifolds that they are well removed from any flammable material, including cylinders of flammable gases. An oxygen manifold should be separated from acetylene generators and cylinders containing combustible gases by at least 15 m (50 ft), or failing that by a fire proof partition.

Only steel or wrought iron pipe should be used for acetylene distribution systems. Acetylene gas should never be brought into contact with unalloyed copper, with which it forms an explosive compound which is easily detonated. Distribution pipework should be designed in accordance with sound engineering standards. In fuel gas distribution systems, a pressure reducing regulator or non-return valve should be installed at every point where gas is withdrawn to supply a torch, etc to prevent back flow.

Regulators or reducing valves must be used on both oxygen and fuel gas cylinders or their manifolds to maintain a uniform and correct supply pressure to the torches. These must be designed for the purpose in hand and made by a reputable manufacturer. Regulators are easily damaged and must be handled carefully. 'Creeping regulators' where the outlet pressure rises when the outlet is closed, should be withdrawn from service for repair. When regulators are connected but not in use, the pressure adjusting device should be released. Cylinder valves should never be opened unless the pressure adjusting device has been released and gas allowed to escape.

The main steps in the procedure of attaching a regulator to a gas cylinder or manifold are:

1. Blow out any dust in the cylinder valve by cracking it open and closing it quickly as described above under cylinders.
2. Check that the regulator threads match those on the cylinder valve, then connect the regulator, after releasing the pressure reducing screw.
3. Open cylinder valve slightly so that the hand on the cylinder contents gauge rises slowly to its steady figure, then close the cylinder valve and check that the contents gauge remains steady. If it falls, there is a leak either on the connection to the cylinder or in the regulator itself. The leak should be located by brushing soapy water onto the regulator and the cylinder connector.
4. Attach oxygen hose, previously blown through with air, to the outlet of the oxygen regulator and the oxygen inlet valve on the torch, and make the corresponding connections with the acetylene hose. Shut both oxygen and acetylene inlet valves on the torch.
5. Set oxygen pressure regulator to give the approximate working pressure, then check the oxygen hose for leaks with soapy water. Do the corresponding things with the acetylene regulator. Check that the outlet pressures on both regulators do not creep.
6. Adjust oxygen and fuel gas pressures by first opening oxygen cylinder valve then torch oxygen valve and adjusting oxygen regulator to give required pressure. Shut off torch oxygen valve. Do the corresponding things to set the fuel gas regulator, but lighting the fuel gas while setting the regulator, taking care that the flame is pointing in a safe direction.
7. Open torch valves and light in accordance with manufacturers recommendations.

6.4.5.3 Hoses and hose connections

Correct hose colours are blue for oxygen and red for acetylene or other fuel gas (ISO standards). The hose should not be much longer than needed for the job, and care must be taken to prevent it becoming tangled or kinked. It must be protected from being run over or otherwise damaged (see *Figure*

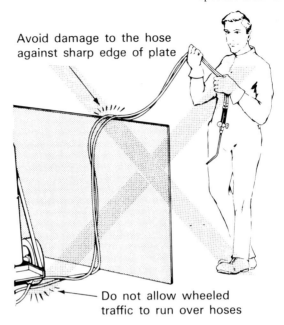

Avoid damage to the hose
against sharp edge of plate

Do not allow wheeled
traffic to run over hoses

Figure 6.4.3   Incorrect use of welding hoses

6.4.3), and it is usually preferable to suspend it overhead high enough to allow unrestricted traffic to pass underneath.

All hoses should be periodically examined and tested for leaks by immersion in water with normal working pressure inside. Leaks should be repaired by cutting the hose and inserting a splice, not be taping the hose on the outside.

Hoses must be protected from flying sparks, hot slag, grease and oil, and should be stored in a cool place. When parallel lengths of oxygen and fuel gas hoses are taped together for convenience, not more than 100 mm in 300 mm (or 4 inches in every foot) should be taped. If a flashback occurs and burns the hose, the burnt section should be discarded.

6.4.5.4 Torches

Only torches made by a reputable supplier should be used. Cutting torches differ from welding torches in having two oxygen valves, one for the premixed heating jet and one for the oxygen cutting jet.

Two types of welding torch are in use, the injector type, in which the acetylene is drawn into the mixing chamber by the velocity of the oxygen, and the medium pressure type where the acetylene is supplied under its own pressure to the mixing chamber.

The following precautions apply to the operation of torches:

1. Make sure that the welding head, tip or nozzle is correctly chosen for the job, and screw it firmly into the torch.
2. Before disconnecting a torch from the hose, shut off the gas at the regulators, never by crimping the hose.
3. To discontinue cutting or welding for a few minutes, shut only the torch valves. When stopping for half an hour or longer close oxygen and fuel cylinder valves, open torch valves to relieve gas pressure in hoses, and release pressure adjusting knob of regulator.
4. When lighting a torch, use a flint, or pilot light, not a match, and point tip in a safe direction away from anyone.
5. Never put down a torch without turning the gases off, and never hang a torch from a regulator or other equipment where it could come in contact with or impinge on a gas cylinder.
6. Follow the makers instructions when extinguishing the flame.

6.4.6 Arc welding and cutting hazards[10,11,12]

Manual arc welding is carried out with currents up to about 500 A and voltages of 15 to 40 V across the arc. The open circuit voltage on both d.c. and a.c. machines should be less than 100 V except when all equipment and circuits are fully insulated and the operator cannot make electrical contact other than through the arc. The a.c. power supply for manual welding should have a voltage reducing control which reduces the open circuit voltage to about 38 V when idling, but automatically restores the higher a.c. voltage when the electrode makes contact with the work, so that the arc is instantly struck.

Special power units are always required for arc welding. When there is adequate mains power, air-cooled transformers are used for a.c. supply and transformers and rectifiers for d.c. supply. Motor generators are used for both a.c. and d.c. supply. In the absence of adequate mains power petrol driven generators are used both for a.c. and d.c. supply.

When portable power units are employed, either of transformer or petrol engine type, adequate ventilation must be provided around them to prevent overheating of transformers or build up of exhaust fumes from petrol engines.

Two welding leads are required (*Figure 6.4.4*), the electrode lead from the machine and the work lead which provides the return circuit from the work to the machine, this being normally earthed (grounded). Sometimes instead of providing a complete return lead, the work may be connected to a steel structure and a connection made from another part of the structure to the power unit. When this has to be done it is essential to ensure that the electrical resistance of the entire return path from the work to the machine is very low. If not, heating will occur at points of poor contact in the return circuit, which can have dangerous consequences. Both leads must be adequate in size, well insulated and maintained.

Loose welding cable should not be left lying around, nor should welding

Bad connection for welding return

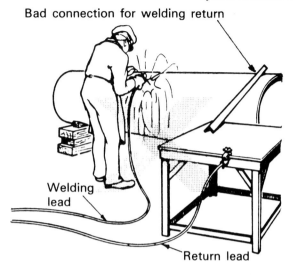

Good connection for welding return

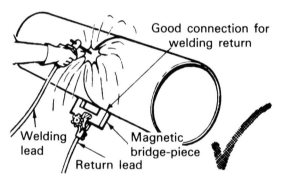

Figure 6.4.4   Good and bad connections for welding leads (All the illustrations in this section are reproduced by permission of the Engineering Industry Training Board)

cables lie in water or oil. Rooms where arc welding is done regularly should be permanently wired with welding cable. Welding cable is best strung overhead over roads and passageways with adequate clearance for persons and vehicles.

Fully insulated electrode holders should be used to reduce the possibility of accidentally striking an arc. Electrode holders should be adequate in size for the work, but if the holder gets too hot, an extra one should be provided for alternate use so that a hot holder can cool down. Dipping holders in water to cool them must not be allowed.

6.4.6.1 Electric shock protection

Although the open circuit voltages employed are not high, they cannot be neglected as a hazard. The work is normally earthed (grounded) and the welder can also easily become earthed. When changing electrodes, setting up work or changing working conditions, the welder can readily be exposed to the open circuit voltage. The danger is greatest when he is hot and sweaty. Welders should aim to keep themselves insulated both from the work and from the electrode and holder. They should never allow the bare metal part of an electrode or holder to touch their bare skin or any wet clothing. Electrodes should never be changed with bare hands or wet gloves or when standing on wet floors or grounded surfaces. In confined spaces, cables should be covered or so placed to avoid contact with sparks.

The frames of portable as well as stationary power units should be grounded. Exposed sections of worn cables should be replaced. Special care must be taken to keep welding cables well away from power supply cables.

6.4.6.2 Hazards of gas shielding

Gas shielded arc welding requires a supply of argon, helium or carbon dioxide which is fed through a torch or gun to surround the electrode. These gases are supplied in cylinders equipped with regulators and fed by hose to the gun as in gas welding. The gas supply is generally controlled by a flowmeter and valve.

Ventilation must be sufficient to maintain a normal oxygen concentration for the welder to breathe. Both argon and carbon dioxide are heavier than air and will settle in pits and excavations displacing air from them. Positive ventilation is therefore needed when such welding has to be done in pits and confined spaces.

6.4.7 Resistance welding hazards[3]

Resistance welding equipment is usually fixed and the operations are automatic or semi-automatic. The main hazards are lack of guards, flying hot metal particles, and unauthorised adjustments and repairs. The secondary voltages applied in resistance welding are low, rarely above 30 V, and there is seldom much electric shock hazard. The cables however are subject to severe conditions, caused not merely by the high currents but by the electromagnetic interactions between them which cause pulsations and metal fatigue. Water cooled coaxial cables are widely used to overcome this problem. They present a hazard if the water supply fails, when steam pressure may build up in the cable and burst out at a weak spot.

Control circuits should operate on low voltages. Stored energy equipment (capacitors) on the primary circuits should be fully enclosed. Back doors of machines and panels should be kept locked or interlocked, and points of operation hazards should be guarded as far as practicable, using the same type of guards as are used for power presses. Flash welding machines should have shields to protect the operative's eyes, and local exhaust ventilation to carry off dust and fume. Safety glass shields may be required to protect employees from flying sparks. Foot switches should be guarded to prevent accidental operation.

6.4.8 Medical examination

All welders and welding operatives should be subject to medical examination, including chest X-ray before appointment, and should be periodically re-examined according to a pre-arranged programme.

A serious attempt should be made to relate any adverse medical conditions found to particular welding operations and hazards, and firm steps taken to remedy the situation.

REFERENCES

1. Fenton, E.A., 'Welding' in *Marks standard handbook for mechanical engineers*, 7th ed., McGraw Hill, New York (1967)
2. Fife, I., and Machin, E.A., *Redgrave's health and safety in factories*, Butterworth, London (1976).
3. National Safety Council, *Accident prevention manual for industrial operations*, 7th ed., NSC, Chicago, 973 (1974).
4. Ross, D.S., *Journal of the society of occupational medicine*, **24**, 125 (1974).
5. Blackadder, E.S., 'Toxic diseases and their biological effects', *Process Industry Hazards Symposium*, Inst. Chem. E. Symposium Series No. 47 (1976).
6. Baird, A.W., 'Lead intoxication in the demolition of railway property', Joint Meeting of Society of Occupational Medicine and Royal Society of Medicine, London (25th March, 1976).
7. Kipling, M.D., *Occupational health*, **25**, 131 (1973).
8. Williams, A., 'Metal toxicity', *The Chemical Engineer*, **323**, 570 (Aug. 1977).
9. Health and Safety Executive, *Guidance Note EH/79. Threshold limit values for 1979*, H.S.E. (1979).
10. Home Office, *Safety measures for the use of oxy-acetylene equipment in factories* (Form 1704) HMSO, London.
11. Sanderson, P.G., 'Welding operations', Chapter 13 in *Industrial Safety Handbook*, editor: Handley, W., 2nd ed., McGraw Hill, Maidenhead, 178 (1977).
12. Ministry of Labour, *Electric arc welding*, HMSO, London (1969).

6.5 GRINDING

Contents

The main hazards of grinding are:

1. Production of airborne dust, often toxic, and resulting lung disease.
2. Production of fast moving particles and possible eye damage.
3. Breakage of wheels when running at speed and personal injuries caused by fast moving fragments.
4. Bruising, abrasions and burns — sometimes worse injuries, resulting when work held in the hand slips when in contact with a grinding wheel.
5. Loose hair, tie or shirt sleeve etc. caught in wheel or workpiece resulting in serious injury.
6. Ignition of finely divided dust from the sanding and grinding of wood, plastics, magnesium and other combustible materials.

Lung disease caused by inhaling toxic dusts (mainly silica from sand stones) was particularly prominent among dry grinders in the late nineteenth century, most of whom as Dr. White noted, died before the age of 32 (see section 3.2.3). The position has improved today through the development of abrasive wheel containing less toxic abrasives (aluminium oxide and silicon carbide), by better ventilation and by the use of dust masks; however, it is surprising how seldom dust masks are worn by grinders even when the local ventilation is poor or non-existent (*Figure 6.5.1*).

The eyes can be fairly simply protected against fast flying particles, e.g. by glasses with side shields, and such protection should always be worn when grinding. Similarly, when dust is likely to be a problem, a mask should be worn.

The use of a proper guard round a grinding wheel should prevent injury to the user should a wheel fracture at speed. Guarding however is seldom as effective on portable grinders as on fixed machines.

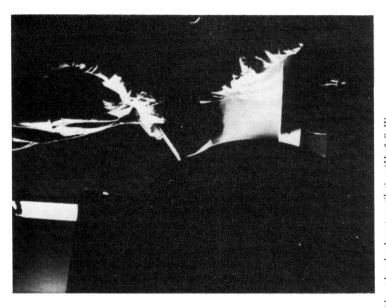

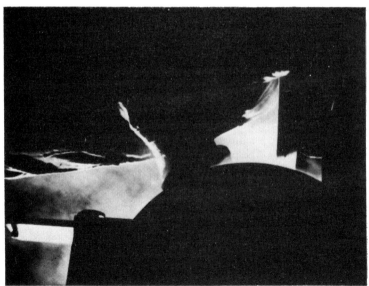

Figure 6.5.1 Pedestal grinder with and without local exhaust ventilation (H. & S. W. Booklet 17)

Proper training in the use of grinding wheels and the employment of sound techniques can do much to reduce accidental breakage of grinding wheels in use, as well as personal injuries caused when hand held work (or a portable grinding tool) slips in use.

Training and supervision are also necessary to ensure that long hair is adequately covered (see section 3.10.6) and that loose or frayed articles of clothing are not worn by those engaged in grinding, buffing, polishing and similar operations. The ignition hazard of finely divided organic dust is largely controlled by good housekeeping.

It is clear that grinding holds a number of pitfalls in store for the unwary, most of which are avoidable by the use of proper equipment training and techniques.

6.5.1 Legal requirements[1]

The various legal regulations applicable to grinding are considered below.

There are several legal requirements with regard to abrasive wheels and their use which are summarised below. The principal ones, contained in the Abrasive Wheels Regulations 1970, are summarised in section 6.5.1.5.

6.5.1.1 The Factories Act 1961, Part IV, Section 65. The Protection of Eyes Regulations, 1974, Regulation 29.

Approved eye protectors, approved shields or fixed shields are required for truing or dressing an abrasive wheel where there is a reasonably forseeable risk of injury to the eyes of any person engaged in the work from particles or fragments thrown off.

6.5.1.2 The Dangerous Occurences (Notification) Regulations 1947, Schedule, Class 1.

This makes it obligatory to report to the (H and SE) Inspector the bursting of a revolving vessel, wheel, grindstone or grinding wheel moved by mechanical power, whether death or disablement is caused or not.

6.5.1.3 The Grinding of Cutlery and Edge Tools Regulations 1925 and The Grinding of Cutlery and Edge Tools (Amendment) Special Regulations 1950

The following notes are a summary of the main requirements:
1. Local dust extraction, with hoods close to origin of dust, ducts and fans, is required for all dry grinding machines when grinding cutlery.
2. Dressing wheels shall only be done in a special room or place where no other work is carried out.
3. Where wet grinding is carried out, there must be adequate local

   ventilation or a stream of clean water supplied to the surface of the
   grindstone, and means of collecting and draining it.
4. All ventilation plant for dust extraction must be inspected, tested
   and maintained where necessary every four months.
5. Wet grinding and glazing (finishing and polishing) operations shall
   be well separated with at least 2.44 m (8 ft) between centres of
   grindstone and glazing appliance.
6. Hacking (an operation involving chipping a grindstone) is not
   allowed unless there is enough water supplied to keep the surface of
   the grindstone wet.
7. Rooms in which wet grinding is done shall be at least 3.05 m (10 ft)
   high, with adequate window area, the windows to be kept well
   cleaned.
8. Rooms in which wet grinding is done shall have smooth impervious
   walls and floors, and the belts, pulleys and shafts shall be well
   protected yet readily accessible for cleaning.
9. Rooms where grinding (wet and dry), racing and glazing are carried
   out shall be cleaned, including belts, pulleys, shafts and fixtures, at
   least once every three months.
10. A register shall be kept of all cleaning carried out with details and
    signature of cleaner. The register is to be available to anyone
    working in the room to which the register applies.
11. No spitting is allowed on floors or walls of grinding rooms.
12. Proper accommodation for clothing is to be provided.
14. There must be at least 1.37 m (4 ft 6 in) between any two
    grindstones or 0.915m (3ft) if the stones are less than 61m (2ft) in
    diameter.
15. All defects in appliances, facilities or accommodation must be
    reported to the owner, occupier or manager or other responsible
    person.

6.5.1.4 The Grinding of Metals (Miscellaneous Industries) Regulations
1925 and The Grinding of Metals (Miscellaneous Industries) (Amendment) Special Regulations 1950

These regulations which are very similar to the Grinding of Cutlery and
Edge Tools Regulations 1925 and Amendment 1950, extend their
application to the grinding of metals generally and the cleaning of castings.
'Rumbling' of castings to free them of sand by rotation in a revolving vessel
and sand blasting are included in their scope. There are a number of
exceptions, notably they do not apply to the intermittent grinding of metals
in a room in which such work is not carried out for more than twelve hours
in a week.
   New provisions include one that not more than one person may be
allowed to perform the actual process of grinding or glazing upon a
grindstone, abrasive wheel or glazing appliance at any one time, and one to
prohibit the use of a rumbling appliance unless an efficient exhaust draught

is applied to prevent any dust evolved from entering the room. The provisions with regard to ventilation are more detailed than those of the Grinding of Cutlery and Edge Tools Regulations.

6.5.1.5 The Abrasive Wheels Regulations 1970

These regulations start with the following definition of an abrasive wheel, namely:

(a) A wheel, cylinder, disc or cone which, whether or not any other material is comprised therein, consists of abrasive particles held together by mineral, metallic, or organic bonds whether natural or artificial;
(b) A mounted wheel or point and wheel or disc having in either case separate segments of abrasive material;
(c) A wheel or disc in either case of metal, wood, cloth, felt, rubber or paper and having any surface consisting wholely or partly of abrasive material; and
(d) A wheel, disc or saw to any surface of which is attached a rim or segment consisting in either case of diamond abrasive particles.

Although an abrasive disc is included in the legal definition of an abrasive wheel, for clarity wheels and discs are distinguished here in section 6.5.2.1. The main points dealt with by these regulations are summarised below:

1. *Speed of wheels (Regulation 6).* No abrasive wheel with a diameter greater than 55 mm other than of metal, wood, cloth, felt, rubber or paper, the surface of which is partly or wholely covered with abrasive material, shall be used unless the maximum permissible speed in r.p.m. is clearly marked on the wheel or its washer.

No similar abrasive wheel with a diameter of 55 mm or less shall be used unless a notice is displayed in the room where grinding is carried out stating the maximum speed in r.p.m. specified by the maker.

No abrasive wheel shall be run in excess of the maximum speed as explained above except where its diameter has been reduced, when the maximum speed may be increased in the same ratio as the original diameter to the new diameter.

2. *Speeds of spindles (Regulation 7).* A notice shall be fixed to every power-driven wheel with a spindle to which an abrasive wheel may be mounted giving its maximum working speed, or for spindles which may be operated at various speeds, the maximum and minimum working speed. No spindle with an abrasive wheel mounted on it shall be operated at a higher speed than the maximum working speed specified for the spindle. The speed of any air driven spindle to which an abrasive wheel is mounted shall be controlled by a governor so that its speed is not in excess of that specified for the spindle.

3. *Mounting (Regulation 8).* Every abrasive wheel shall be properly mounted.

4. *Training and appointment of persons to mount abrasive wheels (Regulation 9).* Nobody shall mount an abrasive wheel unless they have been trained in accordance with the Schedule to the Regulations, is competent for the job and has been appointed in writing by the occupier of the factory to work with that particular type of abrasive wheel.

5. *Provision of guards (Regulation 10).* A guard shall be provided and kept in position at every abrasive wheel in motion, except where this is impracticable.

6. *Construction, maintenance, etc of Guards (Regulation 11).*
   (a) Guards shall be designed and constructed to contain every part of the wheel should the wheel break in motion.
   (b) Guards shall be properly maintained and secured to prevent displacement in case of breakage of the wheel.
   (c) Guards shall enclose the whole of the wheel except such parts that need to be exposed for work done at the wheel.

7. *Tapered wheels (Regulation 12).* Where the work done necessitates that the exposed arc of the wheel shall exceed 180°, then where practicable the wheel shall be tapered from its centre towards its periphery by at least 6% on either side and shall be mounted between suitable protection flanges. These protection flanges between which any abrasive wheel is mounted shall be of substantial construction and shall have the same degree of taper as the wheel. Diameters of the protection flanges are stated. These are usually equal to at least half the wheel diameter.

8. *Selection of wheels (Regulation 13).* Wheels suitable for the work to be done should be selected so as to reduce risk of injury to workers.

9. *Machine controls (Regulation 14).* Any machine on which an abrasive wheel is mounted shall have an efficient device for starting and stopping which can be readily and conveniently operated by the person using the wheel.

10. *Rests for workpieces (Regulation 15).* Where there is a rest for supporting the workpiece the rest must be:
    (a) properly secured,
    (b) adjusted so that it is as close as possible to the exposed part of the wheel,
    (c) of substantial construction and properly maintained.

11. *Cautionary notice (Regulation 16).* The approved cautionary notice about the dangers of abrasive wheels and the precautions to be observed shall be placed in a position where it can easily be read by persons employed

in grinding or cutting in every room where grinding or cutting by abrasive wheels is carried out.

12. *Condition of floors (Regulation 17).* The floor surrounding every fixed or portable machine on which an abrasive wheel is or is intended to be mounted shall be maintained in good condition, kept free of loose material and prevented from becoming slippery.

13. *Duties of employees (Regulation 18).* No employee shall wilfully misuse or remove any guard, protection flange, rest for workpiece or other appliance provided under these regulations and every employee shall make proper use of guards, protection flanges, rests for workpieces, etc. If he discovers any defect in this equipment he shall report it to the occupier, manager or appropriate person.

14. *Training Schedule (Referred to in Regulation 9).* The training shall include suitable and sufficient instruction in the following matters in relation to each class or description of abrasive wheel in respect of which it is proposed to appoint the person being trained, that is to say:
   1. Approved advisory literature relating to the mounting of abrasive wheels.
   2. Hazards arising from the use of abrasive wheels and precautions which should be observed.
   3. Methods of marking abrasive wheels as to type and speed.
   4. Methods of storing, handling and transporting abrasive wheels.
   5. Methods of inspecting and testing abrasive wheels to check for damage.
   6. The functions of all components used with abrasive wheels, including flanges, washers, brushes and nuts used in mounting and including knowledge of the correct and incorrect methods of assembling all components and correct balancing of abrasive wheels.
   7. The proper methods of dressing an abrasive wheel.
   8. The adjustment of the rest of an abrasive wheel.
   9. The requirements of these Regulations.

These regulations taken together give a good picture of the hazards, precautions and training of those working with abrasive wheels.
A few additional points follow to give a more complete picture.

6.5.2 Selection and examination of abrasive wheels[2] [3]

Grinding wheels consist of abrasive particles embedded in a bonding medium. The abrasive particles are generally either aluminium oxide, denoted by A or silicon carbide denoted by C. Aluminium oxide is used mainly for grinding and cutting steel and materials of high tensile strength. Silicon carbide is harder but more friable than aluminium oxide and it is more suitable for grinding hard materials such as glass or porcelain as well as materials of low tensile strength such as copper.

There are four main types of bonding media:

Vitrified media, denoted by V, are glasses or porcelains formed during the final firing of the wheel. They are used for precision grinding wheels.
Resinoid media, denoted by B, are comprised of synthetic resins and fillers. They are used mainly for heavy duty grinding wheels.
Rubber media, denoted by R, are used for fine grinding wheels which give a smooth finish.
Shellac, denoted by E, is also used for special finishing wheels.

Besides the type of abrasive particle and the bonding media, the grit size, the structure and the grade are important. The grit size is given as the size of mesh (openings per inch) which retains the particles. Grit sizes range from 8 (very coarse grinding) to 1200 (super finishing).

The structure is a measure of the grit concentration and ranges from 1 (a very dense structure) to 16 (an open structure with low grit concentration). Grades of wheels range from E (soft) to Z (hard). This refers to the tenacity of the bond between the particles and the bonding medium.

6.5.2.1 Abrasive wheels and discs[4]

It is useful to distinguish between abrasive wheels, mounted on the spindle or arbor of a grinding machine, of which only the circumference is designed for grinding, and abrasive discs, mounted on the machine face of a grinding machine. Only the exposed flat side of an abrasive disc is designed for grinding.

6.5.2.2 Inspection of wheels[4]

Wheels and discs should be unpacked on receipt, and loose packing material removed with a brush. They should first be checked carefully to ensure that they have not been damaged in transit. Abrasive wheels are fragile and easily damaged.

Next the abrasive wheels should be subjected to the 'ring' test (see *Figure 6.5.2*) as follows. A light disc or wheel should be suspended from its hole on a small pin or the finger; a heavy one should be placed vertically on a hard floor. The wheel or disc should then be tapped gently with a light wooden object — a screwdriver handle or a mallet in the case of heavy wheels or discs. The wheel or disc should be tapped at a point 45° from the vertical centreline and 25 to 50 mm from the periphery depending on the size. A sound wheel produces a clear 'ping', whereas a cracked wheel or one in poor condition produces a dull sound. The wheel or disc should then be rotated 45° and the test repeated.

Resin and rubber bonded wheels and discs do not give the same clear 'ping' as do vitrified wheels.

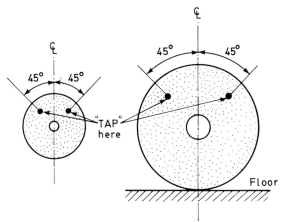

Figure 6.5.2 Tap points of an abrasive tool for the ring test (National Safety Council, U.S.A.)

6.5.3 Handling and storage of abrasive wheels

Care must be taken to prevent abrasive wheels and discs being dropped or bumped; they should never be rolled. Wheels which cannot be carried by hand should be transported by truck, taking care they are properly supported.

Abrasive wheels should be stored in a dry place which is free from frost and away from excessive heat, using racks, bins or drawers. Plain and tapered wheels of appreciable thickness are best stored on edge, preferably in a cradle to prevent them rolling, and with frequent partitions to prevent them falling over. Thin resin, rubber or shellac wheels should be laid flat and stacked on a horizontal surface, such as a steel plate, with no washers between them.

Cylinder wheels and large straight cup wheels may be stored on flat sides with cushioning material between them. Tapered cups should be stored in stacks, not more than six high, with adjacent faces matching.

Abrasive wheels and discs may deteriorate in prolonged storage and should be taken out in rotation. Manufacturers should specify the safe storage life of their abrasive wheels, and their advice on this should be followed. Wheels removed from store where they have been a long time should be given the ring test before being issued for use.

6.5.4 Mounting abrasive wheels (*Figure 6.5.3*)

Many accidents with abrasive wheels have been caused by faulty mounting, and it is now illegal for anyone who has not undergone the required training to mount an abrasive wheel. The following short list of points to be watched

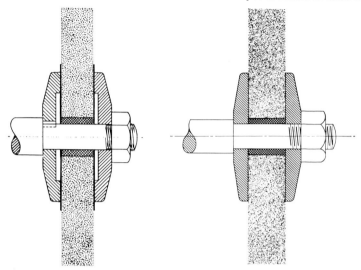

Figure 6.5.3 Correct and incorrect mounting of abrasive wheels (left) correct; (right) incorrect (H. & S. W. Booklet 4)

is intended only as a guide to safety specialists. It is far from comprehensive, and detailed mounting instructions should be available and studied for any particular wheel-machine combination.

1. A wheel should only be mounted on a grinding machine designed to take that particular wheel. The speed of the wheel and of the machine should be checked.
2. Both the wheel and the flange plates should be examined for damage, distortion and foreign matter, and the wheel should pass the ring test.
3. Where a bush is used, it should not extend beyond the face of the wheel.
4. The machine should be isolated electrically before fitting the wheel.
5. When mounting plain wheels, washers of blotting paper or thin rubber or leather, slightly larger in diameter than the flanges, should be fitted between the wheel and the flange to allow for unevenness of wheel or flange. This does not apply to tapered wheel flanges.
6. Protection flanges for taper sided wheels should have the same degree of taper on wheel and flange and should fit closely.
7. The wheel should fit freely but not loosely on the spindle or wheel arbor. This is most important.
8. The wheel flange should be tightened just enough to grip the wheel and drive it without slipping. Where there are a number of screws, those diametrically opposite should be tightened in sequence to ensure even pressure.

9. Where applicable, the correct balancing and dressing procedures should be followed until the wheel is running true and ready for grinding.
10. It must be checked that the overhang is correct for the speed, size of wheel and diameter of spindle.
11. The wheel guard and where applicable the work rest must be secured and adjustable.

6.5.5 Guards and work rests

Guards serve first, to contain the wheel in the event of a burst and secondly, to protect the operator from contact with the wheel. There are several types of guard.

Band type guards which cover the whole periphery are used for abrasive discs. Guards for abrasive wheels may be of cast or fabricated steel but not cast iron or any similar, brittle material; such steel guards have both peripheral and side members, with a cut away sector which exposes the wheel.

Guards must be strong enough to contain the force of a wheel bursting at maximum speed, and recommended thicknesses of peripheral and side members for various wheel sizes and speeds are given in reference 3. They should be adjustable, so that the minimum amount of wheel required for the work is exposed. The correct wheel exposure can be maintained with an adjustable tongue or a movable guard (*Figure 6.5.4*). Readjustment is necessary as the wheel wears and its diameter is reduced so that the maximum distance between the outside of the wheel and the tongue or end of the peripheral band never exceeds 6 mm (¼ in).

The maximum angular exposure varies with the type of grinding. Some examples are given in *Table 6.5.1* and *Figure 6.5.5*.

Table 6.5.1 MAXIMUM EXPOSURE ANGLES FOR VARIOUS GRINDING OPERATIONS

Operation	Maximum Exposure Angle °		
	Above centreline of spindle	*Below centreline of spindle*	*Total*
Bench & floor stand grinders	65	25	90
Bench & floor stand grinder with contact below centreline of spindle	65	60	125
Top grinder	60	—	60
Swing frame grinder	—	180	180
Cylindrical grinder	65	115	180
Surface grinders and cutting-off machines	150	—	150

688

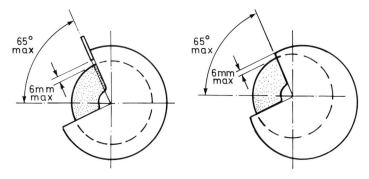

Figure 6.5.4   Correct exposures of grinding wheels maintained by (left) adjustable tongue or (right) a movable guard (National Safety Council, U.S.A.)

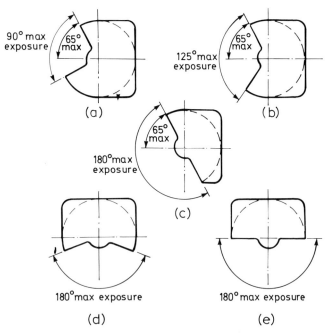

Figure 6.5.5   Angles of exposure of wheels for various grinding operations. (The broken line indicates an alternative guard profile frequently employed)
(a) and (b) Bench and floor stand grinders
(c) Cylindrical grinders
(d) Surface grinders and setting off machines
(e). Swing frame and portable grinders (National Safety Council, U.S.A.)

Safety guards should cover exposed arbor ends. Work rests must be strongly made and securely clamped not more than 3 mm from the wheel. The position of the work rest should be checked frequently but it should never be adjusted while the wheel is in motion

Many injuries have occurred because of too wide a gap between the work rest and the wheel which allowed work to become wedged between the rest and the wheel. Guides should be used to hold the work when slot grinding or in similar operations to prevent twisting the work.

6.5.6 Truing and dressing abrasive wheels[4,5]

Abrasive wheels require occasional truing and dressing. In truing, a diamond supported on a steel rod held rigidly in the machine is passed over the face of the wheel to remove enough material to give it its true shape. Dressing is a more severe operation in which an abrasive stick or wheel is moved over the wheel face to remove loading and glazing and give a sharp face consisting of newly fractured crystals.

Wheel dressing tools should be equipped with guards over the tops of the cutters to protect the operator from flying pieces of wheel or cutter. Only tools approved for the job should be used; lathe cutting tools should not be used. A face shield should be worn by the operator. Wheel edges should be rounded off with a hand stone before and after dressing to prevent them chipping.

When dressing bench and stand grinders, a work rest should always be used to support and guide the tool, and a tool holder used if possible. Moderate pressure should be applied slowly and evenly, moving the dresser across the face of the wheel. Diamond dressers should be applied at or slightly below the centre line — never above it. Wheels should be periodically tested for balance and rebalanced where necessary. Out of balance wheels which cannot be balanced by truing and dressing should be scrapped.

6.5.7 Additional grinding precautions

Many wheel failures occur during the initial run up to operating speed. Operators should therefore place themselves in a safe position while a wheel is coming up to speed and allow it to run for a minute at its working speed before applying work to it. When a grinding fluid is used, this should not be turned on until the wheel has reached operating speed, and it should be turned off before the spindle drive is switched off. Heat stress can quickly build up in a cold wheel when the periphery heats up when work is applied. This applies specially to dry grinding. The difference in temperature between the inside and outside creates stresses which can crack the wheel. The work should therefore be applied lightly to a cold wheel and pressure built up only slowly, as the whole wheel warms up. When dry grinding with vitrified wheels, continuous heavy pressure must be avoided.

The importance of speed on wheel stresses is not always appreciated. The stress in a wheel increases as the square of the speed. Strict precautions must be taken to ensure that safe wheel speeds are not exceeded, particularly with variable speed machines.

6.5.8 Polishing and buffing wheels, wire brush wheels[4]

Buffing wheels are made of discs of felt or fabric sandwiched between flanges, and coated on the outside with rouge or other mild abrasive. Sometimes the working surface is too hard and has to be softened. This should never be attempted while the wheel is rotating, but the wheel should be removed from the machine, placed on a flat surface, and the edges of the wheel pounded with a mallet.

Polishing and buffing wheels should be mounted on substantial stands and be provided with hood guards to protect the operator's hands or clothing from contact with protruding nuts or the ends of the spindles, and exhaust hoods are also needed to catch particles thrown off by the wheels. When variable speed motors are used strict precautions must be taken to ensure that the safe operating speed is not exceeded.

Gloves should not be worn by polishers and buffers as they may catch dragging the operator's hand against the wheel. Small pieces of work should be held in a simple jig against the wheel, not by the operator's bare hands.

Wire brush wheels are used to remove burrs, scale and other materials. They must be held rigidly in place by flanges and nuts, and the same conditions applied as to buffing wheels. The work rest should be adjusted so that it is about 3 mm from the brush wheel. Heavy personal protective equipment is required because of the tendency of wires to break off. Face shields should be mandatory, as well as leather or heavy canvas aprons and leather gloves.

REFERENCES

1. Fife, I., and Machin, E.A., *Redgrave's health and safety in factories,* Butterworths, London (1976).
2. Southwell, K.B., 'Grinding', Chapter 10 in *Industrial Safety Handbook*, edited Handley, W., 2nd ed., McGraw Hill, Maidenhead (1977).
3. Health and Safety Executive, *Safety in the use of abrasive wheels,* Booklet 4, HMSO, London.
4. National Safety Council, *Accident prevention manual for industrial operations,* 7th ed., National Safety Council, Chicago, 841 (1974).
5. Baumeister, T., *Marks standard handbook for mechanical engineers,* 7th ed., McGraw Hill, New York, 13-100 (1967).

6.6 HAND TOOLS AND PORTABLE POWER TOOLS[1, 2]

Contents

About 7% of all industrial injuries in the UK result from hand tool accidents. Minor injuries — cuts, abrasions and contusions — caused by contact with a tool or a chip are very common. More serious injuries — loss of eyes, severed fingers or tendons and deep puncture wounds — too often occur.

Most hand tool accidents can be attributed to a few simple basic causes:

1. Purchase of inferior tools which break in use.
2. Use of tools in poor condition, e.g. screwdrivers with broken handles, hammers with loose heads and electric tools with broken plugs.
3. Using the wrong tool for the job e.g. pliers instead of a wrench, or a file or screwdriver instead of a crowbar.
4. Incorrect use of tools, e.g. fitting a pipe as lever to the handle of a wrench or spanner, or applying a screwdriver to an object held in the other hand.
5. Leaving tools in dangerous places, e.g. a spanner on a building beam from which it may be dislodged and fall on someone, or a sharp unsheathed knife among a bundle of tools which may be grasped by an unwary hand.

In many works, fitters, carpenters and others who use hand tools have their own tool sets, which they keep in locked tool boxes and drawers. Only the less commonly used tools and consumable items such as drill bits and saw blades are regularly issued from the tool store.

The personally owned tool kit has its advantages and disadvantages. In its favour, it encourages workers to take good care of their tools. On the other hand, it is often difficult for each employee to keep a full range of tools needed for his work, especially when he moves from one job to another. Also a personal tool kit makes inspection of tools for safe condition and tool maintenance more difficult to achieve.

Where possible, it is probably sounder for tool kits to be issued by employers on loan to workers who require them, and for these to be called in for maintenance and replacement on a regular basis. In this way it is possible to ensure that only the correct tools for the work are supplied, and that unsuitable tools or ones in an unsafe condition are not used. There are, however, many difficulties, particularly where contract labour is much used or where labour turnover is high.

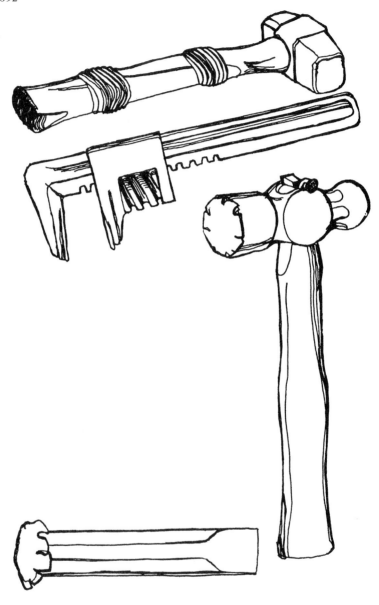

Figure 6.6.1   Defective hand tools. Many accidents occur from the use of badly worn tools, such as those illustrated. Cold chisels are particularly dangerous when they reach the condition shown in the diagram (From 'Safety in Industry' published by RoSPA and reproduced by permission)

A clear policy over the supply, use, storage and maintenance of hand tools which pays careful attention to these five causes of accidents can do much to reduce them. Job specifications should include precise and accurate statements of the tools required for each job. These statements should be checked and amended where necessary by supervisors (in consultation with those who use the tools), and should form the basis for the firm's tool purchasing, inspection and maintenance programmes.

Supervisors must have experience and understanding of the selection of tools for every job under their control, and of their proper use, and all employees must be properly trained in these points. This seldom receives the attention it deserves, and it is all too frequently assumed that 'everyone knows how to use hand tools properly'. Space limitations preclude more than passing treatment of the subject here. An excellent discussion is given by the American National Safety Council,[1] and a shorter but lucid treatment is given by RoSPA.[2]

6.6.1 Analysing hand tool accidents

Different types of hand tool accidents predominate in different works, depending on the nature of the work and the tools used. Hand tools accidents (even seemingly trivial ones) should be recorded and analysed as described in section 1.5.4.

Such an analysis should be used to alert all concerned to the principal hazards at the time and to plan effective action to reduce them.

6.6.2 Carrying, handling and keeping hand tools

Sharp hand tools should be carried in a box or strong bag, never in a pocket with the sharp or pointed end uppermost. The point or cutting edge should be protected, e.g. by a sheath. Where a knife or other sharp tool must be carried in a belt, it should be sheathed and fastened to the belt above the back of one hip. Tools should not be carried in the hands when climbing a ladder, but raised from the ground and lowered in their bag or box by a cord. Wood chisels, knives and other sharp and pointed tools should be kept in special racks or clips above the back of the workbench.

Tools should never be thrown but passed from one man to another with the handle offered, except in the case of axes or hammers when the head is offered. Special care must be taken when carrying large tools on the shoulder to keep well clear of other workers, door posts and lintels, and other obstructions.

6.6.3 Inspection and maintenance of hand tools

Managers of shops and departments where hand tools are used should be responsible for their regular inspection, and have a proper system for withdrawing defective tools from use for repair or scrapping. Proper facilities (including a furnace and baths for hardening and tempering) and experienced personnel must be available if hand tools are to be repaired successfully in the works. Otherwise they should be sent out for repair to a firm with proper facilities.

The heads of hammer-struck tools such as chisels, punches and wedges, require careful heat treatment to ensure they are hard enough to avoid splaying, yet not so hard that they chip. The edges of heads that start to mushroom should be promptly ground off and radiused to about 3 mm. The cutting edges or striking surfaces of hammer struck tools must be kept keen and sharp and in proper temper. Metal cutting tools can generally be sharpened on an emery wheel, taking care to avoid overheating.

Screwdriver tips and axes are best shaped or sharpened with a file to avoid drawing the temper, first securing the screwdriver or axe in a vice. Wood cutting chisels and plane irons are best shaped on a wet grindstone using plenty of water, and whetted on an oil stone mounted securely on a bench to produce a sharp cutting edge. The stone must not be held in one hand and the chisel in the other.

Handles of hammers, axes and mallets should be of straight grained wood, perferably hickory, free from splinters. These invariably shrink and work loose in time, and must be tightened by an experienced man. Plastic or wood handles of screwdrivers, chisels and files must be kept tight and any cracked handles replaced.

6.6.4 Hazards of metal working tools

Many hand injuries occur when using hammer struck tools such as *cold chisels* and *punches*, and eye injuries from flying chips also occur. Hand injuries can be reduced by using a holder, or by fitting a sponge rubber ring over the head of the tool above the left hand which is holding it (*Figure 6.6.2*). Eye protection is needed when chipping, and nearby workers should be protected by a screen.

The work must be held securely in a vice when *taps and dies* are used, and the proper tap wrench must be used — never an adjustable spanner. Freshly cut threads can be sharp and may cut hands and arms.

Figure 6.6.2   Typical sponge-rubber hand protector for hammer-struck tools. Combination hand grips and shields are available for some hammer-struck tools (National Safety Counci, U.S.A.).

The correct blade for the job should be used with *hacksaws* — soft metal requiring a coarse pitch and thin sheet metal and tubing a fine one. Blades must be correctly tensioned.

Files must always be used with a handle which is firmly held in one hand while the point of the file is guided by the thumb and forefinger of the other. Special care is needed when using a file on work rotating in a lathe. The file must never be held in a cloth or rag, which can get caught and wound in. A firm handle must be used, if necessary an offset one, and the hands and arms must be kept well clear of the chuck.

Files are brittle and must not be hammered or used as levers, or struck against a hard object in an attempt to clean them. Special cleaning cards should be available and used for file cleaning.

The main hazard of *tin-snips* is cutting a hand on a sharp metal edge. Gloves should be worn and the work well supported, especially at the end of a cut. The main hazard of *wire, rod and bolt cutters* is an offcut flying and hitting the worker in the eye. Eye protection should be worn. Tin snips and cutters must be heavy enough for the work in hand and the jaws kept tight and well lubricated. It should be possible to use them without strain or rocking with one hand only, with the other free to help support the work. Knives and cutter jaws must have the right hardness, and care is needed to prevent them getting nicked. Snips and cutters must not be used as levers. The work must be properly supported or held in a vice. Handles should be checked that they cannot trap and pinch the fleshy part of the hand when they close suddenly. Modification may be needed to prevent this.

6.6.5 Hazards of woodworking tools

Injuries are sometimes caused by splinters from splayed wooden handles of *wood chisels* or by a handle breaking in use. Handles must be sound and well fixed, and the end should be protected by a metal band from splaying under a mallet blow. The work should be free of nails and the cutting edge of the chisel kept sharp and in good condition. The work should always be held in a vice or jig when a chisel is used and not by hand due to the danger of injury to the hand caused by the work or the chisel slipping.

The most common injuries from *wood saws* are caused by the saw teeth to a hand holding the work when the work or saw slips. Bruised knuckles to the hand holding the saw occur when the saw sticks then moves suddenly when pressure is increased. Splinter wounds are common.

Wood saws should be kept sharp and the teeth well set, and the correct saw always selected for the job. Work must be well supported and securely held. When not in use saws should be hung in a safe place, and the teeth covered if there is any danger of contact with parts of the body.

Injuries from *axes, adzes and hatchets* can occur from several causes — a loose head flying off and hitting someone, a chip hitting a worker, or the swing being deflected by some obstacle, causing the cutting edge to hit a foot or leg. They should only be used by trained workers wearing safety shoes and shin guards. Blades should be kept sharp and covered when not in use. Handles must be securely fixed. Workers should take care they have sufficient room in all directions to swing these tools safely.

Injuries through the use of *wood planes* generally arise when the work is not securely held and the work, plane or worker slip or sometimes even fall. Work must be adequately fixed when planing. When not in use a plane should be stowed in such a way that the cutting edge is not damaged. Planes must be kept sharp and properly adjusted.

Knives are a frequent cause of injury, often through faulty use by cutting towards the body instead of away from it. If the nature of the work makes it essential to cut towards the body suitable body protection should be worn. Knives should always be sheathed or put in a safe rack when not in use — not left with other tools or covered with odds and ends. Knives should be wiped clean with a cloth with the sharp edge pointed away from the hand holding the cloth. Knives should only be used for cutting when the object can be cut without much pressure — other cutting tools should be used for tough materials. Knives should not be used as screwdrivers or tin openers, and care must be taken when knives are used to open cartons not to damage the contents (e.g. plastic containers).

If a *bradawl* is started with its edge in line with the grain, the wood frequently splits. It must enter the wood at right angles to the grain, and be turned as it is pushed in.

6.6.6 Hazards of torsion tools

Screwdrivers are the most abused of tools and suffer broken handles, bent shafts and worn or twisted tips through being used for the wrong purpose or through attempts to use one of the wrong size. Only screwdrivers with insulating handles should be used for electrical work.

The part to be worked on should not be held in the other hand, but supported on a firm surface or clamped. The tip of a screwdriver should be filed with a square end and parallel sides to fit the screw head. Several screwdrivers of different sizes should be at hand so that the right one for the job can be selected. Cross slot screws and screwdrivers should be used wherever possible.

Many injuries are caused by slipping *spanners*, through using the wrong size spanner or open-ended spanners that have splayed apart or through packing a gap with shims that slip out. Spanners should be placed on nuts and bolt heads so that the handle is pulled towards the body, making sure that when it is used the hand will not hit an obstruction. Box spanners and ring spanners are preferred to open ended spanners, and an adjustable spanner should only be used if a rigid spanner of the right size is not available. Extension handles should not be used. Where a frozen nut must be loosened, a striking face spanner should be used after applying penetrating oil to the end of the thread. Where the tension on a bolt or joint is critical, a torque wrench should be used.

Figure 6.6.3 shows the correct use of a wrench. The wrench should be tightly gripped and, in order to protect the hand, it should be pulled and not pushed.

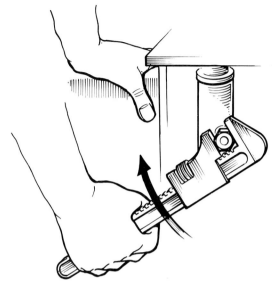

Figure 6.6.3   Showing the correct use of a wrench

6.6.7 Hazards of miscellaneous hand tools

Makeshift bars should not be used to prise joined timber apart or to lever up the edge of a heavy object, as they are liable to slip or bend. A proper *crowbar* of the right size and shape with a sharp toe should be used.

Injuries from *rakes* often occur when someone steps on the prongs of a rake which have been left pointing upwards. A rake should be left on a rack when not in use.

Foot injuries often occur when digging with unsuitable shoes or if the arch of the foot is used to try to force the blade into hard ground. Strong shoes or boots should be used, and the ball of the foot used for pressing the shoulder of the spade or shovel. *Spades, shovels* and *forks* should be examined before use to ensure they are in good condition, and the blades and prongs kept clean.

Tongs used for holding pipes and hot objects should be checked that they do not pinch the hand when closed. If necessary the end of one handle may be turned through 90 degrees to form a stop.

Injuries often occur through misuse of *pliers* as wrenches, when they slip. Pliers should only be used for gripping and cutting, and the handles of electrician's pliers must be insulated.

Spark resistant tools are needed where highly flammable vapours may be present.

6.6.8 Hazards of portable power tools

Injuries have occurred through failure to disconnect the source of power when changing an accessory, and through failing to replace a guard properly. Foot injuries are caused through dropping power tools for various reasons — shock, insecure hold or precarious position, and body injuries are caused through a fast moving tool coming into contact with it. Only power tools which have been approved for safety should be purchased and used, and those using them should be properly trained in their use to appreciate their limitations and to recognise faults and malfunctions.

The electrical hazards of portable electric power tools are discussed in section 5.7, and the explosion hazards of cartridge operated tools in section 4.2.

Portable power tools need frequent inspection and regular maintenance and cleaning. Each tool should be numbered and a maintenance record card kept for it. Portable power saws and grinding wheels have the same hazards as fixed tools — discussed in sections 6.5 and 6.7 — but in a more acute form, since they are hand held and more difficult to guard.

Skin burns are a common hazard of *disc sanders*. Operators should be thoroughly trained in their use. The sanding motion should be away from the body, and no loose clothing may be worn. A vacuum dust collector should be used with them if possible, and dust formed frequently removed. The nature of the dust formed should be checked before starting, and a suitable respirator or dust mask used if the dust is toxic.

Burns and sickness through inhaling lead fumes are the commonest hazards of *soldering irons*. Holders should be insulated and non-combustible, and they should be designed to protect workers against accidentally picking up the hot bit of the iron.

Accidents with *air-powered portable tools* include:

1. Tripping over air hose.
2. Hose becoming detached from the tool and whipping about (the end of the hose may be secured to the tool housing with a short chain),
3. The air hose being disconnected and used for cleaning, causing dust hazard and eye injuries. This should not be allowed; only brushes or vacuum cleaning should be used.
4. Air motor overspeeding; a well maintained governor is required.

Pneumatic rivetting guns and hammers cause noise and vibration hazards described in sections 3.4 and 3.5. They should have a trigger-operated valve on the air supply line so that the tool only operates when the trigger is depressed, and a retaining device which prevents the tool being discharged accidentally. Flying chips are also a hazard, and eye protection and screens are required.

REFERENCES

1. National Safety Council, *Accident prevention manual for industrial operations*, 7th ed., NSC, Chicago, 977-1005 (1974).
2. Creber, F.L., *Safety for industry*, RoSPA, 56, et seq (1967).

6.7 PRESS AND MACHINE TOOL HAZARDS

Contents

A wide variety of different operations and equipment are used in the shaping of metals. It is not possible here to deal with the hazards of the more specialised operations, even when they are practised on a very large scale. These include rolling mills producing bars, rods, coils, sections, rails and plates, the drawing of wire and tube, hot forging, swaging, extrusion and powder compacting. Some of the general hazards and the principal methods of machine guarding have been discussed briefly in section 5.5.

Presses, particularly power presses, are discussed in this section because of the large number of injuries which result from their use. Machine tools which are, in effect, cutting tools are included because their use is so widespread throughout industry.

There are four types of widely used power presses — the mechanical power press, the hydraulic power press, the power press brake and the power shear. Hand and foot operated presses are also still in common use and are dealt with in this section rather than under the heading of 'hand tools.'

The main types of machine tools are lathes, turret lathes, screw, boring, drilling, reaming, threading, milling and gear cutting machines, planers and shapers, broaching, cutting-off, grinding and polishing machines.

6.7.1 Presses

Power presses are one of the few types of machine to be subject to specific regulations over and above those sections in the Factories Act 1961 which deal with dangerous machinery.[1] These are the Power Presses Regulations 1965[2] (SI 1965 No. 1441, as amended by SI 1972 No. 1512).

It should be noted that these regulations only apply to mechanical power presses or power press brakes which have a fly wheel and clutch mechanism. This is clear from paragraph (2) of regulation (2) which states:

'Power press' means a press or press brake which in either case is used wholly or partly for the working of metal by means of tools or for the purpose of die proving, being a press or brake which is power driven and which embodies a flywheel and a clutch mechanism.'

Hydraulic and pneumatic presses are thus excluded (unless they contain a clutch or flywheel), as are hand and foot presses. The regulations also exclude power presses used for working hot metal. Certain other types of mechanical power presses, described below, are also specifically excluded from the regulations:

1. Guillotines
2. Combination punching and shearing machines, turret punch presses, and similar machines where in any case the machine is constructed and used solely for one or more of the following operations, that is to say, punching, shearing or cropping.
3. Machines, other than press brakes, designed and used solely for bending steel sections.
4. Straightening machines
5. Upsetting machines
6. Heading machines
7. Riveting machines
8. Eyeleting machines
9. Press-stud machines
10. Zip fastener bottom stop attaching machines
11. Stapling machines
12. Wire stitching machines

The regulations which are quite brief cover the following main points:

Regulation 4. Appointment of persons to prepare power presses for use. This in effect says that only a trained competent person who has been specially appointed by the factory occupier and who has reached the age of eighteen may install, set, re-set or adjust the tools or safety devices on a press covered by the Regulations, or inspect or test any safety device required by Regulation 7.

Regulation 5. Examination and testing of power presses and safety devices. No power press covered by the Regulations may be used unless all the required safety devices have been fitted and both the press and the safety devices have been inspected and tested in accordance with Regulation 4.

Regulation 6. Defects disclosed during a thorough examination and test. If the examination referred to in Regulation (5) has revealed any defect which is an immediate danger to people using it, the press may not be used until the defect has been remedied. If the examination reveals a defect which may become a danger to people using it, the defect must be remedied within a specified period, and if this is not done within the period, the press must be taken out of service.

Regulation 7. Inspection and test of safety devices. The safety devices fitted to power presses covered by these Regulations must be tested as prescribed in the previous Regulations every day (or shift) during which the press is in use during the first four hours of the day or shift, and a signed test certificate for the safety devices must be issued showing the date and time, listing the safety devices tested and giving sufficient information to identify the press and safety devices tested.

Regulation 8. Defects disclosed during an inspection and test. This reiterates Regulation 6.

Regulation 9. Identification of power presses and safety devices. This requires every power press and safety device covered by the Regulations to be distinctly and plainly marked.

Regulation 10. Indication of speed and direction of flywheel. The maximum speed and direction of rotation of the flywheel are to be conspicuously marked on every press, and the flywheel must only be driven in that direction (except in an emergency when it may be reversed), and not above the maximum speed.

Regulation 11. Preservation of registers. The registers in which the results of the inspections are tested and recorded must be kept in the factory available for inspection by an inspector for at least two years after the date of the last entry.

Regulation 12. Exemption certificates. The Chief Inspector of Factories may at his discretion exempt any power press or safety device from these Regulations if he is satisfied that nobody is thereby endangered. A copy of the exemption certificate is to be kept posted in the factory where it may be conveniently read by employees.

The Regulations include brief particulars of the training required of the persons appointed under Regulation 4. They emphasise certain hazards of power presses and the need for safety devices. Their main aim is to prevent serious injury to fingers and hands through placing them between the tool and the die when the tool is about to enter the die.

HM Factory Inspectorate have issued five test cards, HSW 15, 16, 17, 18 and 23 which detail the minimum safety tests which must be carried out daily for various types of safety device[3]:

HSW 15 is for interlocked guards on positive-clutch presses.
HSW 16 is for automatic guards.
HSW 17 is for fixed guards.
HSW 18 is for photo-electric safety devices on press brakes.
HSW 23 is for interlocked guards on friction clutch presses.

Having outlined the legal requirements of these Regulations, the various types of press and some of their hazards are next discussed. Besides posing the hazard of involuntary amputation, many power presses are very noisy and present the further hazard of involuntary loss of hearing. Other hazards are those common to the handling of heavy, sharp and sometimes hot objects.

6.7.2 Hand and foot presses

Hand and foot presses, despite mechanisation, are still used for various blanking, forming and other operations, mostly on metal, but also on leather, plastics and pastry. These, according to an official booklet[4] were causing 650 reportable injuries per year for hand presses and 200 reportable injuries per year for foot presses in the UK in the early '70s.

Many of these press operations require the operator to place one or both hands near the tool and die at some point in the operating cycle (i.e. to place work in the press or remove it); although the operator is normally able to control his movements so that he does not trap his fingers, there are several reasons why this is not always so:

The ram may descend when the operating handle is released because of slackness in the guides and insufficient frictional resistance.
The operator may be jostled by a passer-by.
The operator's seat may slip or wobble.
The operator's attention may be distracted.
The work may jam in the die, causing an upset in the rhythm of work.

The booklet explains and illustrates a number of safety devices for hand- and foot-operated presses (see *Figure 6.7.1*) which are based on one or more of five methods:

1. Closed tools, fixed guards and restricted stroke.
2. Interlocked guards.
3. Pendulum and sweep-away devices.
4. Automatic devices which arrest the ram.
5. Two hand control.

This is not to suggest that all five methods are equally effective. Sweep away devices which are activated by the press with the object of sweeping the operator's hand out of the way of the tool or ram are deemed ineffective and at times hazardous by American experts.[5] Their action is either too violent or too late, and in many cases they can be avoided.

In addition to providing guards, the following points should also be checked:

1. The press is securely mounted on a firm bench or stand so that neither the press nor stand move when the handle or pedal is turned with the maximum force employed.
2. Presses must be spaced far enough apart or protected by screens to prevent other workers coming into accidental contact with moving parts.
3. Presses must be regularly inspected and maintained, and the handle in particular should be examined for cracks and defects.
4. The press should be so adjusted that the ram cannot descend through the action of gravity.
5. Presses should be so placed and operators so selected that the

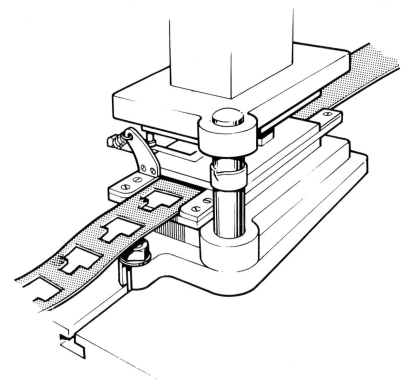

Figure 6.7.1   Closed tools for a strip feed blanking operation. The ram stroke is such that the upper tool does not emerge from the lower one. The location of each blank is facilitated by a trigger stop arrangement which ensures equal spacing between the blanks and so avoids waste (H. & S. W. Booklet 3)

operator never has to raise his hands above head level to hold the handle.
6. Press operators, particularly youngsters, should be carefully trained and supervised.

6.7.3 Power presses

The hazards of mechanical power presses are discussed in great detail in HSE booklet No. 14[6]. Hydraulic and pneumatic presses whose injury potential is quite as great have received less attention in official UK publications.

Safety aspects of other press operations are discussed in several H & SE Technical Data Notes.[7]

No. 30 deals with the hazards of rotating-table pie and tart machines, which are mechanical presses of low power yet sufficient to provide the consumer with an occasional protein bonus in the form of a finger tip of an incautious operator.

No. 33 deals with overrun and fall back devices. It was prepared to help to identify and locate these devices on presses made by different manufacturers when checks are made to see that they are fitted.

No. 50 deals with the hazards of riveting and eyeletting machines, usually clutch operated mechanical power presses, though exempted from the Power Press Regulations 1965.

The technical features, hazards and protection required for a wide range of power presses are broadly described in a US publication.[5] The four main types of power press have been referred to earlier, and some of their features and hazards are now discussed briefly.

6.7.3.1 Mechanical power presses

There are two basic types:

The C type (from its shape) with a gap frame.
The H type with a straight side frame.

C type presses are generally lighter.

Whilst mechanisation of handling operations has reduced injuries caused by mechanical power presses. They still arise on machines which are manually loaded and unloaded. The clutches of these machines are of two types, the full revolution clutch which can only be disengaged at one point in the press cycle, and the part revolution clutch which can be disengaged at any point in the cycle. The latter are naturally generally safer to operate.

Some full revolution clutches for single-piece production with manual die loading are hazardous as they are subject to repeat strokes without warning, and to unintended or premature strokes. It is now compulsory in the USA for such machines to incorporate a single stroke device with an anti-repeat mechanism.

6.7.3.2 Hydraulic and pneumatic presses

These have similar operating characteristics to the part revolution clutch press, and can be stopped by cutting off the power supply. One danger is the sudden dropping of the slide and tool caused by power, mechanical or electrical failure.

Another danger is that residual hydraulic or pneumatic pressure may be accidentally locked in the machine when a power valve is closed. This may cause involuntary and unexpected movement of the machine later.

6.7.3.3 Press brakes

A press brake may be mechanically or hydraulically operated. It differs from an ordinary power press in having a long narrow ram and bed set in front of the gapped frame, and between which various dies can be placed. Press brakes can be started and stopped at will, and are mainly used to form or bend pieces of sheet metal — often on a 'one off' basis.

The variety of jobs which may be carried out on a press brake, each with sheet metal of different shape and size, makes guarding quite difficult, since a guarding system suited to one job may make it impossible to carry out the next one. Two or three alternate guarding systems, which can if necessary be used in combination, may thus be required.

6.7.3.4 Power shears

These are available with mechanical power with either a full or part revolution clutch brake drive or with hydraulic or pneumatic drives. They are used only for shearing metal. Operators are usually protected by fixed guards designed to admit the work to be sheared whilst preventing the operator's fingers coming near the point of operation.[8]

A type of shear known as the alligator which chops rather than cuts and works continuously, is used in salvage yards. It has caused many injuries through difficulties in guarding and lack of guards and operators tend to underestimate its danger because of its low operating speed. The action of shearing often causes the other end of the stock (which an operator may be trying to hold) to fly up and hit him. Some form of retaining or hold down bar is needed to prevent this.

The guarding of power presses is a complex and highly technical subject, and success depends largely on choosing the most appropriate method for the job and in attention to detail. Much useful information is given in the references quoted. Further expert help on particular problems is available through the Health and Safety Executive (Dept F1/7C) 25 Chapel Street, London, N.W.1 5DT.

6.7.3.5 Noise in press rooms

Noise levels in many press rooms are continuously above 90 dBA. Noise originates partly from the drive mechanism and partly from the point of operation, impact of the tool or upper die against the work. Methods which may reduce the noise include:

Design of dies and tools to give a shearing rather than a punching action.
Staggering the entry of the punch points.
Use of polyurethane cushions in place of springs for cushioning.
Use of solid rather than perforated point of operation fixed guards.
Use of mechanical rather than pneumatic ejection from die, or close attention to design of pneumatic system to reduce noise.

6.7.4 Machine tools

Machine tools generally fulfil one or more of five basic functions — turning, boring, milling, planing and grinding. Grinding is dealt with in section 6.5. Most machine tool operations depend on cutting, although spinning, which is performed on a lathe, is a forming operation.

Fewer injuries are reported from machine tool operations than from power presses; nevertheless they constitute a serious industrial safety problem. Most of these injuries result from unsafe practices, faulty procedures, use of defective equipment, absence of guards, poor lighting and poor housekeeping. Most injuries could have been avoided by better training, supervision and guarding. Machine tools are subject to the general safety provisions given in sections 12-16 of the Factories Act 1961 which relate particularly to fencing.[1] Section 21 of the same Act incorporates 'The Dangerous Machines (Training of Young Persons) Order 1954, which among the prescribed machines includes 'Milling machines in use in the metal trades'.

The hazards of milling machines are further emphasised by 'The Horizontal Milling Machines Regulations 1928'[9] which are discussed under section 6.7.4.4.

General rules which apply to the safety of all machine tools are briefly as follows:[10]

1. Only experienced trained personnel should operate, adjust or repair machine tools.
2. A safe working procedure should be established for every operation.
3. Supervisors must enforce points (1) and (2).
4. Careful attention must be given to the hazards of every machine tool when it is purchased.
5. New equipment must be thoroughly examined and checked before being used by operators.
6. A tool rack with all tools needed for operation and adjustment should be provided with every machine tool. Only the proper tools should be used.
7. Every machine tool must have a power isolating switch which can be locked in the 'off' position.
8. Maintenance should only be allowed when the isolating switch has been locked and tagged in the 'off' position.
9. Machine tools must not be left running unattended.
10. Operators must be safely dressed and should wear eye protection and safety shoes.
11. Where circulating coolant is employed, care must be taken not to contaminate it with sputum and rubbish.
12. Machines must be stopped before adjusting or gauging work with calipers.
13. Swarf and chips must not be removed by handtools, brushes or vacuum equipment should be used.
14. The use of compressed air for cleaning is best avoided, but if this is impossible, it should be carefully controlled, using only low pressure

air jets designed for safe operation. Compressed air should not be applied to operator's clothing or hair to remove particles.

Specific hazards relating to different types of machine tool are discussed below.

6.7.4.1 Lathes

Contact with projections on the rotating work, facepieces and chucks are responsible for most injuries. A chuck key left in a chuck when the machine is started is a common hazard. Safety chuck keys which can only be held in a chuck are available (*Figure 6.7.2*).

When a file has to be used on work in a lathe, it is recommended that the handle be held in the left hand and the thumb and forefinger of the right hand used to steady the tip. This reduces the risk of contact between the left arm and the rotating chuck.

Every lathe should have a safe and effective means of braking. Braking by hand should not be allowed.

Wiping rags should not be applied by hand to revolving parts. Long continuous turnings which may cause hand and arm injuries should be prevented by a chip breaker. This may take the form of a small step ground into the tool, or a seperate piece of metal securely attached to it.

Operations on turret lathes should be carefully studied for hazards and safe procedures devised to prevent injuries. Proper devices should be provided to lift and hold heavy face pieces, chucks and stock while these are being attached to or removed from the lathe.

Spinning lathe operations need special study to eliminate hazards such as the swarf from trimming work forming a long coil of sharp edged ribbon. Safe work procedures must be established and supervised, and chucks and tools regularly inspected and maintained.

6.7.4.2 Drills

Injuries commonly result from contact with the spindle or drill, breaking a drill, being struck by insecurely clamped work, failure to replace pulley or gear guards, and many of the hazards noted earlier with lathes.

Health and Safety at Work Booklet 20 *Drilling machines, guarding of spindles and attachments*, describes a number of guards and tripping devices in detail, with numerous illustrations which include safety chuck keys which fall out of the chuck unless held in position (see *Figure 6.7.2*).

Tripping devices which stop a drill in an emergency (such as when an operator's clothing is caught by a revolving drill) are described in Health and Safety Executive's Technical Data Note 38,[11] *Tripping devices for radial and heavy vertical drilling machines*. This contains lists of suppliers of tripping devices, braked motors and of drilling machines which incorporate these features.

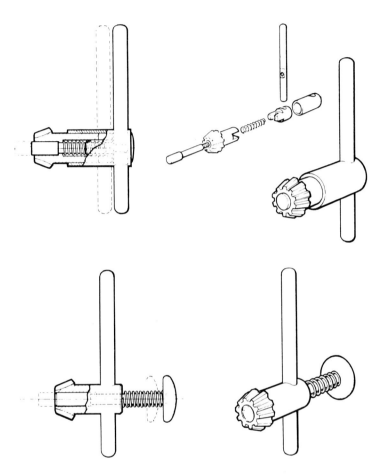

Figure 6.7.2   Spring-loaded safety key for chuck which releases itself when not being held in position. A chuck key is normally located by means of a peg which continues to hold the key in the chuck when the key is released by the operator. These keys are spring-loaded so that the peg extends to the position shown in broken lines only when the operator depresses the end of the key. When pressure is released the peg is withdrawn by the spring into the body of the key, thus removing the support. If the operator fails to remove the key it will fall from the chuck. The upper illustration shows a design in which the spring is enclosed within the body of the key. A simpler design is shown in the lower illustration (H. & S. W. Booklet 20)

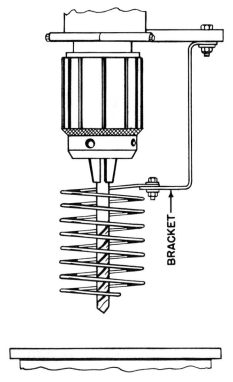

Figure 6.7.3   Spring safety guard which compresses as drill cuts into metal in order to contain metal slivers and chips (National Safety Council, U.S.A.)

6.7.4.3 Borers

Borers are subject to many of the same hazards as lathes and drills. There is also the danger with some borers of falling against revolving work.

Each machine should be carefully studied to assess its individual hazards, safe procedures worked out, and any special guards or other protection needed designed, procured and fitted.

6.7.4.4 Milling machines

Milling machines include a variety of machines where metal is brought into contact with a rotating multi-edged cutter. They include horizontal and vertical milling machines, gear hobbers, profiling machines, circular and band saws. Most accidents occur when the machines are being unloaded or adjusted.

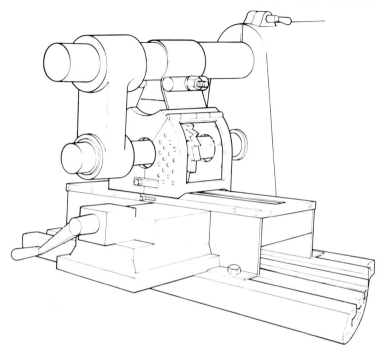

Figure 6.7.4 Fixed guard and false table for horizontal milling machine (H. & S. W. Booklet 42)

The Horizontal Milling Machines Regulations 1928[12] are summarised below.

Regulation 1. The floor around every machine must be kept in good, even and non-slippery condition and clear of loose material.

Regulation 2. The machines must be well lit and the lights so placed that they do not shine into the eyes of the operator.

Regulation 3. The cutters of every machine must be properly fenced.

Regulation 4. Every machine must have an efficient stopping and starting appliance placed where it can be readily used by the machine operator.

Regulation 5. Where cutting lubricants are used, suitable means must be provided to apply the lubricant and remove the swarf.

Regulation 6. Guards must be maintained in an efficient state and always in position when the milling cutter is in motion except when the tool setter is setting up the machines.

There are a number of exemptions to Regulation 3 when milling cutters are used for particular purposes, i.e. for making tools or jigs for use in the factory, for internal milling, for end milling other than face milling, for automatic gear cutting, for automatic hobbing, for automatic profiling and for thread milling.

Health and Safety at Work Booklet 42 *Guarding of cutters of horizontal milling machines*[13] gives a useful analysis of accidents with contributory factors and explains various methods of guarding with copious illustrations. About two-thirds of accidents with these machines occur when removing, inserting or adjusting the workpiece; 15% occur when removing swarf; accidents also occur when adjusting or cleaning the coolant pipe, when cleaning with a rag, during general work when a worker slips, and during gauging.

Special contributory factors to these accidents are:

Cutter not stopped.
Job not withdrawn to safe distance.
Shape of work holder prevents close adjustment of guard.
Closing device for work holder in dangerous position.
Exposure of cutter below guard.
Suitable cleaning brush or appliance not available or not used.
No facilities provided to clear or adjust coolant pipe safely.
Rag used near moving machinery.
Floor in bad condition.

General contributory factors include:

Guarding irregularities (no guard or guard ineffective).
Loose clothing.
Bad lighting (including glare).
Piecework incentives undermining safety precautions.

Most accidents with horizontal milling machines occur with repetition jobs for which there is no excuse for failing to provide proper guarding and protection. Often a single guard is insufficient to prevent finger access to all parts of the cutter, then additional features such as a false table should be provided to ensure that the cutter is always inaccessible. The false table is usually an extension of the vice or fixture which covers the lower part of the fixture when the cutter is withdrawn (see *Figure 6.7.4*).

Further discussion of different types of milling machines and their hazards is given by the National Safety Council.

6.7.4.5 Planing machines

Planers employ a stationary cutting tool above the work which is moved underneath it on a sliding table. Shapers, also classed as planing machines, employ a moving cutting tool while the work is held stationary. Common causes of injury are:

Hand or fingers caught between the tool and the work.
Bare hand cut on sharp metal edge.
Measuring work while machine is running.
Work or tool insecurely clamped.
Insufficient clearance for work or tool, which may project beyond table or guard.

Planers or shapers are generally guarded by rails, but since these cannot completely prevent operators reaching moving parts and nip points, careful training and supervision of operators is required.

6.7.4.6 Job hazard analysis

It is impossible in this short survey to discuss all machine tools and their hazards. Safety specialists are recommended to study in detail the operation and maintenance of machine tools for whose safety they are responsible in detail with the operators and supervisors concerned. They should also prepare a written job hazard analysis for each tool and type of operation, as discussed in section 2.9.

REFERENCES

1. Fife, I., and Machin, E.A., *Redgrave's health and safety in factories*, Butterworths London, 65,83,84 (1976).
2. Reference 1 pages 1373 to 1383.
3. HM Factory Inspectorate, Power presses — minimum safety tests, HSW 15, 16, 17, 18 and 23, HMSO, London.
4. Health and Safety at Work Booklet 3, *Safety devices for hand and foot operated presses*, HMSO, London.
5. National Safety Council, *Accident prevention manual for industrial operations*, 7th ed., NSC, Chicago, 857-896 (1974).
6. Health and Safety at Work Booklet 14, *Safety in the use of mechanical power presses*, HMSO, London.
7. Health and Safety Executive, Technical Data Notes:
   30. *Safety in the use of rotating table pie and tart machines.*
   33. *Power press mechanisms; overrun and fall back devices.*
   50. *Guarding of riveting and eyeleting machines.*
8. Health and Safety at Work Booklet 33, *Safety in the use of guillotines and shears*, HMSO, London.
9. Reference 1, pages 1373 to 1383.
10. Reference 5, pages 828 et seq.
11. Health and Safety Executive, Technical Data Note 38, *Tripping devices for radial and heavy vertical drilling machines*, H & SE, London.
12. Reference 1, pages 1093 to 1096.
13. Health and Safety at Work Booklet 42, *Guarding of cutters of horizontal milling machines*, HMSO, London.

6.8 WOODWORKING MACHINERY

Contents

Many injuries arise through the use of woodworking machines. In 1970 it was stated[1] that approximately 3000 accidents in the UK were being reported annually to HMFI, these being apportioned as follows:

Circular saws of various types	40%
Planing machines	25%
Vertical spindle moulding machines	11%
Band sawing machines	6%

Woodworking is done in three ways, either entirely manually with hand tools or mechanically by the use of hand-fed machines with powered tools, or fully automatically by machines which feed the work to the tool and remove it after machining. It is with the second type of operation that most accidents occur; this is understandable as workers are here most exposed to injury by powerful fast moving tools. A high proportion of the accidents arise through human contact with sharp cutters moving at high speed.

Wood, being soft, is presented by hand to the cutters at a high feed rate. Wood varies greatly in texture because of knots, twists in the grain, etc which may cause the material to jump or twist when it meets the cutters. These variations in the material are often hidden in the timber, so that the machinist is unable to adjust the movement of the hands and the timber to allow for the changes. It is also difficult to guard hand fed woodworking machinery fully and effectively when the cross section of the timber has similar dimensions to those of fingers, arms or hands.

The use of Woodworking Machinery in the UK is covered by the quite extensive Woodworking Machines Regulations, 1974.[2]

These stem from earlier regulations dating from 1922 and revised in 1927, 1945 and again in 1974. The changes and amendments made in 1974 are explained in detail by J.A. Hayward.[3]

The 1974 regulations are arranged in ten parts:

Part I. Application, interpretation and exemptions.
Part II. All woodworking machines — general.

Part III. Circular sawing machines.
Part IV. Multiple rip-sawing machines and straight-line edging machines.
Part V. Narrow band sawing machines.
Part VI. Planing machines.
Part VII. Vertical spindle moulding machines.
Part VIII. Extraction equipment and maintenance.
Part IX. Lighting.
Part X. Noise.

Many of the machines covered in these regulations (particularly Parts IV, V, VI and VII) are of rather specialised concern to factories which produce wooden articles. But some machines, especially circular saws are more widely used, e.g. in maintenance shops or packaging departments of many factories. Since a high level of training is itself required under these regulations for all people involved with woodmaking machinery, it is unnecessary to describe here in detail hazards with the more specialised machinery. Rather we deal with the general hazards of woodworking machines, and discuss in detail only those machines which are widely used throughout industry.

6.8.1 General hazards and precautions required[1],[4]

The following paragraphs deal with guards, electrical precautions and the protection of operatives.

6.8.1.1 Machine selection, design, construction and guarding

Woodworking machines should be carefully selected for the job, and should not be employed on work for which they were not intended.

All moving parts, belts, shafts, gears, etc should be fully enclosed. Guards at points of operation must be movable or adjustable to accommodate the wood, strong enough to protect the operator and positioned so that they keep his hands away from the point of operation without impeding the work.

Adequate jigs and pusher bars should be provided and used to avoid the need for operators to risk their fingers. Machines should be so designed that it is impossible to mount in them tools larger than the largest for which they were designed.

Bearings for spindles must be securely housed and there should be no play in the spindles. Woodworking machines should run at full and idle speeds with a minimum of noise and vibration.

The working surfaces of all woodworking machines should be at a height most convenient for operators, and all accessories and feed tables should be at the same height.

6.8.1.2 Electrical features

All metal frameworks as well as the motor should be earthed (grounded). Every machine should have its own isolation switch which can be locked in the 'off' position for maintenance inspection, etc. Every machine should have at least one conveniently located stop switch, and an emergency foot stop switch if there is any doubt whether the hand operated stop switch could be used in an emergency. Every machine should have a manual reset device which must be reset after the machine has stopped, before it can be restarted.

6.8.1.3 Layout, installation and services

The layout of operations involving woodworking machines should be planned to give a smooth flow of materials without crossing. There must be ample working space round each machine which does not encroach on the aisles.

Every machine should be securely fixed to the floor. Floors should be free from holes and irregularities, and the working area should have a non-slip surface. Machines and their layout should be studied and planned to require the minimum of movement by the operator.

Provision should be made for the removal of sawdust and scrap, and a system provided to keep the working surface clear of scrap. An automatic vacuum system is preferred. The point of operation of all woodworking machinery must be very adequately illuminated and operators protected against glare from any source or reflecting surface.

There is a legal requirement to maintain the temperature above 13°C in any building where woodworking machines are used, except parts of factories used as sawmills where the minimum temperature is 10°C.

6.8.1.4 Machine operators, inspections and protection

All operators of woodworking machines should be thoroughly trained not merely to operate them but also to inspect them. Before starting work on a machine, the operator should check the sharpness of the cutting edges, the safety devices, the power drives and operating controls. When using a machine operators should be able to recognise incipient trouble by a strange noise or change in note, and be trained in what action to take to avoid more serious consequences.

Every woodworking machine should be regularly inspected by a competent person and the results entered in a register. The inspector should have the authority to bar the use of any machine which he judges to be in a hazardous condition.

The noise level to which operators of woodworking machines are exposed should be monitored. Every effort should be made to reduce these levels to below 85 to 90 dBA. For exposure at these levels ear protectors should be provided and their use made obligatory for continuous exposure above 90 dBA (see section 4.2).

Dust entering the atmosphere from woodworking machinery should be checked for concentration and health hazard, and dust masks or respirators provided where necessary. Safety specialists should, in particular, be on the alert for dust from timber treated with toxic preservatives. These include compounds of copper, arsenic, chromium, creosote and chlorophenols. The use of woodworking machinery on asbestos board should be avoided if possible and special precautions including the use of resperators suitable for asbestos dust should be taken should it ever become necessary.

All operators of woodworking machinery and persons working near them should wear safety glasses or eye shields, and close fitting clothing with no loose ribbons, sleeves or jewelry which could become entangled with the machine. Those with long hair should wear hair nets or caps. Gloves may be worn only where there is no chance of their being caught by machinery or rotating stock.

Safety shoes should be worn when handling heavy material and when there is any danger of foot injury.

6.8.2 Circular saws

Any machine which uses a circular blade to cut wood may be termed a circular saw. Circular saws are usually bench mounted and are designed for ripping, deep cutting or cross cutting — sometimes all three as in the general-purpose circular saw. Less commonly cross cutting circular saws are mounted above the bench and moved towards the wood.

The saws discussed here are primarily those where the saw blade is mounted on a spindle below the table. With some of these the position of the spindle is fixed in relation to the table, so that the projection of any blade above it is fixed, but with others either the table or the saw spindle may be raised and lowered by a screw mechanism. Most circular saws allow blades with different teeth to be fitted to suit the type of work, the wood and the required finish of the cut surfaces. Some small machines allow work to be cut at an angle by means of a canting table, a canting fence or a canting saw spindle.

Fixed circular saws are nearly always driven by an electric motor, but mobile saws used on building sites are often engine driven.

Hand held circular saws are built in a range of sizes with blades up to 1200 mm in diameter. Special mechanically-fed saws are used for converting logs into planks or sections, for ripping wide sections into narrow pieces, and for ripping blocks into thin boards for boxes and crates. These special saws are not discussed further here.

Two kinds of accident predominate with circular saws:

1. Injuries to the hands of operators and their helpers caused by contact with the saw blade.
2. Body injuries to operators feeding circular saws resulting from work being caught by the blade and flung back at the operator (known as a kickback). These are often fatal.

Before discussing preventative measures, the common causes of both types of injury are discussed briefly.

6.8.2.1 Blade injuries

Operators may be injured by a hand slipping when pushing work into the saw, by holding the hands too close to the blade, by slipping on the floor, or by someone else bumping into the work whilst it is being cut. Such injuries can be prevented by careful attention to the riving knife, and crown or hood guard, by the use of jigs and pusher sticks or blocks, by sound operating methods and proper maintenance.

6.8.2.2 Kickbacks

Kickbacks may result from the following typical causes:

1. Work handled accidentally touches the crown of an unguarded saw, and is violently ejected.
2. The stock is wet or unseasoned, out of square, improperly planed or cut too close and grips the blade. As the leading end nears the up-running teeth at the rear of the blade, the saw lifts the wood and thrusts it back with a force which the operator cannot control.
3. Similar effects to those described in (2) can occur if the saw is used without a riving knife, if the blade is in poor condition, or if the blade and the guide or fence are improperly aligned so that work is nipped between the rear side of the blade and the fence or guide. They can also be caused if the work is too large for the table and gets knocked, or by a jolt caused by the blade hitting a knot or a nail, etc in the wood. Other causes are the application of force to the cut-off piece, or allowing the cut-off piece to be nipped between the saw blade and the fence when ripping or the end-stop when cross-cutting.
4. When deep cutting, friction may heat the blade causing distortion and slowing down the saw. The timber is withdrawn to ease the load. As the saw picks up speed it imparts an increasing horizontal force to the timber flinging it back towards the operator. Kickbacks can be minimised by the proper use of riving knife, guard and anti-kickback dogs, and by careful selection of timber. Operating practices should be closely reviewed should a kickback nevertheless occur.

6.8.2.3 Circular saw guards

Three aspects of circular saws require guarding:

1. The part of the blade below the bench.
2. The riving knife above the bench behind the saw blade.
3. The crown and front of the blade.

1. *Guards below bench.* Most machines either have a box-frame or have the saw and table mounted on a pedestal. A box-frame is easily enclosed. With the pedestal type the blade below the table is usually enclosed by a shaped hood with an outlet connected to a dust extraction system. The main danger in guarding the blade below the bench arises when clearing sawdust, offcuts or other small objects which have fallen inside the frame or hood.

On a box-frame machine with no exhaust system, the only openings in the enclosure should be cutouts near ground level. A special tool should then be provided to remove sawdust and other objects. It should be a strict rule that nobody puts a hand inside the frame without first stopping and isolating the saw. On machines where sawdust is removed by an exhaust system, care is needed in the design of the hood and outlet to ensure that the outlet cannot become blocked with small offcuts.

Supplementary guards must be provided for saws with a rising table to prevent parts of the blade under the bench from being exposed when the table is raised.

2. *The riving knife or spreader.* The riving knife or spreader is a sickle-shaped blade of high-grade steel which is set behind and in line with the saw (see *Figure 6.8.1*). It prevents the sides of the cut from closing on the rear up-running edge of the saw, and affords some protection to a

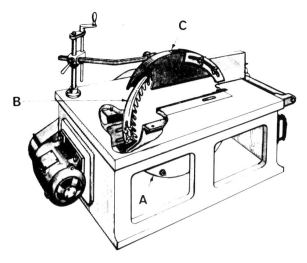

Figure 6.8.1   Typical protection for circular saw
A. *Under-bench guards*
B. *Riving knife*
C. *Guard for crown and front of saw*
        (H. & S. W. Booklet 41)

Figure 6.8.2 Riving knife attachment on circular-saw (Wadkin Ltd.)

worker removing off-cuts, etc from the back of the saw bench. The riving knife should be at least 10% thicker than the saw plate, and it should extend upwards to a height not less than 25 mm below the top of the saw or for large saws to at least 225 mm above the table. The radius of the arc of the inside edge of the riving knife should be not less than that of the largest saw used in the bench.

The knife must be rigidly mounted and its position adjustable so that the distance between its leading edge and the saw blade can be set at 12 mm for any saw which may be used. It must be attached so that it will remain in true alignment with the saw blade even when the table or spindle is tilted. The leading edge of the riving knife should be rounded, not sharp. If several saw blades of different sizes are used, it may be necessary to provide one or more alternate riving knives of different sizes.

3. *Guards for crown and front of saw.* The guard should cover the top of the saw with flanges on either side which are deep enough to cover the roots of the saw teeth. The guard must be strong yet easily adjustable, and extend from the top of the riving knife to a point as low as practicable on the cutting edge of the saw. Bench circular saws manufactured in the USA usually have a hood which adjusts itself to the stock and rides on the stock. A circular saw table used for ripping should be provided with anti-kickback dogs attached to the guard or hood. The points of the dogs ride on the stock and oppose any tendency of the saw to force the work upwards and towards the operator.

6.8.2.4 Jigs and fixtures

On rabbeting and dadoing work it is impractical to use a riving knife and sometimes impossible to use a standard hood or guard. For such operations a jig which slides in the grooves of the transverse guide should be used. The work is locked in the jig and the operator's hands are kept well away from the cutting head. A special jig should be made if a suitable one is not available, since the work is very hazardous without one. If much rabbeting or dadoing work is done, it is best to set a machine aside specially for it.

Sometimes feather-boarding can be used to guard operations where a standard guard cannot be used. It should press against the stock at a point between the saw and the operator, and bear against the stock at an angle of about 45°.

6.8.2.5 Operating practices

Hands should always be kept out of the line of the cut. When ripping with the fence close to the saw, a push-stick should be used to keep fingers away from the saw. A set of push-sticks and blocks of various sizes and shapes should always be kept handy (see *Figure 6.8.3*).

Figure 6.8.3 Push-stick in use on circular saw (H. & S. W. Booklet 41)

Stock should always be held against a guide or fence — never sawn freehand, which endangers the hands and may allow the work to bind on the saw. When ripping stock with narrow clearance on the guide side, a filler board should be clamped to the table between the guide and the saw, thus making it unnecessary to remove the guard or hood.

To provide the maximum force to hold stock on the table, and thus reduce the chance of a kickback, the saw blade should be as high as possible. This however increases the risk of hand injuries from the blade.

When ripping, the operator should stand out of line of the stock to reduce danger from kickbacks. A heavy leather apron should be worn for additional protection.

The correct saw blade should be used for the job; crosscut saws should not be used for ripping nor vice versa.

Long stock should not be crosscut on a table saw as it is difficult to guide long pieces, and ends of the stock are liable to be struck by other people or may overbalance. Work which can be done on special or power fed machines should not be done on hand fed general purpose machines.

All circular saws should be fitted with an effective brake (electrical or mechanical). The operator should always bring a saw to rest before leaving it—it is not enough merely to switch off the power and leave it spinning.

All adjustments to the fence, and guards must be made with the saw stationary. The top of the table should be marked with a permanent distinctive line in line with the saw in front and behind it. This enables the operator to set the guide fence without removing the guards.

6.8.2.6 Maintenance of circular saw blades

Saw blades are designed, built and tensioned to cut at a rated speed without distortion. Any changes in operating speed, tensioning or balance are liable to affect their safety. Some common faults with saw blades are:

1. *Out of roundness.* The largest teeth then do most of the work causing unequal strains on the saw, heating up, warping and running out of line.
2. *Out of straight.* Warps should be checked with a straight edge across the diameter of the blade.
3. *Improper hook or pitch of teeth.*
4. *Incorrect or uneven set of teeth.* The correct set or swage of the teeth is given by bending alternate teeth to left and to right, or by spreading the point, so that the kerf is somewhat wider than the thickness of the blade.
5. *Cracked blades.* A saw blade should be inspected for cracks whenever the teeth are filed or set. A recognised method of non-destructive testing (e.g. the magnetic particle method) should be used to check for cracks invisible to the naked eye. Any blade with a visible crack should be removed from service. Blades should only be repaired and retensioned by the manufacturer or a recognised sawsmith who has all the proper equipment.

The following precautions prevent cracking:
1. The saw blades must be tensioned for the speed at which they will operate.
2. Teeth must have sufficient clearance to prevent the work binding and burning.
3. Blades should be perfectly round and in balance.
4. Blades must be kept sharp at all times.

A saw blade in good condition and running at the correct speed should cut easily. Care must be taken that the collar which secures a saw blade on the mandrel is not misshapen or distorted, as this will bend or buckle the blade when it is tightened. Only the outer edge of the collar should come into contact with the saw when it is tightly clamped in position. When a loose collar is fastened in place, the saw should be tested with a straight edge.

6.8.2.7 Other types of circular saw

Several special types are used for cross-cutting. Only two are mentioned here:
1. *The Pendulum Saw or Overhead Swing Saw.* The blade is carried at the lower end of an arm pivoted at its upper end. The saw is drawn forward by hand and follows the arc of a circle.
2. *The Pullover Cut Off Saw.* The saw and its motor move along a horizontal arm, either manually or hydraulically.

Both types are liable to cause hand injuries, and require adequate guarding. One type of guard is hinged to cover the lower half of the saw when it is not cutting and ride on top of the stock when it is cutting.

A device such as a counterweight, which must be securely fastened, should be used to return the saw automatically to the back of the table when released. A limit device is also needed to prevent the saw swinging beyond the front of the table, and another to prevent it rebounding from its idling position. Stop and start buttons should be located for quick and easy access.

A guard may be required to protect the body when the saw blade is extended to the full length of the support arm.

If the saw is pulled by a handle, it should be attached to one or other side of the saw, not in line with it, and the operator should stand on the handle side and use the hand closest to it.

Usually the saw can be ordered with either right or left hand handles. On a new installation a saw should be ordered with the handle on the side from which the stock is to be pulled. If the handle is on the right side of the saw, boards should be pulled from the right with the right hand, and the saw pulled with the left hand.

To measure a board, the end should be placed against a gauge stop, but if a scale has to be used the board should be moved away from the saw blade first.

After completing a cut the operator should ease the saw back to the idling

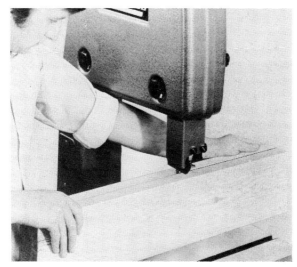

Figure 6.8.4   Bandsaw with adjustable guard and guide (Wadkin Ltd.)

position and wait until it has stopped swinging or bouncing before putting his hand on the table.

The operator should be able to use the saw at his own natural rhythm. No automatic or constant stroking saws should be allowed where the operator is liable to lose a hand if he fails to maintain the rhythm.

6.8.3 Band saws[1,4]

The usual cause of injuries, which are less severe than from circular saws, is that hands come into contact with the saw blade.

Since the operator's hands must come close to the blades, it is important that the saw table be well lit and free from glare. Although the point of operation cannot be fully guarded, an adjustable guard which encloses the saw on the front and sides should be set as close as practicable to the table (*Figure 6.8.4*).

Wheels and all parts of the blade other than that in use should be solidly encased. A bandsaw should have a device to show the tension in the blade and some means of adjusting the tension. At least one wheel should have a positive brake to bring the saw to a halt quickly when work is complete or in case of a band breaking. A special jig or fixture should be used for cutting small pieces of stock.

6.8.4 Hand fed surface planers[1,4]

Most injuries arise through contact between hands and fingers and the

*Figure 6.8.5 Correct method of using a bridge guard on a surface-planing machine.
when plotting and edging (H. & S. W. Booklet 41)*

rotating knives. All planers should have cylindrical heads. The throat in the
cylinder should not be deeper than 11 mm nor wider than 16 mm. The
openings betwen the table and head should be just large enough to clear the
knife. Deep cuts should be avoided because of the danger of kickbacks. The
table opening on the working side of the gauge should be covered by a
guard which adjusts itself as the work rides against it or over it. A shaped
guard that swings sideways is useful for edge jointing but for surfacing work
a guard that rises and rides on top of the work gives better protection. The
unused part of the head on the far side of the gauge should always be covered.

A planer should be no wider than strictly required, as the danger
increases with the width. For surfacing work the operator should have both
hands on top of the stock — never over the front or rear edges (*Figure
6.8.5*).

Push sticks or blocks should be used for surfacing short lengths of stock
when a guard that swings sideways is used.

6.8.5 Vertical spindle moulding machines

A vertical spindle moulding machine or shaper consists of a spindle which
passes up through a hole in the centre of the machine table and carries a

cutter or cutters. The main dangers are contact between the hands and the revolving knives, and broken knives thrown from the machine. When a knife breaks or is thrown from the collar, the other knife is usually thrown as well, so that several pieces of sharp heavy steel are projected with enough speed to kill a man.

The use of solid cutters which fit over the spindle in place of knives greatly reduces the risk of flying knives. These machines, like other woodworking machines, require expert guarding, operation and maintenance.

The use of vertical spindle moulding machines is too specialised to warrant detailed discussion here.

REFERENCES

1. Department of Employment and Productivity, *Safety in the use of woodworking machines*, Booklet 41, Health and Safety at Work Series, HMSO (1970).
2. Fife, I., and Machin, E.A., *Redgrave's health and safety in factories*, Butterworths, London, 1519-1538 (1976).
3. Hayward, J.A., 'The construction industry', Chapter 27 in *Industrial Safety Handbook*, edited Handley, W., 2nd ed., McGraw Hill, Maidenhead (1977).
4. National Safety Council, *Accident prevention manual for industrial operations*, 7th ed., NSC, Chicago, 810-826 (1974).

6.9 RADIATION HAZARDS (1) ELECTRICALLY EXCITED SOURCES

Contents

Radiation hazards may be classified in various ways, e.g.

By the broad type of radiation — pressure waves or electromagnetic waves, whether the radiation is natural or man-made.
By the broad classification of emitter, whether electrically excited or radioactive.

Pressure waves are dealt with in section 3.4.

X-rays and gamma radiation from radioactive sources (whether natural or artificial) have much in common and are often treated together under the heading 'ionising radiation'. There is however one fundamental difference between them. X-rays are emitted from special electrically excited devices which can be switched on and off at will, whereas rays from radioactive sources cannot be simply switched off, but continue at a slowly decreasing rate as the source decays into some other (often harmless) material.

The radiation hazards arising from electrically excited sources and from radioactive ones in industry today are essentially man-made and are brought into man's working environment by man himself. Although radioactive materials may be common in other places in the universe, it is probably true to say that life and man himself could not have developed on the surface of planet earth until they had for the most part died out. Now man is himself reintroducing them under controlled conditions for special purposes.

Unlike most hazards, ionising radiations are not usually apparent to human beings through any known sensory channel. Special instruments and devices such as photographic film are necessary for their detection and measurement.

They are primarily health hazards rather than accident causes; their effects on inanimate matter and machines are generally far less than those

726

on man himself. But the damage they can inflict on man is usually delayed, taking sometimes ten or twenty years from the time of exposure to develop, or even appearing in the next generation. Monitoring these hazards and safeguarding against them requires highly specialised skills and equipment, while the cost of protection is generally a high proportion of the cost of the machines which create the hazards.

6.9.1 The electro-magnetic spectrum

The spectrum of electro-magnetic waves known to or used by man is shown diagrammatically in *Figure 6.9.1* and ranges from wavelengths of 10^5 km (10^8m) (comparable to the earth's diameter) down to 10^6Å (10^{-16}m) (less than the size of a proton).

Some hazards of electromagnetic radiation in order of decreasing wavelength are given below.

6.9.1.1 Low frequency waves

These waves with wavelengths of 1 m upwards, include those used in power transmission (transformers), induction heating, radio and television broadcasting and radar. If sufficiently intense they can produce general body heating, but rarely are they powerful enough for this to become dangerous.[1]

Radio, television and radar waves with wavelengths similar to the dimensions of large steel structures (oil tankers, offshore and onshore oil installations) can induce electrical voltages in the latter. This leads to the possibility of sparking when the structure is near to the transmitter, and possible ignition of flammable vapours present. Two accidents, one fatal, have been definitely ascribed to this phenomenon[2,3] which has only recently received detailed attention[4,5].

Particularly at risk are oil tankers where a high-power h.f. antenna is strung above the deck, below which are highly flammable vapour/air mixtures. It seems that a major naval radio station in Scotland may have to be relocated due to its proximity to the newly built natural gas onshore terminal with its treating and drying installation at St. Fergus, near Peterhead.

6.9.1.2 Microwaves

Microwaves with wavelengths of 1 mm to 1 m are used in radar, communications, cooking and for diathermal treatment in physiotherapy. They are often intense enough to cause pain and localised tissue damage to anybody exposed to them. They may also vapourise and ignite organic materials inside exposed objects. Microwaves of the longer wavelengths are more dangerous than those of shorter wavelengths as they penetrate deeper and can produce temperature rises in deep tissues where

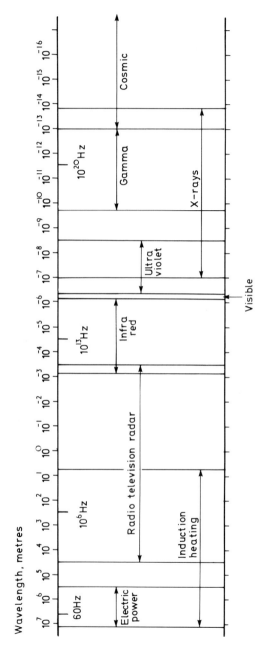

Figure 6.9.1 The electromagnetic spectrum (National Safety Council, U.S.A.)

there are no temperature sensing nerve ends to detect the rise.

Permissible exposures of workers to microwaves and other forms of electromagnetic radiation have been formulated as 'threshold limit values' and published by various authorities, notably the American Conference of Governmental Industrial Hygienists (1973).[6]

6.9.1.3 Infra-red radiation

Infra-red radiation with wavelengths from 7×10^{-6} to 10^{-3} m (7 microns to 1 mm) is given off by hot objects and is used for heating and heat treatment. It affects only the surface of the skin where the rise in temperature is sensed by the nerve endings which cause reflex withdrawal.

The main health hazard of infra-red radiation lies in damage to the rear surface of the lens of the eye. This is known as 'heat cataract' and is caused by prolonged exposure. It has been known among glass blowers for nearly a century.[1]

6.9.1.4 Visible radiation

All visible radiation lies in the narrow waveband of 4000 to 7000 Å (4 to 7 microns). Reflex action causes the eyes of a person exposed to high radiation rates in this range to take evasive action, and eye protectors with suitable colour filters are prescribed for those exposed to excessive visible radiation.

Radiation in the invisible ultra violet range and the use of lasers are more prolific causes of eye damage; recommended exposure limits for all three types of radiation are given in references 1 and 6.

6.9.1.5 Ultra-violet radiation

Near ultra-violet radiation extends from a wavelength of 1000 Å (10^{-7} m) to visible purple (4000 Å)—but the short ultra-violet range extends to 100 Å (10^{-8} m) or less, overlapping with soft X-rays.

Most ultra violet radiation is absorbed by window glass. Welding, UV lamps and direct sunlight are the commonest sources of damaging ultra-violet radiation. The skin as well as the eyes may be affected, and excessive exposure can cause tumours. Skin creams which contain compounds which absorb UV radiation provide some protection.

6.9.1.6 Lasers

These are narrow beams of electromagnetic radiation, mainly in the visible spectrum, but available also in the IR and UV ranges. A laser has a single wavelength and is usually emitted in pulses.

Lasers are generated by special equipment and their industrial uses are quite specialsed, i.e. for cutting cloth and paper and welding plastics. Three types of laser generators are employed, solid state, gas and injection lasers. Solid state lasers which are the most powerful present the greatest hazards and can cause permanent eye damage and skin burns. Injection lasers have the lowest power outputs.

Not only is direct viewing of laser light highly dangerous, but reflected laser light can also cause eye damage. This makes the use of lasers more hazardous since reflections are not always predictable. Great care must, therefore, be taken to contain laser beams, and electrical interlocks should be included in the circuits to ensure that the beam can only be switched on when the protection is in position. Warning notices should be displayed whenever lasers are used. Eye protection for lasers was discussed in section 3.10.3. TLVs for laser exposure are given in references 1 and 6.

Most laser sources employ high voltages which require special electrical protection for the operator. High-powered laser generators can cause ionisation of the atmosphere and formation of ozone, and require good ventilation to prevent this from becoming a health hazard. Some laser sources require the use of liquid nitrogen or helium, which can produce skin burns. Powerful lasers can also present an ignition hazard, especially for flammable vapours.

Laser beams outside the visible spectrum, especially in the UV region, are the most dangerous because they are invisible. The eye is the most vulnerable organ to damage because the cornea and lens of the eye focus and concentrate the laser beam (which may be reflected and invisible) on a small spot on the retina.

6.9.1.7 Ionising radiation

Electromagnetic radiation with wavelengths ranging from 10^{-3} Å to 2 Å (10^{-13} to 2×10^{-10} m) is produced both by X-ray machines and by radio-active isotopes. The resulting electromagnetic rays are termed 'X-rays' or 'gamma rays' depending on the type of source, but there is really no difference between X-rays and gamma rays when they have the same wavelength and energy. Both are used industrially and often in competition for the same purpose.

The rays are known as ionising radiation because they have the power of separating electrons from their atomic orbits and producing charged nuclei or ions and free electrons. These ions may then react with other atoms or molecules in ways which are not normal in living organisms. This can kill living cells, cause tumours and cancers, and mutations in genes — hence the health hazard.

In practical application, the X-ray machine has the advantage that it only produces radiation when it is switched on. Radio-isotopes, on the contrary, give off radiation continuously. They may be either sealed in some type of strong container which is unlikely to get broken, lost or overlooked, or they may be unsealed, when the greatest care must be taken to prevent them being dispersed into the atmopshere, washed down the sink, swallowed or otherwise allowed to contaminate the environment. Radio-isotopes may also give off other types of radiation, e.g. alpha particles, beta particles and neutrons as well as gamma rays.

Although the hazards of X-ray machines and radio-isotopes are often treated together, it is felt advantageous to treat them separately here, dealing first with X-rays. This section forms a useful introduction to the following section (6.10 Radioactive sources).

6.9.2 X-rays

X-rays, discovered by Rontgen in 1895 are formed when a stream of electrons in an evacuated tube strike a cooled metal anode which is maintained at a potential difference of several kilovolts relative to a cathode in the tube. In modern tubes the electrons are generated by passing a current through a hot wire which forms part of the cathode.

The penetrating and ionising power or 'hardness' of X-rays depend on their wavelength which is primarily determined by the voltage between anode and cathode. The intensity of the beam depends on the current flowing. Operating voltages of most X-ray sets lie between 100 kV and 400 kV, although they can range from 30 kV (for research work on crystal structure) up to 2 MV.

X-rays are also produced adventitiously by any thermionic valves (valve rectifiers and valves used in radio, radar and TV transmitters) which operate at voltages of a few kV. UK factory regulations apply to adventitious X-ray sources operated at more than 5 kV, except for TV receivers which are exempt up to 20 kV.

X-ray tubes used industrially should be shock proof and incorporate integral protection in their housing which limits harmful radiation other than that of the beam itself. The limits of the beam are closely defined and should be restricted to the minimum practical size required for the work.

Protection against the harmful effects of X-rays is achieved by shielding, limiting personnel exposure in time, by monitoring and above all by training and selection of personel.[7,8,9]

6.9.2.1 Units of radiation

The unit of exposure to an X-ray beam is the rontgen (R), which is based on its capacity to ionise air. A rontgen is the amount of radiation producing 2.1×10^9 pairs of ions in a volume of 1 cc of air.[8]

As only a portion of the energy to which a body is exposed is actually absorbed, there is a second unit, the rad, which is the absorbed dose. This is equal to an energy absorption of 100 ergs per gram of material exposed, irrespective of the material or the radiation.

The biological effect of the radiation depends partly on the energy absorbed and partly on the type of radiation. This is defined by a quality factor Q, and other modifying factors. The effective dose is measured in rems, where

Rems = rads × Q × other modifying factors.

6.9.2.2 Regulations[6]

The principal UK regulations which affect the use of X-ray machines and personnel exposed to them are the ionising Radiations (Sealed Sources) Regulations 1969. These cover both X-ray machines and sealed radio-active sources. These regulations should be studied in conjunction with the Ionising Radiations (Unsealed Radioactive Substances) Regulations 1968. Other regulations and codes of practice dealing mainly with X-ray machines and apparatus are:

Code of Practice for the protection of persons exposed to ionising radiations in research and teaching (1968).[11]
Code of Practice for the protection of persons against ionising radiations arising from medical and dental use (1964).[12]
ILO Manual of Industrial Radiation Protection, Part IV, Guide on protection against ionising radiations in industrial radiography and fluoroscopy.[13]
Department of Employment and Productivity, Booklet 13, ionising radiations, precautions for industrial users (1969).[7]

In addition, the International Commission on Radiological Protection (ICRP) and the International Atomic Energy Agency (IAEA) have published a number of recommendations, guides and regulations. More detailed bibliographies are given in references 5 and 7.

6.9.2.3 Limits of exposure

Maximum exposure doses are defined as rems/year and rems/quarter for parts of the body and for the whole body. Maximum permissible doses for factory employees for X-rays and external radiation from radioactive sources are given in *Table 6.9.1*.[8,9] The average permissible dose for the whole body is 5 rems per year.

6.9.2.4 Supervision, training and classification of workers[7,8]

In factories where ionising radiations are used, one or more 'competent persons' must be appointed to supervise their use and assist in enforcing observance of the regulations. The competent persons must have been suitably trained; the extent of the training required depends largely on the hazards involved and the precautions which are therefore necessary.
    Advice on training may be obtained from the Advisory Unit of the

Table 6.9.1 MAXIMUM PERMISSIBLE DOSES FOR FACTORY EMPLOYEES IN UK (X-RAYS AND EXTERNAL RADIATION FROM RADIOACTIVE SOURCES)[6,7]

| Organ or parts of body | Radiation | All employees | | Further restrictions for women |
		Yearly limit (rems)	Quarterly limit (rems)	
Hands, forearms, feet, ankles	All	75	40	—
Lens of eye	All	15	8	—
Other parts of body	All (except X, Y, neutrons)	30	15	—
	X, Y, neutrons	5 (N-18)*	3	1.3 rems in a quarter.
For pregnant women	X, Y, neutrons	—	—	1 rem for rest of pregnancy.

*N= age of employee in years.

Department of Employment and Productivity and from the National Radiological Protection Board, Harwell.

In addition to the 'competent persons' various authorised persons may be appointed to carry out special tasks e.g., entry into radiography enclosures and dose rate monitoring.

Both 'competent' and 'authorised' persons are liable to be registered as 'classified workers' who are effectively defined as those employed:

1. In work involving the storage, manipulation, maintenance, operation, use or installation of a sealed source or the operation or use of an X-ray machine or apparatus emitting ionising radiations, and
2. In work at some time in a radiation area, that is, an area where the dose rate exceeds 0.75 mille-rem per hour.

Careful records must be kept of the doses to which all employees are exposed. This is particularly important for classified workers. These records must be preserved through any changes of employment and retained for 30 years after the last entry. When a classified worker leaves employment, he must be given a transfer record containing copies of the relevant entries in his radiation dose record. His employer must at the same time send a copy of this record to the local Health and Safety Inspector.

6.9.2.5 Shielding

The Regulations require all sources of ionising radiation, where reasonably practicable, to be adequately shielded. The definition of this is:

'Adequate shielding in relation to any source of ionising radiations means having provided and properly maintained around that source of ionising radiation, shielding or a demarcation barrier, being shielding, or a barrier outside which the radiation dose rate averaged over any one minute does not exceed 0.75 mille-rems per hour, or where only classified workers are affected, 2.5 mille-rems per hour.'

The dose rate is so stipulated to cover machines which emit pulses of radiation over fractions of a second. Many measuring instruments automatically average the dose rate, but in some cases a calculation is necessary.

Shielding involves interposing some material between the source and the person to attenuate the intensity of the radiation to a safe level before it reaches the person. For shielding against X-rays and gamma radiation, dense materials, such as lead or steel or considerable thicknesses of concrete or brick are used. References to the use of rubber gloves and aprons containing lead compounds and eye protectors with lead glass for use by workers on X-ray machines were made in section 3.10. Because of the problems concerned with the use of such protective clothing and eye protection the authorities now insist that fixed shielding and protection should be adequate, without reliance on clothing and personal protective devices.

6.9.2.6 Protection by distance and limitation of exposure time

The intensity of gamma and X-radiation emitted from point sources decreases with the square of the distance from the source. This does not, however, apply to narrow parallel beams of X-radiation, which are attenuated to a much lesser degree by distance.

Obviously the shorter the time a person spends exposed to certain ionising radiation, the smaller the dose he will receive. Whilst distance and limitation of exposure time are important factors in personnel protection, the specification of safe combinations of distance and time is a job for the expert.

6.9.2.7 Monitoring and personal dosemeters

Since ionising radiation is not detectable by any of our senses, except at very excessive and damaging dose rates, its presence has to be detected and measured indirectly by instruments known as monitors. These are usually battery-operated and must be tested regularly and repaired as necessary by a qualified person.

Details of tests must be kept in a register. Both simple routine testing as well as calibration of monitoring instruments using radiation sources of known strength are required.

Personal dose meters,[14] usually film badges, should be worn (with the number facing outwards) by persons exposed to ionising radiation during the entire working period in which they are exposed. Film badges and other approved personal dose meters are available from the National Radiological Protection Board, Harwell, and other approved laboratories. A film badge consists of a film mounted in a holder with a window and various filters. The dose is assessed by the degree of darkening of the film. X-radiation and gamma radiation are monitored by film badges with appropriate filters. Neutron doses can also be measured by an extension of this technique. Film badges are worn on the outside of normal working clothing between the waist and the shoulder.

For operations where the hands or other parts of the body are subject to higher doses than the chest, thermoluminescent finger tip dose meters should be worn. These are used for work with X-rays where exposure to narrow but intense beams is possible, and also for the manipulation of radioactive sources. Thermoluminescent dose meters contain substances which when heated after being irradiated by ionising radiation emit an amount of light which is proportional to the dose received.

Other types of dose meters are available which give an immediate reading of high radiation dose rates which are used during very short periods as in site radiography.

All classified workers who work in radiation areas are required to wear a film badge or other approved dose meter.

6.9.2.8 Medical supervision and overdoses

All classified workers must be medically examined by the appointed doctor and certified fit before starting work. Workers should be re-examined at intervals not exceeding 14 months.

When any person receives a dose in excess of the maximum permitted by the regulations, the occupier of the factory must notify the appointed doctor and the inspector for the district, hold an investigation or arrange for one to be made, and keep a record of the circumstances. Depending on the severity of the dose, the appointed doctor is required to examine the person concerned without delay.

6.9.2.9 Industrial application of X-rays

At one time during the commercialisation of X-ray apparatus, many wide-ranging applications were envisaged, such as discriminating between real and artificial gems, detecting defective golf balls, examining the fit of boots and shoes, and discriminating fraudulent paintings from 'old masters'.

As the hazards of ionising radiation became better appreciated, some of these applications have disappeared. The principle now employed is that 'Adequate justification should be required for the employment of ionising radiation on however small a scale.

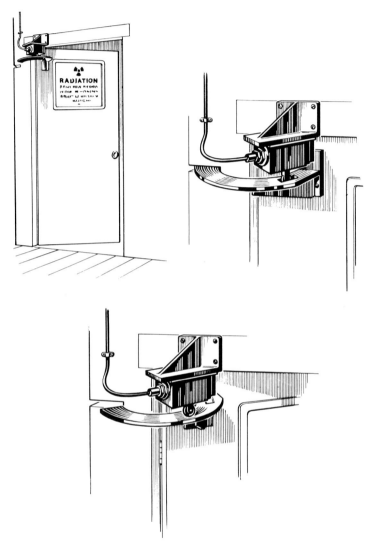

Figure 6.9.2 Electrical interlock on the hinged door of a walled enclosure used for radiography (DEP Booklet 13)

The main industrial applications of X-rays are:
X-ray radiography — especially for inspection of welds.
X-ray fluoroscopy.
X-ray spectroscopy and crystallography, used primarily as analytical and research tools in laboratories where special precautions can be provided.

6.9.2.10 X-ray radiography

X-ray radiography of welds, castings and a variety of manufactured articles and components is carried out in walled enclosures of ample area provided with adequate screening. Special attention must be given to floors and roofs and possible exposure of persons under or over the walled enclosure, as well as observation windows and breaks and discontinuities in the enclosure.

Protection of X-ray radiography machines involves the use of suitably designed electrical interlocks (see *Figure 6.9.2*) and warning signals which ensure that:

No X-ray tube can be energised while a door to the enclosure is open.
Any person accidentally shut inside an enclosure can escape without delay and/or can de-energise the machine.
Adequate warning is automatically given to any persons liable to be exposed to X-rays before a machine is activated; this includes drivers of overhead travelling cranes which are liable to pass over the enclosure.

In circumstances where the use of a walled enclosure is not reasonably practicable or justified, X-ray sets should be used only within suitably marked areas demarcated by solid railings, with suitable warning notices displayed at the boundaries of the areas. X-ray beams should be pointed away from areas where people may be present, and protective lead screens (which may be mounted on wheels) should be placed around the work.

The safety of personnel at all positions round the work should be frequently checked by monitoring instruments or film badges held or mounted in suitable positions while the tube is energised.

Adequate protection must be provided against high voltage circuits and adequate ventilation must be provided to prevent the accumulation of harmful concentrations of ozone and nitrous fumes.

6.9.2.11 X-ray fluoroscopy

Fluoroscopy is used for the routine inspection of many manufactured articles. The following points are important:

1. The lowest practical beam intensities and tube apertures should be used.
2. The screen should be viewed indirectly by the use of inclined mirrors or closed circuit television. Direct viewing is dangerous since the beam is directed towards the face.

3. The X-ray apparatus should be completely contained in a cabinet or walled enclosure provided with adequate screening which should be checked by monitors.
4. The screen should be adequately shielded by a plate of lead glass.
5. Effective interlocks which prevent the machine being used while any access doors are open should form an integral part of the control circuits.
6. Articles to be examined should be inserted and positioned by mechanical means or devices which prevent any part of the operator's arm or body entering the beam.
7. Faulty articles disclosed by the fluoroscopic examination should either be marked or pushed off a conveyor belt into a special container by a device remotely operated from outside the cabinet.
8. Openings for conveyor belts into the X-ray chamber should be shielded by flexible curtains made for example of strips of leaded rubber
9. The viewer should be located in a darkened and well ventilated room.

6.9.2.12 Crystallography and other analytical and research applications

Apparatus designed for these purposes generally employs soft X-rays and is used by well-trained scientists who should have adequate appreciation of the hazards. Familiarity, however, sometimes breeds contempt. The main danger of this type of apparatus is that of narrow X-ray beams which extend beyond the limits of the apparatus. The dangers are increased by the use of make-shift modifications carried out to extend the use of the apparatus beyond that for which it was designed.

The design and construction of all such modifications should be carefully checked and monitored before the apparatus is put into use. These hazards apply especially to the camera and diffraction apparatus.

6.9.2.13 High energy apparatus — linear accelerators

High energy X-ray machines operating at a million volts or more should only be used inside specially designed walled enclosures provided with comprehensive protection and interlocks. Specialist advice on their safety, protection and use should be obtained before installation.

REFERENCES

See References at the end of section 6.10.

6.10 RADIATION HAZARDS (2) RADIOACTIVE SOURCES

Contents

Unlike electrically-excited radiation which can be switched on and off at will, radiation emitted by a radio nuclide continues naturally at a decreasing rate throughout the life of the nuclide. Radioactive emission is accompanied by transformation of the original nuclide into a different one which may or may not be radioactive. The life span òf a nuclide is expressed as the 'half life' which is the time taken for half of the nuclide present initially to be transformed into emission (particles or radiation) plus a different and usually lighter nuclide.

If the half life of a radio nuclide is 1000 years when 50% of the original nuclide will remain, then the amounts remaining after 2000, 3000, 4000 and 5000 years will be 25%, 12½%, 6¼%and 3.125% respectively (*Figure 6.10.1*).

(*Note.* A nuclide is defined as 'a species of atom having a specific mass number, atomic number and energy state).[15]

Until the development of nuclear fission in the late 1940s nearly all radionuclides used by man were found in minerals. The main repository of natural radio nuclides lies in the earth's deposits of natural uranium and thorium, heavy elements of low radioactivity which have half lives of 4.5×10^9 and 1.41×10^{10} years respectively. These materials predate the earth itself and were probably formed in the sun or some other star before or during the birth of the solar system.

Active natural radio nuclides used by man such as radium, have been formed as intermediates in the decay of uranium and thorium, which lead eventually to stable isotopes of lead. Radium has a half life of 1620 years. Nearly all the radium now present in the earth has been formed since the beginning of human civilisation.

Nuclear fission produces many radio nuclides as by products, most of which are unknown in nature. Most of the radio nuclides used today in industry, research and medicine are separated from fission products.

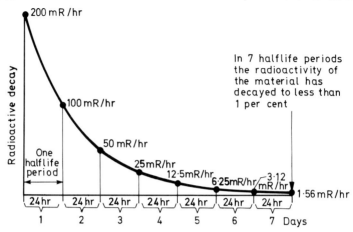

Figure 6.10.1 Decay of radio-active material with 24 hour half life (National Safety Counci, U.S.A.)

6.10.1 Types of emission

There are four main types of emission from radio nuclides, alpha particles (the positively charged nuclei of helium atoms), beta particles (electrons), gamma rays and neutrons which are neutral particles with the mass of a hydrogen atom. Every radio nuclide has its characteristic type of emission which has a characteristic energy, measured in million electron volts (MeV). One MeV is roughly the energy used by an ant in putting one leg forward in walking.[10] Most radio nuclides are either alpha particle emitters or beta particle emitters. Most beta particle emitters also emit gamma radiation, but some radio nuclides emit gamma radiation alone.

Neutrons are emitted by uranium 235 and plutonium, but spontaneous neutron emitters are rarely employed in industry. Where a source of neutrons is required a high voltage accelerator is used to bombard tritium (itself radioactive) with deuterons, which causes neutrons to be emitted. These machines pose difficult safety problems and require the highest standards of protection. A type of radiation similar to gamma rays is produced when beta particles pass through heavy materials. This is called bremsstrahlung.

Alpha particles are the least penetrating and are stopped by cellophane film or the outer layer of the skin. Beta particles are more penetrating than alpha; the most powerful require 6 mm or more of aluminium to stop them.[7] Gamma rays like X-rays are very penetrating and require the same kind of heavy shielding materials.

Neutrons are also extremely penetrating, but they require shielding materials containing a high concentration of hydrogen atoms such as paraffin wax, polyethylene or water. Fast neutrons striking the human body cause severe cell damage.

6.10.2 External and internal exposure. **Sealed and unsealed sources**

When considering the possible effects of different radio nuclides on the human body, it is necessary to discriminate between cases where the nuclide is outside the human body, and those where it is inside. The latter may be caused by inhaling radioactive dust or gas, swallowing food contaminated by radio nuclides or by the deliberate use of radio nuclides internally for medical purposes.

Alpha particles striking the body from outside are relatively harmless, but once an alpha particle emitter is swallowed serious damage may occur, particularly where the nuclide (e.g. strontium 90) is absorbed by some part of the body such as bone and not excreted. Side by side with this distinction it is necessary to discriminate between sealed sources and unsealed sources of radioactivity.[8]

Sealed sources, as the name implies, contain the radio nuclide inside some hard and not easily destructible object which is readily identified and unlikely to enter the human body by accident. With unsealed sources the radio nuclide is present just like any other material and stringent precautions are necessary to prevent contamination of the immediate environment.

Unsealed sources are present during the mining, ore treatment and extraction of radioactive materials, in nuclear power plants when dissolving spent fuel rods and recovering radio nuclides from them. They are also used in luminising dials and in tracer work in chemical research and industry.

6.10.3 Nuclides used in industry

A list of nuclides which are commonly used in industry is given in *Table 6.10.1*.

Table 6.10.1 RADIO NUCLIDES USED IN INDUSTRY[7]

Nuclide	Half-life (years or days)	Energies of principal emissions in MeV	
		Beta	Gamma
Tritium (hydrogen 3)	12.26y	0.018	
Carbon 14	5760y	0.155	
Cobalt 60	5.3 y	0.31	1.17 and 1.33
Krypton 85	10.6 y	0.15 — 0.67	0.51
Strontium 90	28 y	0.54 2.2 *	
Caesium 137	30 y		0.66
Promethium 147	2.6 y	0.22	
Iridium 192	74.4 d		0.31 — 0.61
Thallium 204	3.9 y	0.77	

*The 2.2 MeV beta particle is emitted not by strontium 90 but by its product (daughter) yttrium 90 which has a much shorter half life.

6.10.4 Units of radioactivity

The unit of radioactivity is the curie (Ci), which is the amount of radioactive material required to give 3.7 x 10 disintegrations per second, corresponding approximately to one gram of radium. Since it is rather large for practical purposes, the mille curie (mCi) and micro curie(μCi) equal to one thousandth and one millionth of a curie respectively are more commonly used. Emissions from the source take place in all directions. Radiation rates and radiation dose rates (rads and rems) were discussed in section 6.9.2.1. and related to the strength of the source given in curies. The most important relation is that between the gamma radiation dose rate and the strength and energy of the source. The following approximate formula[9] may be used in the absence of more precise tabulated data:

rems/hour at 30 cm distance = 6 CE

where

C is the strength of the source in curies, and
E is the energy per disintegration in million electron volts.

Specialist advice should be sought for calculating dose rates from alpha and beta particles and neutrons and is usually advisable for gamma rays and bremsstrahlung. The use of several different units, rads, rems, curies and, in the case of X-rays, roentgen, can be quite confusing. The essential points to grasp are that:

1. Curies and roentgen define the strengths of radioactive and X-ray sources respectively.
2. Rems, millirems and microrems define the effective doses received by the human body from both radioactive sources and X-rays, on a scale which is directly related to the biological damage which is expected to occur.

6.10.5 Industrial applications of sealed radioactive sources and protection against exposure (see also section 6.9.2)

The principal applications of sealed radioactive sources are found in:

1. Gamma radiography.
2. Thickness gauges.
3. Level gauges.
4. Density gauges.
5. Static eliminators.
6. Moisture content and soil density gauges.

Gamma radiography and beta particle thickness gauges are used mainly in engineering inspection to check the soundness of welds and to measure the thickness of the metal walls of pipes and vessels that have been subject to corrosion or erosion.

Level and density gauges are used on silos and process plants as part of the control instrumentation required to assist operations.

Static eliminators employing alpha particle sources are used in factory atmospheres where static charges may build up as a result of the operations carried out in order to obviate ignition and nuisance hazards caused by the static charges (see section 5.6).

Moisture and soil density gauges use neutron sources for which special protection is essential. The monitoring of radiation from sealed sources employs the same methods as the monitoring of X-rays which were discussed in section 6.9.2.

Radioactive sources must be kept in a secure store provided with adequate shielding, and as close to the area of use as possible. The use of pits with walled sides and padlocked covers should be considered.[7] Such a pit may be located inside a walled enclosure in which the sources are to be used.

6.10.5.1 Gamma radiography[7,9,13]

Gamma radiography serves the same purpose as X-radiography, namely the inspection of welds and manufactured and fabricated articles. The main difference is that in place of a large X-ray machine, the radiation source consists of a small metal capsule inside which the radioactive material is sealed. The capsule is fitted into a holder which is kept in a larger protective container. The holder must be specially designed to ensure that the capsule cannot be dislodged and knocked out of it. Radiography sources must never be picked up by the hands.

It is often difficult to produce a container which provides adequate shielding and yet is readily portable. The requirement that the radioactivity on the surface of the container should not exceed 2.5 m rem/hour is therefore sometimes relaxed to comply with the following standard:[7]

1. The average dose rate at 1 m from the surface should not exceed 2 m rem/hour and the maximum dose rate should not exceed 10 m rem/hour.
2. The average dose rate at 5 centimetres from the surface should not exceed 20 m rem/hour and the maximum dose rate should not exceed 100 m rem/hour.'

Containers of three main types are available:

1. Those from which the source is not removed, but from which a cover is removed to expose a beam of radiation.
2. Those from which the source is removed on a long handling rod.
3. Those from which the source is removed by mechanical, electrical or pneumatic means.

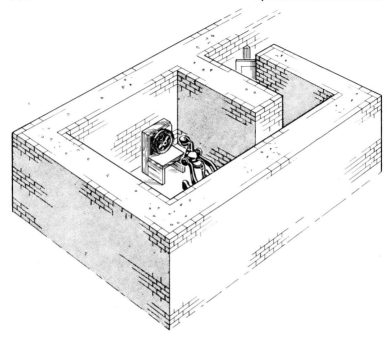

Figure 6.10.2 Labyrinth entrance for a walled enclosure used for gamma radiography (DEP Booklet 13)

The third method whereby the source is moved by remote control is preferred for panoramic exposures and becomes almost essential where high energy sources are employed. Remote automatic handling methods may be combined with time controlled mechanisms so that the source is removed from the container and exposed in each required position in turn for predetermined times and then returned to the container. Such operations may be performed at night in the complete absence of workers. Absence of body is more important for the safe use of radioactive sources than presence of mind.

Gamma radiography requires the use of walled enclosures in the same way as X-radiography, thicker walls are generally needed. If a door into the enclosure cannot provide enough shielding, then a walled labyrinth entrance should be provided (*Figure 6.10.2*).

Warning notices and lights and a permit to work system should be used as required to protect workers who may be incidentally exposed to gamma radiation. Where the sources are operated manually and electrical interlocks cannot be provided, doors of enclosures and stores should be locked.

6.10.5.2 Thickness gauges

Most thickness gauges use beta radiation for which the problems of protection are less severe than for gamma radiation. Thickness gauges may work either by measuring the attenuation of radiation transmitted through the part measured, or by measuring the back-scattered radiation. The source may be a radioactive foil or plaque. Sources must be properly shielded when not in use, and strict precautions must be taken to prevent exposure of operator's fingers and hands. The gamma rays and bremsstrahlung associated with the beta ray source can be a more serious hazard than the beta rays themselves.

Some thickness gauges do in fact utilise gamma radiation, and others use X-ray tubes, which are generally safer than the use of radio nuclides.

6.10.5.3 Level and density gauges

These operate on the same principle as thickness gauges but generally employ a weaker source. The detectors generally operate with a radiation level of 1 m rem/hour or less and require no special precautions, but the sources themselves must be secure and adequately shielded.

Consideration must be given as to whether the plant where the source is used could be destroyed by fire or explosion, and also what extra hazards the sources would cause during subsequent rescue and rehabilitation.

6.10.5.4 Static eliminators

These are similar in design to thickness gauges but generally employ alpha emitting sources. The source must be protected against mechanical damage, i.e. by a stout wire mesh grille, and a shutter should be fitted over or around the source to attenuate the beam when not in use and to protect workers during maintenance or cleaning.

Cleaning of radioactive sources in eliminators, e.g. after a fire, requires extreme care as well as speed, and should only be done by a specially trained person. Sources involved in a fire should be tested afterwards for escape of radioactive material.

6.10.5.5 Neutron gauges

The housing and carrying cases of neutron gauges must be adequately shielded and appropriate cleaning devices for dirty gauges must be provided to protect the users.

6.10.5.6 Breakage and leakage of sealed sources

Whilst sealed sources are specially designed to reduce risks of breakage and leakage, there is always a possibility that this may occur, and sealed sources must be installed and used in ways which minimise these risks. Sealed sources should be tested for leakage at least every twenty six months by a qualified person using an approved method of testing. The following precautions must be taken when a sealed source is damaged or broken:

1. The source should be placed in a leak-proof container for repair or disposed of by properly authorised persons.
2. The area round the source should be roped off.
3. The competent person should be notified and the affected area surveyed and decontaminated by a person with suitable training, experience and equipment.
4. All practicable steps should be taken to prevent the dispersal of the radioactive material to the environment or down the drains.
5. A survey should be carried out of contaminated surfaces both before and after cleansing to ensure that the cleansing has been effective.

6.10.6 Work with unsealed sources[7,9,13]

Special requirements for work with unsealed sources are specified in the Ionising Radiations (Unsealed Radioactive Substances) Regulations 1968. The main aims of these regulations are to restrict external exposure, as discussed for sealed sources, and to prevent inhalation, ingestion or absorption of the substances. The latter relies on three types of precaution:

Avoidance of direct contact between the worker and the radioactive substances.
Maintenance of surface and personal contamination below specified maximum levels.
Avoidance of concentrations of airborne or gaseous radioactive significant quantities to be inhaled by workers.

6.10.6.1 Area classification

Area division forms an important tool in achieving these objectives. The areas of the premises fall under five categories:

Active area.
Tracer area.
Decontamination area.
Total enclosure.
Radiation area.

Many factors need to be considered in limiting the boundaries of these areas; this should only be done by a person with the necessary training and experience, who is familiar with all the operations to be carried out. The classification should be checked by monitoring when the areas are in use.

1. *Active area.* This is defined as 'part of a factory other than a tracer area or the inside of a total enclosure or of a fume cupboard in which any operation involving the manipulation or use of any radioactive substance is carried on as a result of which there is, or under normal conditions is liable to be:

   1. Contamination in excess of the levels specified for Category D in Schedule 2 to the Regulations; or
   2. Airborne or gaseous radioactive substances in the atmosphere to such an extent that persons employed in the area are likely to inhale, ingest or absorb a significant amount.

Only classified workers (see section 6.9.2.4) should be allowed to work in specified areas, and stringent precautions should be taken to minimise risks.

2. *Tracer area.* Unsealed radioactive substances are sometimes used to investigate the operation of machines or processes, for production control or to acquire design data. This is called tracer work. An area in which the only work carried out with radioactive substances is tracer work is called a tracer area and is subject to provisos (1) and (2) noted above for active areas. Since tracer work is usually only carried out for the purpose of special investigations which last for a limited time, an area will remain a tracer area only for so long as the risk remains.

Precautions required in a tracer area are less stringent than those required for active areas. Examples of these are:

   1. Full particulars identifying tracer areas do not have to be recorded.
   2. A worker who works in tracer areas for less than 14 days in a year and would not otherwise be a classified worker need not be classified.
   3. Boundaries of tracer areas need be marked only with warning signs. Barriers are not required.
   4. It is not necessary to carry out tracer work in total enclosures or fume cupboards.
   5. The construction of tracer areas and the furniture employed in them do not need to be specially designed.
   6. Special washing facilities are not required.

3. *Decontamination area.* An area where contaminated articles are present and being decontaminated may be classified as a decontamination area provided the level of contamination of the surroundings does not exceed the maximum permissible level specified for category D in Schedule 2 of the Regulations. However, if highly radioactive articles have to be decontaminated and the level specified above is liable to be exceeded, then the area must be classified as an active area.

6.10.6.2 Monitoring for radioactive substances

Monitoring for the presence and quantity of radioactive substances plays an important part in work with unsealed sources. Such monitoring is broken down into three categories:

1. Surface contamination.
2. Air contamination.
3. Body burdens.

Surface contamination is further sub-divided depending on whether the surface is part of an inanimate object or of the human skin.

1. *Surface contamination.* Monitoring for surface contamination is carried out by scanning the surface with a probe and by wiping the surface with an absorbent material, and measuring the activity of the absorbent. To test whether the contaminant is fixed to the surface, a probe should be used and the surface should then be wiped clean and the probe re-applied. Floors, ceilings and walls should be sub-divided into areas not larger than 100 cm² and average contamination levels determined for each area.

Probes should be sufficiently sensitive to measure contamination down to

10^{-5} μCi/m² for alpha emitters, and to
10^{-4} μCi/cm² for beta emitters.

Where the only radioactive substance present is a luminising compound, a suitable ultra-violet lamp used under appropriate conditions of darkness may be used to detect the presence of radioactive contaminants.

Skin contamination may be measured in the same way as that of inanimate objects. Where luminising work is being carried out, an ultra-violet lamp should be located near to the washing facilities to enable workers to check their hands and clothing.

2. *Air contamination.* Regular air monitoring is needed either

Where gaseous or volatile radioactive materials (radon, krypton 85, tritium and iodine 131) are handled in quantity,
Plutonium is being produced or processed, or
radioactive materials are handled under conditions leading to frequent and heavy contamination of the workplaces.

Monitors for radioactive gases and dusts work on different priciples, so that it is important that the type of monitor used is appropriate to the contaiminant.

Great care is required when sampling air for contamination to ensure that the air sampled corresponds to the intake of individual workers. Personal air samplers are available which give an accurate picture of the concentration of radio nuclides in the air inhaled.

3. *Body burden.* Despite all precautions, there is sometimes a need to monitor personnel for internal contamination. This is done in one of two ways either

1. By measuring the radiation emitted from the body, or
2. By measuring the radioactive content of the breath, urine or faeces.

Urine monitoring is the most common method used, but other methods may sometimes be necessary, e.g. after an accident involving radioactive material. Radon in breath measurements have been used to determine the quantity of radium deposited in luminisers working with radium activated material.

6.10.6.3 Precautions for work with unsealed sources

Precautions to be taken for work with unsealed sources include:

1. Containment, e.g. by the use of glove boxes.
2. Cleanliness, e.g. by the use of specially designed work areas which are readily decontaminated, and by the use of very high standards of housekeeping.
3. Special cleaning and waste disposal.
4. Special changing and wash rooms.
5. Personal protective equipment including breathing apparatus.

1. Containment. Work which involves the handling of unsealed sources may where appropriate be carried out inside glove boxes or sometimes inside fume cupboards. Glove boxes are carefully sealed enclosures provided with a window, illumination, rubber gauntlet gloves whose open ends are sealed into the sides of the box, and means of bringing the radioactive substances into and out of the box in sealed plastic bags or other suitable containers (*Figure 6.10.3*).

Glove boxes should be designed on ergonomic principles and the interiors should be smooth and easily cleaned. Provision should be made to enable the gloves to be changed without opening up the box.

Glove boxes should be maintained under a very slight negative pressure to prevent materials escaping outwards through leakage, and a suitable filter should be placed in the exhaust ventilation pipe which itself should discharge to the open air.

No person should be allowed to open up or enter a glove box or similar enclosure until it has been thoroughly decontaminated and until a special permit has been issued.

2. Cleanliness. Rooms in which unsealed sources are to be used should be specially designed. Surfaces should have a smooth finish free from cracks or crevices. Joints between floors, walls and ceilings should be filled in where necessary and rounded.

Fittings for pipes and electrical conduit should be either flush and sealed or supported clear of the surface so that the sides and back are readily cleaned. All joints should be as tight as possible. Floors should be waxed, so

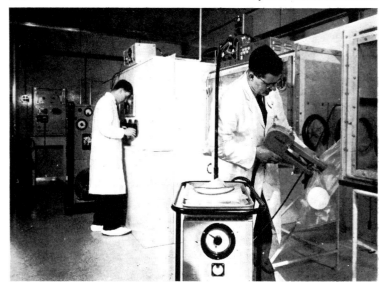

Figure 6.10.3   Use of a plastic bag to remove articles from a glove box (DEP Booklet 13)

that in the event of a spillage the wax coating can be removed. Walls should be painted with a high gloss paint. Work benches should be protected with a sheet of disposable material such as polyethylene or PVC, secured in place by adhesive tape.

Laboratories in which unsealed sources are handled are graded A, B or C by the International Commission on Radiological Protection.[9] Grade A laboratories are specially designed for work with high levels of radioactivity. Grade B laboratories are top quality laboratories which have been well provided with ventilated fume cupboards and other special features for work with unsealed sources.

Most modern conventional chemical laboratories qualify as Grade C laboratories with only minor modification. Only work involving low levels of radioactivity is allowed in Grade C laboratories.

3. *Cleaning and waste disposal.* Cleaning should be done frequently by methods which avoid the spread of radioactive materials; wet methods are preferred. Cleaning articles should be kept in a separate cupboard and used for no other purpose. Radioactive waste and contaminated articles should be placed in specialised receptacles with an inner lining which extends over the top of the sides of the receptacle, so that the receptacle itself does not become contaminated. Authorisation and advice for the disposal of radioactive waste should be obtained from the DoE.

4. *Changing and wash rooms.* Changing and wash rooms should be arranged in accordance with the most stringent precautions recommended

in section 3.11. Separate lockers are required for protective clothing and equipment and those not worn during working hours. Soap, towels and soft nail brushes should be provided. It is advisable that all access to an active area should be via a 'change room'. Employees must be allowed adequate time to wash and monitor themselves for contamination.

5. *Personal protection and protective devices.* No eating, drinking, smoking, snuff taking or the application of make-up should be allowed in active areas, tracer areas or decontamination areas. Disposable paper handkerchiefs should be provided and used in place of normal ones. Nothing must ever be put in the mouth. Any cuts, wounds or abrasions must be treated immediately by an authorised person, and any such wounds must be kept covered when working in active areas.

Protective clothing should not only protect the worker, but serve also to prevent the spread of contamination outside the working area. Where contamination is heavy complete suits with boots, gloves and helmets connected to air supply lines may be required.

6.10.6.4 Applications of unsealed sources

The most common uses of unsealed sources outside atomic energy establishments and associated factories are for luminising and tracer work. The general principles of protection have already been given.

1. *Luminising.* Luminising is carried out using compounds activated by radium, tritium or promethium 147. For hand luminising a glove box is recommended (see *Figure 6.10.4*), but brushes should not be employed to apply the luminous compound. Machines are available for illuminising on a continuous basic, which may be carried out in a fume cupboard type enclosure. Tritium and promethium 147 are weak beta emitters and only a small amount of shielding is necessary. Radium luminising compounds require rather more shielding.

2. *Tracer work.* A radioactive source is introduced into the material being processed at one point in the plant, and measurements are then made further along the production line or in the product to check on the performance of the process. The radio nuclides used should have half lives only slightly longer than the duration of the experiments. The use of beta emitters is preferred where possible, though sometimes gamma emitters have to be used (i.e. where the material measured is in a large volume or inside enclosed plant).

Typical radio nuclides for tracer work are:

Sodium 24	half life 15 hours
Potassium 42	half life 42 hours
Manganese 56	half life 2.58 hours
Lead 212	half life 10.6 hours
Bromine 82	half life 36 hours

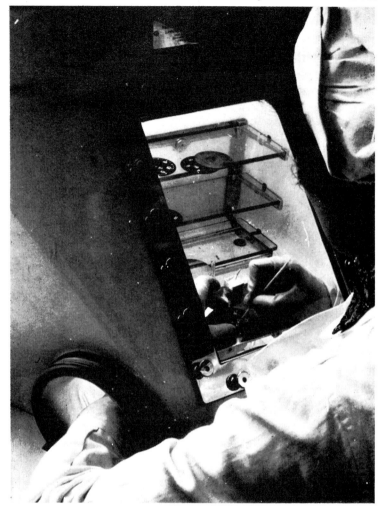

Figure 6.10.4   The use of a glove box for luminising (DEP Booklet 13)

Tongs should be used when pouring radio nuclides into the material.

Suitable notices must be displayed at the boundaries of a tracer area while the work is in progress. All persons in the area and its vicinity should receive appropriate warning and instruction.

6.10.6.5 Incidents involving escape of radioactive substances

Emergency procedures should be drawn up in anticipation of possible incidents and all those likely to be involved should be trained in their duties. The main action necessary is that of minimising the spread of radioactive material. If all the equipment required cannot be kept on the premises, prior arrangements should be made with a specialist organisation such as the National Radiological Protection Board.

The measures most likely to be needed are:

1. Immediate reporting of the incident to a competent person.
2. Segregation of the area by closing doors and erecting barriers.
3. Excluding all workers from the area, except those authorised to deal with the situation.
4. Monitoring and where necessary decontaminating workers who may have become contaminated.
5. Monitoring and where necessary decontaminating any articles which have to be removed from the area.
6. Removing any gaseous contaminants that have been released by ventilation through the use of exhaust fans.
7. Monitoring of the contamination, especially of surfaces in the area.
8. Declaration by the competent person that conditions in the area are satisfactory before work is resumed.

REFERENCES FOR SECTIONS 6.9 AND 6.10

1. National Safety Council, *Accident prevention manual for industrial operations,* 7th ed., National Safety Council, Chicago, 1056 (1974).
2. *Judicial Report of the Tiergarten Court,* West Berlin No. (204) 59 MS 81/55 (301/55) (Jan 1956).
3. 'Radar radiation causes fire', Approach (*US Naval Aviation Safety Review*) 44-45 (April, 1958).
4. BS 4992:1974, *Guide to protection against ignition and detonation initiated by radio frequency radiation,* British Standards Institution.
5. Excell, P.S., Butcher, G.H., and Howson, D.P., 'A generalised basis for the determination of the hazard of ignition of flammable gas mixtures by radio-frequency radiation' *Proc, 2nd European Electromagnetic Compatability Symposium,* Montreaux, Switzerland, 59-64 (1977).
6. American Conference of Governmental Industrial Hygienists, *Threshold limit values for physical agents, 1973,* ACGIH, PO Box 1937, Cincinnati, Ohio 45201 (1973).
7. Department of Employment and Productivity, *Ionising radiations: precautions for industrial users,* HMSO, London (1969).
8. Fife, I., and Machin, E.A., *Redgrave's health and safety in factories,* Butterworths, London (1976). See The Ionising Radiations (Sealed Sources) Regulations 1969 and The Ionising Radiations (Unsealed Radioactive Substances) Regulations 1968.
9. Bennellick, E.J., 'Ionising Radiation', Chapter 5 in Handley's *Industrial Safety Handbook,* 2nd ed., McGraw Hill, London (1977).
10. Denney, R.C., *This dirty world,* Nelson, London, 148 (1971).

11. Department of Employment and Productivity, *Code of Practice for the protection of persons exposed to ionising radiations in research and teaching*, HMSO, London (1968).

12. Department of Health and Social Security, *Code of Practice for the protection of persons against ionising radiations arising from medical and dental use*, HMSO, London (1972).

13. International Labour Office, *Manuals of Industrial Radiation Protection. Parts I to VI*, ILO Branch, 87 New Bond Street, London (1963 to 1968).

14. BS 3385: 1961, *Specification for direct reading personal dose meters for X and gamma radiation*, British Standards Institution.

15. BS 2597:1955, *Glossary of terms used in radiology*, British Standards Institution.

APPENDICES

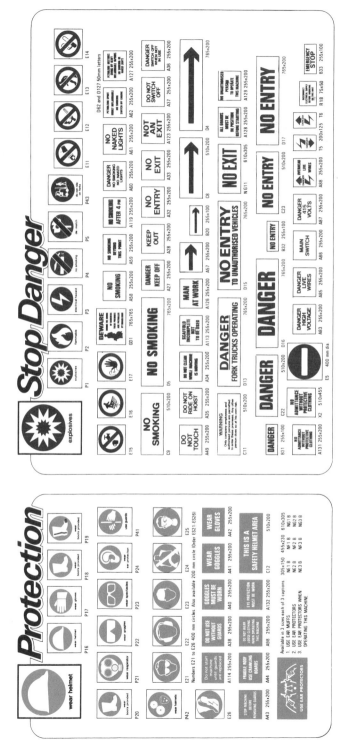

British and European international signs relating to protection and danger (Courtesy Focal Displays Ltd)

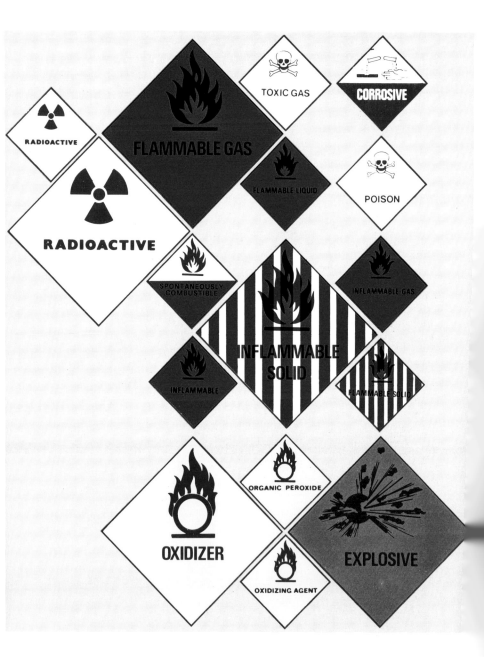

Selection of warning diamonds used in the Hazchem identification system (Courtesy Focal Displays Ltd)

Appendix A

Hazard warning and identification signs

Contents

A.1 INTRODUCTION

Hazard warning and identification signs depend for quick recognition on the use of various methods of coding. Most signs employ two or more coding methods, which are compared in *Table A.2*.

A.2 INTERNATIONAL STANDARDS

Whatever codes are employed, it is important that they be standardised and it is also desirable that they should apply internationally. Recommendations for international standards for several identification codes have been made by the International Standards Organisation; some of these have been adopted by the majority of member countries and have become international codes. The more important of these recommended and adopted International Standards are listed in *Table A.1*. Those standards which are of special importance to industrial safety are marked* and are summarised in the following paragraphs.

Table A.1. INTERNATIONAL CODES FOR HAZARD WARNING AND IDENTIFICATION (R = RECOMMENDED)

Code No.	Title	No. of pages
ISO 12-1976	Aerospace pipelines — identification scheme	5
ISO 32-1977	Gas cylinders for medical use. Marking for identification of content.	2
ISO/R 408-1964	Safety colours*	2
ISO 443-1965	Marking of aircraft gas cylinders	3
ISO/R 508-1966	Identification colours for pipes conveying fluids in liquid or gaseous conditions in land installations and on board ships*	5
ISO 513-1975	Application of carbides for machining by chip removal — designation of the main groups of chip removal and groups of application	3
ISO/R 557-1967	Symbols, dimensions and layout for safety signs*	6
ISO 1769-1975	Laboratory glassware — pipettes — colour coding	2

Table A.2 COMPARISON OF CODING METHODS

Method	Max. number of different possibilities to give 95% or better accuracy of recognition	Evaluation	Comment
Colour	11	Good	Requires little space. Requires good illumination.* Requires adequate colour vision.* Quickly recognised.
Numbers and letters	Unlimited	Good	Requires little space if contrast is good. Recognition slower than colour.
Geometric shapes	About 15	Good	Requires little space if resolution is good.
Size	5	Fair	Requires considerable space. Recognition time longer than for colour or shapes.
Number of dots	6	Fair	Requires considerable space & easily confused.
Orientation of line	12	Fair	For special purposes
Length of line	4	Fair	Clutters display with many signals
Brightness	4	Fair	Reduces visibility of other signals.
Stereoscopic Depth	Unknown	Fair	Poor. Confuses other methods with which used Complex display & viewing equipment needed.

A.2.1 ISO/R 408-1964 Safety colours

This recommendation was circulated to all ISO members in 1962 and was approved by twenty-nine countries including the UK and the USSR, but was not approved by the USA, which has a somewhat different code, ANSI Z 53 1-1971.

The purpose of the ISO/R 408 is to define the meaning and application of a limited number of safety colours *and geometrical forms* in order to prevent accidents and meet certain emergencies. It is intended to give a rapid indication of danger. The code defines the use and meaning of three main safety colours — red, yellow and green, one auxiliary colour — blue and two contrasting colours — white and black to improve visibility.

The meaning of the main colours is given in *Table A.3.*

Table A.3 MEANING OF MAIN SAFETY COLOURS IN ISO(R 408-1964)

Colour	Meaning	Examples
Red	Stop, harmful activity	Stop signals Fire fighting equipment and its location Emergency stop devices
Yellow	Attention, danger	Warning of danger Caution signs Inside of machinery guards
Green	Safety	Escape route and refuges Clear or go signals for persons and vehicles First aid and rescue stations and equipment and their location, other than those authorised by the International Red Cross Society

Blue may be used as a signalling colour as an auxiliary to the main colours for organisation, instruction or information.

Black and white are employed in combination with the above colours to provide contrast and improve visibility of signs.

The main and auxiliary colours, with or without contrasts, may be applied directly on objects or on safety signs so as to be always clearly visible but generally on a limited scale in order not to interfere with the colour scheme of the surroundings.

Safety signs are of three shapes:

Circular. For prohibitions
Triangular (equilateral, with apex upwards). For warning
Rectangular. For information

Symbols may be used inside the signs. Written text may be inside the signs or outside and adjacent to them.

A.2.2 ISO/R 557 1967 Symbols, dimensions and layout of safety signs

This recommendation was circulated to all ISO members in 1964 and approved by twenty-three countries including the UK and USSR, but opposed by Belgium, France and India. The recommended symbols are described in *Table A.4*. The use of macabre or horrifying designs should be avoided as far as possible.

Table A.4 SYMBOLS RECOMMENDED IN ISO(R 577-1967)

Notion of symbol	Symbol
First aid equipment	Greek Cross (except for Arab Countries (crescent) Iran (Lion & Sun) & Israel (Star of David))
Escape route, direction to emergency exit, first aid and fire fighting equipment	Arrow
Protection of eyes	Safety goggles
Protection of head	Helmet
Protection of respiratory organs	Gas mask
Protection of hands	Glove(s)
Flammable material	Flame
Explosive material	Exploding bomb
Toxic material	Skull and cross-bones
Corrosive material	Corroded hand — or drops from test tube on corroded hand with optional corroded plate
Radioactive material	Trefoil — see ISO/R 361
Electricity	Flash of lightning or figure of man struck by flash of lightning
Suspended loads	Load suspended on hook
Falling objects	Falling stone, brick or hammer
Dangerous temperatures	Thermometer showing high or low temperature
Risk of loss of balance slipping or falling	Person falling backwards
Smoking	Burning pipe or burning cigar, with or without burning match
Open fire	Burning candle or flame

A.2.3 ISO/R 508 1966 Identification colours for pipes

This has been approved by 27 countries including the UK and the USSR. Its scope provides:

(a) Basic identification.
(b) Additional code indications where more precise determination is important.

British Standard No.	International equivalent	Title of British Standard	Notes on contents
BS 349:1973	= ISO/R 448	Identification of contents of industrial gas cylinders	Marking and colour to denote contents
BS 349:C:1973		Colour chart for above	
BS 837:		The construction of road traffic signs and internally illuminated bollards	
Part 1 1970		General traffic signs	
Part 2 1973		Miscellaneous signs	
Part 3 1970		Internally illuminated bollards	
Part 4 1973		Road studs	
BS 1319:1976		Chart of colours for the identification of the contents of medical gas cylinders	
BS 1710:1975	± ISO/R 508	Identification of pipelines	Colours for identifying fluids conveyed in pipes on land and marine installations
BS 1843:1952		Colour code for twin compensating cables for thermocouples	To indicate nature of the thermocouple and polarity of conductor
BS 3510:1968	= ISO 361	A basic symbol to denote the actual or potential source of ionizing radiation	
BS 4099:		Specification of colours of indicator lights, push buttons, annunciators and digital readouts	
Part 1 1976		Colours for indicator lights and push buttons	
Part 2 1976		Flashing lights, annunciators and digital readouts	
BS 4159:1967		Colour marking of plastics pipes to indicate pressure ratings	Relates to pipes carrying fluids and independent of composition and method of manufacture of plastics
BS 4410:1969	= IEC:1969	Connections of flexible cables and cords to appliances	Describes colours of cores of flexible cables and cords
BS 4610:1970		Colours for high visibility clothing	Specifies colour including visual performance and fasteners for high visibility clothing
BS 4765:1971		Safety signs to denote the actual or potential presence of a dangerous level of radio-frequency or other non-ionizing radiation	Exclusive symbol showing shape, colour and dimensions
BS 5378:1976		Specifications for safety colours and safety signs	Four safety colours with interpretation, contrasting colours and geometric forms to be used.
BS 5383:1976		Specification for material marking and colour coding of metal pipes and piping system components in steel nickel alloys and titanium alloys	Gives simple colour code to identify groups of materials. This reduces dangers resulting from installations of item in materials different from that intended.
AU 7:1968		Chart and colour code for vehicle wiring	
AU 47:1965		Advance warning triangle to indicate temporary obstructions	
PD 2379:1962		Register of colours of manufacturers identification threads for electric cables and cords	

The basic identification colours are:

Green	Water in liquid state
Silver grey	Steam
Brown	Mineral, vegetable and animal oils, combustible liquids
Yellow ochre	Gases in gaseous or liquefied condition except air
Violet	Acids and alkalis
Light blue	Air
Black	Other liquids

The basic colours may be applied to pipes in one of three ways:

1. By painting over whole surface.
2. By painting as bands with a length of about 150 mm according to the pipe diameter.
3. By applying as adhesive bands round the pipes.

The basic identification colour should be applied at all junctions, at both sides of valves, service appliances, wall penetrations and at other places where identification of the fluid is necessary.

Valves may be painted with the basic identification colour unless the pipeline carries water or steam for fire fighting. In this case the valves should be painted red.

The safety colours given in ISO/R 408 may be used in conjunction with the basic colours with the following meanings:

Red.	For fire fighting
Yellow, with black diagonal stripes.	For warning of danger
Auxiliary blue in conjunction with ̣asic green.	Denotes pipes carrying fresh water, potable or non-potable.

If the safety colour is used, it should be applied as follows:

1. By painted bands over the basic colour, if the colour is applied over the whole surface.
2. By painted bands alternating with painted bands of the basic colour.
3. By adhesive bands alternating with adhesive bands of the basic colour.

Further identification (words, formulae or abbreviations according to a code established for the works) should be written on the pipe in black or white (to give the best contrast) either on the basic identification colour or next to the basic identification colour band.

The direction of flow should be indicated by a black or white arrow on the basic identification colour.

A.3 OTHER IMPORTANT CODES USED INTERNATIONALLY

It was noted earlier that ISO/R 408 1964 is still only a recommendation. Whilst it contains the basic and commonly agreed essentials it is not accepted by the USA. The American code ANSI Z 531 1971 gives more detailed guidance on colour marking. It is important as many of its features are seen in other countries. Fortunately it is in fact broadly in agreement on most basic matters with ISO/R 408 1964.

A.3.1 United States Code ANSI Z 531 1971 Safety Colour Code for marking physical hazards

This code differs from ISO/R 408 in using five colours, red, orange, yellow, green and purple, in addition to black and white. Blue was formerly used under this code for designating control equipment (electrical controls, valves, breaks and switches), but the use of blue has been discontinued in the latest edition of the code. Red and green have basically the same meaning in this code as in ISO/R 408, but orange plays a role intermediate between those of red and yellow. Purple is used exclusively for radiation hazards.

The use of the different colours in this code is summarised below.

1. *Red — shall be the basic colour for the identification of:*

   1. Fire protection equipment and apparatus.
   2. Danger.

   The following suggestions are made for the use of red:

   1. *Fire protection:* alarm boxes, blanket boxes, buckets or pails, extinguishers, hose stations, hydrants, pumps, sirens, and sprinkler piping.
   2. *Danger:* safety cans and other containers for flammable liquids (excluding shipping containers), red lights at barricades and temporary obstructions, and danger signs.
   3. *Stop:* stop bars on hazardous machines, stop buttons on electrical switches.

2. *Orange — shall be a basic colour for designating dangerous parts of machines or equipment such as:*

   The inside of movable guards for gears, chains and pulleys and machinery.
   Safety starting buttons.
   Exposed parts (edges only) of pulleys, gears, rollers and cutting devices.

3. *Yellow — shall be the basic colour for designating caution and for marking physical hazards (striking, stumbling, falling, tripping and crushing). Solid yellow, yellow and black stripes, yellow and black checkers shall be used interchangeably, using the combination which will draw most attention in the particular environment.*

The following suggestions are made for the use of yellow:

Construction equipment (or parts of them), such as bulldozers and tractors.
Material handling equipment, (or parts of them) trucks, trailers and overhead cranes.
Locomotives (or parts of them).
Corner markers for stacks and storage piles.
Covers or guards for guy wires.
Exposed and unguarded edges of platforms.
Pits and walls.
Fixtures suspended from ceiling which extend into operating areas.
Handrails, guardrails and upper or lower treads of hazardous steps or stairs.
Lower pulley blocks, cranes.
Markings for projections, doorways, low beams and pipes, elevator frames and gates.
Pillars, posts or columns liable to be struck.
Vertical edges of horizontally sliding doors.
Pipes containing dangerous materials.

Waste containers for explosive or very combustible materials should be marked with a yellow band whose height is not more than one third of the height of the can. Wording to indicate the contents such as EXPLOSIVE or COMBUSTIBLE should be painted in large red letters on the band.

Yellow shall also be used for warning signs against starting equipment under repair, and these shall be prominently located.

4. *Green — shall be used as the basic colour for designating safety and the location of first aid equipment such as:*

Safety bulletin boards.
Gas masks.
First aid kits and dispensary.
Stretchers.
Safety showers.

5. *Purple — shall be the basic colour used to designate radiation hazards.*

6. *Black, White – or combinations of both shall be used for the designation of traffic and housekeeping signs. Solid white, solid black, single colour striping, alternate stripes of black and white or black and white checkers shall be used in accordance with local conditions.*

The following suggestions are made for the use of black and white.

1. *Traffic:* dead ends; location and widths of aisles and passageways; stairways (direction and border limit lines); direction signs.
2. *Housekeeping:* location of refuse cans; white corners for rooms or passageways; drinking fountains and location of food dispensing equipment; clear floor areas around first aid, fire fighting or other emergency equipment.

The colours which are defined in detail in the specification correspond to the designations and block numbers on the ISCC/NBS code in *Table A.6.*

Table A.6. DESIGNATIONS AND BLOCK NUMBERS ON THE ISCC-NBS CODE

Safety colour	*Designators Colour description*	*Block numbers*
Red	Vivid red	11
Orange	Vivid orange	48
Yellow	Vivid yellow	82
Green	Brilliant green	140
Purple	Vivid reddish purple	236
White	White	263
Black	Black	267

A.4 BRITISH STANDARD SIGNS AND COLOUR CODES

The British Standards Institution has drawn up and published a number of standards on signs and colour markings for identification and safety. Perhaps the most basic of these are the 'Kitemark' and 'Safety Mark' shown in *Figure A.1.* Safety specialists should know of these standards and be familiar with those relevant to their work. A list of the main ones are given in *Table A.4.* Many of these standards cover the same ground as international codes, although they may differ in the colours or signs employed. The international codes to which they are equivalent are shown in *Table A6* with the following symbols in front of them:
= identical to.
≠ technical equivalent to.
± covers same subject matter as.

The Kitemark   The Safety Mark

Figure A.1 BS Kitemark and Safety Mark

A.5 EXAMPLES OF COLOUR WARNING SIGNS

A number of coloured warning signs used in the UK are shown in the colour plates.

REFERENCES

1. Morgan, C.T., *Human engineering guide to equipment design*, McGraw Hill, New York (1963).
2. *The International Standards Organisation, Geneva, Switzerland.*
3. *The American National Standards Institute, New York.*
4. *The British Standards Institution, London.*

Appendix B

First Aid

Contents

B.1 INTRODUCTION

First aid is exactly what it says, i.e. skilled help given *immediately* and *on the spot*. It should not be confused with subsequent professional medical treatment.

First aid lies on the borderline between safety and medicine. The organisation of first aid services, first aid boxes, ambulance rooms, record keeping and the various statutory requirements are dealt with in booklet 36[1] in the Health and Safety at Work Series. Usually when someone is injured at work they will be assisted until the ambulance arrives by an employee who has taken and passed a recognised course in first aid. Sometimes, especially on shift work, there may not be anyone available with training in first aid, and there may be considerable delays before an ambulance arrives.

This appendix which is based mainly on the standard first aid textbook[2] of the two ambulance associations and British Red Cross Society is intended to provide helpful information for just such a situation. It is intended to supplement, but in no way to supplant, recognised first aid training courses.

Whenever a serious injury occurs, an ambulance should be called immediately, so that the injured man can be removed speedily either to the works medical centre or to hospital for full treatment.

The treatment given by the nurse or attendant at the works medical centre is not dealt with here. The treatment given at the place of the injury should not cause any interference with or delay to that given at the medical centre or hospital. Even simple treatment given immediately may, however, be more effective than more elaborate treatment given five or ten minutes later at the medical centre. If the victim has stopped breathing, resuscitation may be achieved by prompt treatment given on the spot, whereas he would be dead by the time he was taken to the medical centre.

Before even simple treatment is given, some diagnosis is needed and rapid answers to eight key questions are needed.

Is the casualty breathing?
Is the casualty conscious?
Is there any bleeding?
Are there any broken bones?
Is the casualty burned?
Has the casualty been electrocuted?
Has the casualty been gassed?
Has the ambulance been called, and is it on its way?

B.2 CASUALTY NOT BREATHING

If the casualty has been electrocuted, gassed, suffocated or drowned, breathing may have stopped. In each case some preliminary action is needed before first aid resuscitation begins, and medical assistance should also be called for promptly.

B.2.1 Electrocution

The casualty must first be removed from contact with the live conductor. The person doing this should ensure that they are not electrocuted (see section 5.7).

B.2.2 Gassing

The casualty must first be removed to a gas-free atmosphere as quickly as possible, the rescuer taking steps to avoid being gassed (see section 6.1).

B.2.3 Suffocation

This may be caused by the casualty being buried in loose material or by entering an atmosphere deficient in oxygen, or by a foreign body lodged in the trachea. If the casualty is suffering from a deficiency of oxygen the first steps are the same as for gassing. If the casualty has been buried, all debris round the nose and mouth should be removed as quickly as possible, and resuscitation should then start at once while others continue to uncover his body.

B.2.4 Drowning

Remove the casualty from the water as quickly as possible, clear any debris from the mouth, and start resuscitation at once. If rescue was by boat, carry out resuscitation in boat.

B.2.5 Procedure for mouth-to-mouth resuscitation *(Figures B.1 and B.2)*

The procedure to be followed for this method is given below:

1. Lay the casualty on his back.
2. Clear the mouth and throat and nostrils of obstruction if possible, but do not delay inflation to do this. Loosen all tight clothing round neck chest and waist. (Note. The brain begins to suffer irrepairable damage if left without oxygen for a few minutes).
3. Tilt the head well back and push the jaw up to open air passages so that they are not blocked by the tongue *(see Figure B.1)*
4. Take a deep breath, open your mouth wide, seal your lips round the casualty's mouth and seal the nostrils with your cheek or by pinching his nose with your fingers.
5. Blow into the lungs and watch for the chest to rise, then remove your mouth.
6. Take another breath and watch the chest fall.
7. Repeat the process. The first six inhalations should be given as quickly as possible and subsequent ones at ten per minute.
8. If you are unable to open the casualty's mouth or, if the casualty has no teeth, then seal the mouth and blow through the nose, using the same procedure as for the mouth-to-mouth method.
9. Continue the process until the casualty begins to breathe, or a medical expert with special oxygen equipment is ready to replace you, or until a doctor has pronounced the casualty to be dead.

B.3 CASUALTY UNCONSCIOUS BUT BREATHING

If the casualty is breathing but unconscious, he should be laid on a flat surface on his front, with one leg out straight and the arm on the same side as his leg pointing down in the same direction as the leg. The other arm and leg should be drawn up and the head should be tipped to the same side and slightly lower than the rest of the body. This helps the air passages to drain and prevents the casualty choking from his own blood or vomit as could happen if he is laid flat on his back. If the casualty is on a stretcher, the correct position can be achieved by tilting it slightly to one side. No pillow should be used for this will tilt the head upwards.

While waiting for an ambulance, the unconscious person may be covered with a blanket, but the casualty should not be overheated. Extra wrappings and hot water bottles should not be added. Shock and loss of blood will turn him cold as his blood gathers at the injured areas, but this is a natural reaction and should not be interfered with. No attempt should be made to arouse an unconscious accident victim, but on no account should they be left alone, since they may wake up and become restless, the breathing may stop or they may begin to bleed.

(a) NORMAL   *(b)* UNCONSCIOUS

(c) UNCONSCIOUS — WITH HEAD BACK

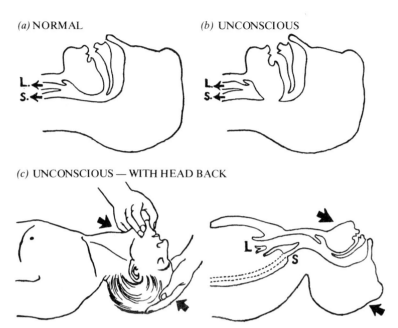

Figure B.1   Mouth-to-mouth method, position of head. (a) In the conscious person, the back of the throat is clear, so that air can pass freely from the nose and mouth to the lungs. L is the wind-pipe leading to the lungs. S is the gullet leading to the stomach. (b) In the unconcious patient, lying on his back, the tongue falls backwards, blocking the air-passage. (c) By bending the patient's head right back, the air passage behind the tongue is once more opened (From 'First Aid in the Factory' (Longmans) by Lord Taylor of Harlow)

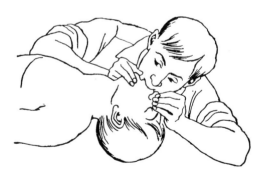

Figure B.2   Mouth-to-mouth method. The operator, having taken a deep breath, applies his mouth to the patient's mouth. He then blows until out of the corner of his eye he sees the patient's chest rise. Throughout the patient's nose must be kept closed by pinching (From 'First Aid in the Factory' (Longmans) by Lord Taylor of Harlow)

B.4 CASUALTY BLEEDING

If the casualty is wounded and bleeding externally, try to stop the loss. One may lose ½ litre of blood without harm, but to lose 1 litre is serious. There may be multiple injuries and further loss of blood through internal bleeding.

Whenever bleeding is suspected, seek to find its location, and remove the casualty's clothing as needed to do this. Do not remove large pieces of embedded matter. Keep pieces of glass to show to the doctor — particularly glass from spectacles. Stop bleeding by direct pressure over the area affected unless a fracture is present or suspected, using the cleanest material available until the contents of the first aid box can be reached.

Cover open wounds as soon as possible with sterile dressings from the first aid box, taking care not to touch them on the inside. If there are loose particles in the wound large enough to be picked out without touching it, do so quickly, but do not attempt to prise out foreign matter or clean away dirt or grit. Once a dressing has been applied, do not remove it, as this will disturb any clotting that has occurred.

If blood seeps through a dressing, apply an additional dressing over it, and bandage it so as to exert pressure, except when there are foreign objects or broken pieces of bone in the wound. But if a finger or other limb has been amputated by the accident, the dressing must be bandaged tightly on to the stump and the limb elevated where possible to prevent serious loss of blood.

If there has been a substantial loss of blood, and there are no chest injuries, raise the patient's legs if possible so that the blood remaining in the body is concentrated about the heart, lungs and brain. But if there are chest injuries, prop up the patient to a seated position.

If the patient has broken bones or internal injuries there is likely to be internal bleeding at these points. This will be shown by swelling and a rise in the pulse rate to 90 beats per minute or more although the strength of the pulse will diminish. It is then vitally important to have him moved to hospital as speedily as this can be done without causing pain.

B.5 BROKEN BONES

The first aider should be able to recognise a fracture, know when to suspect one and to distinguish its type so as to be able to decide whether early care or early removal to hospital is most important. There are several types of fracture, simple, compound, complicated, comminuted, impacted, stellate.

In a compound fracture, there is an associated wound through which the broken bone may appear. There is risk of infection and the wound should be treated like any other and covered with a sterile dressing.

If the patient is conscious, ask them what happened, whether there is pain with movement, whether the site of the injury is tender and whether they have lost the use of any limb.

Whether the patient is conscious or not, ask any witnesses what

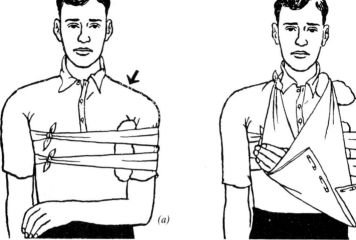

Figure B.3   Fixing a broken collar bone. (a) A pad of cotton wool is placed in the armpit, and the upper arm bound to the side of the chest by two folded triangular bandages. (b) The forearm is supported in a sling at an angle of 45 degrees, and a large cotton-wool pad placed between the sling and the injured shoulder. Alternatively, a clove-hitch may be used or, by those who have learnt its use, the special St. John sling. The object of (a) is to lever out gently the point of the shoulder. The object of (b) is to take the weight of the arm off the injured bone (From 'First Aid in the Factory' (Longmans) by Lord Taylor of Harlow)

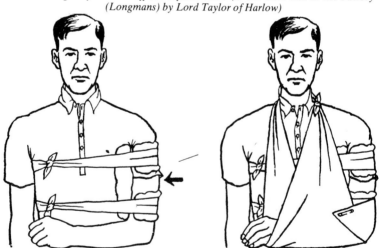

Figure B.4   Fixing a broken humerus. (Left) The side of the chest is used as a splint. A large pad of cotton wool is placed between the arm and the chest and cotton wool placed round the arm above and below the break. The arm is bound to the chest with two folded trinagular bandages. (Right) The forearm is supported in a sling at a right-angle (From 'First Aid in the Factory' (Longmans) by Lord Taylor of Harlow)

happened and look for deformity, unnatural position and swelling at the site of the injury. Listen for the grating of broken bone ends if the patient tries to ease their position by moving, but do not deliberately move the patient to do this.

From the information found in these ways, the location of a suspected fracture must be diagnosed. The history of the accident is often more important for making a diagnosis than are the symptoms. The first treatment of all broken bones is to immobilise them. Broken limbs should be moved gently, slightly stretched, and the tension should not be released until the limb is in the correct position.

B.5.1 Broken fingers or toes

Open wounds should be dressed. Shoes should be left on, but shoe-laces removed.

B.5.2 Broken collar bone

A collar bone *(Figure B.3)* is easily broken by a heavy fall on the hand or a blow on the point of the shoulder. The casualty can travel upright with his arm supported in a triangular sling.

B.5.3 Broken upper arm *(Figure B.4)*

This is generally the result of a direct blow on the arm. Suspend the wrist from a bandage round the neck and allow the weight of the injured arm to provide tension.

B.5.4 Fracture of elbow or near elbow

Transfer the casualty to a stretcher. Lay the arm out straight, padded, and secured to the body with bandages knotted on the other side opposite the injury. Arrange for the casualty to be taken to hospital on the stretcher, because of risk of damage to the circulation if the artery is damaged.

B.5.5 Fracture of wrist or forearm *(Figure B.5)*

Support the broken limb in a sling taken round the neck. The patient may travel to hospital sitting up.

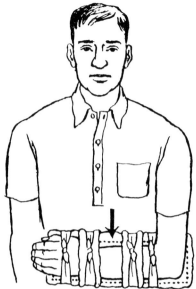

Figure B.5   Fixing a broken radius and ulna. A simple splint padded with cotton-wool extends from the elbow to the knuckles along the palm surface of the forearm and hand. Note the cotton-wool pads under the bandages on each side of the fracture. The forearm is carried in a right-angle sling (not shown)

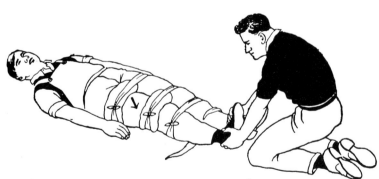

Figure B.6   Splinting a broken hip thigh or shin. For simplicity, the first-aider doing the bandaging is not shown. The arrow indicates the position of the suspected break. An assistant exerts a steady pull on the foot, without bending or turning it while, the bad leg is tied to the good leg with folded triangular bandages. Plenty of cotton-wool is placed around the injured limb before the two bandages are applied just above and below the fracture (From 'First Aid in the Factory' (Longmans) by Lord Taylor of Harlow)

B.5.6 Broken pelvis

Place the casualty on a stretcher and make comfortable with padding. Ask the casualty to try to stop any voluntary flow of urine. Tie the legs together with bandages at the knee and ankle to prevent unnecessary movement. Arrange for the patient to be transferred as quickly as possible to hospital as there is certain to have been internal bleeding.

B.5.7 Broken thighbone (*Figure B.6*)

Treat as for broken pelvis, but with two more bandages added above and below the injury.

B.5.8 Fracture of lower leg

When the shin bone breaks it is liable to form a sharp point. Being close to the shin, it may easily penetrate it thus converting a simple fracture into a compound one. Great care in handling is therefore needed. Place the casualty on a stretcher. Place padding between the legs and secure the injured leg to the sound one at the knees and ankles and above and below the injury.

If both legs are broken, make two long wooden splints running from armpit to ankle and bandage round them.

B.5.9 Fractured ankles

If a fractured ankle is suspected, treat the casualty as a stretcher case and support the ankle on a cushion.

B.5.10 Fractured spine

If injury to the spine is suspected *(Figure B.7)*, either from the casualty's pain, the nature of his injury, or the position, do not move the patient, but arrange for a doctor or qualified ambulance personnel to come to the casualty.

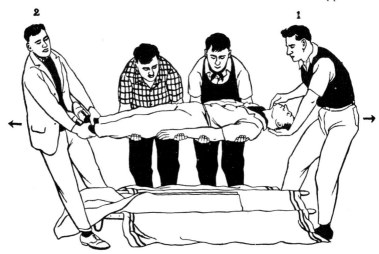

Figure B.7   The right way to move a patient with an injured back. The patient's spine must be kept absolutely straight and head and foot traction applied at the same time. The first man places one hand under the patient's neck with the head bent back. The second puts his hands around the patient's heels. All four men must move very slowly and carefully and in complete unison. It is best, however, not to move the patient at all, but to wait for the expert ambulance staff (From 'First Aid in the Factory' (Longmans) by Lord Taylor of Harlow)

Any movement of a broken back may add damage to the spinal cord and result in death or paralysis. If the patient has to be moved, at least four people should do this. They must take great care to see that this is done without altering the position of the back.

B.5.11 Broken ribs

Broken ribs can be serious if a lung is pierced. This may be indicated by the casualty coughing blood, or by the presence of an open wound large enough to allow air to be sucked into the chest. Plug any such wound quickly with a sterile dressing held in position with plaster. If this is not done, the lungs may cease to function.

If conscious the casualty may travel sitting up. If the patient must recline, the injured side should be lowest. The patent should be supported sitting up if both sides are affected.

B.5.12 Head injuries

Any casualty who has received a severe head blow should be examined promptly by a doctor and his breathing should be continuously checked until he is delivered to the doctor. All head injuries should be treated as serious until seen by a doctor.

B.6 CASUALTY BURNED

Burn casualties require prompt treatment with cold drinking water. Rapid cooling reduces the effect of the burn and relieves pain. A burned limb should be dipped in a bucket of water, while body burns may be gently soaked.

An aerosol burn spray is strongly recommended and should be kept in every first aid box. This should be applied over the burn following the cold water treatment, or in place of it if there is no ready supply of cold water available. The emphasis here is on promptness.

The aerosol spray cools as well as sterilises the burn and brings speedy relief. If charred, fused or scorched clothing is adhering to the skin at the point of the burn, it should not be removed by the first aider, but left for expert treatment in the works medical centre or hospital.

Burns should be lightly covered by sterile burn dressings after application of the burn spray. If there are not enough burn dressings available, dry, non-fluffy cotton materials such as handkerchiefs or strips of sheets may be used to cover the burn. Overlapping strips are preferred to large sheets.

Burn casualties may be given a drink of water, tea or milk every 10 minutes, but no alcohol should be given. No other class of seriously injured casualty should be allowed anything to drink.

Burn blisters should not be burst and burns should not be touched with the fingers.

B.7 CASUALTY ELECTROCUTED

Rescue of electrocution casualties and their resuscitation if they are not breathing has been dealt with in section 5.7 and B.2.1.

When the casualty is breathing, or earlier if possible, while resuscitation is taking place, attention should be given to any other injuries caused by the electrocution. These often include burns and injuries resulting from a fall after the electric shock.

B.8 CASUALTY GASSED

The most common cause of gassing is through carbon monoxide, formed as a result of incomplete combustion. It is particularly dangerous since it cannot be detected by smell, and its casualties rarely recognise its symptoms in themselves in time to escape. If the casualty is unconscious or unable to walk, he should only be treated by someone equipped with proper breathing apparatus (section 3.10). All casualties should be removed to fresh air and made to rest.

If the casualty is not breathing, artificial respiration by the mouth-to-mouth method (section B.2.5) should be given and cardiac massage should be applied if unconscious. Medical assistance should be obtained as quickly as possible.

There are, however, many cases of persons who are suffering from carbon monoxide poisoning without being totally incapacitated, and it is important to recognise their symptoms. The principal symptoms are:

Pinkness of the face.
Rapid pulse rate.
Headache.
Sickness.
Weakness of the limbs.
Dizziness.

The casualty should not be allowed to make unnecessary movements, since any exertion increases the toxic effect of carbon monoxide. Oxygen should be given wherever possible through the use of a simple oxygen resuscitation apparatus which the patient can use alone.

Whenever carbon monoxide poisoning is suspected, this should be checked and confirmed by chemical analysis of the atmosphere. Any concentration above 50 parts per million is dangerous and requires immediate action by the plant manager and engineer to ascertain the cause and prevent its further occurence.

B.9 OTHER FIRST AID DUTIES

Plant foremen, supervisors and other workers trained in first aid have responsibilities additional to those of immediate treatment of the casualty. These include:

Calling for the ambulance and explaining briefly the nature of the casualty.
Looking after the kit, clothing and any possession left by the injured person at the works.
Informing the injured person's family.
Completing an injury report.
Arranging for the injured person's work to be carried on by some other suitable person.
Taking immediate steps to deal with the cause of the injury, so that others do not suffer the same fate.
Restoring the morale of other workers.

B.10 MAIN LEGAL REQUIREMENTS

This section outlines the legal requirements which all industrial premises should comply with.

B.10.1 First aid boxes

The requirement to provide and maintain first aid boxes is contained in the Factories Act 1961, section 61.

All factories are required to provide and maintain at least one first aid box or cupboard, and where more than 150 persons are employed, an additional box or cupboard is needed for each 150 persons.

Each first aid box or cupboard is to be kept in charge of a responsible person whose name must appear on a notice fixed in every workroom for whose workers the box is provided. (The box or cupboard may, however, be outside the workroom.)

The person responsible for the first aid box in a factory where more than 50 persons are employed shall be trained in first aid treatment and shall always be available during working hours. They should keep a record of all cases treated or report them to the person keeping such records. It is also their duty to see that the box is properly stocked and kept in a clean and serviceable condition.

Boxes should be placed so that injured workers can get treatment in the best conditions as quickly as possible. If possible the box should be placed near a wash basin or sink. It should be in a good light with a strong chair close at hand. If the box is in an office the office should not be locked.

Standards (including minimum contents) of first aid boxes are given in the First aid Boxes in Factories Order 1959 (SI 1959), details of which are given in the H and SE booklet 36 referred to earlier.

B.10.2 First aid training

The standard of training required is given in the First Aid (Standard of Training) Order 1960 (SI 1960 No. 1612). This requires the person to hold a qualifying certificate of training which is valid for three years. At the end of that time the person must pass a further examination in first aid, after first taking any additional training necessary.

B.10.3 Ambulance rooms

The name 'ambulance room' is used in the legislation for what is usually referred to as the first aid room, works surgery or medical centre. Certain works and factories with more than a stipulated number of employees (from 100 to 500) are obliged by law to have their own 'ambulance rooms' where immediate treatment can be given.

The size, location and equipment provided in an ambulance room depends on the number of workers, the type of work done in the factory and on its layout. The person in charge of an ambulance room should not only hold the ordinary certificate of training in occupational first aid, but where possible, a full-time registered nurse with special training in occupational health nursing should be employed. Where a factory is too small to justify employing its own full-time nurse, it is suggested that the employer explores the possibility of sharing the services of a nurse with one or more neighbouring factories.

B.10.4 Records of first aid treatment

Every accident treated at a first aid box should be recorded. The record should include:

1. The name and clock card of the injured person.
2. The nature and site of the injury and the times both when the accident occurred and when the injury was first treated.
3. An indication of the alleged cause of the injury and where it occurred.
4. Treatment given and by whom.
5. A note on action taken after treatment, e.g. return to work, sent to hospital or personal doctor.

More extensive information on the planning, layout and use of works medical centres is given in the references quoted.

The use of first aid records in investigating causes of accidental injuries was discussed in section 1.5.

REFERENCES

1. Health and Safety Executive, *First aid in factories*, HMSO, London (1976).
2. The St. John Ambulance Association and Brigade, The St. Andrew's Ambulance Association and The British Red Cross Society *First aid* 3rd ed. (1975)
3. Cameron, J.D., 'Medical services: first aid and casualty treatment *Industrial Safety Handbook*, editor Handley, W., 2nd ed., McGraw Hill (1977).
4. Hall, M.H., 'Medical services: liaison with the hospital emergency department', also from Industrial Safety Handbook.
5. Lord Taylor of Harlow, *First aid in the factory*, Longmans, London (1973).

Appendix C

Some accident statistics

Most so-called accident statistics are in fact injury statistics, according to the definitions given in section 1.2. Selected statistics of injuries of different types and those found in particular occupations have been given in the various sections of this book. Most of these statistics were derived from the extensive data which appeared up to 1974 in the Annual Reports of HM Inspector of Factories and subsequently in the annual *'Health and Safety, Industry and Services (later Health and Safety, Manufacturing and Service Industries')* published by the Health and Safety Executive.

The figures given in this Appendix are abstracted partly from the Pearson Report, which compares accidental injuries caused at work with those on the road and in the home, etc, partly from *'Health and Safety, Industry and Services 1975* and partly from *'Health and Safety, Manufacturing and Service Industries, 1979'*. The latter have been rearranged to allow quick comparisons to be made between injury incidence rates of different industries and between the frequencies of different types of accident.

For more information the sources referred to should be consulted.

C.1 DATA FROM THE PEARSON REPORT

An overall view of all accidental injuries serious enough to lead to four or more days off work or an equivalent degree of incapacity for those not in work, is given in the Pearson Report, from which the following passage and figures have been taken.

'We estimate that no more than 3 per cent of men and 8 per cent of women will escape injury, as we have defined it, during an average lifetime. Injuries happen in the course of every kind of human activity — at work or at school, when driving, when playing games, in the street or at home. In infancy and in old age, accidents at home predominate; in the school years, accidents at school; and in the middle years, accidents at work. Until around retirement age, and even in infancy, males are more susceptible to injury than females. Manual workers and their families are more subject to injury than non-manual workers.'

Figure C.1   Graph showing relationship between injuries and time off work

Table C.1 gives the number of injuries and deaths following injury each year in the UK from 1973 to 1975, caused at work, through motor vehicles and other causes. This shows a number of interesting points. The most striking is that injuries at work although more than twice as numerous as motor vehicle injuries, produce only about a fifth as many deaths.

Figure C.1 shows the distribution of periods off work resulting from an accidental injury throughout British industry. 50% of injured workers return to work after a little more than two weeks. Over 98% return within 6 months. These figures will of course be quite different for road accidents, and they do vary considerably from one industry to another, as shown in section C.2.

C.2 DATA FROM 'HEALTH AND SAFETY, INDUSTRY AND SERVICES 1975'

Table C.2 gives the incidence rates of fatal accidents serious (group 1) accidents and total accidents for various types of industry in the years 1970 and 1979. There is approximately an eight-fold range in the incidence rates both of serious and total accidental injuries from 830 accidental injuries per 100 000 at risk (per annum) in clothing and footwear to 6900 in metal manufacture. It is clear from this table that light industries such as clothing and footwear, instrument engineering and electrical engineering are much safer than construction and heavy industries.

Table C.3 gives the total accidental injury frequency rate for the entire range of industries (according to Standard Industrial Classification) throughout the country. These show an even wider span. The table has been divided into three parts. Part 1 gives branches of industry with low accidental injury frequency rates, up to 2000. Several concerned with clothing have frequency rates below 500. Part 2 gives those branches of industry with medium frequency rates, taken as 2000 to 4000.

Part 3 gives branches of industry with high frequency rates, above 4000. Coke ovens, iron foundries, brickworks and the production of refractories have the highest incidence rates, as might be expected from the combinations of hazards present — material handling, burns, gassing and falls of persons and things. Some branches of industry have higher incidence rates than one might expect, brewing and malting are examples.

One possible use of this table is to enable managers and safety specialists to compare the injury incidence rate in their own works with the national average for the same branch of industry.

Table C.4 gives the numbers of total and fatal accidental injuries throughout manufacturing industry for various types of accidents. The predominance of injuries caused by manual handling, falls and machinery is clear from this table.

Table C.5 gives a similar breakdown of accidental injuries and fatalities in the construction industries. Here, as one might expect, falls of persons have just edged manual handling out of the top place in the table.

Table C.6 gives a breakdown of the falls of persons in construction industries listed in *Table C.5* but for 1975 only. The high proportion (40%) of these falls which occurred on the flat seems a little surprising. Most of the rest are from ladders and scaffolding.

REFERENCES

1. *The Royal Commission on Civil Liability and Compensation for Personal Injury.* Chairman: Lord Pearson, 3 volumes. HMSO (1978).
2. The Health and Safety Executive. *Health and Safety Industry and Services* (1975) HMSO.

Table C.1 INJURIES AND DEATHS FOLLOWING INJURY IN UK 1973-1975

	All injuries			*Males 15 & over*		
	Injuries	*Deaths following injury*	*% of injuries leading to death*	*Injuries*	*Deaths following injury*	*% of injuries leading to death*
Injuries at work	720 000	1 300	0.18	630 000	1 270	0.20
Motor vehicle injuries*	290 000	7 220	2.49	160 000	4 330	2.71
Other injuries	2 040 000	12 900	0.63	660 000	5 100	0.77
All injuries	3 050 000	21 420	0.70	1 450 000	10 700	0.74

	Females 15 & over			*Children under 15, students*		
	Injuries	*Deaths following injury*	*% of injuries leading to death*	*Injuries*	*Deaths following injury*	*% of injuries leading to death*
Injuries at work	90 000	30	0.03	—	—	—
Motor vehicle injuries*	80 000	2 080	2.60	50 000	810	1.62
Other injuries	620 000	6 540	1.05	760 000	1 260	0.17
All injuries	790 000	8 650	1.40	810 000	2 070	0.26

Table C.2 MANUFACTURING AND CONSTRUCTION INDUSTRIES (GROUPED BY SIC ORDER) IN ORDER OF INCREASING SERIOUS ACCIDENTAL INJURY INCIDENCE RATE

Averages from 1971 to 1975 compared with 1979. Fatal accident incidence also included

SIC order	Industry	Incidence rate per 100 000 at risk			
		Serious Group I accidental injuries		Fatal accidents	
		1971–1975	1979	1971–1975	1979
XV	Clothing and footwear	124	120	—	0.3
VIII	Instrument engineering	246	230	—	—
IX	Electrical engineering	340	370	—	0.6
XIV	Leather, leather goods and fur	432	320	—	—
XI	Vehicles	448	490	2.1	0.8
XVIII	Paper, publishing and printing	476	360	—	2.8
XIII	Textiles	512	510	2.6	1.1
V	Chemical and allied industries	626	510	9.0	5.0
VII	Mechanical engineering	698	640	3.6	2.7
III	Food, drink and tobacco	706	750	3.3	3.1
XII	Metal goods not elsewhere specified	714	630	3.2	1.8
XX	Construction	756	500	18.8	11.7
XVI	Bricks, pottery, glass, cement, etc	806	580	7.6	5.1
XVII	Timber, furniture, etc	816	550	4.7	2.6
IV	Coal and petroleum products	818	720	—	24.6
X	Shipbuilding and marine engineering	856	900	13.8	8.3
VI	Metal manufacture	1022	930	13.7	10.2
XIX	Other manufacturing industries	486	460	—	3.4

Note: Too few incidents for statistically reliable averages

Table C.3

ACCIDENTAL INJURY INCIDENCE RATES OF TOTAL REPORTED ACCIDENTS FOR
DIFFERENT INDUSTRIES PER 100000 AT RISK 1975-1977 AVERAGES AND 1979

Part 1. Low Incidence Rates (below 2000 in 1971-73)

Min list heading	Industrial branch	1975-1977	1979
446	Hats, caps and millinery	340	330
443	Women's & girls' tailored outerwear	480	450
893	Dry cleaning, job dyeing, carpet beating	470	530
445	Dresses, lingerie, infants' wear	670	740
892	Laundries	620	480
353	Surgical instruments & appliances	720	710
441	Weatherproof outerwear	880	830
366	Electronic computers	820	740
367	Radio, radar & electronic capital goods	690	940
449	Dress industries not elsewhere specified	710	750
444	Overalls and men's shirts & underwear	860	1010
354	Scientific & industrial instruments & systems	1090	1030
417	Hosiery & other knitted goods	1150	1150
485	Printing and publishing of newspapers	800	680
486	Printing and publishing of periodicals	1230	1150
442	Men's & boys' tailored outerwear	1090	940
432	Leather Goods	980	860
450	Footwear	1150	1300
474	Shop and office fitting	1100	990
433	Fur	1230	720
489	Other printing & publishing, bookbinding, engraving, etc	1450	1450
390	Engineers small tools & gauges	1790	1520
365	Broadcast & sound reproducing equipment	2050	2200
363	Telegraph & telephone apparatus & equipment	1420	1250
422	Made up textiles	1600	1410
421	Narrow fabrics (not more than 30 cm wide)	1980	2380
396	Jewelry and precious metals	1580	1520
495	Miscellaneous stationer's goods	1670	2290
894	Motor repairers, distributors, garages and filling stations	1910	1850

Table C.3
Part 2. Medium Incidence Rates (2000 to 4000 in 1971-1973)

Min. list heading	*Industrial branch*	*1975-1977*	*1979*
221	Vegetable & animal oils & fats	2660	2770
493	Brushes & brooms	1840	2000
273	Toilet preparations	3420	3090
352	Watches & clocks	2270	2220
418	Lace	2860	2370
380	Wheeled tractor manufacturing	2010	1710
383	Aerospace equipment manufacturing & repair	2310	2140
484	Manufacturing of paper & board not elsewhere specified	2270	2620
483	Manufactured stationery	2130	1540
369	Other electrical goods	2730	2470
232	Soft drinks	3160	2690
499	Miscellaneous manufacturing industries	2550	1880
473	Bedding, etc	2000	2350
472	Furniture and upholstery	2310	2230
494	Toys, games, children's carriages and sports equipment	1820	1840
215	Milk and milk products	3150	4040
338	Office machinery	2100	1990
368	Electric appliances primarily for domestic use	2310	2040
393	Bolts, nuts, screws, rivets	3140	3000
482	Packaging products of paper, board and associated materials	2490	3200
392	Cutlery, spoons, forks & plated tableware	2770	2440
212	Bread, flour and confectionary	3640	3480
413	Weaving of cotton, linen & man-made fibres	3690	3460
364	Radio and electronic components	2020	1650
335	Textile machinery and accessories	3130	3210
496	Plastics products not elsewhere specified	2800	2770
279	Other chemical industries	2980	2570
381	Motor vehicle manufacturing	2970	3120
464	Cement	3140	4150
277	Dyestuffs and pigments	3260	3390
361	Electrical machinery	2950	2650
462	Pottery	3120	3300
414	Woollen and worsted	3410	3010
416	Rope, twine and net	2660	2220
333	Pumps, valves and compressors	3760	3100
219	Animal and poultry foods	2760	2810
272	Pharmaceutical chemicals and preparations	3990	3980
423	Textile finishing	3360	3130
274	Paint	3840	4300
431	Leather (tanning & dressing) & fell mongery	3750	3270

(cont.)

Min. list heading	Industrial branch	1975-1977	1979
500	Construction	3430	3130
412	Spinning & doubling in the cotton & flax systems	3730	4010
332	Metal working machine tools	3780	3040
213	Biscuits	3680	3690
411	Production of man-made fibres	3990	4140
349	Other mechanical engineering not elsewhere specified	3690	3260
337	Mechanical handling equipment	3940	3400
217	Cocoa, chocolate & sugar confectionery	3830	3770
271	General chemicals	4210	4550
341	Industrial (inc. process) plant & steelwork	3880	3470

Table C.3
Part 3. High Incidence Rates. Over 4000 in 1971-1973

Min. list heading	Industrial branch	1975-1977	1979
399	Metal industries not elsewhere specified	3860	3610
429	Other textile industries	3810	3650
382	Motor cycle, tricycle and pedal manufacture	5050	·4550
334	Industrial engines	4600	4210
262	Mineral oil refining	4150	5300
278	Fertilizers	2980	2810
339	Other machinery	4230	3840
391	Hand tools and implements	4500	3310
479	Miscellaneous wood and cork manufactures	4340	3770
229	Food industries not elsewhere specified	4610	4450
263	Lubricating oils and greases	3940	3600
419	Carpets	4090	3450
491	Rubber	4430	4250
276	Synthetic resins & plastics materials & synthetic rubber	4280	4780
475	Wooden containers and baskets	4180	4080
218	Fruit and vegetable products	4920	5160
239	Other drink industries	5150	5880
395	Cans and metal boxes	5000	4780
336	Construction and earth moving equipment	4130	3550
492	Linoleum, plastics floor-covering, leather cloth, etc	4190	3980
469	Abrasives & building materials not elsewhere specified	4690	4430
471	Timber	4720	4640
351	Photographic & document copying equipment	6080	4780
323	Other base metals	6220	7090
362	Insulated wires and cables	4670	4570
312	Steel tubes	3990	3820
221	Grain milling	2660	2770
216	Sugar	4550	4250
321	Aluminium & aluminium alloys	5600	5040
342	Ordnance and small arms	4970	5160
322	Copper, brass & copper alloys	5810	5690
463	Glass	5260	5260
415	Jute	4830	4520
214	Bacon curing, meat and fish products	6720	6930
385	Railway carriages & wagons & trains	6390	5000
370	Shipbuilding & marine engineering	6120	4960
231	Brewing and malting	6250	6140
311	Iron and steel (general)	6310	5850
331	Agricultural machinery (except tractors)	5920	4920
384	Locomotives & railway track equipment	8050	6190
394	Wire and wire manufacture	6490	6140
481	Paper and board	7240	7120
461	Bricks, fireclay & refractory goods	7040	6180
313	Iron castings, etc	8920	8250
261	Coke ovens & manufactured fuel	10530	9260

Appendices

Table C.4

Arranged in decreasing order of total accidental injuries in 1979

Accident classification	Total reported accidents		Fatalities	
	1975	1979	1975	1979
Handling goods (not otherwise specified)	58 460	56 380	3	8
Falls of persons	32 049	37 125	39	28
Machinery	31 913	23 145	63	38
Stepping on or striking against objects or persons	17 127	13 893	3	1
Transport other than rail	14 118	13 492	42	47
Use of hand tools (non-powered)	13 244	10 682	nil	2
Struck by falling objects	12 248	8 941	18	8
Due to electricity	881	567	10	12
Rail transport	489	303	6	7
Poisoning and gassing	417	474	7	12
Other accidental injuries including fires and explosions	23 322	24 279	40	16
Total reported accidental injuries	204 268	189 281	231	179
Fires and explosions of combustible material	663	—	13	—
Explosions of pressure vessels	417	—	9	—
Explosions of molten metal	337	—	11	—

Table C.5

TOTAL REPORTED ACCIDENTAL INJURIES AND FATALITIES IN CONSTRUCTION
PROCESSES 1975 AND 1979

Accident classification	Total reported accidents		Fatalities	
	1975	1979	1975	1979
Falls of persons	9 892	9 020	67	56
Handling goods (not elsewhere specified)	9 019	8 814	nil	nil
Falls of materials	2 743	1 829	23	10
Stepping on or striking against objects	2 662	2 487	nil	nil
Hand tools (not power or cartridge operated)	2 349	2 112	nil	1
Non-rail transport	2 052	2 156	26	14
Machinery, power and non-power (not lifting)	1 612	686	5	2
Lifting equipment	635	300	22	17
Excavations	221	192	11	8
Electricity	188	135	10	5
Fires and explosions	144	N.R.[1]	2	nil
Tunnelling	77	2	nil	2
Rail transport	55	12	nil	nil
Poisoning and gassing	16	40	11	4
Other	2 497	3 221[2]	4	nil
Totals	34 162	29 006	181	119

Notes (1) Not reported separately
(2) Includes fire and explosion

Table C.6 FALLS OF PERSONS. ALL CONSTRUCTION PROCESSES. 1975

Accidental injuries by type of fall and % of each type in decreasing order

Type of fall	*Total*	*%age*
Total falls on flat	3 972	40.16
From ladders or step ladders	1 503	15.19
From scaffolding (including those during erection or dismantling, or due to collapse of scaffolding)	1 325	13.39
Through openings in floors or walls, or downstairs	814	8.23
Into excavations	391	3.95
Through fragile roofs	257	2.60
From sloping roofs	169	1.71
From structural frameworks during erection	63	0.64
From building structures during demolition	48	0.49
From cradles, boatswains chairs, skips, etc	31	0.31
From hoists or into hoistways	20	0.20
During work on other sloping surfaces	18	0.18
Into water	9	0.09
Other falls from heights	1 272	12.86
Total falls from heights	5 920	59.84
Total falls	*9 892*	

Appendix D

Polyurethane foam

Contents

D.1 INTRODUCTION

Since this book first went to press, the dangers associated with flexible polyurethane foam have received considerable publicity. In May 1979 there was a large fire at an F. W. Woolworth store in Manchester, in which ten people were killed. Polyurethane foam in furniture upholstery was considered to be a substantial factor contributing to the severity of this fire. Although not an isolated incident, the extent of the damage with its estimated loss of around £2½m and the deaths (and possibly because everybody knows Woolworth's), this particular fire received very wide publicity.

In May 1980, following the recommendations made by the Central Fire Brigade Council sub-committee which investigated the Woolworth's fire, the Home Secretary forecast new regulations to control the display of furniture containing pu foam. In the same week, the Minister for Consumer Affairs promised new measures on the flammability of furniture upholstery.

Flexible pu foam may not generally be considered to be an industrial hazard in the same category as, for example, electrical and some mechanical hazards. However, as it is commonly used both in our homes and our workplaces, the authors and the publishers thought it worthwhile to include the following notes on this subject. In the time and space available we can only give a general picture of the problems and point the interested reader in the direction of the sources of authoritative information.

D.2 THE MATERIAL

The term polyurethane foam is often used loosely to mean flexible polyurethane foam although the generic term also covers the rigid foams which are used structurally and which are made from different raw materials. This appendix concentrates on the flexible foams with a brief section at the end on rigid foams. Where the term 'pu foam' is used it refers to the flexible variety.

The Health and Safety Executive, in its publication *HS(G)1—Polyurethane Foam,*[1] defines flexible pu foam as 'a material produced by the reaction between toluene di-isocyanate or diphenyl methane di-isocyanate and a polyol of molecular weight 3000 or above, together with other ingredients, resulting in a flexible cellular product having a predominantly open-cell structure'.

Pu foams are widely used in upholstery in all of its forms, domestic, commercial and automotive. In addition it has many industrial applications including packaging, acoustic and thermal insulation and other more specific uses. Accordingly, it is possible that the vast majority of industrial and commercial premises have some pu foam somewhere on site. Because of this, and the specific problems associated with these materials, particularly in a fire, all industrial safety specialists need to be able to identify problem areas within their organisations, and to ensure that all the necessary safety precautions are taken.

The widespread use of pu foams has arisen because they are cheap and convenient. Cheap, that is, in comparison with the so-called 'traditional' materials they replace. Being based on petrochemicals, however, this situation could change. Pu foams are convenient in that they are comparatively simple to make and to handle. For some insulating purposes they can be sprayed directly into position. They are also very light in weight with densities typically falling in the range $22.5–60 \, kg/m^3$.

D.3 THE PROBLEMS

The problems associated with the manufacture and use of pu foams fall into two categories: toxicity and flammability.

D.3.1 Toxicity

One of the basic raw materials in the manufacture of pu foams is an isocyanate and in the majority of cases this is toluene di-isocyanate (TDI). Isocyanates are very toxic and great care is needed in their handling. However, this is sufficiently well known within the industry so that specific precautions for the storage and handling of TDI and related compounds are (or should be) universally observed.

The second toxic problem involves the products of combustion of pu foams. When these materials are involved in a fire they produce large quantities of black smoke containing, amongst other things, oxides of nitrogen, carbon monoxide, carbon dioxide, hydrogen cyanide and, possibly, free isocyanates. Of these, the presence of hydrogen cyanide has received considerable publicity, much of it emotional and ill-informed. Any organic nitrogen-containing material is likely to produce hydrogen cyanide on combustion. In fact wool, nylon and acrylic fibres all produce more hydrogen cyanide on combustion than do pu foams.[2] The British Rubber Manufacturers Association, in its publication *Flexible Polyurethane Foam its Uses and Misuses*[3] discusses these various toxic products in more detail. This publication will help to counter some of the very misleading statements which have appeared in the press.

D.3.2 Flammability

Flexible polyurethane foams ignite easily and burn readily, like a large number of other natural and synthetic materials. However, pu foams rapidly produce large quantities of dense, black smoke.

Because of the rapid production of this smoke, which as noted above, will contain some toxic materials, people caught in a fire which involves any quantity of pu foam may only have a very short time to reach an exit. Thus, in any situation where pu foam is present in quantity, the standard for travel distance in the event of a fire, as laid down in the Home Office/Scottish Home and Health Department's publication *Guide to the Fire Precautions Act 1971; No 2. Factories* may be inadequate.[1] If there is any doubt as to the adequacy of the provision of escape routes and means of escape, the Fire Prevention Branch of the local Fire Brigade should be consulted.

In general, uncovered pu foams will rarely be encountered outside those industries which handle them as raw materials. More usually the pu foam will be protected by some form of covering material, and the fire properties of this covering material have a vital effect on the overall fire hazard.

Over the years, various attempts have been made to improve the fire characteristics of pu foams. The first approaches used chemical additives, and whilst these certainly improved the ignition properties of the foams (i.e. made them more difficult to ignite), once they were involved in a fire, they released more smoke than the untreated foams, and in some cases they burnt more rapidly.

A more recent approach, which has the same effect of making the foam more difficult to ignite, is used in the 'High Resilient' and 'Neomorphic' foams. These have been formulated so that they begin to decompose at, or below, ignition temperature.

D.4 CONCLUSIONS

Flexible pu foams are very widely used and are likely to be found in any industry. They are flammable, and when involved in a fire they rapidly produce quantities of black smoke and toxic gases. For this reason, in areas where pu foams are present in quantity, great care should be taken to ensure that all the people present can be evacuated quickly and safely.

Where the foam is covered, the covering material should be chosen so as to inhibit the ignition of the foam as much as possible.

In the case of pu foam used as a thermal insulating material, wherever possible a thermal barrier should be used to protect the foam in a fire environment.

D.5 RIGID URETHANE FOAMS

These are almost exclusively manufactured from diphenyl methane di-isocyanate (MDI), and are closed-cell in structure. They are less easy to ignite than the flexible foams but, once alight, will burn as readily. They are used, inter alia, as thermal insulating materials, and in some cases may be foamed *in situ*.

During manufacture the same problems of toxicity occur as with flexible foams, although a toxic concentration of MDI is less likely than a toxic concentration of TDI in the atmosphere because MDI is less volatile.

In use, rigid foams are less likely than flexible foams to be the starting point of a fire. However, they will burn rapidly and spread a fire once they are involved. This is particularly worrying in some installations where they are used as building insulating materials. Under these conditions of use one should not rely on the standard 'surface spread of flame' tests when assessing the suitability of a coating material. If there is any doubt about fire safety, the practice required in the USA, of protecting the urethane foam with a thermal barrier, should be adopted. Advice on rigid foams can be obtained from the British Rigid Urethane Foam Manufacturers Association (Manchester).

D.6 FURTHER INFORMATION

As explained in the introduction, this Appendix can only touch briefly on the subject of pu foam. Organisations involved in the manufacture, use and problems of pu foams have done and are doing a lot of work to overcome these problems. For example the British Rubber Manufacturers' Association (in London) has published booklets and organised more than seventy meetings to discuss the fire problems of these materials. The Fire Research Station (at Boreham Wood) has conducted detailed research into the ignition and combustion of pu foams over several years, and has published its results in a series of papers. Other organisations active in this field include The Rubber and Plastics Research Association (Shawbury) which has carried out a lot of research into the fire properties of pu foams. This list is by no means complete, but it should provide a basis for further investigation.

REFERENCES

1. Health & Safety Executive, *Polyurethane foam*, HS(G)1 HMSO (1978).
2. 'The fire threat from plastics furnishings', *Fire Prevention*, 1973, August, No 99, pp 6–11, 21.
3. Simpson, B., *Flexible polyurethane foam, its uses and misuses*, The British Rubber Manufacturers' Association.

Appendix E

Key to Abbreviations

Note: Names of societies, companies, etc. are British except where otherwise stated

Abbreviation	Full Name
a.c.	*alternating current*
atm	*atmosphere (pressure unit)*
ACGHI	American Conference of Governmental Industrial Hygienists
AMOCO	American Oil Company
ANSI	American National Standards Institute
API	American Petroleum Institute
ASH	Action on Smoking and Health
ASME	American Society of Mechanical Engineers
BASEEFA	British Approval Service for Electrical Equipment in Flammable Atmospheres
BCF	Bromochlorodifluoro methane
BLEVE	Boiling liquid expanding vapour explosion
BOAC	British Overseas Airways Corporation
BP	Boiling point
BS	British Standard
BSC	British Safety Council
BSI	British Standards Institution
BSSR	British Society for Social Responsibility in Science
BTM	Bromotrifluoro methane
C	Capacity
CD	Capacitor discharge
CENELEC	European Committee for Electro Technical Standardisation
CIA	Chemical Industries Association
CM	Condition monitoring
CP	Code of Practice
d.c.	direct current
dB	decibel
dBA	weighted decibel (to correspond to same noise intensity sensed by human ear over entire frequency scale)
DB	Dry bulb temperature
DD	Draft for development
DHSS	Department of Health and Social Security
DOE	Department of Employment

Abbreviation	Full name
E	Energy
EEC	European Economic Community
EEL	Emergency exposure limit
EEUA	Engineering Equipment Users Association
EOT	Electrically operated overhead travelling (crane)
EPEL	Emergency population exposure limit
ERA	Electrical Research Association
FAFR	Fatal accident frequency rate
FDT	Fractional dead time
FITC	Foundry Industry Training Committee
FPA	Fire Protection Association
g	gauge
GKN	Guest Keen and Nettlefolds
GT	Globe thermometer temperature
H & S	Health and Safety
HASAWA	Health and Safety at Work, etc. Act
HMFI	Her Majesty's Factory Inspectorate
HMSO	Her Majesty's Stationery Office
HSC	Health and Safety Commission
HSE	Health and Safety Executive
Hz	Hertz (Frequency, times per second)
IAEA	International Atomic Energy Agency
IATA	International Air Transport Association
ICI	Imperial Chemical Industries
ICRP	International Commission on Radiological Protection
IEC	International Electrotechnical Commission
IES	Illuminating Engineering Society
IP	Institute of Petroleum
ISO	International Standards Organisation
kg	kilogram
kV	kilovolt
kVA	kilovolt amp
LC	Level controller
LD	Lethal dose
LEL	Lower explosive limit
LPG	Liquefied petroleum gas
LS	Level switch
mR	milli Rontgen
MDI	Diphenyl methane di-isocyanate

Abbreviation	Full name
MESG	Maximum experimental safe gap
MeV	Million electron volts
MIC	Minimum ignition current
MPPD	Maximum probable property damage
NFPA	National Fire Protection Association (USA)
NSC	National Safety Council (USA)
pf	picofarad
pu	polyurethane
PERA	Production Engineering Research Association
PO	Post Office
PVC	Polyvinyl chloride
Q	Probability of malfunction $= I - R$
R	Reliability (probability that a piece of equipment or component will function as intended for a given working life in a stipulated environment) or, Rontgen (unit of radiation)
RoSPA	Royal Society for the Prevention of Accidents
RP	Recommended practice
SI	Statutory Instrument
SIC	Standard Industrial Classification
SMRE	Safety in Mines Research Establishment
TDI	Toluene di-iso-cyanate
TLV	Threshold Limit Value
TNT	Tri-nitro toluene
V	volt or potential
WB	Natural wet bulb temperature
WBGT	Wet bulb globe temperature
FR	Failure rate (time per year)
μJ	Microjoules

—

Index